Catecholamines and behaviour

Catecholamines and behaviour

STEPHEN T. MASON
Department of Psychology, Brunel University, London

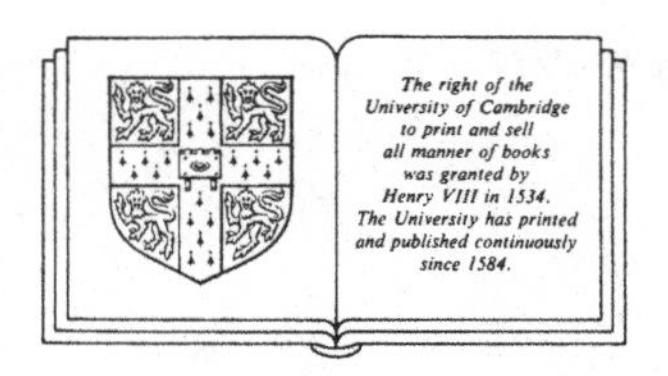

CAMBRIDGE UNIVERSITY PRESS
Cambridge
London New York New Rochelle
Melbourne Sydney

CAMBRIDGE UNIVERSITY PRESS
Cambridge, New York, Melbourne, Madrid, Cape Town, Singapore, São Paulo

Cambridge University Press
The Edinburgh Building, Cambridge CB2 8RU, UK

Published in the United States of America by Cambridge University Press, New York

www.cambridge.org
Information on this title: www.cambridge.org/9780521249300

First published 1984

A catalogue record for this publication is available from the British Library

Library of Congress Catalogue Card Number: 83-7722

ISBN 978-0-521-24930-0 hardback
ISBN 978-0-521-27082-3 paperback

Transferred to digital printing 2008

Contents

Preface

There are many strange wonders, but nothing more wonderful than man.

Sophocles, *Antigone*.

Research on the function of the catecholamine neurotransmitters in the central nervous system has been proceeding apace for some twenty years. Discoveries have perhaps been few. We know where the catecholamine neurones are to be found (anatomy) and how they synthesise, release and inactivate their chemical transmitter substances (biochemistry), but for all this we have only a few hints as to what the systems actually do (behaviour). Or to put it another way, what happens when they do *not* function normally (psychopathology). Some might suggest that it is premature then even to contemplate a volume entitled *Catecholamines and behaviour*. Is such a project hubris, and do the Furies await? No pretence is made that this book offers the definite answer to the behavioural functions of catecholamine systems. Where a consensus of evidence exists this is presented to the reader, but where controversy reigns, or data are absent, the reader is not shielded from these facts of scientific life. In this fashion the experimental process of science is laid bare and some appreciation is achieved of the balance of knowledge in this field and the probable ways forward for the future.

Why not wait until all the answers are in before writing this compendium? Perhaps one day neuroscience will reach the status of the Homeric epics – handed down unchanged for generations.[1] Would this not be the time for such an undertaking? Alas, progress in science is not, it seems,

[1] At least from the putative Peisistratean recension or the Alexandrian scholiasts. Myers, J.L. (1958). *Homer and his critics*, Chap. 2. Routledge & Kegan Paul: London.

linear but dances to the dictates of fashion. Research interest has moved from the classic catecholamine systems to the much-in-vogue peptides. The wave has crested and the bare bones of what has been learned over the last twenty years of work on catecholamine neurotransmitters are laid exposed on the sands. The purpose of this book is to gather them up for safekeeping before the winds of altered research priorities scatter them and they are lost. For in time, in the ebb and flow of scientific endeavour, workers will return to this field. Realising how much remains to be learnt, what potential still exists, the fashions of research funding will carry our latter day Odysseus back once more to these shores. When that happens, it is hoped that this book will be of profit by having systematised what has been learned to date, by pointing out questions which are of interest and amenable to current techniques, and perhaps by preventing repetition of already explored or unprofitable avenues.

The plan of the text opens with a survey of the anatomy of catecholamine systems in the brain and moves on to the techniques of pharmacology available for investigating their behavioural function. Very little background is assumed beyond a grasp of the neuronal structure of the brain. Terms are defined as they are introduced and there is an Appendix of commonly used behavioural paradigms and terminology. The main body of the text then examines a series of behaviours such as learning, reward, motor output, cognitive processes and ingestive/vegetative behaviours. First, the early work implicating catecholamines but failing to differentiate noradrenaline from dopamine is reviewed. Then, the relative contributions of noradrenaline and subsequently of dopamine are assessed. The work closes with an examination of the role of catecholamines in pathological behaviour as exemplified by the human clinical conditions of schizophrenia, depression, epilepsy, Parkinson's disease and senile dementia.

I am grateful to many different people for help in the preparation of this book; if I do not list them all by name they will nonetheless know who they are. I am especially grateful to Drs S.D. and L.L. Iversen and to the Provost and Fellows of King's College, Cambridge, who together gave me the initial opportunity to embark on the study of this fascinating subject. Whether that study has been fruitful the readers must judge for themselves.

April 1983

S.T. Mason
Sheffield, England

Permissions

Permission to quote copyright material is gratefully acknowledged from Penguin Books, The 'Introduction' by A. R. Burn to Herodotus: *The Histories* translated by Aubrey de Selincourt (Penguin Classics, revised edition 1972); Weidenfeld & Nicolson Ltd, *Periclean Athens* by C. M. Bowra (1971); Oxford University Press, *Greek lyric poets from Alcman to Simonides* by C. M. Bowra (2nd edn 1961); Weidenfeld & Nicolson Ltd, *The Greek experience* by C. M. Bowra (1957).

Abbreviations

6-ADA	6-aminodopamine
AMPT	α-methyl-*para*-tyrosine
BOL	2-bromo-*d*-lysergic acid diethylamide
cAMP	cyclic adenosine monophosphate
CRF	continuous reinforcement
CSF	cerebrospinal fluid
DBEE	dorsal bundle extinction effect
DBH	dopamine-β-hydroxylase
DDC	diethyldithiocarbamate
5,7-DHT	5,7-dihydroxytryptamine
DITA	3′,4′-dichloro-2-(2-imidazolin-2-yl-thio)-acetophenone
DMI	desipramine
DMT	*N*,*N*-dimethyltryptamine
DOPA	dihydroxyphenylalanine
DOPAC	dihydroxyphenylacetic acid
DOPS	3,4-dihydroxy-*ortho*-phenylserine
DRL	differential reinforcement of low rates of responding
DSP4	*N*-(2-chloroethyl)-*N*-ethyl-2-bromobenzylamine
ECS	electroconvulsive shock
EEG	electroencephalogram
FI	fixed interval
FR	fixed ratio
GABA	γ-aminobutyric acid
5-HIAA	5-hydroxyindoleacetic acid
HVA	homovanillic acid
ICSS	intracranial self-stimulation
LC	locus coeruleus
LSD	lysergic acid diethylamide

MAO monoamine oxidase
MAOI monoamine oxidase inhibitor
5-MeODMT 5-methoxy-*N*,*N*-dimethyltryptamine
MHPG 3-methoxy-4-hydroxyphenylglycol

(−)NPA (−)-*N*,*n*-propylnorapomorphine

6-OHDA 6-hydroxydopamine

PGO ponto-geniculate-occipital
PREE partial reinforcement extinction effect
PRF partial reinforcement

REM rapid eye movement
RLA Roman Low Avoider rat strain
RSF reward summation function

SWS slow wave sleep
STP 2,5-dimethoxy-4-methylamphetamine

UV ultraviolet

VI variable interval
VMN ventromedial nucleus of the hypothalamus
VR variable ratio

1

Anatomy of the catecholamine systems in the brain

'It is my principle that I ought to repeat what is said;
but I am not bound always to believe it.'
Herodotus, *The histories.*

Introduction

The initial demonstration of cell bodies in the brain which contained catecholamines was by Dahlstrom & Fuxe in 1964. This was based on the Falk–Hillarp histofluorescence technique in which brain tissue was reacted with formaldehyde gas in order to convert the catecholamines noradrenaline and dopamine into derivatives which fluoresce with a bright green intensity when exposed to ultraviolet (UV) light under a microscope. Other neurotransmitters fail to form these fluorophores with formaldehyde and hence are not seen under the UV microscope. The related indolamine, serotonin, forms a derivative which has a yellow fluorescence and which fades rapidly. It can hence be distinguished from dopamine and noradrenaline. However, since both dopamine and noradrenaline form green-fluorescing derivatives they cannot be separated one from the other by the Falk–Hillarp technique. Further, this technique by itself is not sufficiently sensitive to yield observable fluorescence in either fibres or terminals of the catecholamine systems. In order to reveal the pathway traversed by the axons of these cell bodies, Ungerstedt (1971) placed lesions at various levels of the brain and observed the build-up of transmitter caused by axoplasmic transport caudal to the cut. This allowed mapping of the projection systems of both the dopaminergic and noradrenergic systems to be accomplished.

Since this technique depended on lesions to raise the level of fluorescence to detectability, inevitably some systems were not detected and others were confused into one larger pathway. The next stage in the mapping of catecholamine systems came with the introduction by Lindvall & Bjorklund (1974) of the glyoxylic acid fluorescence technique, which visualised both terminals and fibres in the normal brain without resort to lesions. This revealed an almost embarrassing plethora of projections systems and pathways.

Other, newer anatomical techniques such as the retrograde transport of horse-radish peroxidase have recently been applied to catecholamine systems with interesting results (Mason & Fibiger, 1979). However, since the behavioural experiments described in the following chapters have generally not made use of the fine distinctions and subdivisions of pathways revealed by anatomical studies, no further mention will be made of them.

Much further investigation of the detailed anatomy of the terminal ramification of catecholamine fibres in the areas that they innervate has also been carried out, especially by Moore and associates (Fallon, Koziell & Moore, 1978; Fallon & Moore, 1978; Moore, 1978). From this it has proved possible on purely morphological grounds to distinguish terminals originating from the locus coeruleus (LC) system from those contributed by ventral systems. However, since the behavioural experiments in the literature have not made use of these anatomical findings, no further mention will be made of them either.

The basic neuroanatomy of the brain catecholamine systems will now be presented. Since the remit of this book is behaviour, not neuroanatomy, only as much detail is included as is required for the understanding of the behavioural experiments reported subsequently. Since most of these have used as their conceptual basis the divisions suggested by Ungerstedt (1971), most emphasis is given to this view of catecholamine neuroanatomy.

Noradrenaline

In Fig. 1.1 is shown the picture of noradrenergic pathways obtained by Ungerstedt (1971) on the basis of lesion and Falk–Hillarp

Fig. 1.1. Noradrenergic pathways in the rat brain. (Redrawn from Ungerstedt, 1971.)

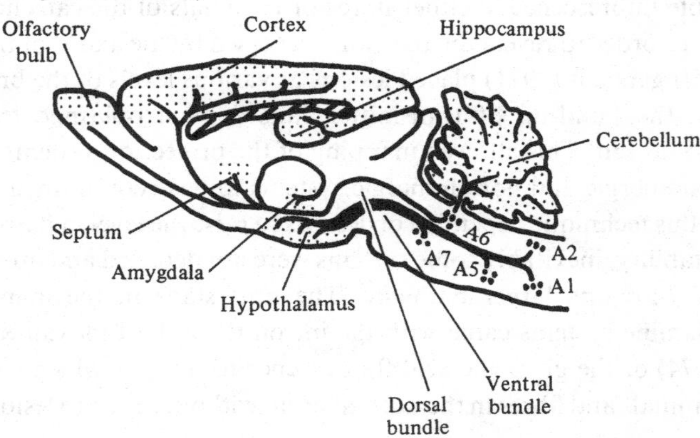

histofluorescence. The features of importance are the two distinct projection systems. The *dorsal bundle* originates in cell bodies in the LC (A6 of Dahlstrom & Fuxe, 1964). It then gives rise to bifurcating axons which run both to the cerebellum and in the dorsal bundle to innervate wide areas of the forebrain such as the cortex, hippocampus, amygdala and septum. A second system, the *ventral bundle* originates in cell bodies scattered throughout the medulla (A1, A2, A5 and A7 in the terminology of Dahlstrom & Fuxe, 1964). These axons collect together to form the ventral bundle, distinct along much of its course from the dorsal bundle, and innervating especially the hypothalamus. This view has formed the basis of virtually all subsequent behavioural studies. In Fig. 1.2 is shown the more detailed mapping achieved by Lindvall & Bjorklund (1974) using the glyoxylic acid fluorescence technique in normal brain. This confirms the basic pattern of Ungerstedt (1971) but adds some refinement. Thus, the dorsal bundle system is subdivided into a dorsal tegmental bundle originating from the LC and a dorsal periventricular bundle arising from cell bodies scattered along its course from the pons to the mesencephalon; the latter bundle innervates the thalamus and hypothalamus. These two bundles run very close to each other and it is most unlikely that any behavioural study has manipulated one without affecting the other. Despite deceptively accurate titles such as 'The dorsal tegmental noradrenergic projection: analysis of its role in maze learning' (Roberts, Price & Fibiger, 1976), both the dorsal tegmental *and* the dorsal periventricular systems were manipulated in tandem. Certainly, no neurochemical evidence was presented to justify the suggestion that only the dorsal tegmental system was affected. More recent literature has reverted to the older term 'dorsal bundle' without subdivision to indicate more accurately the nature of the manipulation.

Other details emerging from the work of Lindvall & Bjorklund (1974) are that the ventral bundle systems are seen rather as a wide plexus of fibres fanning out in the tegmental area than as a discrete bundle. They are thus called the central tegmental tract. These ventral systems are now known to contribute rather more to forebrain areas than was previously realised. Thus, some 20 % of amygdala noradrenaline and as much as 50 % of septal noradrenaline come from the ventral systems. The cortex and hippocampus are still viewed as pure dorsal bundle projection areas. However, at least three distinct input pathways to the hippocampus have been visualised from the dorsal bundle. Conversely, a small contribution to hypothalamic noradrenaline is seen to come from the LC itself.

Finally, descending projections to the spinal cord have been detected

(Nygren & Olson, 1977). These seem to come about primarily from LC but also with contributions from medullary cell groups.

Although the state of knowledge of noradrenergic systems has now progressed well beyond that of Ungerstedt (1971), virtually all the behavioural work reported in the ensuing chapters, with the notable

Fig. 1.2. (*a*) Dorsal periventricular noradrenergic system (medial parasagittal view). (*b*) Dorsal tegmental and central tegmental noradrenergic systems (more lateral parasagittal view). (*c*) Enlargement of (*b*). (*d*) Coronal view showing relation of the three main noradrenergic pathways. (Redrawn from Lindvall & Bjorklund, 1974.)

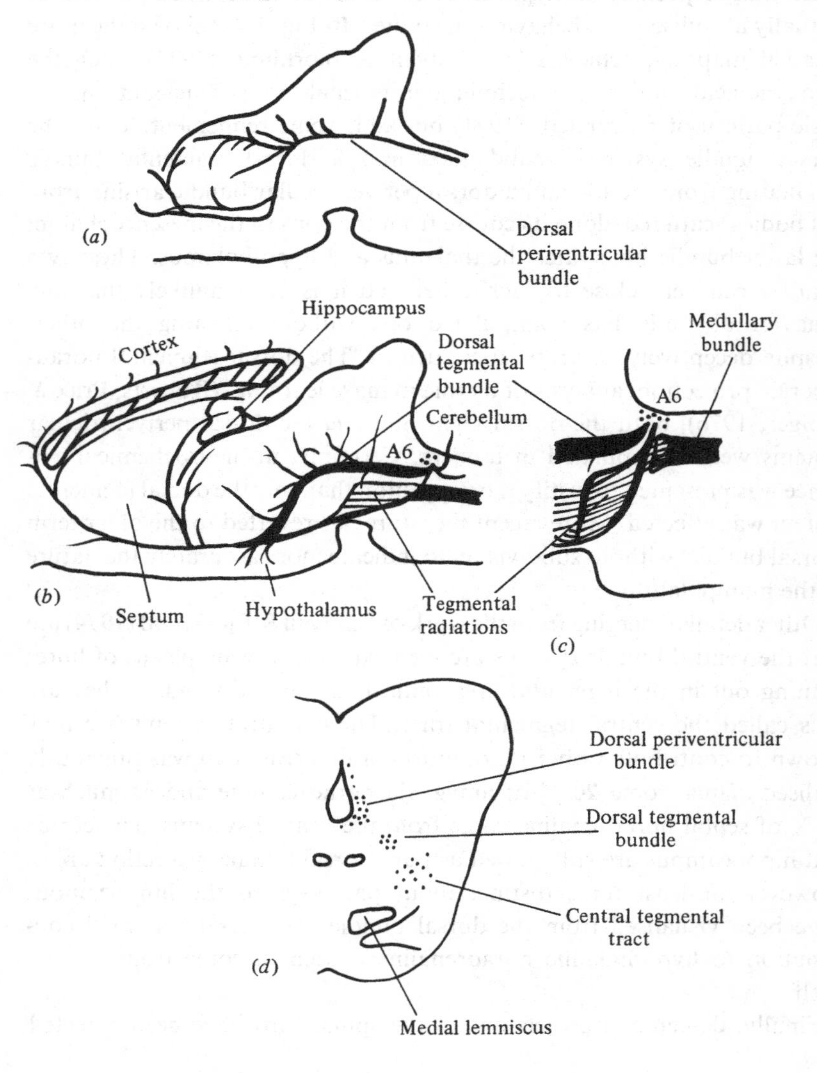

exception of that by Leibowitz on food and water intake (see Chapter 7), has been based on no more sophisticated a view than the dichotomy between dorsal bundle and ventral bundle. In part this reflects an absence of methodology for manipulation of the subcomponents of these systems in isolation.

Dopamine

In Fig. 1.3 is shown the conceptualisation of the dopaminergic system from Ungerstedt (1971). The main distinction, after the hypothalamic and pituitary dopaminergic pathways, is between the *nigrostriatal* pathway, originating in cell bodies in the substantia nigra (A9 of Dahlstrom & Fuxe, 1964) and innervating the caudate nucleus (striatum in the rat), and the *mesolimbic* pathway, originating in cell bodies in the ventral tegmental area (A10) and innervating the nucleus accumbens and olfactory tubercle. To the latter system has been added dopaminergic innervation to the *medial and sulcal prefrontal cortex*. Dopaminergic cells are also found in the retina and in the olfactory bulb, where they seem to act as interneurones of one form or another. Many of the behavioural experiments have failed to distinguish nigrostriatal from mesocorticolimbic dopaminergic systems since they lesioned the ascending dopaminergic bundle which contains all the fibres from both systems. More recent studies have, however, investigated this question either by localised lesion of A9

Fig. 1.3. Dopaminergic pathways in the rat brain. (Redrawn from Ungerstedt, 1971.)

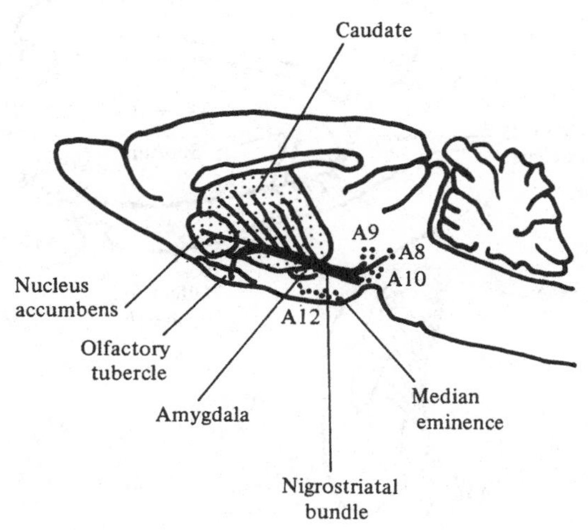

versus A10 or by destruction of specific dopaminergic terminal regions, such as the nucleus accumbens. In Fig. 1.4 is shown the more detailed mapping studies of Lindvall & Bjorklund (1974). From this and subsequent work it has been realised that there is some crossover between the nigrostriatal and the mesolimbic systems, with some striatal innervation coming from A10 and some accumbens–tubercle dopaminergic fibres arising from medial regions of the substantia nigra (A9). Further developments have included the demonstration of a small dopaminergic input to the traditionally noradrenergic hippocampus and a quite significant descending dopaminergic projection, possibly from hypothalamic cell bodies, to the spinal cord. Conversely, a traditionally dopaminergic area, the caudate, has been shown to receive a small noradrenergic projection from both ventral and dorsal bundle systems.

Again, possibly because of the lack of methodology for separate and fine manipulation, behavioural experiments to be described subsequently have tended to highlight the dichotomy between the *nigrostriatal* system and the *mesolimbic* pathway. The cross-innervation of these systems suggests that this has perhaps been a false dichotomy.

Fig. 1.4. (*a*) Parasagittal view of ascending dopaminergic pathways of the mesolimbic system. (*b*) Horizontal view of (*a*). (Redrawn from Lindvall & Bjorklund, 1974.)

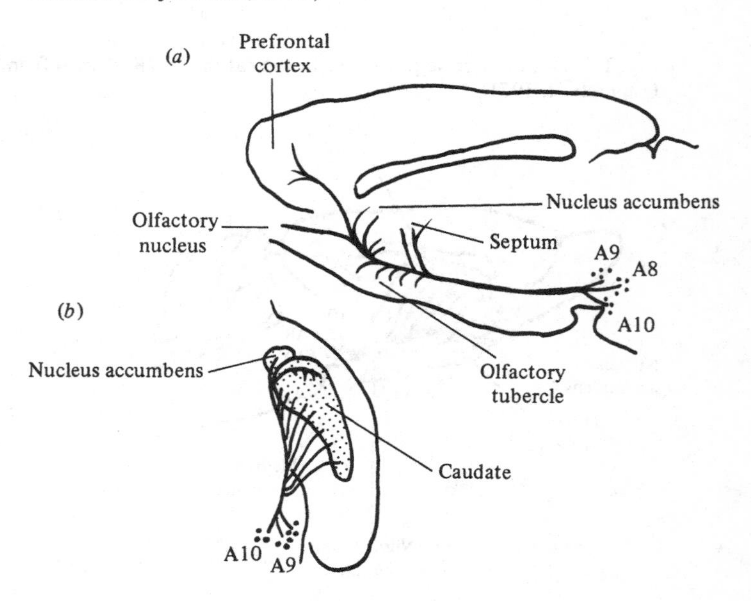

Further reading

For those desirous of a more detailed coverage of catecholamine neuroanatomy use may be made of the following:

Moore, R. Y. & Bloom, F. E. (1978). Central catecholamine neuron systems: anatomy and physiology of the dopamine systems. *Annual Review of Neuroscience*, **1,** 129–69.
Moore, R. Y. & Bloom, F. E. (1979). Central catecholamine neuron systems: anatomy and physiology of the norepinephrine and epinephrine systems. *Annual Review of Neuroscience*, **2,** 113–48.

Reviews of the changing picture of catecholamine neuroanatomy and highlights of some of the more important results for noradrenergic systems may be found in:

Mason, S. T. (1981). Noradrenaline in the brain: progress in theories of behavioural function. *Progress in Neurobiology*, **16,** 263–303.
McNaughton, N. & Mason, S. T. (1980). The neuropsychology and neuropharmacology of the dorsal ascending noradrenergic bundle – a review. *Progress in Neurobiology*, **14,** 157–219.

Detailed consideration of the evidence that noradrenaline in the LC system functions as a neurotransmitter may be found in:

van Dongen, P. A. M. (1981). The central noradrenergic transmission and the locus coeruleus: a review of the data, and their implications for neurotransmission and neuromodulation. *Progress in Neurobiology*, **16,** 117–43.

References

Dahlstrom, A. & Fuxe, K. (1964). Evidence for the existence of monoamine containing neurons in the central nervous system. I. Demonstration of monoamines in the cell bodies of brain stem neurons. *Acta physiologica scandinavica, suppl.*, **232,** 1–55.
Fallon, J. H., Koziell, D. A. & Moore, R. Y. (1978) Catecholamine innervation of the basal forebrain. II. Amygdala, suprarhinal cortex and entorhinal cortex. *Journal of Comparative Neurology*, **180,** 509–32.
Fallon, J. H. & Moore, R. Y. (1978). Catecholamine innervation of the basal forebrain. III. Olfactory bulb, anterior nuclei, olfactory tubercle and piriform cortex. *Journal of Comparative Neurology*, **180,** 533–44.
Lindvall, O. & Bjorklund, A. (1974). The organization of the ascending catecholamine neuron systems in the rat brain as revealed by the glyoxylic acid method. *Acta physiologica scandinavica, suppl.*, **412,** 1–48.
Mason, S. T. & Fibiger, H. C. (1979). Regional topography within noradrenergic locus

coeruleus as revealed by retrograde transport of horse-radish peroxidase. *Journal of Comparative Neurology*, **187,** 703–24.

Moore, R. Y. (1978). Catecholamine innervation of the basal forebrain. I. The septal area. *Journal of Comparative Neurology*, **177,** 665–84.

Nygren, L. G. & Olson, L. (1977). A new major projection from locus coeruleus: the main source of noradrenergic nerve terminals in the ventral and dorsal columns of the spinal cord. *Brain Research*, **132,** 85–93.

Roberts, D. C. S., Price, M. T. C. & Fibiger, H. C. (1976). The dorsal tegmental noradrenergic projection: analysis of its role in maze learning. *Journal of Comparative and Physiological Psychology*, **90,** 363–72.

Ungerstedt, U. (1971). Stereotaxic mapping of the monoamine pathways in the rat brain. *Acta physiologica scandinavica, suppl.*, **367,** 1–49.

2

Pharmacological manipulation of catecholamine systems

Introduction

The reader with a background in pharmacology may omit this chapter, since it is designed to take those with little or no experience of central neurochemistry to the point where they can appreciate the detailed manipulations described in subsequent chapters. The following sections cover the common techniques used in the experimental investigation of brain catecholamine systems. Some mention is made of the main limitations or shortcomings of the techniques, but, since this is rather diffuse without specific examples, most critical discussion of technical and conceptual shortcomings is postponed to the individual chapters dealing with catecholamines and specific behaviours.

Because of the different purpose of this compared to subsequent chapters, it has been thought inappropriate to include detailed citations to original papers and so the barest, and it is hoped, the clearest of factual coverage is given.

Lesions and synthesis inhibition

The rationale behind this approach is that study of what goes wrong with behaviour when a specific brain system is permanently or temporarily put out of action may yield insights into the function of that system in the intact brain. Objections to this approach include the possibility that other brain structures may compensate for the loss of the lesioned area. Further, a given failure of behaviour (say, movement) may occur through more than one mechanism (thus, loss of motor initiation, loss of motivation to move, loss of detection of sensory components which usually trigger a movement, etc.). These should be borne in mind throughout the subsequent chapters and explicit reference will be made when the problem is particularly acute.

Tyrosine hydroxylase inhibitors

Catecholamines in the brain are synthesised from the precursor tyrosine by the enzyme tyrosine hydroxylase to form dihydroxyphenylalanine (DOPA). This occurs in both noradrenaline- and dopamine-containing neurones and is believed to be one of the rate-limiting steps in the synthesis of these neurotransmitters. It is thus a particularly good point to interrupt their production. DOPA is then converted into dopamine by the enzyme aromatic amino-acid decarboxylase and here the process ends in dopaminergic neurones. In noradrenaline neurones a further step, catalysed by the enzyme dopamine-β-hydroxylase (DBH), converts dopamine into noradrenaline. This step, since it takes place only in noradrenergic neurones can be used to disrupt the synthesis of noradrenaline without affecting dopamine.

A classic drug used in the early study of the functional role of amine systems is α-methyl-*para*-tyrosine (AMPT). This inhibits the enzyme tyrosine hydroxylase and so prevents the synthesis of both dopamine and noradrenaline. It cannot, by itself, distinguish the contribution of dopamine from that of noradrenaline to the behaviour in question. Some attempts have been made to answer this by the combination of AMPT with 3,4-dihydroxyphenylserine (DOPS) which is converted by a non-physiological pathway into noradrenaline and hence replenishes noradrenaline stores without restoring function to dopaminergic systems. A further attempt at delineating the role of dopamine, as distinct from noradrenaline, after AMPT synthesis inhibition has been to seek to reverse the behavioural deficit by injections of *either* noradrenaline *or* dopamine. However, noradrenaline can be taken up not only into noradrenergic neurones, but also into dopaminergic ones and may, upon release, weakly activate dopamine receptors. Thus, AMPT, even in combination with other manipulations, offers only weak potential for distinguishing between a functional role of noradrenaline and of dopamine.

DBH inhibitors

More recently a new generation of synthesis inhibitors has become available which act at the DBH step. Thus, synthesis of noradrenaline is prevented but that of dopamine may continue unchecked. Deficits caused by these drugs should then be ascribable specifically to noradrenergic systems. Four caveats are needed. First, there exist systems in the brain which synthesise and release adrenaline as their neurotransmitter. In analogous fashion to AMPT and dopamine/noradrenaline so DBH inhibitors will prevent the synthesis of adrenaline, which is normally brought about by the enzyme phenylethanolamine-*N*-methyltransferase acting on noradrenaline. Secondly, the question of behavioural specificity assumes great importance. DBH inhibitors seem to have marked sedative effects and a somnolent animal will fail to press a lever merely because it is asleep, not necessarily because its brain reward system is blocked for example. This is expounded in greater detail in future chapters. Thirdly, compounds used to inhibit noradrenaline synthesis include diethyldithiosulphate, disulfiram, fusaric acid, FLA 63, FLA 157 and U 14,624, amongst others. Many of these compounds act by copper chelation which will inhibit many enzymes other than merely DBH. Thus, diethyldithiosulphate has been reported to inhibit liver aldehyde-dehydrogenase, intestinal indolamine-2,3-dioxygenase, liver D-glucuronolactone, mixed function oxygenase, superoxide dismutase and hepatic microsomal and plasma esterases. The nonspecificity of neurochemical action is a further drawback to the use of DBH inhibitors.

Finally, since these drugs are most often administered systemically they penetrate to all brain areas and inhibit noradrenaline synthesis in both the dorsal and the ventral bundle systems. They deplete noradrenaline from the ascending, descending and cerebellar projections of the dorsal bundle system. They thus cannot localise the brain area of importance for a given behavioural effect. Such administration will also deplete noradrenaline from the peripheral nervous system, totally outside the brain as such.

Electrolytic lesion

To deplete noradrenaline from the dorsal bundle system without altering ventral bundle noradrenaline, small electrolytic lesions of the source of the dorsal bundle system (the locus coeruleus, LC) have been attempted. These, by definition, lack neurochemical specificity since the passage of electrical current destroys all tissues in its vicinity and not merely noradrenergic neurones. In the rat the LC is purely noradrenergic and so an electrolytic lesion restricted to this nucleus might claim a degree of neurochemical specificity. However, in the cat and some primate species nonnoradrenergic cells are also present (very markedly so in the cat). Additionally, nonnoradrenergic fibres of passage may run through the region of the LC even in the rat. Finally, the shape of the LC makes it very difficult to restrict an electrolytic lesion to the noradrenergic cells without encroaching on adjacent nonnoradrenergic tissue (See Fig. 5.3 for graphic demonstration).

6-Hydroxydopamine

Perhaps the best lesion techniques available today make use of 6-hydroxydopamine (6-OHDA), a compound which is taken up into catecholamine neurones and then destroys them. This yields a neurochemically selective and permanent loss of catecholaminergic elements. Intraventricular administration can deplete both noradrenaline and dopamine together, while pre-treatments may enhance the dopamine loss or afford virtually complete protection to the noradrenergic neurones, thus yielding depletions specific for one or other amine. Local tissue injection into the dopaminergic or the noradrenergic fibre bundles can cause loss of only dopamine or only noradrenaline, often in just one part of the projection system. Finally, 6-OHDA injection into terminal areas (such as the nucleus accumbens) can permit loss of one amine in a specific and localised brain area and thus represents the acme of the technique. Further details of the doses, routes of administration and ancillary drug pre-treatments will be found in the following chapters as use is made of them in behavioural situations. Objections to the 6-OHDA techniques have generally focussed around the neurochemical specificity of this agent. It is now generally agreed that a region of nonspecific necrosis will be found at the tip of the injection cannula for local tissue injections. However, this constitutes less than 10 % of the volume of tissue affected, the remaining 90 % being selectively depleted of catecholamines. An example of the very profound loss of noradrenaline, with virtually complete sparing of other neurochemical systems, which can be achieved by 6-OHDA injection into the dorsal bundle system is given in Table 2.1. It has also been suggested that functional reorganisation of the brain may occur at long periods after the initial lesion and this might complicate interpretation of the behavioural findings. Both these points, of possible nonselective damage and functional reorganisation, should be borne in mind in the following chapters and are explicitly mentioned when the problem becomes particularly acute. Control measures exist for both of these objections and are discussed in context.

Stimulation

The rationale for this approach is that exogenous activation of a brain system involved in the normal expression of behaviour might cause that bit of behaviour to be emitted soon after onset of stimulation. Objections to this approach include the difficulty in separating functional centres of cell bodies from fibres of passage merely running through this area on their way to and from totally different brain regions. A second objection to the stimulation approach is that the pattern of natural neuronal activity is unlikely to be elicited by the stimulation. Thus, in a nerve bundle it may be that some neurones increase their firing while others decrease their activity during a natural behaviour and that it is this patterning of activity, rather than a general increase or decrease by all

Table 2.1. *Post-mortem assays on control and dorsal bundle 6-OHDA lesion rats*

	Control ($n=10$)	*Lesion* ($n=10$)	%
Noradrenaline			
Cortex	289±14	9±6	3
Hippocampus	305±18	15±3	5
Hypothalamus	2380±110	860±90	36
Amygdala	408±5	61±4	15
Septum	954±12	435±34	46
Cerebellum	219±12	271±8	124
Spinal cord	255±6	307±12	120
Dopamine			
Striatum	13570±660	1284±230	95
Amygdala	88±9	66±27	75
Septum	650±70	500±20	77
Hypothalamus	452±19	421±26	93
Serotonin			
Cortex	377±21	382±19	101
Hippocampus	493±38	485±41	98
Hypothalamus	1206±110	1156±96	96
Choline acetyltransferase			
Cortex	13.4±0.7	13.1±0.6	98
Hippocampus	15.0±0.5	14.9±0.6	99
Hypothalamus	11.9±1.68	10.2±1.14	86
Septum	15.1±0.8	13.7±0.4	91
Amygdala	18.4±1.5	19.6±1.2	106
Glutamic acid decarboxylase			
Cortex	12.1±0.4	11.3±0.4	93
Hippocampus	10.8±0.4	11.0±0.5	101
Hypothalamus	25.0±1.2	23.6±1.5	91
Septum	16.7±0.9	16.3±1.3	98
Amygdala	10.2±0.2	11.0±0.8	108

Values are means with standard error of the mean. Noradrenaline, dopamine and serotonin values are in nanograms of amine per gram wet weight of tissue and choline acetyltransferase (an index of cholinergic systems) and glutamic acid decarboxylase (an index of GABA minergic systems) are micromoles per 100 mg protein per hour. The last column represents values of lesion groups expressed as percent relative to control values.

neurones, which is important for elaboration of behaviour. Such a micropatterning of neuronal activity could not be mimicked by exogenous stimulation. Finally, the possibility exists to elicit behaviours by electrical or chemical stimulation of systems which are never physiologically active in the elaboration of normal behaviour.

Precursor administration

To increase the functioning of a given neurochemical system it might be possible to increase the amount of neurotransmitter synthesised by the brain by administering its precursor substance. In the catecholamine field this has most often been done by administration of L-DOPA, the precursor of both noradrenaline and dopamine. As such, it fails to distinguish between the two amines in the elicited behaviour. This may be overcome in part by combined administration of L-DOPA with a DBH inhibitor so that only dopamine can be formed. This faces the problems cited with DBH inhibitors above, namely of behavioural and neurochemical nonspecificity. Precursor administration has been most useful historically in demonstrating a role for catecholamine systems in a class of behaviour, while other approaches have been required to establish the particular catecholamine of importance.

Direct receptor agonists

Drugs, or the natural neurotransmitters themselves, which directly activate dopamine or noradrenaline receptors have been used extensively in the study of catecholamines and behaviour. As such, they figure prominently in the following chapters. Since there is a wide diversity of drugs with different patterns of selectivity for receptors (some postsynaptic, some presynaptic, some α-, some β- and so on), they will best be dealt with as they occur in specific behavioural experiments. However, a word about the application of the neurotransmitters themselves is appropriate. Neither noradrenaline nor dopamine cross the blood–brain barrier in appreciable amounts when injected peripherally and thus must be injected directly into the brain, either into the ventricles or into local brain areas. As such the degree of spread from the area of injection to adjacent regions must be determined. This can be done by radio-tracer studies and as such should be *de rigueur* for all local injection work. A three-dimensional grid of injections around the primary area should also be carried out to demonstrate that these surrounding areas *fail* to give rise to the behaviour ascribed to the area of primary injection. Too often these vital controls are not included in published studies. Further problems with the natural neurotransmitters are that noradrenaline may be taken up, not only into noradrenergic neurones and activate noradrenaline receptors, but also into dopaminergic neurones and weakly activate dopamine receptors. To control for this it is necessary to demonstrate that the effects of noradrenaline on behaviour are blocked by noradrenaline receptor blockers but not by dopamine antagonists. This has the further benefit of determining the type of receptor (α- or β-, α_1 or α_2- etc.) by which noradrenaline was acting. Use of the natural neurotransmitters is thus the start, not the end, of a long programme of investigation. It has perhaps most usefully been employed in the study of noradrenaline and feeding behaviour (see Chapter 7).

Electrical stimulation

Electrical stimulation via chronically implanted electrodes lacks neurochemical specificity, since all neurotransmitter systems at the electrode tip will be activated, not only those using noradrenaline or dopamine. Some improvement may be effected if stimulation of pure noradrenergic or dopaminergic nuclei is used, but even here nonnoradrenergic fibres of passage may be being activated rather than aminergic cell bodies. Further, more detailed mapping studies may indicate that the behavioural effect was not being obtained from the aminergic nucleus but in fact from closely adjacent areas. Considerable work is needed to detect this and artefacts may thus linger in the literature for many years (see section on Noradrenaline and ICSS, p. 28). A further problem with electrical stimulation in branching systems is the possibility that it is not the terminal area of the branch being stimulated which is of importance but the antidromic conduction of electrical activity back to the cell bodies with either dendritic transmitter release or further proliferation of the electrical activity down another branch of the bifurcating system to a totally distant terminal area. Such considerations are even now only starting to enter the critical literature.

Recording

The rationale here is that emission of a piece of behaviour should have correlates in the change in electrical or chemical activity of the brain areas involved in the causation of that behaviour. Problems here are the correlative nature of the approach. That is, more than one brain area may change its activity during ongoing behaviour, perhaps as the consequence of changed sensory input, altered motivational states and so on. Which area is causative of that behaviour, and which merely reflecting the occurrence of that behaviour caused by a totally different brain area, cannot be determined by passive recording methods. Interventionist approaches such as lesion or stimulation are required. Some brain areas may be excluded as showing changes in activity *after* behaviour has already started, not preceding behaviour emission as would be required if they had a causative role in that behaviour.

Single units

Recording of electrical activity in single cells of catecholaminergic nuclei is only in its infancy. The technical problems of holding a single cell for long enough to determine changes in firing rate while the animal is moving around and engaging in vigorous behaviour are formidable. However, some preliminary work in both noradrenergic and dopaminergic cell body regions has been reported and this technique would appear to hold great promise. It is, of course, necessary to demonstrate that the cell which was recorded was indeed catecholaminergic since nothing in its electrical activity *per se* will necessarily guarantee this. A combination of fluorescence histochemical methods with post-mortem demonstration of recording site appears promising. Specific instances of single unit recording in the catecholaminergic systems in relation to behaviour are mentioned (see, for example, Catecholamines and sleep, pp. 316–29). A further problem of single unit recording lies in the sampling bias associated with the technique. Thus, larger cells are more likely to be hit by the electrode and hence their activity monitored than smaller cells. Further, in a mixed population of cells some of which increase and some of which decrease their activity during ongoing behaviour, a clear conclusion as to the functioning of the overall nucleus might be difficult.

Release and push–pull

To add a further dimension of neurochemical specificity to recording techniques, studies on the levels of neurotransmitters and their release during or at the completion of behaviour have been carried out. Much will be made of these in the appropriate sections in subsequent chapters. Detailed analysis of many brain areas is possible with this approach and an idea of the dynamics of transmitter release can be achieved with the addition of synthesis inhibitors. The major problem at present seems to be to achieve sufficient behavioural sophistication to assign the observed neurochemical changes to a single element of behaviour, rather than to the many sensory, motivational and motor components all inevitably present in even a simple behavioural task. Only a few studies reach this exalted level. Push–pull studies in which transmitter release during, rather than after completion of, behaviour can be measured would appear to offer greater investigation of behavioural parameters and their temporal relationship to neurochemistry but the invasive nature of the push–pull technique may limit its applicability (see, for example, Chapter 7).

Conclusion

As may be seen from the preceding account, no attempt at a comprehensive catalogue of pharmacological manipulations has been made. Such an endeavour, as well as yielding a monograph in its own right, would be spectacularly barren and arid in the absence of their actual use in behavioural experiments. Thus, most critical discussion of the pharmacological techniques is tightly integrated with their behavioural use, description of which follows in subsequent chapters. The foregoing serves merely to give the flavour of the repast which will be served up in more extended courses in later sections.

Further reading

Those wishing a more detailed, yet still comprehensible, exposition of catecholamine pharmacology may consult the following:

Cooper, J. R., Bloom, F. E. & Roth, R. H. (1978). *The biochemical basis of neuropharmacology*, 3rd edition, especially Chapters 6 and 7. Oxford University Press: New York.

3

Catecholamines and intracranial self-stimulation behaviour

Introduction

The paradigm

Intracranial self-stimulation (ICSS) refers to the administration of electrical stimulation to the brain as a result of a voluntarily emitted response on the part of the animal. Olds & Milner (1954) observed that rats electrically stimulated in the septum would return, with increasing regularity, to the corner of the box in which stimulation happened to be delivered. It appeared that the electrical stimulation had reinforcing properties, since it would increase the likelihood of the response which had immediately preceded its onset. In a typical ICSS paradigm nowadays the rat is allowed to press a lever which will deliver short bursts of electrical stimulation (typically 100 ms or so) for each lever-press it makes.

Animals are implanted with, typically, bipolar electrodes made up of twisted wire insulated along its length except at the immediate cut end of the tip. This electrode is positioned in the brain at a particular structure with the help of a stereotaxic instrument and fixed to the rat's head by dental cement which is anchored to three or four screws driven into the skull (see Fig. 3.1). When the dental cement has dried the skin is sown together around the mound of dental cement and protruding electrode, a connector fitted to the end of the electrode and the animal allowed to recover from the anaesthetic. Each day the animal is taken from its home cage and connected via a swivel to wires in the chamber containing the lever which run to electronic apparatus which delivers the electrical brain stimulation (of the order of 10–100 μA). As the rat approaches the lever it receives a burst of electrical current from the experimenter. This is called shaping and is repeated until the rat makes contact with, and ultimately presses, the lever. For some electrode placements, typically the hypothalamus, the rat may simply be left in the box to stumble across the lever for itself. For other placements, typically the locus coeruleus (LC) or caudate, anything up to 10 days of patient shaping by the experimenter may be required to get the rat to press the lever by itself. After this learning phase the rat is given, typically, a 30 min session each day in the test apparatus. Sometimes the session is initiated with a free, experimenter-delivered burst of current. This is called priming. It was once very popular but appears to be declining in vogue. Animals will usually emit up to a thousand lever-presses or more in a 30 min session with hypothalamic electrodes. Other placements, such as the LC, elicit much more sedate lever-pressing. Other responses, such as nose-poking, tail-moving or alley-way running, can also be rewarded by electrical brain stimulation. Other, so-called 'rate-free' paradigms of ICSS have also been utilised. A drawback of the use of the rate of lever-pressing to quantify ICSS behavior is that it requires a high rate of motor output from the animal. It is thus particularly sensitive to disruption of motor mechanisms, such as happens with a great number of drugs. This motor effect may occur in the absence of any direct effect on the rewarding nature of the electrical brain stimulation *per se*. In order to try to minimise this possible confound, paradigms requiring only very low rates of motor response have been developed. These are generally called rate-free paradigms. However, this is a slight misnomer since they will require some rate of responding, albeit a much lower one than is the case with lever-pressing.

One rate-free paradigm greatly used is a shuttle box. Here the animal is placed in one half of the box and the response consists of running, often over a low barrier, to the other side of the box. There is some evidence that the running response is less motorially demanding than that involved in lever-pressing. Often the rat receives ICSS for a period while it remains in one side of the box. Shuttling to the other side terminates the ICSS. Thus,

Fig. 3.1. Typical implantation arrangement for self-stimulation in the rat. (Redrawn from Milner, 1970, p. 45.)

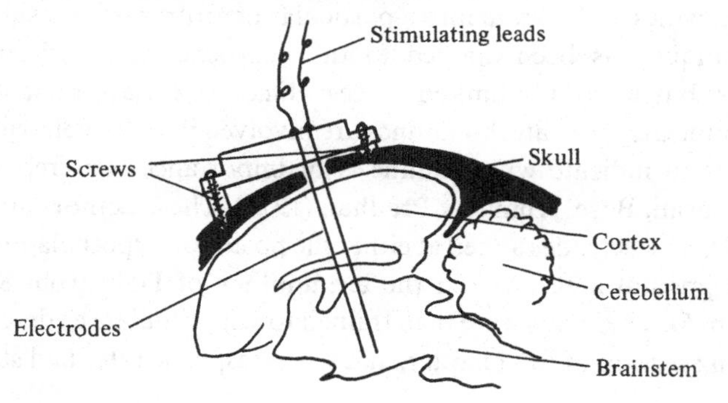

latency to escape ICSS can be determined. Further, the ICSS periodically shifts sides and the rat has to track it. Thus, if the animal remained immobile on one side of the box it would no longer receive ICSS. Every two or three minutes, as the ICSS shifts side, so too has the rat in order to continue to get ICSS. Thus, a latency to initiate ICSS can be measured from the time that the ICSS shifted to the other side of the box to the time that the rat eventually shuttled over to the other side to follow it. Since the ICSS only shifts side every two or three minutes, only a very low rate of responding is required compared to a lever-press paradigm. Nonetheless, some responding is required and the term 'rate-free' considerably overemphasises the advantages of this paradigm.

Its significance

The great theoretical interest in ICSS is that by electrically stimulating the brain we seem to be tapping into a system which mediates the central effects of reward. It is presumed that this same system is activated by natural reinforcers, such as food to a hungry rat or water to a thirsty one. If we can find out the brain mechanisms and transmitter systems involved in ICSS we may have gone a long way to determining those systems involved in natural reward.

Starting about the mid 1960s a series of lines of evidence seemed to suggest that the systems in the brain mediating ICSS might be catecholaminergic in nature.

Early evidence for catecholamine involvement

Precursor administration

As described above (p. 14), one technique available to investigate the role of catecholamines in a given behaviour is that of precursor administration. Here, an attempt is made to increase the amount of neurotransmitter available for release in the brain by injection (often directly into the ventricular spaces) of compounds which will be synthesised by enzymes in the brain into a particular neurotransmitter substance. This technique has been applied to the neurochemical basis of ICSS behaviour, but with only limited success. One typical experiment which serves to indicate that catecholamines are involved in ICSS behaviour, but which fails to indicate which amine is of importance, was reported by Nimitkitpaisan, Bose, Kumar & Pradhan (1977). These authors implanted rats with ICSS electrodes aimed at either the posterior hypothalamus or the ventral tegmental area (A10 in the terminology of Dahlstrom & Fuxe, 1964, see p. 5). These rats were then trained to self-stimulate as described in the opening section of this chapter, and when response rates had stabilised were injected intraperitoneally with various doses of L-DOPA. This substance is converted in the brain into dopamine in dopaminergic neurones; but it is also converted into noradrenaline in noradrenergic neurones. So right from the start this technique cannot tell us *which* amine is critically involved in ICSS. It should also be noted that in this study L-DOPA additionally acted to decrease serotonin levels in the brain.

The effects of L-DOPA treatment were biphasic. Initially, a decrease in ICSS was observed. This lasted for some 40–60 min and then a much longer increase in ICSS rates was observed. This is shown in Fig. 3.2, for a dose of L-DOPA of 150 mg kg^{-1}. Similar but smaller effects were seen with lower doses of L-DOPA. Identical effects were seen for both posterior hypothalamus and A10 electrodes. Measurement of the neurochemical results of L-DOPA administration in another group of rats showed that dopamine concentrations in the brain were increased between 200 and 1000 % while serotonin concentrations were decreased some 50 %.

What can be concluded from these results? Clearly, increasing catecholamine concentrations in the brain has an effect on ICSS behaviour.

Fig. 3.2. Effect of 150 mg kg^{-1} L-DOPA on ICSS rates from A10 and posterior hypothalamic electrodes ($n=4$) during 2 h session. Neurochemical changes in caudate dopamine content at various times after L–DOPA injection are also shown. (Redrawn from Nimitkitpaisan *et al.*, 1977.)

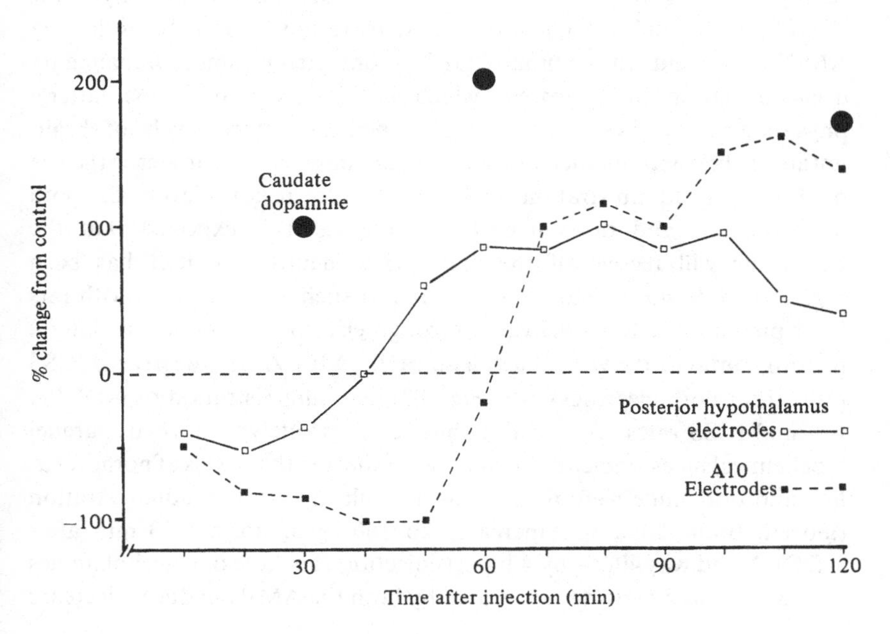

However, we cannot from these data say whether dopamine, noradrenaline or even serotonin is the important amine. Further, we cannot specify the direction of the involvement, i.e. whether amine X increases ICSS or decreases it, since both effects were seen at different times after L-DOPA treatment.

Synthesis inhibition

Cooper, Black & Paolino (1971) used a rate-free measure of ICSS in a shuttle box (see p. 19) and examined the effects of α-methyl-*para*-tyrosine (AMPT) on it. AMPT blocks the enzyme tyrosine hydroxylase, which is responsible for converting tyrosine into DOPA and thus permitting the synthesis of both noradrenaline and dopamine (see p. 10). Cooper *et al.* found that ICSS from both septal-forebrain and from lateral-hypothalamic electrodes decreased after 200 mg kg^{-1} or 600 mg kg^{-1} AMPT. While the authors chose to conclude that this was evidence for a role of noradrenaline in ICSS, because of the known depletion of dopamine as well as noradrenaline following tyrosine hydroxylase inhibition, I would conclude that this experiment certainly indicates that catecholamines are involved in ICSS but again fails to differentiate between dopamine and noradrenaline.

Attempts have been made to extend the use of the AMPT synthesis inhibition technique so as to distinguish between the two amines. This involves the administration of *dl*-threo-3,4-dihydroxy-phenylserine (DOPS) to animals which have ceased to respond for ICSS following AMPT treatment. In the brain DOPS is converted into noradrenaline by decarboxylation in a process which is not involved in the latter's physiological synthesis. Thus, it is hoped to restore levels of brain noradrenaline without increasing brain dopamine. A similar idea is the use of L-DOPA administration after AMPT to restore levels of both noradrenaline and dopamine. In both cases it is expected that the behaviour will recover if the appropriate neurotransmitter has been replenished. Stinus & Thierry (1973) report such an experiment with rats lever-pressing for ICSS delivered through electrodes in either the lateral hypothalamus or the ventral tegmental area (A10). AMPT decreased ICSS rates from both electrodes (see Fig. 3.3) and administration of L-DOPA restored ICSS rates after AMPT, but not completely to normal. Parallel biochemical measurements on other rats indicated that levels of noradrenaline and dopamine were restored as a result of L-DOPA administration (indeed, brain dopamine increased to 160 % of normal 30 min after L-DOPA and was 402 % by 4 h post-injection. The role of catecholamines in ICSS is thus clearly demonstrated by both the AMPT-induced decrease in ICSS rates and the subsequent restoration of the behaviour by L-DOPA reinstatement of brain catecholamine levels. DOPS, when given after AMPT, restored ICSS from ventral tegmental electrodes to near normal rates. It was, however, ineffective in animals with lateral hypothalamic placements. From this it might be tempting to conclude that lateral hypothalamic reward in this study was being mediated by a dopaminergic, rather than a noradrenergic, substrate (i.e. depletion of dopamine and noradrenaline by AMPT decreased ICSS, restoration of both dopamine and noradrenaline by L-DOPA reinstated the behaviour, but restoration of only noradrenaline by DOPS was ineffective). Indeed, given the clear effect of L-DOPA in replenishing brain dopamine this is not an unreasonable conclusion. The other half of the interpretation is more complex. It might additionally be concluded that the ventral tegmental electrodes were tapping into a noradrenergic substrate (i.e. depletion of dopamine and noradrenaline by AMPT decreased ICSS, restoration of both dopamine and noradrenaline by L-DOPA reinstated ICSS *and* restoration of noradrenaline alone by DOPS *also* reinstated ICSS). However, in parallel biochemical studies on other rats, DOPS did *not* increase noradrenaline levels after AMPT. Eight hours after AMPT, brain noradrenaline levels

Fig. 3.3. Effects of inhibition of catecholamine synthesis with 150 mg kg^{-1} AMPT (α-methyl-*para*-tyrosine) on ICSS rates from lateral hypothalamic and ventral tegmentum electrodes. L-DOPA (200 mg kg^{-1}, orally) and dihydroxyphenylserine (DOPS) (400 mg kg^{-1}, orally) were administered after the fifth 30 min test session. (Redrawn from Stinus & Thierry, 1973.)

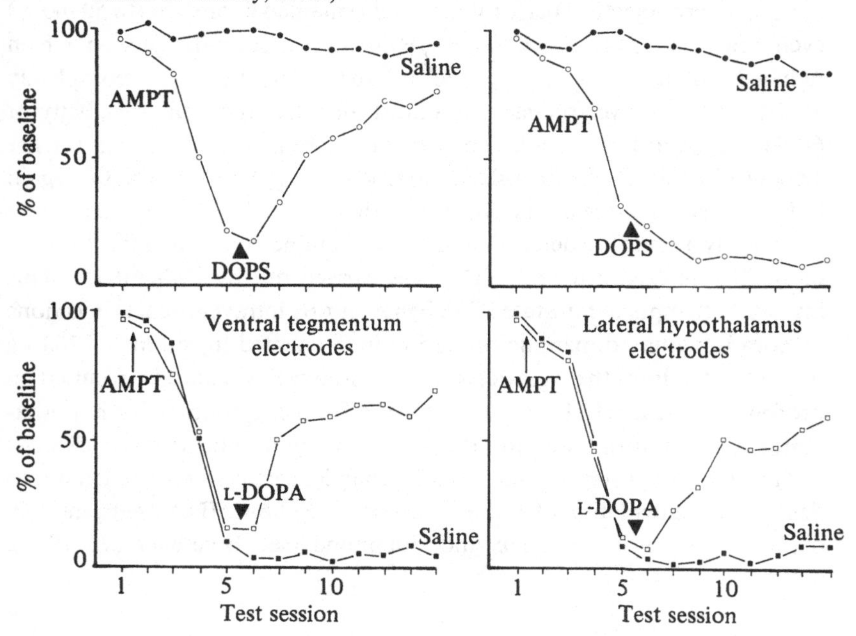

were 39 % of normal following placebo injection and 35 % of normal after DOPS injection. This latter was 4 h after DOPS injection but, even 1 h after DOPS, brain noradrenaline was still only 46 % of normal. Although DOPS restored the behaviour there is no evidence that it did so by increasing brain noradrenaline.

Thus, the above attempt to give additional specificity to AMPT depletion of catecholamines by subsequent DOPS treatment failed to indicate which amine was of critical importance.

Neurotoxic lesions

The use of 6-hydroxydopamine (6-OHDA) to destroy brain catecholamine systems started around 1970 (see p. 11) and was soon applied to ICSS behaviour. Initially, techniques to deplete only one amine were not available. Early 6-OHDA administration usually depleted noradrenaline and dopamine together. However, subsequent modifications are now available for depletion of each amine independently.

Breese, Howard & Leahy (1971) reported a typical early application of 6-OHDA to ICSS behaviour. Rats trained to self-stimulate through electrodes in the lateral or posterior hypothalamus were injected intracisternally with 200 μg of 6-OHDA following an intraperitoneal injection of pargyline. Depletions were confirmed at the end of the experiment on the same animals and showed an 83 % decrease in both noradrenaline and dopamine. ICSS rates immediately after the 6-OHDA treatment dropped to 20 % of pre-injection performance and remained depressed (to 50–60 %) even after 35 days. A second 6-OHDA intracisternal injection which reduced dopamine by 92 % and noradrenaline by 93 % completely abolished ICSS lever-pressing in almost all rats. Given the specificity of 6-OHDA towards catecholamine neurones (see p. 13) this is then further evidence for catecholaminergic involvement in ICSS behaviour, but again fails to dissociate one amine from the other.

An early attempt to determine the critical amine involved in the decrease of ICSS rate seen after 6-OHDA was carried out by Olds (1975). This involved attempts to reinstate ICSS behaviour by intraventricular injections of noradrenaline, dopamine or serotonin. Repeated injections of 250 μg of 6-OHDA into the ventricles via an indwelling cannula resulted in profound decreases in ICSS lever-pressing from electrodes in the hypothalamus, some pontine sites and the substantia nigra, but with less effect at midbrain or other pontine sites. No biochemical assays were carried out to determine the depletion of amines, but the dose of 6-OHDA used generally depletes both noradrenaline and dopamine (see Breese *et al.*, 1971).

Intraventricular infusion of noradrenaline was effective in reinstating the ICSS behaviour in rats with hypothalamic electrodes but *not* in rats with substantia nigra or pontine electrodes. Dopamine was ineffective at all sites and serotonin was effective only with lateral hypothalamic electrodes.

The decline in ICSS rates after 6-OHDA again implicated catecholamines in ICSS and served to highlight the possibility that not all ICSS sites in the brain use the same neurotransmitter. Thus, hypothalamic electrodes were affected by 6-OHDA catecholamine depletion, but some in the pons and midbrain were not. The possibility of noncatecholaminergic ICSS sites must not be overlooked. Reinstatement of ICSS behaviour with noradrenaline suggests that this is a critical amine – but only for the hypothalamic electrodes, not for the substantia nigra where it was without effect. However, even the noradrenergic nature of hypothalamic sites is questioned by the finding that serotonin was also effective in reinstating ICSS behaviour at these same sites. Is the transmitter involved noradrenaline, or is it serotonin – or is it both?

It seems that attempting to reinstate ICSS behaviour by injection of transmitters after amine depletion has not clarified the picture much. A better approach is to use the selective depletions of amines by 6-OHDA which are now available.

Summary

The early experiments described above using precursor administration, synthesis inhibition with AMPT or lesion with 6-OHDA have certainly demonstrated the involvement of catecholamines in ICSS behaviour, but have not allowed us to identify the amine of critical importance. Even replacement therapy with DOPS or the natural neurotransmitter substance have not teased out this question. Many more recent experiments have been carried out specifically to answer this question and are covered in the next sections.

Noradrenaline and ICSS

Mapping studies

Dorsal system. One of the lines of evidence which became available quite early to support a role for noradrenaline in ICSS was the correspondence of sites in the brain which would yield ICSS behaviour with the known distribution of noradrenergic fibre systems and cell bodies. If the release of noradrenaline is rewarding, and this release is actually being stimulated by the ICSS electrode in order to maintain ICSS behaviour, then it would be expected that the best sites in the brain in which to implant electrodes to obtain ICSS behaviour would indeed be those containing noradrenergic fibres, terminals or cell bodies. This correspondence between maps of the brain showing the distribution of ICSS-positive sites and other maps drawn up independently to show the distribution of noradrenergic pathways in the brain was first stressed by Crow (1971). He implanted rats with an electrode in the midbrain region and deliberately varied the stereotaxic coordinates so that a wide spread of electrode positions was obtained. The animals were then tested for ICSS behaviour with every effort being made to shape them to lever-press so as to stand the very best chance of detecting ICSS if indeed that site would yield it. This raises the first possible shortcoming of the technique. If ICSS behaviour is indeed obtained from that site, then the result is unambiguous. But suppose that one fails to get ICSS behaviour. Is that simply because one has failed to teach the rat that lever-pressing yields a 'pleasurable' sensation – i.e. that site was indeed ICSS positive but for one reason or another the rat did not make the connection and undergo the necessary learning? Or is that site indeed ICSS negative and no matter how you tried to shape or persuade the rat it would never find electrical stimulation of that part of the brain 'pleasurable' and so would never lever-press to receive it? Thus, negative results – failure to find ICSS from a given site – must be treated with some caution. Most experimenters use a standard shaping technique to maximise the likelihood of detecting ICSS – and indeed they do obtain it from many sites, thus making the negative ones most suggestive. However, failure to detect ICSS from a given site is not as compelling as a positive demonstration of that behaviour from a site.

In the experiment by Crow (1971) cited above, ICSS was obtained from electrodes in the region of the dopaminergic cell bodies of the substantia nigra and A10 and also from a position ventrolateral to the central grey which, as Crow (1971) points out, corresponds to the course of the noradrenergic fibres of the dorsal bundle as they ascend from the LC to innervate the forebrain (Fig. 3.4). Looking at the scatter of points tested in the midbrain, many of which were 'negative', it does seem impressive that the positive ones fall so tightly into two discrete groups. That catecholaminergic fibres of cell bodies are also to be found in the same brain regions would immediately suggest a connection; however, coincidence cannot be accepted as proof until more tests have been applied.

If ICSS acts to stimulate noradrenergic neurones (as seen with the ascending fibres of the dorsal bundle) then it should be possible to obtain ICSS from the cell bodies of origin of that bundle, namely in the pontine nucleus, the LC. Crow, Spear & Arbuthnott (1972) carried out a study of this possibility and found that electrodes in the vicinity of the LC did indeed yield ICSS behaviour, although it was less vigorous than that typically seen from the lateral hypothalamus or A10 areas. More extensive shaping was also required to reveal it. It also differed from A10 in that mouth movements, rather than excited sniffing, accompanied the ICSS behaviour (Anlezark *et al.*, 1973). Electrodes placed lateral or medial to the LC or ventral to it generally failed to yield ICSS, although one site lateral and one anterior to the LC proper were ICSS-positive. The other 12 ICSS-positive sites, however, were classified as being within the LC by the authors. Inspection of the published histological micrographs (Crow *et al.*, 1972) suggests that it was not such an easy job to determine if the electrode tip was indeed in the LC. Most of the published placements appear to be very slightly outside the LC, but certainly much closer to it than the sundry ICSS-negative sites. This point has been made by more than one author previously (Amaral & Routtenberg, 1975). Doubtless, it may be argued that electrical current would spread to the LC even from electrodes just outside it, but nonetheless Crow, *et al.* (1972) felt unable to rule out conclusively the possible involvement of the adjacent mesencephalic root of the Vth cranial nerve (*ibid.*, p. 284). This will assume greater importance

Fig. 3.4. Distribution of electrode placements which yielded ICSS (□) and those which did not (■) throughout the rat mesencephalon. The black areas on the right-hand side of each diagram indicate the location of the A9/A10 dopamine cell bodies, while the shaded area shows the position of the ascending noradrenergic fibres of the dorsal bundle. (Redrawn from Crow, 1971.)

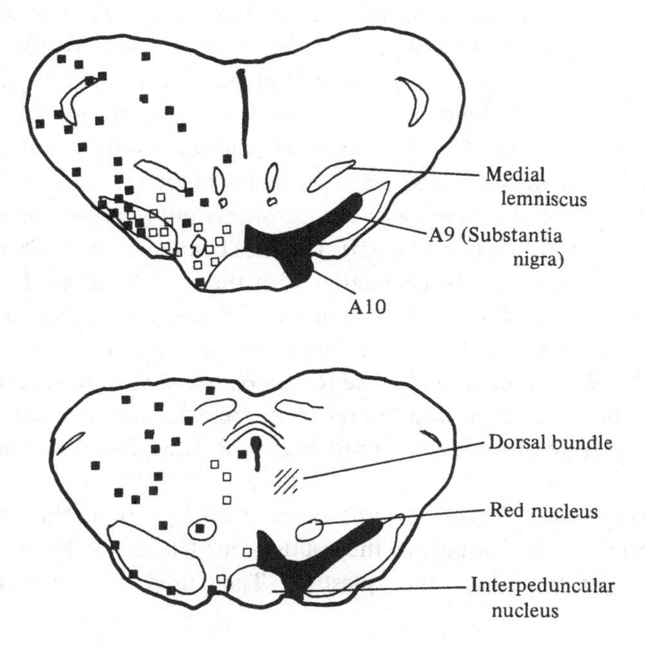

than it seemed to at the time in light of subsequent more detailed mapping studies (Corbett & Wise, 1979).

Since 1972, other authors have also reported ICSS behaviour from electrodes in the LC (Ritter & Stein, 1973). Although here the published histology reveals a few electrode tips more clearly in the LC than the previous study (Crow *et al.*, 1972), more than 10 of the 39 positive ICSS electrodes were definitely out of the LC proper, even if not by much. These authors based their conclusions, of the noradrenergic nature of the ICSS, more on the fact that amphetamine increased and AMPT decreased ICSS rates. As stated in the previous section, AMPT affects both noradrenaline and dopamine and so too does amphetamine. Such conclusions are seen, with the benefit of hindsight, as being unfounded.

The most detailed mapping study, available to date, of ICSS sites in and around the LC is that of Corbett & Wise (1979). These authors used a moveable ICSS electrode which enabled them to test several different sites for ICSS within the same rat. Altogether, they examined 425 sites in and around the LC. Although 140 of the sites were ICSS positive, none of these was directly in the LC proper. Indeed, Corbett & Wise presented in detail two sites which were clearly within the LC but, despite 17 sessions or more of behavioural shaping, could not be made to yield ICSS. These authors also used histofluorescence techniques to reveal the location of catecholamine-containing cell bodies and fibre tracts. All previous authors had used conventional stains to reveal the LC, which need not necessarily correspond to the noradrenaline-containing cell bodies, and further did not reveal the noradrenaline-containing fibre pathways. Corbett & Wise (1979) found five animals with ICSS-positive electrodes in noradrenergic fibre pathways but were unable to show that the closer the electrode was to the pathway the better the ICSS behaviour (Spearman $\rho = 0.03$, not significant). The overall map of ICSS sites in the region of the LC showed that the positive sites corresponded not with the LC proper but were predominantly anterior and lateral to it (see Fig. 3.5). They conclude that the ICSS sites in this pontine region are not correlated with the noradrenergic LC. The findings, that electrodes clearly in the LC did not yield ICSS, that no correlation between the closeness to noradrenergic fibre bundles and the rates of ICSS was found and that the ICSS-positive sites were actually out of the LC, being anterior and lateral to it, seem to undercut the whole previous literature of a role for noradrenaline in ICSS based on mapping studies.

If electrodes near the LC do not cause ICSS by stimulating the LC noradrenergic system, what are they acting on? Clavier & Routtenberg (1976*a*) sought to answer this question. They used the Fink–Heimer staining method which can reveal degenerating axons and terminals to see which pathways were close to the tip of an ICSS-positive electrode in the region of the dorsal noradrenergic bundle. To do this they passed a current large enough to cause a lesion through the ICSS electrode after behavioural testing was complete and then sacrificed the rat and processed the brain with the Fink–Heimer technique. They detected two descending pathways and a number of ascending ones, all of which were running in the vicinity of the so-called 'dorsal noradrenergic bundle' electrode. The descending ones seemed to originate from limbic–midbrain structures and from the prefrontal cortex. The ascending ones, in addition to the dorsal tegmental noradrenergic bundle, included the mammilary peduncle system which originates in the nonnoradrenergic nuclei of Gudden, the superior cerebellar peduncle system and the medial longitudinal fasciculus system, both of which are nonnoradrenergic. A second noradrenergic system, in addition to the dorsal tegmental bundle, was identified as the more medially located tegmental radiation of Weisschedel which contains the dorsal periventricular noradrenergic system as well as other nonnoradrenergic fibres. So the question is not so much what system other than the dorsal noradrenergic bundle is there in this region of the brain to mediate so-called dorsal bundle ICSS, but which of the many other nonnoradrenergic (and one different noradrenergic pathway) is actually crucial?

Perhaps the most important aspect of the above study by Clavier & Routtenberg (1976*a*) is the demonstration that a whole host of pathways are to be found in any small area of the brain. Whereas one may choose to focus on, say, a noradrenergic system which happens to run in that area,

Fig. 3.5. Distribution of electrode placements which yielded ICSS (▲) in the vicinity of the locus coeruleus and those which did not (△). Abbreviations: LC, locus coeruleus; IV, fourth ventricle; DTG, dorsal tegmental nucleus of Gudden; Mes V, mesencephalic nucleus of the trigeminal nerve. (Redrawn from Corbett & Wise, 1979.)

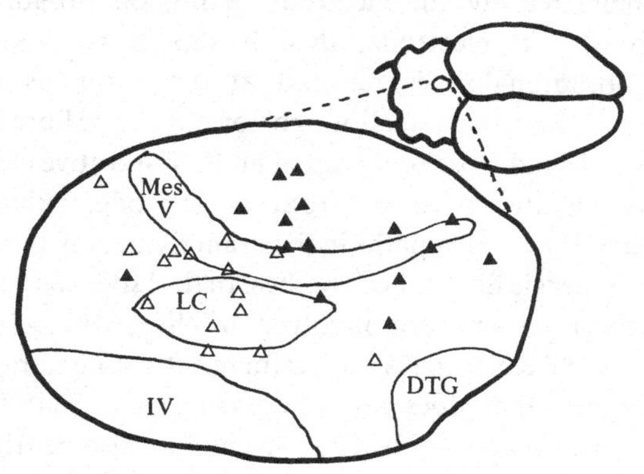

there are in fact many other fibres and pathways any of which could be critical in preference to the rather arbitrarily selected noradrenergic one. The apparent correspondence of ICSS-positive sites with the known distribution of noradrenergic systems cannot be taken to prove a role for noradrenaline.

Ventral systems. Ritter & Stein (1974) found that 11 of 71 electrodes in the vicinity of the ventral noradrenergic bundle, posterior to the dopaminergic cell bodies in the ventral tegmentum, would support ICSS. Very little behavioural shaping was carried out, which may serve to explain the rather low success rate. However, the authors did not use histofluorescence techniques to visualise the ventral noradrenergic fibres directly. Since the ventral bundle, unlike the dorsal noradrenergic bundle, is now known to be a diffuse plexus of fibres rather than a discrete bundle (Lindvall & Bjorklund, 1974) the correspondence between ICSS sites and noradrenergic fibres remains unproven. Certainly, the ventral noradrenergic fibres run somewhere in the vicinity of the ICSS-positive sites but, without direct histofluorescence visualisation relative to the electrode tip, no further connection can validly be drawn. In this respect it is suggestive that Anlezark *et al.* (1972, 1974) were completely unable to obtain ICSS from electrodes in A1 or A2, the cell bodies of origin of the ventral fibre system. Further, the region of the ventral bundle itself more anterior to the cell bodies failed to yield ICSS. Similarly, Clavier & Routtenberg (1974) failed to find ICSS-positive sites in A1, A2 or A5 or in the ventral bundle itself. Although a word of caution about failure to obtain ICSS and its interpretation has been given above, it seems uncertain whether the ventral noradrenergic fibre system sustains ICSS or not.

Noradrenaline release during ICSS

If the rewarding properties of ICSS are being mediated by the electrical current delivered by the electrode acting on noradrenergic systems at the tip of that electrode, then it should be possible to demonstrate that noradrenaline is released at the terminals of that noradrenergic system. It may be that, although noradrenergic fibre systems or cell bodies are to be found close to the tip of an ICSS-positive electrode, they are not actually activated by current from the electrode. Perhaps their threshold is too high or the current paths in the brain tissue sink the current away from the noradrenergic fibres or cell bodies. In the latter case then, the noradrenergic system is clearly not involved in the ICSS behaviour. However, even if the ICSS electrode is successful in activating noradrenergic fibres or cell bodies, that does not necessarily mean that ICSS is mediated by the noradrenergic system. The current may also be triggering another system in addition to any noradrenergic one. It has been seen in the previous section how many systems are to be found in even an apparently 'small' area of the brain. So the conclusions to be drawn from the following studies are limited from the start. If ICSS does not activate noradrenergic systems, they can be excluded from a role in ICSS; if it does, however, this is merely consistent with the model while in no way proving it conclusively.

Push–pull tracers. The first experiment to examine these questions was carried out by Stein & Wise (1969). They injected radioactively labelled noradrenaline (^{3}H-, ^{14}C-; or, in some cases, ^{14}C-labelled precursors such as DOPA) into the lateral ventricles through an indwelling cannula in the freely moving rat. After waiting some 45 min to 1 h they then allowed the rat to press a lever for electrical brain reward delivered through electrodes implanted in the region of the medial forebrain bundle. Any labelled noradrenaline released as a result of this ICSS was collected by a push–pull cannula in either the amygdala or the hypothalamus.The push–pull cannula perfused a small region of brain tissue by injecting artificial cerebrospinal fluid (CSF) solution into the brain via the inner cannula of a concentric double cannulae system and withdrawing fluid at an equal, slow rate from the outer cannula. Substances released by synaptic transmission in the region of the cannula tip will be picked up by the perfusion and carried into the withdrawal cannula, whence they can be detected by their radioactivity and analysed chemically. Stein & Wise (1969) found that ICSS led to an increase of release of radioactivity in perfusates of both amygdala and hypothalamus. The ICSS-induced release was some three to five times that observed under resting conditions without applied electrical stimulation. The increase often took 15 min or so to occur after the initiation of ICSS, suggesting that some diffusion from distant synapses might be required. It also failed to continue at its high level during the whole course of ICSS. Inevitably, after 15 or 30 min of ICSS the release of radioactivity declined to control, pre-stimulation levels despite the fact that electrical current continued to be applied. The authors suggest that this reflected the depletion of the radiotracer from the brain. This result was replicated by Holloway (1975), who also found an increase in the release of ^{3}H-labelled serotonin from the amygdala after medial forebrain bundle ICSS. This questions the specificity of noradrenaline for ICSS if other amines are also released.

Numerous additional technical problems apply to this approach. The push–pull cannula, being a double concentric device, is very large in diameter and from the post-mortem histology can be seen to have inflicted quite considerable damage to the region of the brain in which it was implanted. Further, the perfusion of fluid through a relatively solid medium such as the brain also causes damage. However, although later authors have turned to different methods of measuring noradrenaline release, the work by Stein & Wise (1969) remains of historical importance as the first demonstration of noradrenaline release during ICSS.

Amine concentrations. A second technique often used to demonstrate noradrenaline release after ICSS is to measure the concentration of the transmitter itself in brain regions. Prolonged application of ICSS current, it is argued, would lead to an eventual depletion of the stores of transmitter in the systems being activated by the current. Thus, a decrease in the concentration of noradrenaline is expected in various brain regions if the ICSS electrode is actually activating noradrenergic systems. An early study by Olds & Yuwiler (1972) illustrates this approach, as well as some of its limitations. Animals were implanted with electrodes in the hypothalamus and were allowed to lever-press for ICSS for 20 min and then immediately decapitated and their brains removed, dissected into regions and assayed for noradrenaline. A 40 % decrease in noradrenaline content of the hypothalamus was found, as well as a smaller decrease in the rest of the brain, compared either to unimplanted controls or to rats implanted with electrodes but not allowed to lever-press for ICSS. One problem of using this study to argue that ICSS has it effects on behaviour by releasing noradrenaline is that a comparable decrease in dopamine was also seen. A second curiosity is that rats which had electrical stimulation of the brain imposed on them, rather than earning it by lever-pressing, failed to show a decrease in hypothalamic noradrenaline. If electrical current from the ICSS electrode were directly activating noradrenergic systems, it is hard to see why it should matter whether the animal pressed a lever in order to turn on the current or whether the current came on without participation by the rat. Certainly, it suggests a more complex route of activation of the noradrenergic fibres than immediate stimulation by the electrical current at the tip of the electrode.

A more detailed approach was used by St Laurent, Roizen, Miliaressis & Jacobowitz (1975), who implanted rats with electrodes in the ventral tegmentum and, after a 3 h ICSS session, sacrificed the animal and dissected the brain into 20 different regions by taking 300 μm slices from which circular punches between 1 and 0.5 mm in diameter were taken. These were then assayed for noradrenaline and dopamine, with the results being expressed per milligram of protein in the sample. This beautifully detailed study yielded unexpected results in that no decreases were seen in any area. Rather, a number of quite large increases in catecholamine content were observed; a doubling of the noradrenaline content of the nucleus accumbens, the olfactory tubercle and the central nucleus of the amygdala was seen, along with a small increase in the noradrenaline content of the dorsal bundle punch and an increase in the dopamine levels in the olfactory tubercle sample. The explanation for an increase in catecholamine concentrations is obscure, but it may be of significance that the rats were sacrificed 3 h after the end of the ICCS session, rather than immediately. Perhaps recovery processes of increased synthesis or decreased release had come into play in order to restore transmitter depleted by the ICSS session, and had actually led to a transitory *increase* in absolute transmitter level. This does, however, serve to highlight the major disadvantage of measures of transmitter concentrations – that they may tell you nothing about the state of functional activity of the neurones themselves. Thus, although increased firing of the neurones might be expected initially to decrease transmitter content because of its release, subsequent feedback responses on the neurone, such as an increase in synthesis, may mask or even reverse this effect. Other authors have sought to overcome these problems by measuring levels of transmitters *after synthesis has been blocked* or by adopting other measures of the functional activity of the neuronal system.

Histofluorescence. Arbuthnott *et al.* (1970) implanted rats with electrodes aimed at the dorsal or ventral noradrenergic bundles, imposed stimulation for 1 h, and then immediately killed the rat and assessed the degree of depletion of brain catecholamines by visual inspection of brain sections after reaction with the histofluorescence method of Falk–Hillarp. To avoid the problem of activation-induced increases in synthesis outlined above, a synthesis inhibitor, FLA 63 or H44/68 (a synonym for AMPT) was administered immediately or 1 h prior to the electrical stimulation. Decreases in the histofluorescence intensity in the cortex and hippocampus were seen after electrical stimulation of the dorsal bundle and decreases in fluorescence in the hypothalamus and septum as well as the lateral neostriatum, were seen after ventral bundle stimulation. This seems to be a clear result, indicating an increased release of noradrenaline from terminal areas as a result of stimulating the ascending fibre system. One limitation is that the histofluorescence technique does not allow one to quantify the degree of depletion beyond very gross estimation, nor to distinguish noradrenaline from dopamine. The efficiency of the histochemical reaction may vary from one slide to another, thus making interanimal comparisons a questionable practice. Further, in this study the electrodes were not demonstrated to support ICSS. They were aimed at areas of the brain known to be ICSS-positive, but no test was made on the actual electrodes themselves. Since not all electrodes in a ICSS-positive area will actually yield ICSS and, since not all electrodes in the above study gave depletions of fluorescence, despite being close to ICSS-positive areas, the connection between the two events is not as clear as would be wished. Other studies have remedied these defects.

Arbuthnott, Fuxe & Ungerstedt (1971) implanted rats with various electrodes and were able to demonstrate ICSS from those in the vicinity of the ventral bundle. These rats were then injected with AMPT and allowed to lever-press for ICSS for 2 h. Many rats ceased to press during this period and the electrical stimulation was then imposed at a rate similar to that elicited by the rat itself. Immediately after the session the animals were sacrificed and histofluorescence using the Falk–Hillarp technique carried out. A marked decrease in histofluorescence in hypothalamic nuclei, such as the supraoptic, the paraventricular and the preoptic, was seen. No decrease in dopaminergic fluorescence in the lateral neostriatum was seen with ventral bundle electrodes, thus improving the evidence for release solely of noradrenaline. Rough quantification on a four-point scale showed about a 50 % reduction in fluorescence in these areas compared to non-ICSS exhibiting control rats. This is a major improvement over the previous study in that the same electrodes which caused a semi-quantified decrease in fluorescence were also shown to support ICSS. However, this study has its peculiarities. One electrode which was right in the middle of the ventral bundle area failed totally to yield ICSS, despite causing a marked decrease in fluorescence in the hypothalamus. Also, two placements in the close vicinity of the dorsal noradrenergic bundle failed to yield ICSS despite again causing a marked decrease of fluorescence, this time in the cortex.

It seems clear that electrodes in the vicinity of the ventral bundle in the mesencephalon which support ICSS do increase the release of noradrenaline from the hypothalamus. However, the failure of most authors to obtain ICSS from the source nuclei of the ventral bundle (A1, A2 and A5), or from the ventral bundle posterior to the level of the LC, suggests that the coincidence of electrode sites yielding ICSS with the anatomical location of the ventral bundle in the more anterior mesencephalon is fortuitous (despite causing an increased release of noradrenaline), and is actually mediated by another, possibly noncatecholaminergic, system which is present in that region of the brain.

Biochemical assays. The difficulties of the histofluorescence technique in quantifying the changes in amine levels found after ICSS can be overcome by the use of biochemical assay to determine actual catecholamine concentrations after AMPT blockade of synthesis following ICSS exposure. Miliaressis, Thoa, Tizabi & Jacobowitz (1975) used this to examine 17 brain areas after ICSS through electrodes in the ventral tegmental area. Rats were allowed to lever-press for ICSS 1 h after injection with AMPT and sacrificed immediately after the 90 min session. Significant decreases in noradrenaline content were found in the piriform cortex and the dorsal bundle region and a small nonsignificant change seen in the vicinity of the LC. Decreases in dopamine in the olfactory tubercle after synthesis inhibition were also detected. The advantage of this approach, as well as quantification of changes, is that relatively small ones can be detected reliably. Thus, the piriform cortex decrease was from 3.96 ng amine mg^{-1} protein in nonstimulating controls to 2.60 ng mg^{-1} in ICSS rats. It is unlikely that this would have been detected by histofluorescence, let alone exactly quantified.

Similar changes in the content of catecholamines after synthesis inhibition with either AMPT or FLA 63 have been seen by Stinus *et al.*, (1973). These authors again used ICSS electrodes in the ventral tegmental area and demonstrated that the same electrodes did indeed support ICSS. Some rats received imposed electrical stimulation through these electrodes using the same parameters of current and duration as yielded ICSS. These rats were treated 1 or 2 h previously with FLA 63 or AMPT. Decreases in noradrenaline content of the hypothalamus, hippocampus, cortex and brainstem were demonstrated after both such treatments combined with imposed ICSS. One limitation of this approach is that the use of AMPT, to prevent synthesis of amine from masking possible changes resulting from increased release, will also cause ICSS behaviour to cease within 30 min to 1 h. Thus, imposed, rather than freely earned, electrical stimulation is used in most cases. Stinus *et al.* (1973) turned to another technique for demonstrating the release of amine transmitters to enable the experiment to be carried out in rats which were actively pressing for ICSS, rather than passively receiving imposed stimulation. They injected ^{3}H-labelled noradrenaline into the fourth ventricle 5 min before an ICSS session of 1 h duration. Following immediate sacrifice the amount of [^{3}H]noradrenaline in various brain regions was determined, as it was for rats similarly treated but not permitted to lever-press for ICSS. A 40 % decrease in the [^{3}H]noradrenaline content of the hypothalamus, hippocampus and cortex was observed in rats which had been actively lever-pressing for ICSS. [^{3}H]Dopamine in the olfactory tubercle was also decreased by a similar amount.

The conclusion to be drawn from the above experiments is that ICSS electrodes in the ventral tegmentum certainly release noradrenaline from both ventral and dorsal bundle projection regions. However, such electrodes also release dopamine from the mesolimbic dopaminergic system in the olfactory tubercle. Thus not only is the amine of causative significance in the ICSS behaviour not determined, but even the role of the ventral versus the dorsal noradrenergic pathways is uncertain.

Electrophysiological assays. An electrophysiological demonstration of the release of noradrenaline by electrodes which would support ICSS was reported by Segal & Bloom (1976). They implanted rats with electrodes in the vicinity of the LC and demonstrated ICSS from some of these placements. The animals were also implanted for electrophysiological recording from the hippocampus using 6 μm nichrome wire semimicroelectrodes. Single-unit activity was recorded from these electrodes during ICSS behaviour. It was found that each electrical shock delivered to the LC as a result of an ICSS lever-press caused a 300–800 ms inhibition of firing of hippocampal single units, starting about 50–150 ms after the LC shock. The inhibition of hippocampal single unit firing by electrical stimulation of the LC was shown to be noradrenergic since diethyldithiocarbamate, a noradrenaline synthesis inhibitor, abolished it in 6 out of 15 cells tested. Further, 6-OHDA injected intracisternally abolished both the ICSS behaviour and the LC-induced inhibition of hippocampal single-unit firing within 24 h. However, in the latter case it must be noted that no biochemical assays of the amine depletion were performed and the dose of 6-OHDA used would almost certainly have caused a loss of dopamine, as well as of noradrenaline.

This study demonstrates that ICSS can be obtained from the LC, although again careful examination of the sites (Segal & Bloom, 1976, Fig. 1, upper section) shows that many were significantly outside the LC proper, being anterior and dorsal. The electrical current delivered by each ICSS lever-press clearly has the effect of inhibiting hippocampal unit firing by a noradrenergic mechanism. Whether this is the basis of the rewarding properties of the ICSS, or a totally unrelated coincidence, is not clarified by this study.

Conclusion. A series of studies have demonstrated that catecholamines are indeed released during ICSS behaviour. Unfortunately, both dopamine and noradrenaline, within both the dorsal and the ventral systems, are liberated. Further, the whole thing may have no causative relation to the behaviour under question. The latter point is simply not approached by the techniques used in the above section.

Noradrenergic drugs and ICSS

If ICSS involves transmission at noradrenergic synapses, it should be possible to modify that behaviour with drugs which alter transmission at such synapses. Some drugs, such as synthesis inhibitors or neurotoxic lesioning agents, will be dealt with in subsequent sections and I will concentrate here on agonists and antagonists of the noradrenergic receptors, and on indirect releasing agents. The first such agent which will modify transmission at noradrenergic synapses is, of course, noradrenaline itself.

Noradrenaline. Olds (1974) reported that intraventricular injection of noradrenaline increased ICSS rates from electrodes in the posterior hypothalamus. Her rats were implanted with ICSS electrodes, and with an indwelling cannula aimed at the lateral ventricle through which small volumes of solution containing neurotransmitters could be injected immediately prior to an ICSS session. Injection of 10 μg of *l*-noradrenaline (the physiologically active isomer) caused a 46 % increase in the rate at which rats would lever-press for ICSS at posterior hypothalamic electrodes. The physiologically inactive isomer, *d*-noradrenaline, had virtually no effect, even at doses two or three times those effective with the *l*-isomer. This would appear to support the idea that noradrenaline is released by electrical brain stimulation to cause reward. Thus, the more noradrenaline present at the synapse the greater the reward and hence the more time spent lever-pressing. However, not all rats showed an enhancement of ICSS; nine increased their responding whereas three decreased (a 75 % decrease, moreover) and one was unaffected. While the one unaffected rat might be explained as a failure of the injected noradrenaline to reach the site of electrical stimulation (blocked cannula, diffusion barriers) the marked decrease in ICSS exhibited by three rats cannot be thus dismissed. A further complication was that increasing the doses of *l*-noradrenaline resulted not in an increase in the enhancement of ICSS but in a progressive weakening of the effect. Thus, 20 μg caused a 32 % increase, 30 μg caused an 18 % increase, while 40 μg caused only a 14 % increase in ICSS rates. This was also seen by other authors who found that *l*-noradrenaline injected into the ventricle during ICSS caused an actual decrease in the rate of ICSS (Wise & Stein, 1969: 5 μg of *l*-noradrenaline caused a 19 % *decrease* in ICSS rates). At the most generous it must be postulated that noradrenaline has two effects: a low dose enhancement of ICSS rates and a higher dose reduction of ICSS, possibly related to a more general sedation (Olds, Yuwiler, Olds & Yun, 1964). A further cautionary note to the above study is that noradrenaline was not the only effective amine. Thus, both dopamine and serotonin increased ICSS rates, but less effectively than *l*-noradrenaline.

The effects of central administration of noradrenaline on ICSS do not yield much convincing evidence one way or the other.

Noradrenergic agonists. Noradrenergic receptors in the brain exist in two classes, each with two subtypes. Thus, α- and β-noradrenergic receptors have been identified on pharmacological grounds and each type further subdivided into α_1 and α_2, β_1 and β_2 respectively. Some of the problems described above concerning the difficulties in demonstrating profound effects of centrally administered noradrenaline on ICSS behaviour might be the result of stimulation of more than one receptor class or subtype. It might be speculated, for example, that activation of α_2 receptors has a sedative effect on the animal, while the receptor involved in the reward component of ICSS might be of the α_1 type. Noradrenaline, the natural neurotransmitter, stimulates both types and hence would produce mixed effects which were difficult to interpret. Clearer results might be obtained with artificial compounds which activate only one of the four types of adrenoreceptor. Not all subtypes are as easily manipulated as others since some agonists fail to cross the blood–brain barrier, or when they do they activate both types 1 and 2 of a given class. However, some relatively selective agonists and antagonists are now becoming available.

Clonidine is believed to be an α_2-agonist (Langer, 1980), although it also possesses effects on histamine receptors (Audigier, Virion & Schwartz, 1976; Sastry & Phillis, 1977) and on adrenaline receptors (Bolme *et al.*, 1974; Cedarbaum & Aghajanian, 1976). When injected intraperitoneally into rats with electrodes in the medial forebrain bundle clonidine serves to decrease ICSS responding in a dose-related manner (Vetulani, Leith, Stawarz & Sulser, 1977). Indeed, it is highly effective in this action, with doses as small as 0.025 mg kg^{-1} producing clear results. Two shortcomings of this study are obvious. First, clonidine at high doses causes a general sedation, so that the decrease in lever-pressing reported above might be no more than a generalised reduction in all forms of locomotor activity, unrelated to a specific effect on the reward component of ICSS. Secondly, clonidine injected intraperitoneally will also markedly affect the peripheral adrenergic system. No proof of a central site of action was provided. Hunt, Atrens, Chesher & Becker (1976) remedied these defects by using a rate-free measure of ICSS from lateral hypothalamic electrodes (see p. 19) and found that clonidine was effective in reducing the rewarding value of ICSS as judged by an increase in latency for the rat to shuttle across a shuttle box to initiate ICSS. That this was not just a block of locomotor ability is suggested by the relative failure of clonidine to alter the rat's latency to shuttle back across the shuttle box in order to terminate (escape from) the ICSS once initiated. Doses of clonidine as small as 0.016 mg kg^{-1} and 0.03 mg kg^{-1} caused 60 and 80 % increases in initiation latencies, respectively (see Fig. 3.6). A proof of the central site of action of clonidine was also offered by these authors with the finding that the peripheral α-agonist, *l*-phenylephrine, which does not cross the blood–brain barrier, was *not* effective in altering ICSS behaviour. Similar depressant effects of clonidine on ICSS have been reported by Herberg, Stephens & Franklin (1976) for electrodes in the LC and substantia nigra, as well as the lateral hypothalamus.

Thus it may be concluded that clonidine clearly depressed ICSS behaviour in a highly potent fashion, probably not related to its additional sedative or hypotensive effects. Whether this is an effect mediated by its action on the α_2-receptor, rather than via a histamine H_2 or an adrenaline receptor, has not been investigated. Further, it is unknown whether the depression of ICSS, *if* α_2 in nature, is the result of the decrease release of noradrenaline caused by activation of presynaptic α-receptors (Langer, 1980), either in the LC itself (Cedarbaum & Aghajanian, 1976) or on the noradrenergic terminals in distant regions, or because of noradrenaline-

Fig. 3.6. Effects of clonidine on mean response latencies to initiate and to escape hypothalamic ICSS in a shuttle box. (Redrawn from Hunt *et al.*, 1976.)

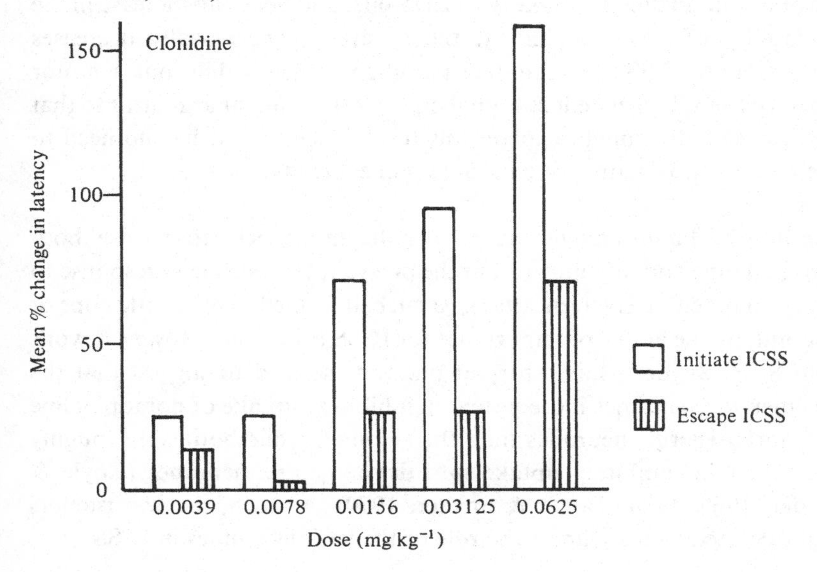

like activation of post-synaptic α_2-receptors in forebrain regions (U'Pritchard *et al.*, 1980). If the last of these is the case, it need not necessarily mean that noradrenaline is antagonistic to ICSS but, as argued by Herberg *et al.* (1976), that ICSS behaviour requires noradrenaline synaptic transmission in a 'packet-like' process. That is, a brief activation of the receptor after each lever-press is necessary for that lever-press to be rewarded and hence to occur again. Continuous activation of the receptor removes the need even to lever-press to obtain reward and hence would lead to a decrease in ICSS behaviour, since the animal is in a continuous state of pharmacological nirvana. Clearly, studies with clonidine have raised more questions than they have solved.

The α-blocking drug, thymoxamine, has also been reported to decrease ICSS (Herberg *et al.*, 1976). Decreases in lever-pressing for ICSS of 79 and 61 %, respectively, were reported for electrodes in the LC and in the substantia nigra after 10 mg kg^{-1} thymoxamine. Blockade of α-receptors by intraperitoneal injection of phentolamine (5 mg kg^{-1}) has also been found to reduce ICSS lever-pressing (Hastings & Stutz, 1973) while blockade of β-receptors by 10 mg kg^{-1} propranolol was without effect. Wise, Berger & Stein (1973) have reported similar results with intraventricular administration of phentolamine and lack of effect with propranolol, thus arguing for a central site of action. However, none of these studies addressed the question of whether the sedative effects of α-blockade could be the cause of the decrease in ICSS lever-pressing, rather than a true blockade of the rewarding properties of the electrical stimulation. Blockade of noradrenaline reuptake with LU5-003, and hence an increase in the availability of noradrenaline at the synapse, paradoxically decreases hypothalamic ICSS in a rate-free paradigm, hence ruling out a motor deficit (Atrens, Ungerstedt & Ljungberg, 1977). A similar argument to that used for clonidine might explain this result since the rat has no need to lever-press if it is getting free noradrenaline already.

The d- *and* l-*amphetamine story*. Amphetamine acts to release both noradrenaline and dopamine from the presynaptic neurone in response to an action potential (Iversen, 1967). As such, it seemed to offer little hope of determining the amine of importance for ICSS behaviour. However, work from S. H. Snyder's laboratory at one time seemed to suggest that the *d*-isomer was ten times more potent in inhibiting uptake of noradrenaline into noradrenergic neurones than the *l*-isomer, while both were roughly equipotent in inhibiting uptake into dopaminergic neurones (Coyle & Snyder, 1969, 1970; Taylor & Snyder, 1970). As such, the two isomers might be used to tease apart the role of the various amines in ICSS.

Phillips & Fibiger (1973) investigated the effects of the two isomers on ICSS lever-pressing from electrodes in the lateral hypothalamus or in the substantia nigra. With lateral hypothalamic electrodes, the *d*-isomer was some seven to ten times more potent in causing an increase in ICSS rate than the *l*-isomer (see Fig. 3.7). On the other hand, the two isomers were equipotent for substantia nigra electrodes. Although the authors were commendably cautious and refrained from spelling it out, this would at that time have seemed to suggest that lateral hypothalamic ICSS may involve a noradrenergic substrate, while that from the substantia nigra reflects dopaminergic effects. A later study by Stephens & Herberg (1975) showed that the hypothalamic area actually consisted of two distinct

Fig. 3.7. Effects of the *d*- and *l*-isomers of amphetamine in increasing ICSS rates from hypothalamus and substantia nigra placements. (Redrawn from Phillips & Fibiger, 1973.)

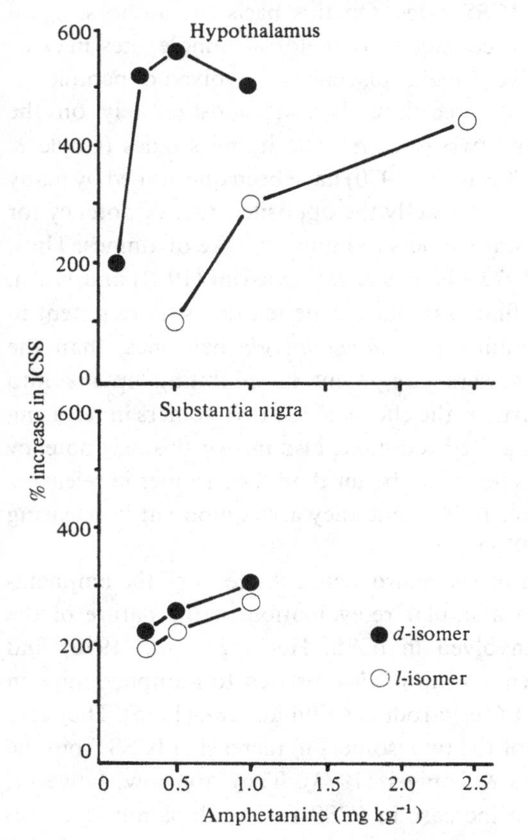

groups of electrode placements. A medial group of electrode positions responded with a much greater increase in ICSS lever-pressing to *d*-amphetamine than to the *l*-isomer, and thus on the above basis were called 'noradrenergic' by the authors. A second group of electrode placements, clustering more laterally than the preceding ones, but still within the hypothalamus, responded more equally to the two isomers and were termed 'dopaminergic' (Stephens & Herberg, 1975). This seemed to make sense, with the known path of the ascending dopaminergic fibres from the substantia nigra being more lateral in the hypothalamus than the course of the ventral bundle and noradrenergic terminal innervation of the hypothalamus.

Ellman *et al.* (1976) investigated the relative effects of the two isomers on ICSS lever-pressing from electrodes in the dorsal noradrenergic bundle or in more medial, periventricular sites. Dorsal bundle ICSS showed a much greater increase in response to *d*-amphetamine than to the *l*-isomer. The periventricular placements responded equally to the two isomers and with a much reduced increase in ICSS rates. On this basis the authors, again expressing enviable caution, concluded that dorsal bundle sites may be noradrenergic whereas periventricular placements involved dopamine.

The interpretation of the preceding data depends entirely on the neurochemical effects of the two isomers. The initial studies (Coyle & Snyder, 1969, 1970; Taylor & Snyder, 1970) have been questioned by many more recent ones which suggest exactly the opposite order of potency for the ability of the amphetamine isomers to inhibit uptake of amines. Thus, Ferris, Tang & Maxwell (1972), Harris & Baldessarini (1973) and Horn, Cuello & Miller (1974) all find *d*-amphetamine ten times more potent in inhibiting the uptake of amine into *dopaminergic* neurones, than the *l*-isomer. The two isomers were equipotent in inhibiting uptake into *noradrenergic* neurones. Further, the effects of the two isomers in releasing amines, rather than blocking their reuptake, also mirror this new potency order. Thus, the *d*-isomer is ten times better than the *l*-isomer in releasing *dopamine* (Chiuch & Moore, 1974) while they are equipotent in releasing *noradrenaline* (Svensson, 1971).

The above re-evaluation of the neurochemical effects of the amphetamine isomers must lead to a similar re-evaluation of the nature of the neurotransmitter system involved in ICSS. Herberg *et al.* (1976) find exactly the same 10:1 potency ratio of *d*-compared to *l*-amphetamine in increasing ICSS rates from LC electrodes as Ellman *et al.* (1976). They also find the same equipotency of the two isomers in increasing ICSS from the substantia nigra as Phillips & Fibiger (1973). They are now, however, forced to conclude that the increase in ICSS rate of dopaminergic sites (substantia nigra electrodes) is caused by *noradrenaline* released by the amphetamine, while the increase in rate at noradrenergic sites (LC electrodes) results from *dopamine* released by amphetamine.

All in all, the use of the isomers of amphetamine to discriminate the neurochemical basis of ICSS at different sites has not proved conclusive.

Synthesis inhibitors

Another class of drugs of great interest in terms of elucidating the neurochemical substrate of ICSS behaviour are the synthesis inhibitors. These act to inhibit the enzyme dopamine-β-hydroxylase (DBH), which is responsible *in vivo* for the conversion of dopamine into noradrenaline. Rats treated with DBH inhibitors are thus prevented from synthesising further noradrenaline and exhaust available stores within a few hours, as judged by the fall in brain noradrenaline concentrations after DBH inhibitor treatment. The great advantage of DBH inhibitors is that they do not greatly affect the synthesis of dopamine and as such can, in theory, be used to separate the roles of dopamine and noradrenaline in ICSS.

The first experiment to use DBH inhibitors to investigate the role of noradrenaline in ICSS was reported by Wise & Stein (1969). They implanted rats with ICSS electrodes in the medial forebrain bundle and with cannulae into the lateral ventricle. Injection of disulfiram intraperitoneally or diethyldithiocarbamate (DDC) intraventricularly reduced ICSS to 20 % of normal within 1 to 3 h of disulfiram administration and almost immediately after intraventricular DDC. Subsequent injection of neurotransmitters into the ventricles showed that only *l*-noradrenaline was effective in reinstating ICSS behaviour; while dopamine, serotonin and the physiologically inactive isomer *d*-noradrenaline were ineffective in the doses used (see Fig. 3.8). The animals were reported to appear 'sedated and. . .disinterested (*sic*)' (Wise & Stein, 1969) after both DBH inhibitors and injections of *l*-noradrenaline 'rapidly produced a state of arousal and alertness' (Wise & Stein, 1969). This will assume greater importance later.

The results of Wise & Stein (1969) appeared to argue strongly for a role of noradrenaline in mediating the reward component of ICSS and were replicated by Shaw & Rolls (1976). These later authors showed that 200 mg kg^{-1} disulfiram injected intraperitoneally into rats lever-pressing for ICSS through lateral hypothalamic electrodes resulted in almost total suppression of this behaviour. The direct receptor agonists, clonidine (0.73–3 μg intraventricularly or 37 μg kg^{-1}–3.0 mg kg^{-1} intraperitoneally), naphazoline or oxymetrazoline (0.9–250 μg intraventricularly) did not restore the behaviour whereas the indirect releasing agent, phenylephrine (15 μg intraventricularly), caused a prompt recovery of ICSS rates. Other indirect releasing agents such as amphetamine or methylphenidate were also able to restore ICSS after disulfiram treatment. This suggests that release of noradrenaline must be impulse-contingent. Thus, a brief packet of noradrenaline released after each lever-press mediates the reward of the electric current. Continuous stimulation of the postsynaptic receptor, not related to the timing of a lever-press, is not effective, and, as argued by Herberg *et al.*, (1976), may actually decrease ICSS rates by rendering the lever-press superfluous to central pharmacologically induced reward.

Other DBH inhibitors have not always yielded the same story. Thus, Lippa, Antelman, Fisher & Canfield (1973) report that FLA 63 failed to disrupt ICSS behaviour despite being given in doses which in other rats led to a profound depletion of brain noradrenaline. Stinus, Thierry & Cardo (1976) found that neither FLA 63 nor U 14,624 decreased ICSS from hypothalamic or ventral tegmentum electrodes. A similar failure of FLA 63 on its own to affect ICSS was reported by Franklin & Herberg (1975). Decreases in ICSS rate of only 14 % were found in rats with either hypothalamic or LC electrodes. However, these authors suggest that when fresh synthesis is prevented by FLA 63, a reserve pool of already existing noradrenaline may be mobilised to permit ICSS to continue. This they claimed to support by the finding that pre-treatment with reserpine, which depletes the reserve pool, greatly enhanced the effects of FLA 63 on ICSS. Now, 3 or 5 days after reserpine pre-treatment, FLA 63 caused 81 % and 56 %

Fig. 3.8. Cumulative record showing the suppression of ICSS from medial forebrain bundle electrodes after 200 mg kg^{-1} disulfiram (DS) and reinstatement by 5 μg of *l*-noradrenaline (*l*-NA), but not by the physiologically inactive *d*-isomer (*d*-NA) or by dopamine (DA). Disulfiram was administered intraperitoneally and the other agents via chronically indwelling ventricular cannulae. The cumulative record was reset after each 100 responses. (Redrawn from Wise & Stein, 1969.)

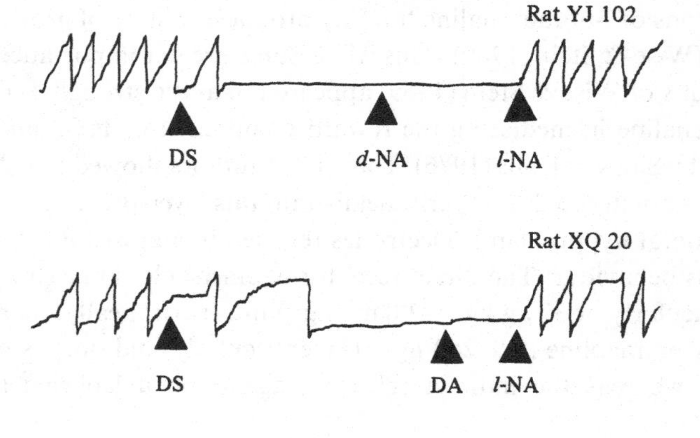

decreases in ICSS rates, respectively (see Fig. 3.9). The conclusion is open to question, since reserpine depletes not only reserve pools of noradrenaline, but also those of dopamine and serotonin. Further, reserpine causes considerable debilitation of the animal, as does FLA 63 to a lesser extent, so a simple addition of nonspecific debilitation might account for these results. Indeed, repeated FLA 63 treatments are successful in reducing ICSS rates but the authors attribute this to the general debilitation of cumulative doses (K. B. Franklin, unpublished observations cited in Franklin & Herberg, 1975). The 'reserve pool' hypothesis also leaves unexplained why disulfiram and DDC *are* effective in suppressing ICSS behaviour, since the reserve pool is intact soon after administration (see Wise & Stein, 1969).

Cooper, Konkol & Breese (1978) found that the combination of the DBH inhibitor U 14,624 with 6-OHDA treatment which caused better than 90 % depletion of whole brain noradrenaline still failed to affect ICSS rates from electrodes in the substantia nigra or the LC.

The mechanism of action of DBH inhibition in causing reduction of ICSS rates has also been questioned. Wise & Stein (1969) observed that their rats were sedated after disulfiram or DDC. Roll (1970) observed a similar cessation of ICSS following disulfiram, but found that if she placed the rat on the ICSS bar it would, in all cases, resume lever-pressing for ICSS. If the long pauses following disulfiram were eliminated from the

Fig. 3.9. ICSS rates in the third hour after FLA 63 (25 mg kg^{-1}) in rats pre-treated with reserpine (2.5 mg kg^{-1}) either three or five days previously. (Redrawn from Franklin & Herberg, 1975.)

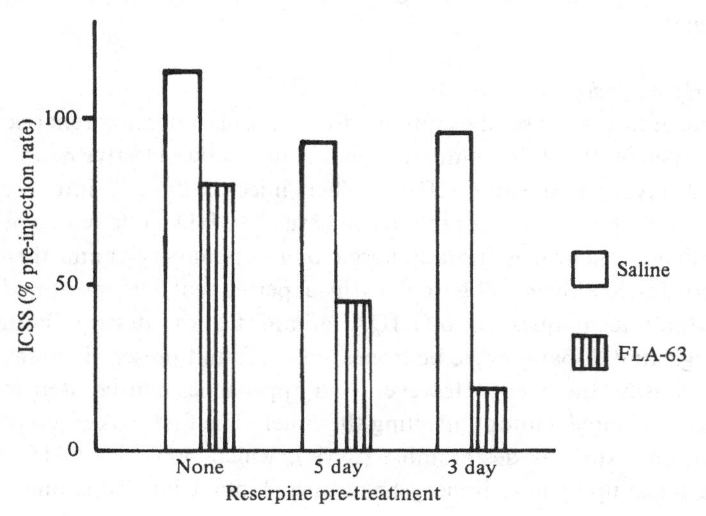

analysis *no* decrease in rate of lever-pressing was found. She interpreted this to mean that when the rat was awake it would press as avidly for ICSS as it would do normally. Cessation of ICSS was caused by the onset of sedation, rather than a decrease in the rewarding value of the electrical brain stimulation. Placing the rat manually on the lever served to wake it up and enable it to resume ICSS at a near-normal rate.

The above conclusion is further supported by studies of Rolls, Kelly & Shaw (1974*a*), who compared the ability of disulfiram to disrupt ICSS lever-pressing with its disruption of other behaviours, such as locomotor activity and rearing. Although 50 % reduction in ICSS rate was observed with 200 mg kg^{-1} disulfiram and a similar effect with the α-receptor blocking agent phentolamine, both agents caused considerably greater disruption of rearing and locomotor activity (see Fig. 3.10*a*). Thus, the effects of noradrenergic depletion or blockade were not specific for the reward element of ICSS but also disrupted many other types of general behaviour. Figure 3.10*b* shows a dose–response curve for disulfiram in terms of its effects on ICSS lever-pressing and locomotor activity. Compared to suppression of ICSS, roughly twice as much suppression of locomotor activity was caused for a given dose. Thus, any decrease in ICSS rates cannot be attributed to a blockade of the reward element of ICSS, but is the result of a general sedative effect of disulfiram on all types of active behaviours.

In conclusion, it seems that not all DBH inhibitors act to decrease ICSS behaviour (FLA 63 and U 14,624). Further, those that do (disulfiram, DDC) probably act to sedate the rat and render it incapable of pressing the lever rather than specifically reducing the reward aspect of the electrical brain stimulation.

6-Hydroxydopamine lesions

One of the most useful techniques for manipulating noradrenergic systems independently of dopaminergic ones employs the selective neurotoxin 6-hydroxydopamine (6-OHDA). When injected directly into the brain, either into the ventricles or into brain tissue, 6-OHDA is taken up by catecholaminergic neurones (noradrenergic *and* dopaminergic) and then proceeds to destroy them. This it does in a permanent and profound fashion. Many techniques of 6-OHDA administration destroy both noradrenergic and dopaminergic neurones – a confound present in many early studies using the agent. However, two approaches can be used to manipulate one amine without affecting the other. The first makes use of uptake blockers, such as desipramine (DMI), which prevent 6-OHDA from being taken up by noradrenergic neurones. Uptake into dopaminergic neurones still occurs and the resulting destruction is of dopaminergic neurones while sparing the noradrenergic ones.

The second technique makes use of injection of small volumes (1–2 μl) of 6-OHDA dissolved in ascorbate–saline into brain areas which contain only noradrenergic or only dopaminergic neurones. Thus, because of the spatial limits of diffusion, only one of the two amine systems will be affected. Typical of this approach is the use of bilateral injections of 6-OHDA into the ascending dorsal noradrenergic bundle in the posterior mesencephalon, where it is separate from both the ventral bundle and the dopaminergic cell

Fig. 3.10. (*a*) Effects of disulfiram (DS: 200 mg kg^{-1}) and phentolamine (PH: 100 mg kg^{-1}) on rearing, locomotor activity and ICSS lever-pressing compared to control (CON) rats. (*b*) Dose–response curve of the suppressant effects of disulfiram on ICSS lever-pressing and on locomotor activity (Loco). (Redrawn from Rolls *et al.*, 1974*a*).

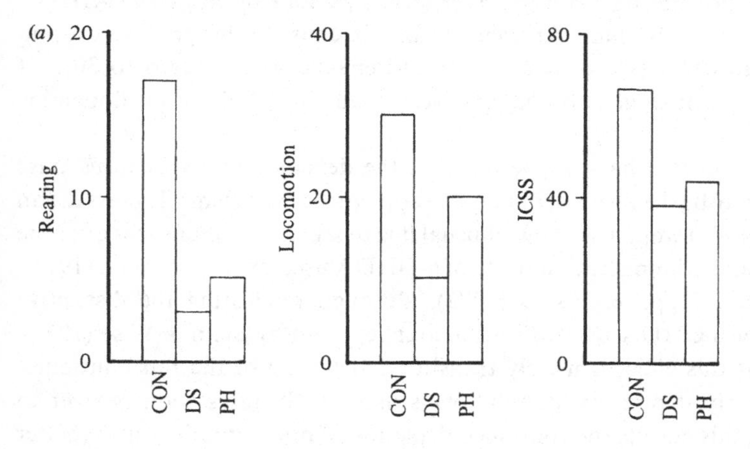

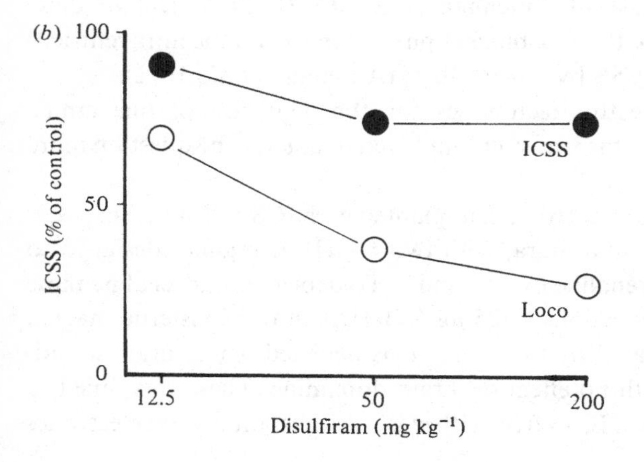

bodies of A9 and A10. A related approach is to make small electrolytic lesions in the same area, or to the cell bodies in the LC which give rise to the dorsal bundle fibres. This has a similar effect in destroying noradrenergic systems without affecting dopaminergic ones, but has the disadvantage of destroying other, noncatecholaminergic pathways which might run in the vicinity of the electrolytic lesion. 6-OHDA injection would *not* affect noncatecholaminergic systems.

Early experiments. The first experiment to make use of 6-OHDA to destroy catecholaminergic systems and examine ICSS behaviour was reported by Stein & Wise (1971). Intraventricular injection of 200 μg of 6-OHDA via an indwelling cannula into the lateral ventricle resulted in a decrease in ICSS rate from that recorded by electrodes in the medial forebrain bundle. Rates fell, after injection, to some 40 % of control and remained depressed at 65–70 % for the next 6 days. Repeated injections of 25 μg of 6-OHDA through the indwelling ventricular cannula caused a progressive drop in ICSS rates to 67 % of control. Noradrenaline was reduced to 30% of normal; surprisingly no changes were found in whole brain dopamine values.

Other authors have suggested that the deficits seen by Stein & Wise (1971) may be the result of damage to dopaminergic systems. Bowers & van Woert (1972) emphasise how unusual it is to see no dopamine loss after the dosage and administration route of 6-OHDA used by Stein & Wise (1971). Antelman, Lippa & Fisher (1972), although replicating the disruptive effects of 6-OHDA on ICSS behaviour reported by Stein & Wise (1971), find that this effect is merely transitory. If testing of the rats continues, particularly if priming stimulation is used, ICSS rates soon recover to normal; this despite the continued depletion of brain noradrenaline (better than 90 % in the case of Antelman *et al.*, 1972). Treatment of these 'recovered' animals with the α-blocker phentolamine via the intraventricular route disrupted ICSS by a mere 10 % (Antelman *et al.*, 1972).

Clearly, more selective techniques for the depletion of one amine independently of the other must be employed to answer the question more convincingly.

Cooper *et al.* (1978) treated rats, implanted with ICSS electrode in either the LC or the substantia nigra, with two 6-OHDA regimes designed to deplete either noradrenaline or dopamine. To deplete noradrenaline these authors used three injections of 25 μg 6-OHDA into the cisterna magna, each injection separated by two days. This depleted whole brain noradrenaline by 70 % with no effect on brain dopamine. These rats failed to show any disruption of ICSS from either LC or substantia nigra electrodes (see Fig. 3.11). Depletion of brain dopamine was achieved by injection of 200 μg 6-OHDA intracisternally following pre-treatment with 30 mg kg^{-1} of the noradrenaline uptake inhibitor DMI. This reduced brain dopamine by 60 % without markedly affecting noradrenaline levels. This treatment did reduce ICSS rates from both electrodes to less than 20 % of pre-operative rates. However, recovery occurred, with normal ICSS rates reappearing within 8–12 days post-operation (see Fig. 3.11). Cooper, Cott & Breese (1974) obtained an identical pattern of results using lateral hypothalamic ICSS electrodes. The noradrenaline-depleted group, when tested 2 days after the 6-OHDA treatment, pressed at pre-operative rates, whereas the dopamine-depleted group fell to less than 30 %. This then seems to implicate brain dopamine in the reduction of ICSS seen after 6-OHDA administration, in contradiction to Stein & Wise's (1971) conclusion of a role for noradrenaline.

One relatively novel study used the neurotoxin 6-aminodopamine (6-ADA) in an attempt to dissociate the role of the two catecholamines in ICSS behaviour. Monasmith, Plotsky, Blank & Adams (1976) injected 100 μg of 6-ADA into the lateral ventricles of rats implanted with ICSS electrodes in the medial forebrain bundle. Unlike 6-OHDA, the neurotoxin

Fig. 3.11. Effects of different 6-hydroxydopamine (6-OHDA) treatments on ICSS rates from locus coeruleus or substantia nigra electrodes. Controls received intracisternal injection of ascorbate–saline, while dopamine depleted (DA↓) animals received a similar injection of 200 μg 6-OHDA 1 h after 30 mg kg^{-1} of the noradrenaline uptake inhibitor desipramine intraperitoneally. Noradrenaline depleted (NA↓) rats received three injections intracisternally of 25 μg of 6-OHDA, each separated by 2 days. 'P' refers to response rate prior to lesion. (Redrawn from Cooper *et al.*, 1978.)

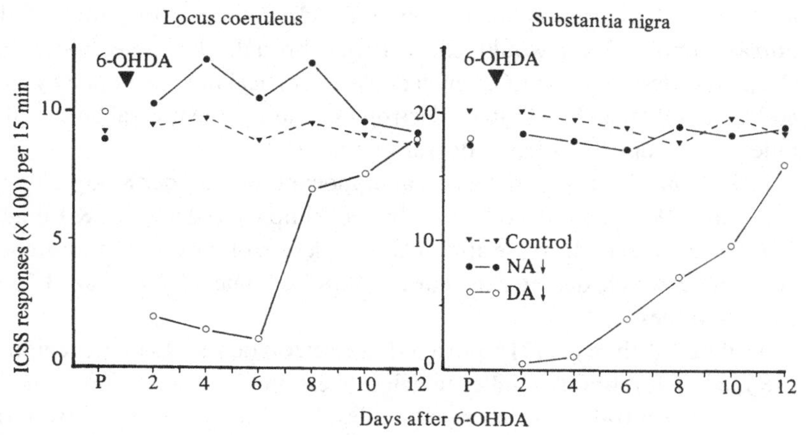

6-ADA depletes noradrenaline without affecting dopamine. Indeed, whole brain noradrenaline was reduced to 50 % of control values by this procedure without any change in dopamine. The ICSS rates of these rats fell to between 0 and 35 % of pre-injection rates immediately after injection but recovered to better than 50 % after 6 days. These data would appear to suggest a role for brain noradrenaline, were it not for the marked toxicity of 6-ADA (Monasmith *et al.*, 1976). Virtually no other experiments have used 6-ADA neurotoxin for this reason.

Noradrenergic lesions. Corbett, Skelton & Wise (1977) found that destruction of the dorsal noradrenergic bundle by bilateral electrolytic lesion failed to affect ICSS rates from electrodes in the LC and actually increased rates from lateral hypothalamic electrodes. Even testing 24 h post-lesion failed to reveal any disruption of ICSS. The validity of the lesion was shown by post-mortem fluorescence histology in which build-up fluorescence could be seen on the caudal side of the lesion in the region of the dorsal bundle, with a disappearance of fluorescence rostral to the lesion. Similar results were reported by Clavier, Fibiger & Phillips (1976), who found that 6-OHDA lesion to the dorsal bundle which depleted forebrain noradrenaline by over 96 % did not affect ICSS from LC electrodes when tested one week post-operatively.

A similar picture seems to apply for ICSS from electrodes in the posterior hypothalamus. Koob, Balcolm & Meyerhoff (1976) found that electrolytic lesions of the LC which depleted forebrain noradrenaline to 20 % of normal caused not a decrease in ICSS rate but a 180 % *increase* in lever-pressing. Clearly, in this case, release of noradrenaline by the electrical current at the tip of the ICSS electrode was not necessary for self-stimulation behaviour.

Clavier & Routtenberg (1976*b*) showed a similar result for electrodes located in the mesencephalic tegmentum close to the trajectory of the dorsal bundle. Rats would show ICSS through these electrodes but electrolytic lesion of the LC, which depleted cortical noradrenaline by over 80 %, did not affect ICSS rates. Electrolytic lesion of the ventral bundle also failed to disrupt ICSS from 'dorsal bundle' electrodes.

ICSS from electrodes in the dorsal hippocampus (van der Kooy, Fibiger & Phillips, 1977) or in the olfactory bulb (Phillips, van der Kooy & Fibiger, 1977) also failed to decrease after 6-OHDA lesion of the dorsal noradrenergic bundle which depleted forebrain noradrenaline to less than 3 % of control values.

Stiglick & White (1977) reported that electrolytic lesions in the region of the medial forebrain bundle, which corresponded to the course of the dorsal and ventral noradrenergic bundles, failed to disrupt ICSS from the lateral hypothalamus. This was true, irrespective of whether the required response to deliver electrical brain stimulation was lever-pressing, tail-moving or alley-way running. Unfortunately, since these studies did not validate their lesions either by post-mortem assay of biogenic amines or by fluorescence histochemical inspection of the amine fibre systems, the negative conclusion is rendered much more tentative. These authors sought to demonstrate that their lesions would be expected to destroy the ascending catecholamine systems since they correspond to the positions marked by Ungerstedt (1971) for these fibre bundles. We have already seen the shortcomings of this approach in the noradrenaline mapping story (see p. 28).

The conclusion of this section is perhaps the firmest in the noradrenaline and ICSS story to date. It is simply that lesioning the noradrenergic systems running at the tip of the stimulating electrode does not impair ICSS from that electrode one iota. This would imply, on the most straightforward model, that release of noradrenaline as a result of the electrical current does *not* play a role in the rewarding element of ICSS. Release certainly happens (see p. 30), but it is unimportant for whatever it is that does mediate the ICSS behaviour from that electrode.

Noradrenergic ICSS from a dopaminergic region? It has even been suggested that ICSS from a typically dopaminergic area (that is, one containing dopaminergic fibres or cell bodies), such as the substantia nigra, was actually the result of activation by the stimulating electrode of noradrenergic fibres of passage. Belluzzi, Ritter, Wise & Stein (1975) found that ICSS from electrodes in the vicinity of the dopaminergic substantia nigra could be disrupted by knife cuts or 6-OHDA injections caudal to this level, which interrupted the ascending dorsal and ventral noradrenergic bundles (see Fig. 3.12). Thus, ICSS fell to very low levels and only started to recover after 6–8 days. Knife cuts, or 6-OHDA, contralateral to the ICSS electrode caused a 1–2 day transitory decrease in ICSS rates with a subsequent rebound to 160 % above pre-lesion rates. Administration of the noradrenaline synthesis inhibitor DDC also suppressed ICSS from nigral electrodes and injection of intraventricular noradrenaline reinstated ICSS behaviour. This the authors interpret as indicating that ICSS from a supposedly dopaminergic area such as the substantia nigra actually involves the release of noradrenaline by the stimulating electrode.

However, Clavier & Fibiger (1977), using a smaller dose of 6-OHDA, found that depletion of forebrain noradrenaline to less than 6 %, and hypothalamic noradrenaline to around 20 %, of normal did *not* disrupt ICSS from electrodes in the substantia nigra. These noradrenaline depletions were, in fact, greater than those obtained by Belluzzi *et al.* (1975) and yet no effect on ICSS was seen. Clavier & Fibiger (1977) suggested that the results of Belluzzi *et al.* (1975) might be the nonspecific effects of the high dose of 6-OHDA which they used.

6-OHDA lesion of the ascending dopaminergic pathway also failed to cause a permanent loss of ICSS from substantia nigra electrodes (Clavier & Fibiger, 1977). Although ICSS rates fell to nearly zero for 2 or 3 days post-operatively they subsequently recovered over the next 6–7 days. A similar time course was seen whether the lesion was ipsilateral or contralateral to the stimulating electrode. If the electrode was activating the ascending dopaminergic fibre bundle to release dopamine as the mediator of electrical brain reward, then destruction of the contralateral bundle would have no effect, while destruction of the ipsilateral bundle

Fig. 3.12. Effects of knife cuts posterior to the substantia nigra, or injections of 6-hydroxydopamine (6-OHDA), on ICSS rates from substantia nigra electrodes. 'C' refers to pre-lesion control response rates. (Redrawn from Belluzi *et al.*, 1975.)

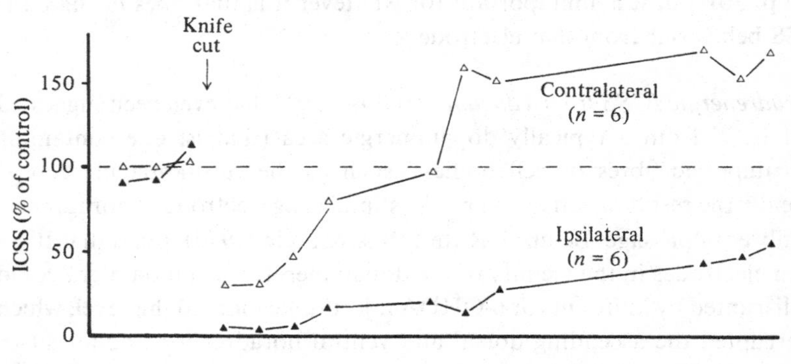

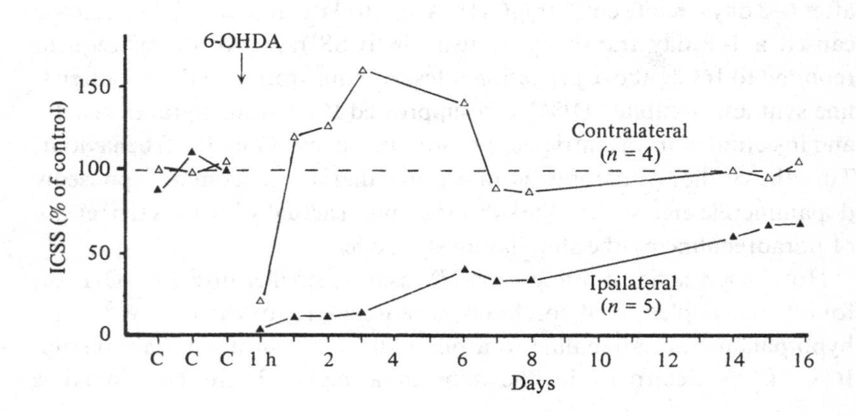

should permanently block ICSS. A confound in the interpretation is that unilateral lesion to the dopamine bundle will also cause a motor impairment (see p. 302). However, this impairment in the motor act of lever-pressing should be the same whether the ipsilateral or contralateral bundle to the stimulating electrode is lesioned. It was this latter pattern of results which Clavier & Fibiger (1977) obtained. They thus concluded that *neither* noradrenaline *nor* dopamine was involved in nigral ICSS.

If neither catecholamine is involved in ICSS from electrodes in the substantia nigra, what is the system actually mediating the reward element of ICSS? Clavier & Corcoran (1976) reported that electrolytic lesion of the sulcal prefrontal cortex caused a 67 % decrease in ICSS rates from nigral electrodes. This decrease, moreover, was permanent throughout the 14 days of post-operative testing. Fink–Heimer histochemistry was used to reveal which neuronal systems had been damaged by the prefrontal lesion and a descending fibre pathway from the sulcal prefrontal cortex running through the vicinity of the substantia nigra was seen. Thus, a noncatecholaminergic basis for nigral ICSS may be possible via this descending system from the sulcal prefrontal cortex.

Conclusion

Briefly summarised the preceding sections have suggested the following:

(1) Early work proposed a correspondence between the noradrenergic systems in the rat brain and ICSS-positive sites. Later work showed, with more detailed mapping, that the two systems were slightly, but significantly, out of alignment. Mapping studies now do not support a role for noradrenaline in ICSS.

(2) Release studies most certainly indicated that ICSS from electrodes near the major noradrenergic systems does release noradrenaline from the terminals of these systems. Whether this release then mediates the rewarding element of that ICSS is not answered by release studies.

(3) Drugs and ICSS studies suggest that α-agonists decrease ICSS by a mechanism separable from simple sedation. The same appears to be true for injection of noradrenaline itself. α-blockers also decrease ICSS but simple sedation has not been ruled out. β-blockers do not affect ICSS.

(4) The *d*- and *l*-amphetamine story, once hailed as a way to dissociate the aminergic basis of ICSS at different brain sites, has collapsed because of neurochemical uncertainty about the potency order of

the different isomers. No conclusive evidence is yielded from the use of *d*- versus *l*-amphetamine.

(5) Disulfiram and DDC inhibition of noradrenaline synthesis decreases ICSS, probably by simple sedation. No other DBH inhibitor reliably decreases ICSS. The synthesis inhibition story fails to provide evidence for a role for noradrenaline in ICSS.

(6) 6-OHDA lesion to the ascending noradrenergic systems reveals that ICSS continues unabated. This is true for ICSS from LC, dorsal bundle, substantia nigra, hippocampus and olfactory bulb following better than 95 % removal of the noradrenergic fibres at the tips of the respective electrodes. The 6-OHDA story strongly suggests that noradrenaline is not necessary for ICSS and probably does *not* mediate the reward element of the stimulating current.

Overall: Despite a number of early suggestions of a role for noradrenaline in the rewarding aspect of ICSS, further experiments have failed to bear this out and research interest has turned to the other catecholamine, dopamine.

Dopamine and ICSS

Mapping studies

I have already mentioned the work of Crow (1971) in demonstrating ICSS from sites in the region of the dopaminergic substantia nigra in the rat mesencephalon (see p. 26; Fig. 3.14). Crow (1972) further defined the self-stimulation obtained from this region, particularly noting that sniffing, licking or gnawing behaviour often accompanied such ICSS, along with some head movements, and movements of the eyes, ears and limbs. Since the same syndrome of sniffing etc. can be elicited by drugs, such as apomorphine or amphetamine, which activate dopamine receptors, either directly or indirectly, Crow (1972) argued that this further indicated a role for dopamine in ICSS from ventral tegmental electrodes and adjacent areas.

The above ICSS-positive sites were mainly in the area of the A9 dopamine cell bodies with a few nearer to the lateral part of the A10 dopamine group. Anlezark *et al.* (1973) reported that electrodes on the midline itself will also elicit ICSS, presumably from the dopamine cell bodies in the medial part of A10.

Clavier & Routtenberg (1974) used the technique of lesioning the amine systems running near the tip of an ICSS electrode by passing a large current through it, after demonstrating ICSS at lower currents. They then processed the brains for fluorescence histochemistry to visualise the relationship of the monoamine systems to their stimulating electrode (see p. 29). They found build-up of fluorescent amine as a result of the lesion in the A9 and A10 dopaminergic fibre systems, as well as in the dorsal noradrenergic bundle. This suggests that the tip of the stimulating electrode was at least somewhere in the vicinity of the dopaminergic systems in virtually all the ICSS-positive cases they examined. Of course, since the lesioning current had to be so much higher than the ICSS stimulating current (1 mA compared to 10 μA) it would spread considerably further and might affect aminergic systems which were actually too far removed to be affected by the lower ICSS current.

Fig. 3.13. Simplified summary diagram of electrode placements yielding high response rates (greater than 20 per minute) in rat mesencephalon. Abbreviations: SNC, substantia nigra, pars compacta – containing dopamine cell bodies; SNR, substantia nigra, pars reticulata; IC, internal capsule. (Redrawn from Corbett & Wise, 1980.)

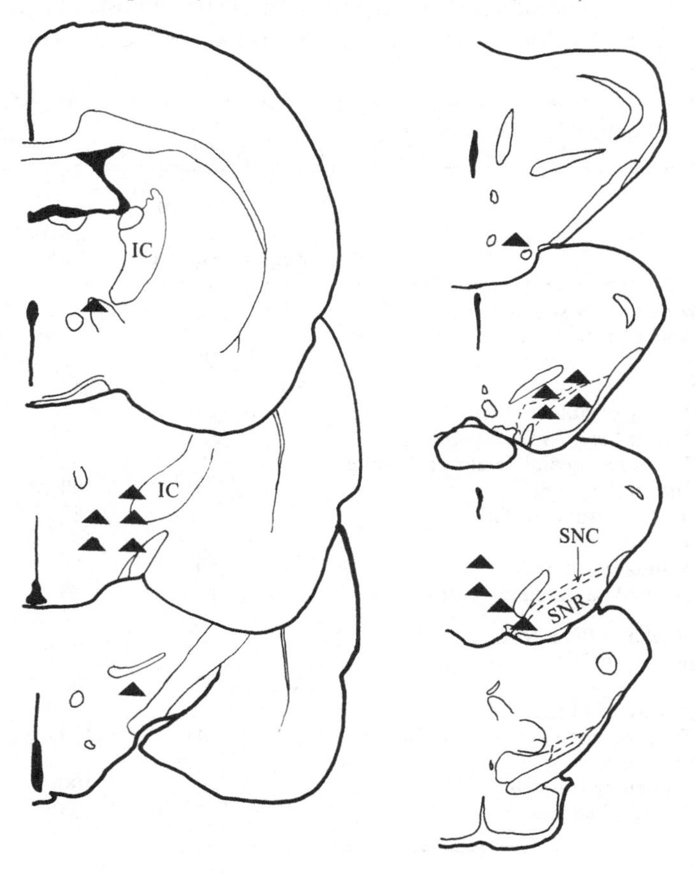

German & Bowden (1974) reviewed the evidence that the dopaminergic systems would support ICSS and concluded that ICSS-positive sites could be found throughout both the A9 and the A10 systems, including cell body areas, regions containing dopaminergic fibres and terminal areas. Table 3.1 is taken from their 1974 review and lists the correspondence between the dopaminergic systems and ICSS-positive sites.

Corbett & Wise (1980) tested a total of 268 sites in the rat mesencephalon for ICSS using a moveable electrode. They then sacrificed the rat and processed the brain for fluorescence histochemistry to permit the direct visualisation of the amine systems relative to the stimulating electrode. They found ICSS-positive sites in the region of the zona compacta of the substantia nigra (A9) and in the ventral tegmental area (A10). Very caudal regions of either group, or the lateral regions of A9, did not yield ICSS (see Fig. 3.13). The zona reticulata of the substantia nigra (a nondopaminergic area) was also convincingly negative for ICSS (all 21 electrodes in the zona reticulata were negative). Further, there was a strong positive correlation between the density of fluorescent dopaminergic cell bodies within a 500

Table 3.1. *Correspondence of ICSS-positive sites with the course of the A9 and A10 dopaminergic systems*

	Region	No. of sites explored	% supporting ICSS
Nigrostriatal A9 system			
Nuclei of origin			
	A9	29	52
	A8	2	50
Pathway of fibres			
Ventral tegmental bundle		53	85
Posterior medial forebrain bundle		15	92
Internal capsule		2	100
Transcapsular radiation		9	83
Terminal areas			
Caudate putamen		50	70
Amygdala, nucleus centralis		12	83
Mesolimbic A10 system			
Nuclei of origin			
	A10	39	82
Pathway of fibres			
External medial forebrain bundle		4	100
Terminal areas			
Olfactory tubercle		2	100
Nucleus accumbens		21	95

μm zone around an electrode tip and the rate of ICSS obtained from it (Spearman $\rho = 0.70$). ICSS was also obtained in more anterior regions close to the fluorescent fibres of the ascending dopaminergic pathways in the zona incerta of the hypothalamus. However, because of the uniformity of dopaminergic fluorescence and ICSS rates, no correlation was obtained for this region (Corbett & Wise, 1980). This last study seems to offer very firm evidence of a correspondence between ICSS-positive sites in the ventral mesencephalon of the rat and some, but not all, parts of the dopaminergic systems (the caudal and far lateral areas of A9 were negative, for example).

In conclusion, the mapping studies have succeeded in demonstrating a close correspondence between ICSS-positive sites and the dopaminergic systems of the rat brain. This contrasts markedly with the same studies of the noradrenergic systems. However, this correlation of ICSS sites with dopaminergic systems does not necessarily prove causation. This has to be addressed by other experiments.

Dopamine release during ICSS

Histofluorescence. Arbuthnott *et al.* (1970) applied electrical stimulation after synthesis inhibition with AMPT to electrodes implanted into areas, which in other rats had been shown to yield ICSS. They then processed the brain for histochemical fluorescence by the method of Falk–Hillarp. The brains were examined for a decrease in the intensity of fluorescence in dopamine-containing areas as a result of the electrically induced release of transmitter. If the ICSS electrodes were in A9, a reduction of fluorescence was seen in all parts of the neostriatum. More medial electrode placements also depleted noradrenaline from the forebrain, and a loss of fluorescence was now seen in mesolimbic terminal zones such as the olfactory tubercle, the nucleus accumbens, the central nucleus of the amygdala and the interstitial nucleus of the stria terminalis. If electrodes were implanted into the A10 group depletion of fluorescence was seen after imposed electrical stimulation in the above mesolimbic terminal zones and often in the neostriatum as well. Since the ventral noradrenergic bundle runs through the lateral part of A10 and the medial part of A9, depletion of noradrenaline fluorescence from the hypothalamus was also seen. However, electrodes implanted in the medial part of A10 decreased fluorescence in limbic terminal zones *without* any depletion in the neostriatum or effect on fluorescence in noradrenaline areas. Thus, depending where the tip of the electrode was located, release (as judged by depletion of amine following synthesis inhibition) was seen either in the nigrostriatal or in the mesolimbic dopaminergic systems, with some activation of the ventral noradrenergic systems as well. The major criticism of this experiment is, of course, that ICSS had not been demonstrated from the same electrodes for which release was shown, only from electrodes 'in the same region as that which yields ICSS'. Further, numerical quantification of the amount of release was not possible with the purely qualitative fluorescence technique.

Amine concentrations. Olds & Yuwiler (1972) rectified some of the shortcomings by using the biochemical determination of amine levels in the brain, following ICSS, through electrodes in the hypothalamus. The animals were trained to lever-press for ICSS and then sacrificed immediately after a 20 min ICSS session. Following such an ICSS session, dopamine levels in the whole brain were lower than in similar rats not allowed to receive ICSS or in unimplanted controls. The shortcoming of this experiment is that it tells us nothing about *which* dopaminergic system is involved in hypothalamic ICSS, only that, generally, dopamine is released.

Biochemical assays. Yuwiler & Olds (1973) made use of the combination of biochemical techniques of synthesis inhibition prior to imposed electrical stimulation and subsequent biochemical determination of amine levels. Rats implanted with electrodes in the hypothalamus or in the caudate nucleus were trained to lever-press for ICSS, given AMPT, and then 16 h later received imposed electrical stimulation for 30 min, in a pattern similar to their usual ICSS stimulation. Whole brain dopamine was reduced as a result of AMPT inhibition of synthesis. However, imposed electrical stimulation did not result in any additional fall, either from hypothalamic or caudate electrodes. This serves to highlight the drawbacks of imposed electrical stimulation. It simply is not the same as ICSS.

Another biochemical technique was used by Arbuthnott, Mitchell, Nicolaou & Yates (1977) to demonstrate increased dopamine release by ICSS from ventral tegmental electrodes. They trained rats to deliver ICSS through these electrodes and then sacrificed the animals immediately after a 30 min ICSS period. Biochemical measurement of the metabolites of dopamine, homovanillic acid and dihydroxyphenylacetic acid, was performed. Those rats in which histology confirmed a placement of the electrode near the A10 cell group showed an increase in dopamine metabolites in the olfactory tubercle. No change in the noradrenaline metabolite 3-methoxy-4-hydroxyphenylglycol was found in the cortex or hippocampus. Thus, at least with some ICSS electrodes, an increase in dopamine release, as judged by an increase in its metabolites, has been demonstrated.

Tracer studies. Stinus *et al.* (1973) injected [^{3}H]dopamine into the cisterna magna via a chronic, indwelling cannula and allowed the rats to lever-press for ICSS from ventral tegmental electrodes for a 5 min session. A decrease in the [^{3}H]dopamine content of the olfactory tubercle relative to unstimulated controls was observed. No change was seen in the neostriatum. Thus, the release of dopamine from at least the mesolimbic system has been demonstrated by this experiment as a result of ICSS from ventral tegmental electrodes.

Mora & Myers (1977) used a push–pull technique to demonstrate release of dopamine during ICSS from the medial prefrontal cortex. A double concentric push–pull cannula was implanted 1 mm caudal to the ICSS electrode in the medial prefrontal cortex. A small amount of [^{3}H]dopamine was injected down the push–pull cannula and then the washout of the radioactivity into the push–pull perfusate circulating through this area was determined by taking 5 min perfusates every 15 min for 2 h. Without stimulation a logarithmic washout curve is obtained. If the rat were allowed to lever-press for ICSS for 15 min and the perfusate of the last 5 min collected, a marked increase in the release of radioactivity was seen. Thin-layer chromatography showed that this radioactivity was made up of dopamine or its metabolites. This increased rate of release of radioactivity returned to the baseline washout rate when stimulation ceased. However, this result was obtained in only 9 out of 16 attempts (Mora & Myers, 1977). This may reflect either the very difficult technical nature of the experiment or that ICSS from the medial prefrontal cortex can sometimes occur without the release of dopamine. In a second study this effect was seen in six out of nine experiments (Myers & Mora, 1977). A further complication was that ICSS from the ventral tegmental area, the undisputed source of the dopaminergic afferents to the medial prefrontal cortex, failed totally to increase the release of [^{3}H]dopamine from the perfused medial prefrontal cortex (Myers & Mora, 1977). Nonetheless, in at least some cases, medial prefrontal cortical ICSS does indeed cause the release of dopamine.

Conclusion. Although the release of dopamine as the result of ICSS from electrodes in A9, A10 or the medial prefrontal cortex seems to have been demonstrated in at least some experiments, it still remains to be shown that this has any causative relationship to the ICSS. It may be an epiphenomenon which has nothing to do with mediating the reward element of ICSS. In order to show the causative connection, other approaches must be adopted.

Dopaminergic drugs and ICSS

Antagonists. The neuroleptic drugs, which block dopamine receptors, have been shown to reduce ICSS in a number of sites in a number of species.

Mora, Rolls, Burton & Shaw (1976) reported in the rhesus and squirrel monkey that intraperitoneal injection of spiroperidol in doses ranging from 0.016 to 0.1 mg kg^{-1} reduced ICSS lever-pressing from electrodes in the hypothalamus, orbito-frontal cortex, and LC (see Fig. 3.14*a*). The orbito-frontal cortical ICSS was particularly sensitive in that doses of 0.016

Fig. 3.14. (*a*) Dose–response curve for response suppressant effects on ICSS of spiroperidol in rhesus and squirrel monkey for hypothalamus (solid circle), locus coeruleus (starred circle), and orbito-frontal cortex (star) electrodes. (*b*) Effect of 0.016 mg kg^{-1} of spiroperidol on ICSS rates from hypothalamus (□), locus coeruleus (●) and orbito-frontal cortex (▲) electrodes in a single rhesus monkey. (Redrawn from Mora *et al.*, 1976.)

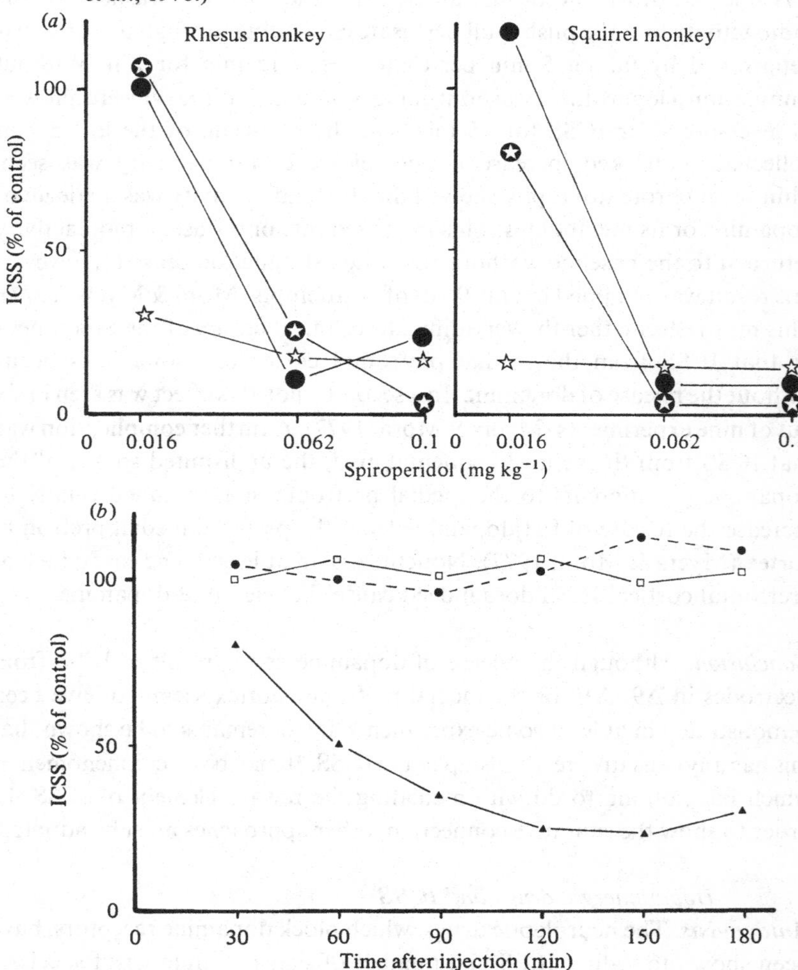

mg kg^{-1} had a profound effect, with much less effect on other ICSS sites (see Fig. 3.14*b*). A dose of 0.062 mg kg^{-1} reduced ICSS to less than 20 % of placebo injection (spiroperidol vehicle,[1] 0.01 mol l^{-1} tartaric acid). The cause of this reduction was not clear since the authors report 'a general effect of spiroperidol on behaviour (such as a motor impairment or sedation) was suspected, since at the higher dose of 0.1 mg kg^{-1} the monkeys looked drowsy' (Mora *et al.*, 1976).

Rolls *et al.* (1974*a*) argued against a motor impairment since spiroperidol reduced ICSS from lateral hypothalamic electrodes in rats far more drastically than it did rearing or locomotor activity (see Fig. 3.15*a*). Note particularly the 0.02 mg kg^{-1} point in Fig. 3.15*b* which reduced ICSS by more than 30 % with no effect on locomotor activity. However, the separation of the potencies in reducing ICSS and other forms of more general activity is not great, so more convincing evidence is required.

Liebman & Butcher (1973) showed that ICSS from electrodes in the lateral hypothalamus was decreased by intraperitoneal injection of another neuroleptic drug, pimozide. Doses of 0.35 and 0.5 mg kg^{-1} were used. Electrodes in the central grey (all but one were near the course of the dorsal noradrenergic bundle; no. 589 was more dorsal) yielded the same effect. Increasing the stimulating current to twice the normal value was effective in reinstating control levels of ICSS responding in the case of the lower dose of pimozide but had virtually no effect with the higher dose. The authors argue that since control rates of ICSS could be reinstated by increasing the current this indicates that the rat was able to emit the motor activity of lever-pressing while under 0.35 mg kg^{-1} pimozide, and hence the decrease in ICSS cannot be attributed to general motor impairment. A direct blockade of the reward component of the ICSS is thus suggested. However, the failure of this manipulation to reinstate responding at 0.5 mg kg^{-1} suggests that the latter did indeed cause a generalised motor impairment. It might be argued from these data that the reduction in lateral hypothalamic ICSS (from electrodes close to the course of the nigrostriatal dopaminergic bundle and with one electrode, no. B-29 actually in the substantia nigra) by 0.35 mg kg^{-1} is caused by blockade of dopamine receptors on which dopamine released by the stimulating electrode normally acts. In the case of the central-grey electrodes the situation is more complex. There are no known dopaminergic cells, fibres or terminals in this region so the electrode

[1] Vehicle, as used here and in subsequent chapters, refers to the solution in which the active substance has been dissolved. It is often physiological saline (0.9 %) but may be, as here, some other solvent or saline containing additional material to facilitate entry into solution or stability once in solution of the active substance.

cannot be directly activating dopaminergic fibres to release dopamine. Since dopamine receptor blockade *did* reduce the reward component of the ICSS, this suggests that the electrode was trans-synaptically activating dopaminergic systems. That is, the electrode directly activated nondopaminergic fibres, or cell bodies, which, via an unspecified number of synapses, projected ultimately to dopaminergic systems where they acted to increase the release of dopamine. This increased dopamine release was then blocked by 0.35 mg kg^{-1} of pimozide to attenuate the rewarding effects of ICSS.

Fig. 3.15. (*a*) Effects of 0.1 mg kg^{-1} spiroperidol (SP) and control injection (CON) on rearing, locomotor activity and ICSS lever-pressing. (*b*) Dose–response curve of effects of spiroperidol on ICSS lever-pressing and on locomotor activity (Loco). (Redrawn from Rolls *et al.*, 1974 *a*.)

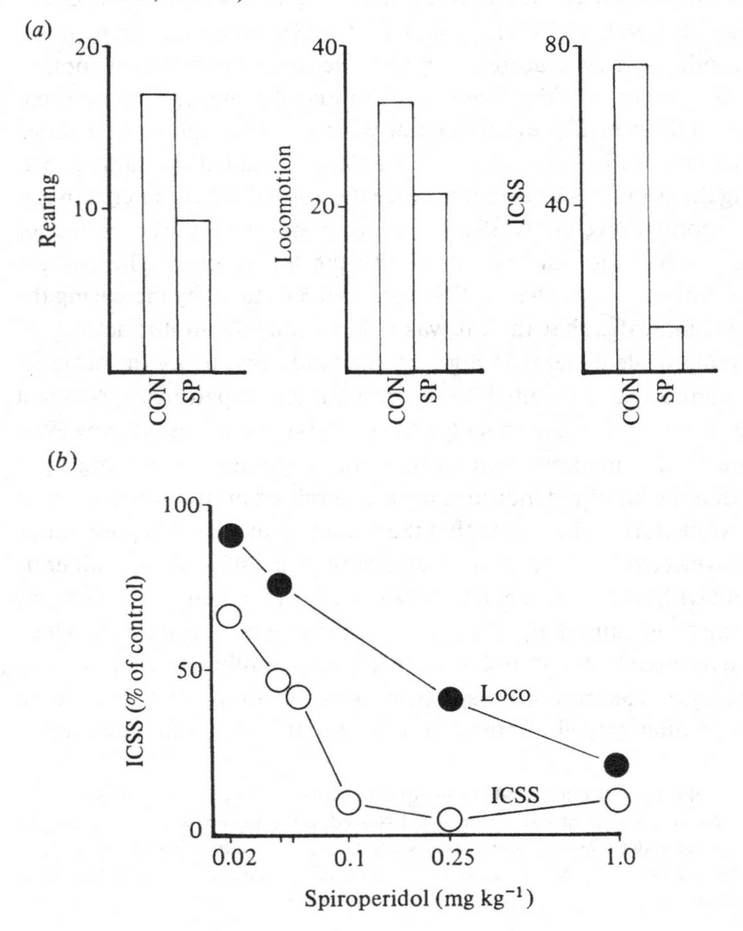

Fouriezos, Hansen & Wise (1978) reported that pimozide in doses ranging from 0.125 to 0.5 mg kg^{-1} attenuated lever-pressing for ICSS from hypothalamic electrodes, with the highest doses virtually suppressing it entirely after some 10 min of lever-pressing. Other dopamine blocking agents such as butaclamol (+)-isomer, (0.1 to 0.4 mg kg^{-1}) produced a similar result. Pimozide also impaired ICSS in a runway paradigm where the rat had to run from a start box along an alley-way to a goal box in order to reach a lever that it could then press to deliver ICSS. Dopamine receptor blockade with pimozide reduced both the lever-press rate and the running speed to get there. More will be made of this experiment in the section on performance versus reward effects (see p. 75).

Wauquier & Niemegeers (1972) found that, as well as pimozide, other neuroleptics such as haloperidol and pipamperone would also block ICSS. A dose effective in suppressing ICSS in 50 % of their subjects (ED_{50}) was 0.22 mg kg^{-1} for pimozide, 0.055 mg kg^{-1} for haloperidol and 19.8 mg kg^{-1} for pipamperone. Low rates of ICSS, as a result of low stimulation current, were more effectively blocked by a given dose of the drug than higher rates maintained by higher stimulation currents.

Rolls *et al.* (1974*b*) reported that spiroperidol in doses from 0.02 to 0.1 mg kg^{-1} would attenuate ICSS not only from lateral hypothalamic electrodes but also from those in the septum, the nucleus accumbens, the hippocampus, the ventral tegmental area and the anterior hypothalamus. These authors found that although feeding and drinking were reduced by spiroperidol, this occurred much less, and required higher doses, than the suppression of ICSS. However, if the rats were required to lever-press to receive food or water, spiroperidol had a much more disruptive effect on responding, equally as severe as that seen on ICSS lever-pressing. The authors thus conclude 'This finding suggests that impairment of motor function accounts for the effects of dopamine-receptor blockade on self-stimulation' (Rolls *et al.*, 1974*b*).

Wauquier & Niemegeers (1976) found that ICSS from the lateral hypothalamus could be wholly or partially reinstated, following pimozide blockade, by a number of drugs which increased transmission at dopaminergic synapses. Thus, ICSS totally suppressed by 0.63 mg kg^{-1} of pimozide could be restored to near normal levels by cocaine or nomifensine, drugs which block the reuptake of dopamine presynaptically. Amphetamine, which increases the impulse-dependent release of dopamine, partially restored ICSS, whereas the direct receptor agonists, apomorphine and piribedil, showed very little activity. This serves to emphasise the need for dopamine to be released into the synapse in an impulse-dependent fashion to maintain the reinforcing relationship of dopamine receptor stimulation to the preceding lever-press. Direct activation of the dopamine receptor may be said to obviate the need to lever-press at all, since the rat is already in a state of pharmacological nirvana (see p. 40 for similar arguments about direct noradrenaline receptor agonists).

Atrens, Ljungberg & Ungerstedt (1976) used a rate-free shuttle box paradigm (see p. 19) to demonstrate that haloperidol increased the latency to initiate ICSS as well as increasing by an equal amount the latency to escape ICSS. Such a pattern cannot be dissociated from that expected to be produced by a simple motor deficit. However, the atypical neuroleptic clozapine increased the latency to initiate ICSS without any effect (except at very high doses) on the escape latency. This, the authors argue, suggests a direct effect on reward processes *per se*.

Using a two-lever reset paradigm (see Appendix), Schaefer & Michael (1980) have compared the effects of five different neuroleptics on ICSS. They found that chlorpromazine and pimozide had little effect on the reinforcement threshold (the current at which the rat pressed the reset lever), but markedly reduced the level of responding on the ICSS lever. Haloperidol and loxapine seemed to have a mixed action, in that they increased the reinforcement threshold slightly while at the same time decreasing the response rate on the ICSS lever. Only clozapine clearly increased the reinforcement threshold at doses which had no effect on the response rate on the ICSS lever. Thus, all drugs tested except clozapine have profound motor effects and only haloperidol and loxapine may additionally have effects on reinforcement mechanisms. Since the last two had severe motor effects at these doses, their elevation of the reinforcement threshold cannot unequivocally be ascribed to a direct effect on reward. Only clozapine, which increased reinforcement thresholds at doses totally without effect on response rate, can be said to have demonstrated clear reward-related action.

In conclusion, it seems clear that the neuroleptic drugs, which on other grounds are believed to block dopamine receptors, act to decrease ICSS behaviour, often in a most dramatic fashion. What is not so clear is whether this is a performance or a reward deficit. That is, does the rat fail to lever-press for ICSS because it is unable to carry out the motor act of lever-pressing (or what ever other response may be required by the experimenter to deliver ICSS) or can the rat perfectly well lever-press but refrains from so doing because it is no longer reinforcing to it? Only if the latter is the case can the neuroleptic-induced decrease in ICSS be adduced as evidence for a role of dopamine in the reward component of ICSS. This question will be taken up in more detail in the section on performance versus reward (see p. 75).

Agonists. Whereas the effect, if not necessarily the interpretation, of dopamine antagonists on ICSS behaviour is clear, even the effect of dopamine agonists on that behaviour is disputed. Some authors have reported an increase, some a decrease, in ICSS lever-pressing and yet others have found no effect. I will now proceed to explore this veritable jungle of conflicting results.

Liebman & Butcher (1973) found a marked decrease in ICSS lever-pressing with hypothalamic electrodes at both 0.75 mg kg^{-1} and 1.5 mg kg^{-1} apomorphine. Now, apomorphine and other direct dopamine receptor agonist drugs cause a behavioural syndrome called stereotypy (see p. 98). This stereotypy is often very intense and seems to lock the animal into repeated bouts of sniffing or gnawing in rather restricted locations in the cage. Thus, decreases in goal-directed behaviour such as lever-pressing might be caused by the predominance of stereotyped behaviour. A rat cannot be sniffing or gnawing a corner of the cage *and* pressing a lever simultaneously. This is called response competition. (However, many rats will solve the problem by gnawing the lever.) So a decrease in ICSS lever-pressing rate under apomorphine need not necessarily reflect any direct effect on reward values of the electrical current. Response competition would also cause a decrease. Liebman & Butcher (1973) observed that only one out of nine rats seemed to show stereotypy under 0.75 mg kg^{-1} apomorphine whereas all the rats showed a decrease in ICSS rates. This, they argue, excludes a response competition explanation of the decline in ICSS rates. With 1.5 mg kg^{-1}, however, clear stereotyped movements were seen.

For an opposite example, take Kadzielawa (1974), who found with posterior hypothalamic electrodes that apomorphine increased lever-pressing rate. At 1 mg kg^{-1} a 45% increase was found, while small increases were seen even at threshold doses of 0.4 mg kg^{-1} (see Fig. 3.16). Amantadine, another putative dopaminergic releasing agent, caused a decrease, however, at doses of 10 to 50 mg kg^{-1}. Similar excitatory effects were also seen with apomorphine by St Laurent, Leclerc, Mitchell & Milliaressis (1973).

Wauquier & Niemegeers (1973, 1974*a*) found a weak response increase in some rats, particularly those maintained at low current levels. This effect was not, however, dose-dependent (Wauquier & Niemegeers, 1974*a*, Fig. 2 *left*). A response decrease was also seen in other rats, irrespective of ICSS current and this effect did get bigger with increasing doses of the drug.

Broekkamp & van Rossum (1974) found that apomorphine increased ICSS lever-press rates in 14 out of 29 rats. However, another 13 rats at this dose (0.2 mg kg^{-1}) showed an inhibition of ICSS. Two rats presumably showed neither change. This dichotomy of response to apomorphine is nicely shown in Fig. 3.17. It did not seem to be dissociable in terms of the site of ICSS. Rats with electrodes in the nucleus accumbens, lateral hypothalamus, A9/A10 area and LC all showed either an increase or a decrease which was consistent for each individual rat, irrespective of electrode placement. Interestingly, a lower dose (100 μg) resulted in either no effect or inhibition. No excitation was reported. This might suggest a biphasic response curve, with low doses causing inhibition and higher doses giving way to facilitation of ICSS rates. Individual responses would then depend on that rat's sensitivity to the drug and thus on what part of the biphasic curve it happens to fall for a given drug dose. Since the dose used to show marked enhancement of ICSS rates (0.2 mg kg^{-1}) is lower than that usually needed to cause stereotyped behaviour, the authors dismiss this as not being the cause of their observed increase in lever-pressing.

Wauquier (1978) again emphasised the role that stimulating current plays, with the finding that high response rate maintained by high current showed a decrease with apomorphine (0.01, 0.63 and 2.5 mg kg^{-1} but strangely no effect at 0.04 mg kg^{-1}). Low response rates, maintained by low ICSS currents, however, showed exactly the opposite – a linear increase with increasing dose of drug, so that a 1100 % increase was seen over baseline rates at 0.63 mg kg^{-1}. Other dopaminergic agonists, such as piribedil, *N*-dipropyldopamine, (–)-*N*,*n*-propylnorapomorphine and 2-(*N*,*N*-dipropyl)amino-5,6-dihydroxytetralin all showed a response sup-

Fig. 3.16. Cumulative record of ICSS responding from hypothalamic electrodes following control or apomorphine (1 mg kg^{-1}) injection. (Redrawn from Kadzielawa, 1974.)

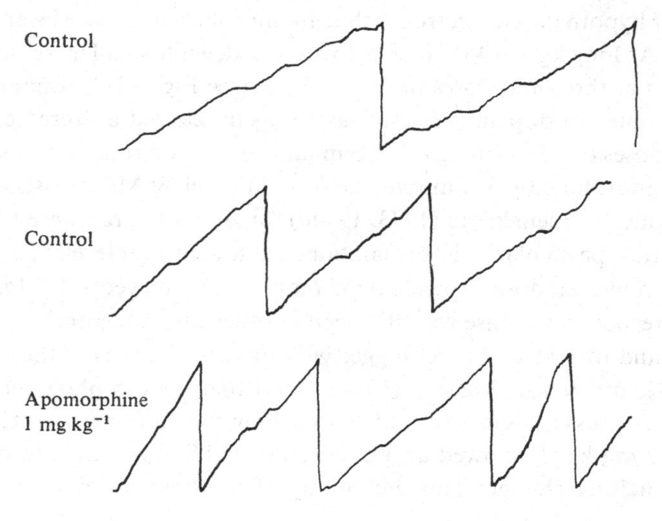

pressant effect with high response baselines. Further, a clear dose-dependency of this suppression was seen (bigger suppression for bigger doses). Only *N*-dipropyldopamine and 2-(*N,N*-dipropyl)amino-5,6-dihydroxytetralin showed a response enhancing effect when ICSS was maintained at low rates by lower currents.

In an attempt to dissociate the effects of apomorphine on the reward component of ICSS from its effects on stereotypy and locomotor stimulation, Atrens, Becker & Hunt (1980) used a rate-free shuttle-box paradigm (see p. 19). Here the drug had very little effect on the latency to

Fig. 3.17. Effects of 200 μg kg^{-1} of (*a*) apomorphine on ICSS rates in 29 rats compared to (*b*) 10 saline-injected controls. Four different electrode placements were used; A9/A10, lateral hypothalamus, nucleus accumbens and locus coeruleus. Both increases and decreases in ICSS rates were seen after apomorphine injection. (Redrawn from Broekkamp & van Rossum, 1974.)

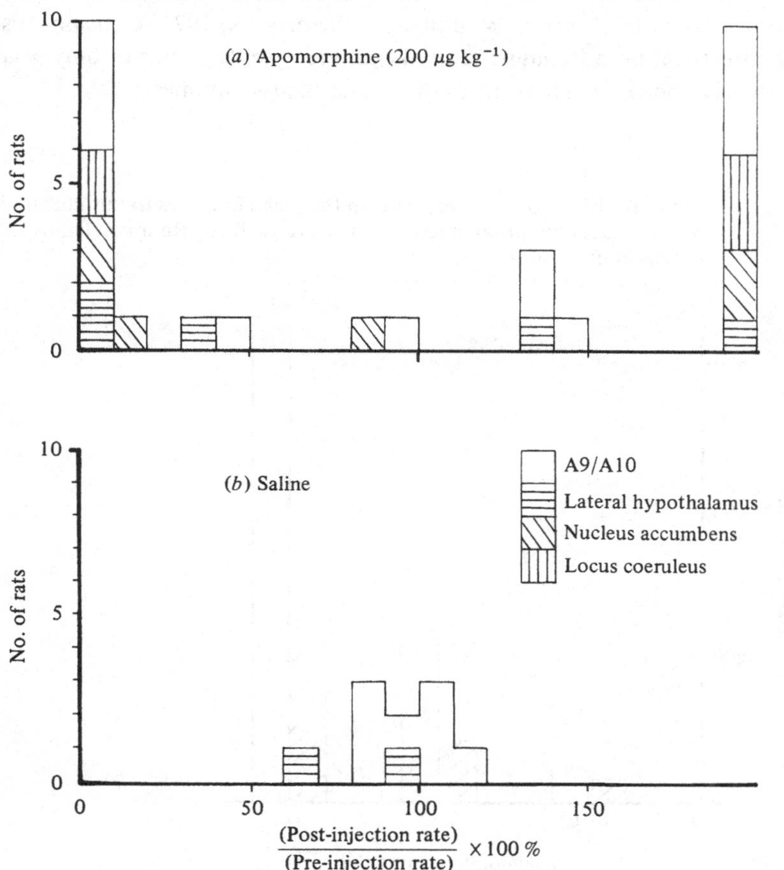

initiate ICSS but caused a dose-dependent increase in the latency to terminate ICSS. This suggests that apomorphine may attenuate the *aversive* component of ICSS (see Fig. 3.18). However, it must be noted that the largest dose of apomorphine, 0.64 mg kg^{-1} appears to have caused a 100 % increase in the latency to initiate ICSS. Although small in comparison to the 700 % increase in latency to terminate ICSS this suggests that the situation is not as clear-cut as it might be (Fig. 3.18).

In summary, the effects of dopamine agonists on ICSS behaviour appear to depend on the baseline rate of responding and on the individual rat. It is hard to extract much support for a role of dopamine in the reward component of ICSS from the above data.

Of passing interest is that the dopaminergic reuptake inhibitor cocaine causes a dose-dependent increase in ICSS rates, especially at low currents of ICSS (Wauquier & Niemegeers, 1974*a*). A 50 % increase in rates was seen at a cocaine dose of 10 mg kg^{-1}. However, as with amphetamine, which also increases ICSS rates (Wauquier & Niemegeers, 1974*b*), cocaine also has effects on noradrenaline reuptake and release. It is thus of only weak persuasive power as an agent to dissociate the two amines.

Fig. 3.18. Effect of apomorphine on the mean latencies to initiate and to escape lateral hypothalamic ICSS in a shuttle box. (Redrawn from Atrens *et al.*, 1980.)

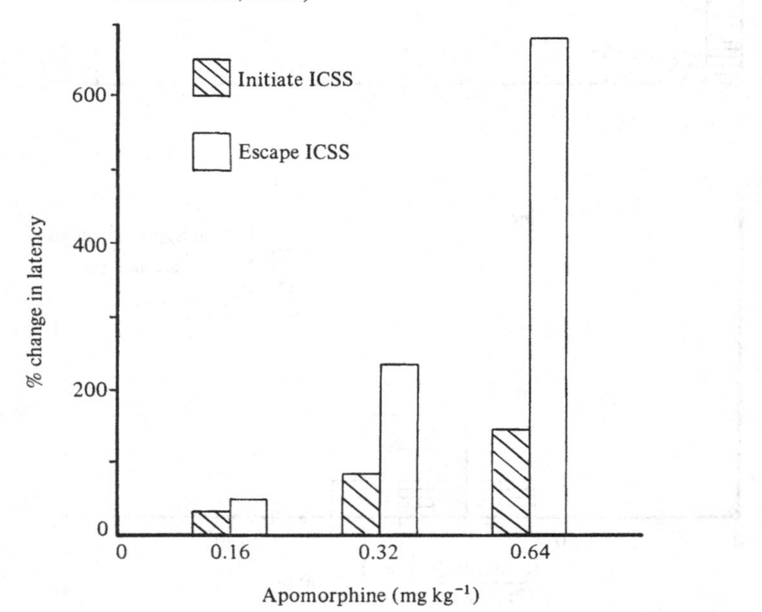

Interactions with other systems

Acute effects. Wauquier & Niemegeers (1976) reported that the anticholinergic agent dexetimide would restore responding for hypothalamic ICSS which had been disrupted by pimozide (0.63 mg kg^{-1}) administration. The authors point out, however, that dexetimide also possesses dopamine uptake blocking ability and hence could be acting in a similar fashion to, say, cocaine, which also restores ICSS rates and which does not act via cholinergic systems.

Stephens & Herberg (1979) were able to restore ICSS from lateral hypothalamic electrodes suppressed by spiroperidol with the anticholinergic agent, scopolamine. This drug does not possess dopamine uptake blocking action and so a clear acetylcholine restoration of a dopaminergically suppressed behaviour was demonstrated. Further, scopolamine on its own (0.1–1.0 mg kg^{-1}) acted to increase the rate of ICSS above saline baseline. Increases of between 10 and 30 % were seen in lever-pressing rates. However, the effect of scopolamine on the spiroperidol suppression was shown by the analysis to be simply subtractive. No interaction was demonstrated. That is, a given dose of scopolamine increased ICSS responding by say 30 %. If ICSS responding had been reduced by 30 % as a result of spiroperidol treatment, then that dose of scopolamine would restore responding to normal. However, if ICSS rate had been reduced by 60 %, that dose of scopolamine would cause some increase (to 30 % less than saline value) in ICSS rates but would not restore them to normal. The effects of one drug simply added (or in this case subtracted) from the effects of the other. Thus, they could be acting through totally separate mechanisms and failed to demonstrate an interaction between cholinergic and dopaminergic systems in the brain in ICSS. Stephens & Herberg (1979) also showed that scopolamine injected directly into the nucleus accumbens was effective in increasing ICSS rates, either without pre-treatment (20 %) or in partially restoring responding suppressed by spiroperidol injection (from 5 % of saline rates up to 35 %). Similar injections into the neighbouring caudate–putamen were ineffective in increasing ICSS. Figure 3.19 shows the two models of the effects of dopaminergic and cholinergic manipulations suggested by Stephens & Herberg (1979); they favour the lower (model *b*), which is additive, rather than interactive, in nature.

Chronic effects. Robertson & Mogenson (1979) examined the effect of chronic injections of spiroperidol, administered 30 min after each daily ICSS session, on the rate of ICSS from a number of electrode placements. Little effect was seen during the 9 days of drug administration (0.1 mg kg^{-1}); but following withdrawal of the drug a marked, and possibly permanent, increase in ICSS rates was seen for electrodes in the prefrontal cortex. In the 3 or 4 days after termination of drug treatment, rates rose sharply until they levelled off at some 200 % of normal. This elevation was still present in rats tested 45 days after termination of drug. No increase in ICSS rates was seen from electrodes in the nucleus accumbens, the caudate–putamen, the ventral tegmental area, the subfornical organ or the supracallosal bundle. Chronic administration of amphetamine (1.5 mg kg^{-1}) 30 min after each daily ICSS session for 9 days resulted in an increase in rates from prefrontal electrodes, which was seen even while the drug was still being administered, and which continued to increase after termination of the drug. By the sixth day after termination of chronic amphetamine administration ICSS rates had reached 200 % of pre-drug values. A suppression of responding from electrodes in the supracallosal bundle was also seen with chronic amphetamine, down to less than 20 % of normal, but ICSS from the nucleus accumbens was not affected. This study indicates that chronic drug administration can drastically change the neurochemical substrates involved in ICSS behaviour.

Shorter treatments with neuroleptics may also be effective in inducing supersensitivity since Ettenberg & Milner (1977) reported enhancement of ICSS 48 h after a 3-day regimen of pimozide (4 mg kg^{-1} twice daily) and Schaefer & Michael (1980) found a decreased reinforcement threshold in a

Fig. 3.19. Two possible models for the restoration of ICSS responding by anticholinergic drugs (scopolamine) following its suppression by antidopaminergic agents (neuroleptics). (*a*) Dopamine (DA) neurones inhibit cholinergic (ACh) neurones, possibly via interneurones. The cholinergic neurones themselves inhibit structures concerned with ICSS. (*b*) DA and ACh systems function independently, producing effects of opposite sign on ICSS.

Model (*a*) is interactive, while model (*b*) is merely additive. (Redrawn from Stephens & Herberg, 1979.)

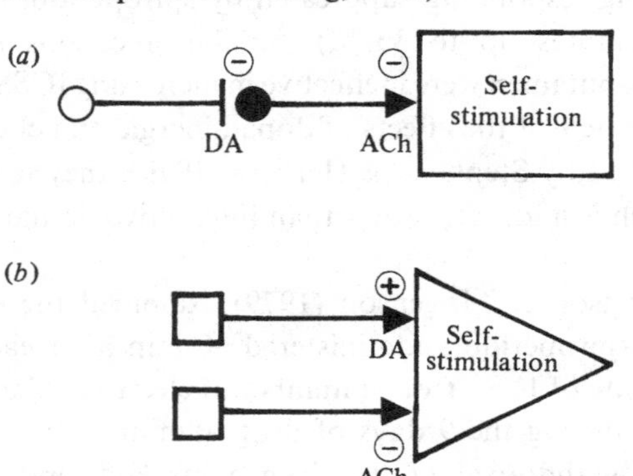

two-lever reset paradism (see Appendix) 24 h after only a single dose of pimozide (1.7 mg kg^{-1}).

The dependence of the effects of chronic amphetamine administration on the particular electrode site used is shown by a study of Liebman & Segal (1976). They administered 1.5 mg kg^{-1} of amphetamine, as well as other lower doses, for 10 days to rats lever-pressing for ICSS from substantia nigra electrodes. Since the drug was given prior to the ICSS session, it acted to increase the rate during that session compared to saline baseline. However, over the 10 days of treatment no additional increase occurred, contrary to that reported by Robertson & Mogenson (1979) at around day 3 or 4. Further, when the animals were removed from amphetamine and tested for ICSS their response rates were no different from chronically saline-injected controls. Whatever it is that is being altered so as to enhance ICSS rates as a result of chronic treatment with either spiroperidol or amphetamine seems to be specific for ICSS from prefrontal electrodes.

The reverse interaction has been shown by Eichler & Antelman (1979). That is, chronic ICSS will alter the subsequent effects of amphetamine. Thus, animals which had experienced daily ICSS sessions for four to seven weeks showed a greater stereotypy response to amphetamine, as well as a greater anorexia in response to the drug, than animals which had not received such chronic ICSS experience. This was true for rats with electrodes in the nucleus accumbens and in the medial prefrontal cortex, but not if the electrodes were in the A9 group.

Whether the neurochemical substrate of either the potentiated response to amphetamine after chronic ICSS, or the potentiated ICSS after chronic amphetamine, involves a dopaminergic element is completely unknown at this time.

It has been suggested that a further example of interaction between transmitter systems involved in ICSS may take place between dopaminergic and noradrenergic systems (Antelman & Caggiula, 1977). Thus, lesion to the LC actually potentiated ICSS from lateral hypothalamic electrodes (assumed by Antelman & Caggiula, 1977, to be dopaminergic) as shown by Koob *et al.* (1976). However, the noradrenergic nature of this latter effect has been questioned by Corbett (1980). He found that ICSS from hypothalamic electrodes was potentiated after lesions to the nonnoradrenergic pontine tegmental taste area, lateral and slightly dorsal to the LC. Lesions of the LC itself did not affect ICSS from hypothalamic electrodes. He also points out that known projections of the pontine tegmental taste area run very close to the dorsal noradrenergic bundle along much of its ascending mesencephalic course, suggesting an alternative, noncatecholaminergic substrate for 'dorsal bundle' ICSS. Further, inhibition of noradrenaline synthesis with FLA 63 has occasionally been reported to increase ICSS from hypothalamic electrodes (Antelman, Canfield & Fisher, 1976). Given the conclusion that noradrenaline certainly does not directly mediate the reward element of the stimulating current at the tip of the ICSS electrode (see p. 54), any such interaction must be more indirect.

The field of transmitter interactions in ICSS behaviour is only in its infancy and we may hope to see additional development in future years.

Intracerebral localisation

The rate suppressing effects of dopaminergic blockade with neuroleptic drugs, such as pimozide or spiroperidol, whilst clear in effect are cloudy in interpretation. One problem, that of performance v. reward, will be dealt with in another section. The second problem, to be covered in the foregoing paragraphs, is that of the site of action of the drug within the brain. Which dopaminergic synapse, in which brain area, is the one blocked in order to reduce ICSS rates? There are dopaminergic synapses in the caudate, the accumbens, the amygdala, the prefrontal and sulcal cortices and in many, many other parts of the brain. To determine the site of intracerebral action, a number of studies have been conducted injecting neuroleptic drugs, and other substances in very small quantities directly into brain regions.

Agonists and releasing agents. Redgrave (1978) used the push–pull technique to administer small amounts of noradrenaline, dopamine or serotonin into either the nucleus accumbens or the caudate–putamen while the animal was lever-pressing to obtain ICSS through electrodes either in the medial forebrain bundle or in the ventral tegmental area. Infusion of dopamine caused a dose-dependent increase in ICSS from either medial forebrain bundle or ventral tegmental electrodes. Effective sites for infusion were found in either the caudate–putamen ($n = 10$) or in the nucleus accumbens ($n = 2$). Infusion of serotonin at the same sites usually led to a decrease in ICSS rate. Infusion of artificial CSF caused no change (control procedure), as did infusion of equal amounts of dopamine and serotonin mixed together in the same perfusate. Infusion of noradrenaline also altered ICSS but in a more complex fashion. Some sites showed a pronounced and long-lasting increase in the rate of ICSS, which eventually returned to baseline rates up to 2 h after the perfusion was discontinued. Other sites, however, exhibited a slight decrease in ICSS rate which remained depressed for long periods after termination of the perfusion with no sign of recovery to baseline response rates. Redgrave (1978) suggested that the latter is nonphysiological, especially since no dose–response relationship could be demonstrated.

Neill, Parker & Gold (1975) and Neill, Peay & Gold (1978) found that application of crystalline dopamine into the ventral striatum (caudate–putamen) would cause nearly 30 % increase in ICSS lever-pressing from lateral hypothalamic electrodes. Similar application to regions in the dorsal striatum were ineffective, whereas application of noradrenaline to ventral sites yielded a significant, but much smaller, increase in rate (about 50 %). The above two experiments lend some indirect support for a role of brain dopamine in ICSS, and further indicate particularly a region of the ventral anterior caudate–putamen as being of importance. It is of interest that noradrenaline was also reported to enhance response rates from this traditionally dopaminergic area of the brain.

Broekkamp, Pijnenburg, Cools & van Rossum (1975) found that bilateral injection of the indirect releasing agent *d*-amphetamine into either the nucleus accumbens or the caudate–putamen served to increase ICSS rates from electrodes in the ventral tegmentum. Up to a 200 % increase was obtained from either injection site, but the nucleus accumbens appeared the more sensitive. Injection into the anterior hypothalamus was without effect. In that amphetamine increases the release of noradrenaline and dopamine together, this fails to discriminate between the two amines but does concentrate attention on the terminal areas (nucleus accumbens and caudate–putamen) rather than on the regions such as the anterior hypothalamus, through which the dopaminergic fibres run without terminating on their way to more anterior sites.

Stephens & Herberg (1977) confirmed these results with amphetamine. Injection into either the nucleus accumbens or the caudate–putamen yielded a 50 % increase in ICSS rates from electrodes in the lateral hypothalamus. Injection of the direct receptor agonist apomorphine had variable effects. All rats showed an initial brief depression of ICSS. A few rats ($n=2$) showed a subsequent further depression, but most rats ($n=8$) then exhibited a sharp increase in ICSS. This effect occurred from either the nucleus accumbens or the caudate–putamen. This is similar to the individual nature of the response to apomorphine reported by other workers after peripheral administration (Wauquier & Niemegeers, 1973; Broekkamp & van Rossum, 1974). Indeed, Stephens & Herberg (1977) showed that if a rat responded in one way to apomorphine when injected peripherally, it would also respond in that way to intracerebral injection of the drug. What determined whether the response should be an increase, rather than a decrease, in ICSS for that particular individual rat, however, remains unknown.

Antagonists. Stephens & Herberg (1977) also found that injection of haloperidol into either the nucleus accumbens or the caudate–putamen sharply decreased ICSS to about 50 % of baseline rates.

Broekkamp & van Rossum (1975) found that microinfusion of haloperidol into the caudate–putamen would depress ICSS from electrodes in the ventral tegmental area. Such infusions were effective whether given ipsilaterally or contralaterally to the stimulating electrode. Since it is supposed that reward processes are localised to the side of the brain of the stimulating electrode, these authors interpret their finding as indicative of a motor effect. A motor impairment, as a result of blocking dopaminergic systems on one side of the brain, would be expected to be equally severe on either side in relation to the stimulating electrode. However, Robertson & Mogenson (1978) found that only ipsilateral injection of spiroperidol at the tip of the electrode would block ICSS from an electrode in the nucleus accumbens. Injection of the drug into the prefrontal cortex was ineffective on either side. This they argue is direct evidence for dopaminergic mediation of reward during ICSS from the nucleus accumbens. A similar experiment, with an ICSS electrode in the prefrontal cortex, yielded a decrease in lever-pressing rates only with higher doses of spiroperidol on the ipsilateral side. Injections into the contralateral prefrontal cortex or into the nucleus accumbens were ineffective. These last experiments are starting to approach the dual criteria required for a convincing demonstration of a role for brain dopamine in ICSS; namely, a dissociation of the performance versus reward confound and a localisation of the particular dopaminergic synapse of importance.

Further localisation of the site of action was obtained by Neill, Peay & Gold (1978), who showed that injections of haloperidol into the ventral anterior striatum, but not into the posterior striatum, were effective in reducing ICSS rates from lateral hypothalamic electrodes. However, here too, the injection was effective irrespective of the laterality of the electrode, suggesting a motor effect (Neill *et al.*, 1978, p. 524).

In conclusion, it would appear that intracerebral injections have highlighted the role of the ventral anterior striatum and the nucleus accumbens in the dopaminergic blockade of ICSS behaviour. Some confusion still remains as to whether this is an effect mediated by a block of the rewarding element of ICSS or by an impairment in the motor ability of the animal to perform the lever-press response. Experiments to tease apart these two processes will be described in the next section.

Performance versus reward

As mentioned before, although the effect of pimozide or other dopamine receptor blocking agents in decreasing the rate of ICSS is clear, the mechanism of this effect is not. Either a block of the rewarding aspect of the electrical current *or* an inability to initiate voluntary movement, and hence perform the response, could explain pimozide's actions. Only if the first case is true does the reduction of ICSS by pimozide constitute evidence for a role of brain dopamine in ICSS reward. A number of attempts have been made to separate the two factors of reward of performance. Few have been totally convincing.

Fouriezos & Wise (1976) examined the pattern of response decrease with time shown after injecting rats lever-pressing for ICSS from lateral hypothalamic electrodes with either 0.5 or 0.16 mg kg^{-1} of pimozide 4 h prior to the test session. These authors concentrated particularly on the responding early in the session, and emphasised that significant lever-pressing occurred for the first 3–5 min, sometimes at higher than usual rates. It was only after the rat had sampled the consequences of lever-pressing that a progressive and precipitous decline occurred. This, they argued, suggested that the rat was physically capable of lever-pressing. It did not suffer from a deficit in the ability to initiate a voluntary movement necessary for the physical ICSS response. The decline which occurred only after a number of lever-presses, they suggested, reflected the rat coming to learn that a lever-press no longer brought central reward since the dopamine receptors were blocked with pimozide. In other words, an extinction curve was seen (Fig. 3.20*a*). Extinction being the behavioural paradigm in which a previously rewarded response no longer brings reinforcement and hence decreases in the frequency of its occurrence. Extinction processes show the phenomenon of spontaneous recovery. If the stimulus situation is changed, or if the animal is removed from the situation briefly, renewed responding occurs when the animal is reintroduced, albeit for only a short period until it once again declines to zero. Fouriezos & Wise (1976) demonstrated spontaneous recovery in one animal by blocking off access to the alcove containing the lever for 10 min in the middle of the ICSS session. Prior to this, the animal had completely ceased to lever-press for ICSS as a result of 0.5 mg kg^{-1} pimozide. When access to the lever was reinstated after this 10 min period, significant lever-pressing occurred, which again showed an extinction-like decline over time (see Fig. 3.20*b*).

The argument of the above authors is essentially that pimozide and extinction (turning off the electrical stimulating current) produce the same pattern of response decrement over time. However, Fibiger, Carter & Phillips (1976) examined the effects of 0.1 mg kg^{-1} of the neuroleptic drug haloperidol on rats which had been trained to lever-press on a variable interval (VI) 60 schedule (see Appendix) for either food or lateral hypothalamic ICSS reward. Equivalent decreases in rates of responding were seen for either food or ICSS. The authors state that the cumulative records failed to show an extinction curve and that responding was

Fig. 3.20. Self-stimulation rates as a function of time. (*a*) Four hours after injection of various doses of pimozide (0.05, 0.16, 0.5 mg kg^{-1} and tartaric acid control). (*b*) Four hours after 0.5 mg kg^{-1} pimozide with access to lever blocked for 10 min after the rat had ceased to respond. This shows spontaneous recovery in the subsequent 10 min period. (Redrawn from Fouriezos & Wise, 1976.)

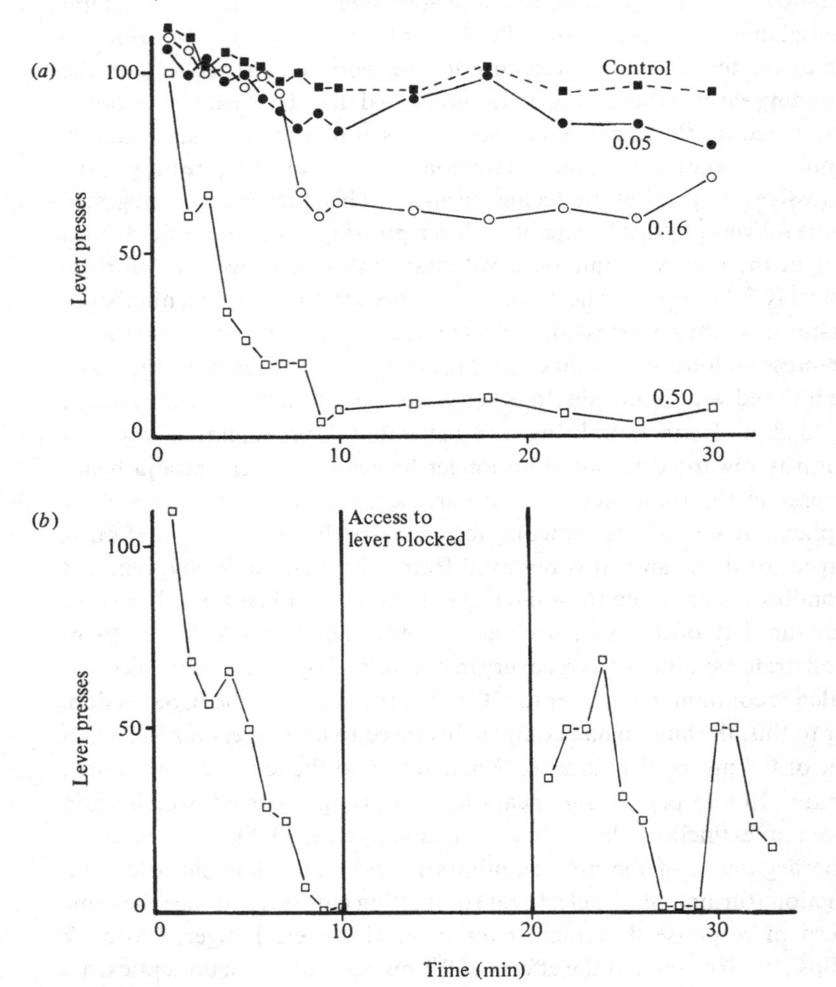

depressed right from the start of the session. However, no sample cumulative records were presented to permit the reader to judge for himself.

Fouriezos, Hansen & Wise (1978) returned to this point concerning the shape of the decline in response rate seen after neuroleptic treatment. Again, rats with electrodes in the lateral hypothalamus showed a progressive decline in ICSS rate when treated with either pimozide or another neuroleptic, (+)-butaclamol. Doses of pimozide from 0.125 to 0.5 mg kg^{-1} and 0.2 and 0.4 mg kg^{-1} of butaclamol showed high rates of responding in the first 3–5 min of the ICSS session, which declined over the next 5 min to much lower values (see Fig. 3.21). A pattern of response decrement expected as a result of a motor deficit (decreased rates right from the start of the session) was seen with the α-blocker phenoxybenzamine, indicating that this experimental paradigm could indeed detect such a motor impairment. A further ICSS response was examined, in which the rat had to run from a start box along a 183 cm long alley-way to reach a goal box in which was a lever which it could press to deliver ICSS via hypothalamic electrodes. The time taken to run this alley-way was measured, as was the number of lever-presses emitted once in the goal box. Turning off the current (extinction) led to a slowing of running speed over three to four trials and a more rapid cessation of lever-pressing (see Fig. 3.22). A very similar, if more delayed, decrease in running speed and

Fig. 3.21. Response rates for ICSS as a function of time following treatment with pimozide, butaclamol, phenoxybenzamine or current reduction. CON represents control injection or current level. (Redrawn from Fouriezos *et al.*, 1978.)

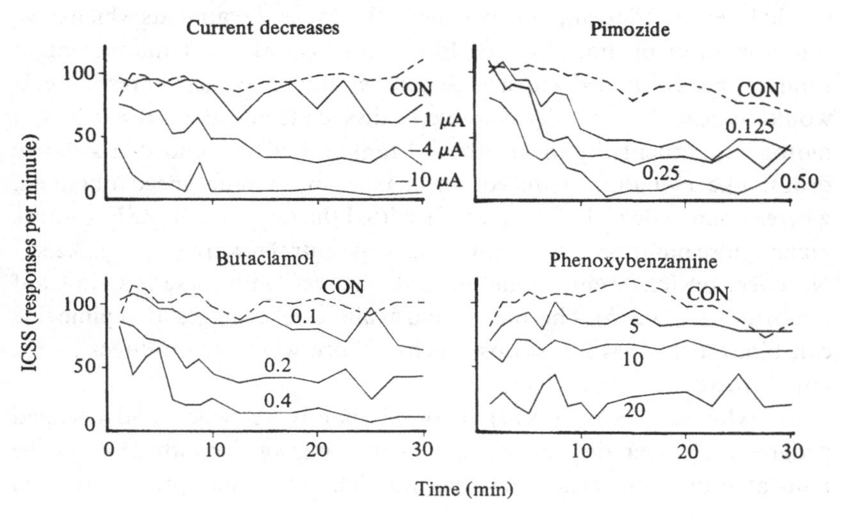

lever-press rate was seen when the rats were treated with 0.5 mg kg^{-1} pimozide. Again, responding did not cease immediately from trial one but required some six or seven experiences of the now-altered consequences of a lever-press before that behaviour ceased to occur. Running speed also remained high until some seven to eight trials had been completed under the drug. Both lever-pressing and alley-way running showed spontaneous recovery under the drug when the rat was removed from the apparatus for 10 min after having ceased to respond totally during the preceding block of 10 trials.

This further demonstration of the similarity of pattern of response decrement over time caused either by extinction (turning off the current) or by the dopamine receptor blocking drug pimozide (see Figs. 3.21 and 3.22) must then be taken as strongly suggestive that pimozide has its effects on responding by directly blocking the reward element of the ICSS, and not merely by inducing a motor impairment which renders the animal unable to respond. The demonstration of spontaneous recovery is consistent with this, without necessarily being very strong evidence *per se*.

Other authors have adopted different approaches to try to tease out the motor effects of pimozide and other neuroleptic drugs from direct effects of reward. Franklin (1978) examined the reward summation function (RSF) of ICSS from hypothalamic electrodes. As the number of pulses delivered as a result of an ICSS response increases so does the vigour of that response. In an alley-way running task a sharp increase in running speed is seen as the number of ICSS pulses delivered at the end of the response increases past a certain critical value. This is referred to as the RSF locus. Treatments which alter the rewarding effects of the electrical stimulation would be expected to alter the placing of this locus. Treatments which have a motor effect do not alter the locus, but instead affect the maximum running speed obtained with a large number of ICSS pulses per response. It would appear theoretically possible to dissociate effects on reward from motor impairments by this method. Franklin (1978) found that sedative drugs, like clonidine, reduced the maximum running speed obtained, whereas pimozide (0.1–0.3 mg kg^{-1}) shifted the rapid rise in RSF towards higher pulse numbers, suggesting a block of central reward (see Fig. 3.23*a*). No effect on maximum running speed was seen with these low doses of pimozide (Fig. 3.23*b*). This then would again seem to suggest that pimozide can block ICSS reward at doses below those which cause direct motor impairments.

Franklin & McCoy (1979) reported that ICSS which had declined progressively over time as a result of 0.25 mg kg^{-1} pimozide could be reinstated by presentation of a flashing light which had previously been associated with ICSS delivery. Since neither a flashing light *not* previously associated with ICSS, nor physically shaking the rat, served to reinstate ICSS the authors argue that this indicates that pimozide acted to decrease ICSS by blocking its rewarding value, rather than by causing a motor impairment. The logic of this falls short of being compelling.

Other authors have attempted to make use of so-called rate-free

Fig. 3.22. Median response data for alley-way following treatment with pimozide (PIM; 0.5 mg kg^{-1}), phenoxybenzamine (PBZ; 10 mg kg^{-1}) or current off (Cur. off). Dashed lines are experimental, solid lines are controls. (Redrawn from Fouriezos *et al.*,1978.)

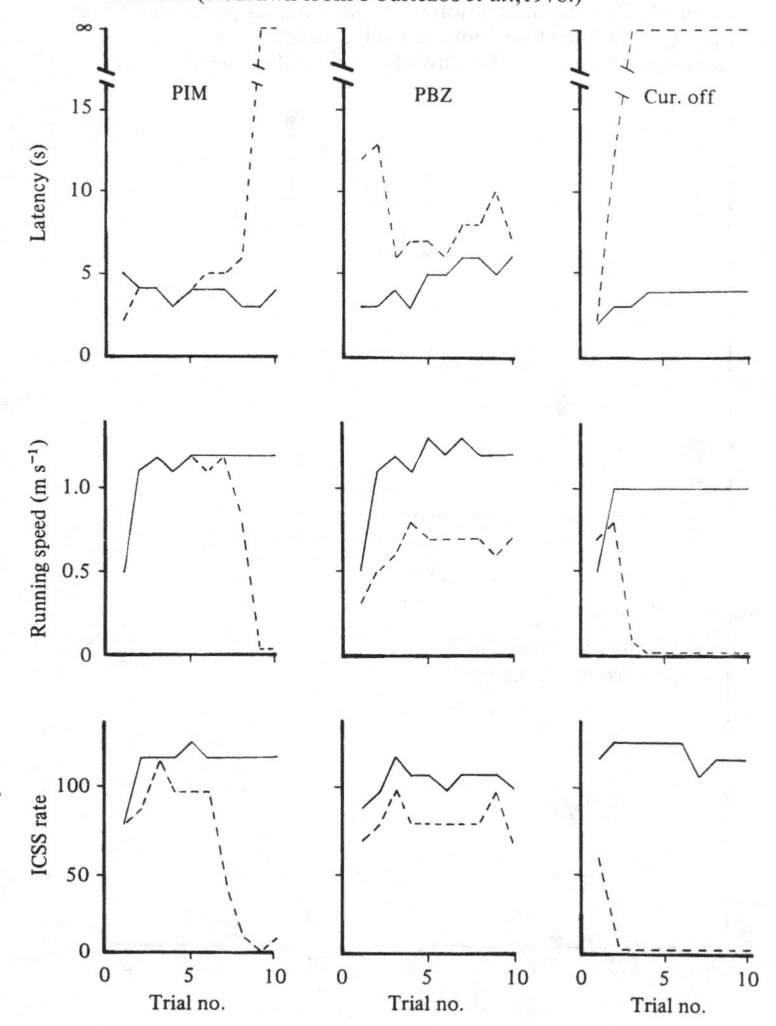

measures of ICSS to overcome the motor deficit problem. Liebman & Butcher (1974) tested the effects of pimozide (0.35 and 0.5 mg kg^{-1}) on ICSS via substantia nigra and hypothalamic electrodes. The rate-free paradigm used (see p. 19) was a shuttle box in which stimulation was delivered to the rat if it was on the 'correct' side of the box. This side changed randomly between 1 and 3 min to the other side. If the rat failed to track over to the other side it did not receive stimulation. Thus, relatively

Fig. 3.23. (*a*) Mean locus of the reward summation functions after different doses of pimozide. Vertical bars show standard errors of the mean. (*b*) Reward summation functions 4 h after pimozide (0.1 or 0.2 mg kg^{-1}). Dashed lines indicate the loci of individual reward summation functions. (Redrawn from Franklin, 1978.)

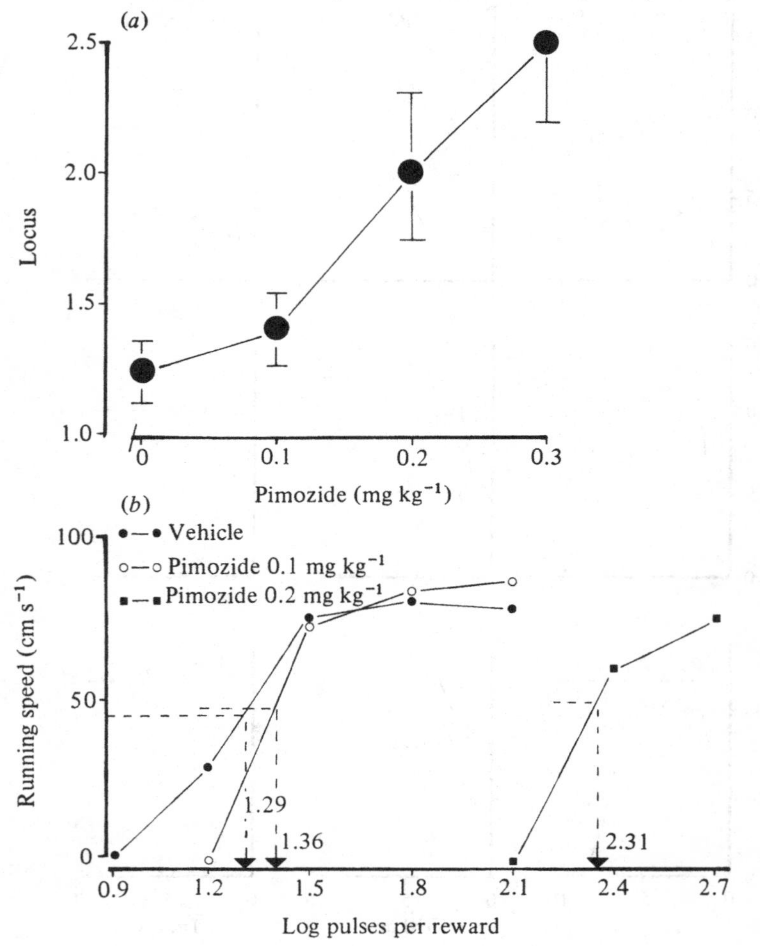

low rates of responding were required for the rat to obtain quite long deliveries of ICSS (an average of 1 shuttle over a 3.5 cm barrier per 2 min). The measure of ICSS received was the time spent on the 'correct' side receiving ICSS, expressed as a percentage of the maximum possible duration of ICSS (that would be if the rat immediately tracked over to the other side and made no further shuttles until the stimulation shifted). Pimozide decreased the ICSS received from both electrodes by a similar and dose-dependent amount. Typically, rats under saline would receive stimulation 80 % of the available ICSS time. Under 0.35 mg kg^{-1} of pimozide this declined to about 60 % and fell slightly further under 0.5 mg kg^{-1} to around 55 %.

Zarevics & Setler (1979) used a procedure in which two levers were present in the operant chamber. Responding on one delivered ICSS via medial forebrain electrodes. This ICSS decreased in magnitude one step after each fifth response. The stimulation intensity would thus progressively decline. To return the intensity to its start value the rat had to press the second lever once. After this the cycle repeated. It was found that 0.1, 0.2 or 0.4 mg kg^{-1} of pimozide increased the reward threshold. That is, resets occurred at higher current levels than in the nondrug condition, suggesting that the current was less rewarding as a result of the drug and so the rat had to reset it sooner and maintain it at an average higher value. Lever-pressing on the nonreset lever, the one which actually delivered the ICSS, was not markedly disrupted by those doses of pimozide (especially at 0.1 mg kg^{-1}). Thus, this again suggests that for ICSS pimozide can attenuate the rewarding aspects of brain stimulation at doses lower than those which cause motor disruption.

The above two 'rate-free' studies, although minimising the effect of motor impairment, do not render it completely irrelevant. Thus a totally immobile animal would fail to track the brain stimulation when the 'correct' side shifted in the paradigm of Liebman & Butcher (1974). This could give rise to the reduced percentages of possible ICSS time. Even though one shuttle response per 2 min is not much, it might still be too much for the drugged rat to execute promptly and as competently as a normal animal. The earlier reset of the stimulating current reported by Zarevics & Setler (1979) appears more hopeful, since it indicates more, not less, motor activity by the drugged animal.

In summary, it seems that no unequivocal single demonstration of a separation of reward from performance effects exists with regard to the effects of neuroleptic drugs on ICSS. However, the several very suggestive ones all combine to make it likely that at some doses neuroleptic drugs indeed affect ICSS by reducing the rewarding aspect of the electrical current. This, then, is the first halfway satisfactory evidence of a role for dopamine in the rewarding nature of ICSS. Even more convincing data have been obtained by selective lesion of dopaminergic systems using 6-OHDA.

6-OHDA lesion

A number of 6-OHDA lesion studies have been carried out using ICSS. Some of the early ones depleted both noradrenaline and dopamine and hence failed to discriminate between the two amines. More recent ones have corrected that deficit by depleting only dopamine. However, they were then beset with the performance versus reward problems. Did the 6-OHDA lesions decrease ICSS by reducing the reward element of the electrical stimulation, or rather by rendering the animal unable to perform the required lever-press because of a motor deficit? The most recent studies have solved even this problem, by the use of unilateral lesions which can dissociate reward versus performance effects.

Early studies. Breese *et al.* (1971) showed that rats lever-pressing for ICSS through either posterior or lateral hypothalamic electrodes dropped in rate to 50 % of pre-operative level after intracisternal injection of 200 μg 6-OHDA following pargyline pre-treatment, which depleted *both* noradrenaline and dopamine by better than 80 %. A second injection of the same dose of 6-OHDA reduced both amines to less than 7 % of control and virtually eliminated ICSS behaviour.

Phillips & Fibiger (1976) found a virtual abolition of ICSS immediately after intraventricular injection of 250 μg of 6-OHDA following tranylcypromine pre-treatment, which reduced noradrenaline to 24 % and dopamine to 19 % of control values. Gradual recovery of ICSS occurred in the 50 days after lesion.

Lippa *et al.* (1973) found a similar decrease in ICSS rates after intraventricular injection of two 200 μg doses of 6-OHDA without pre-treatment. Depletion of noradrenaline by 90 %, and dopamine by an estimated 50–60 %, was seen with ICSS rates recovering to near normal after the fifth or sixth post-operative day. These authors suggested a primary role for dopamine in ICSS.

Selective amine depletions. Cooper *et al.* (1974) were amongst the early authors to effect a selective depletion of brain noradrenaline, and independently brain dopamine, using 6-OHDA techniques. Depletion of brain noradrenaline to 40 % of control by three intracisternal injections of 25 μg, separated by 3 days each, caused no effect on ICSS from lateral hypothalamus electrodes. Depletion of brain dopamine by a single intracisternal injection of 200 μg of 6-OHDA following pre-treatment with the noradrenaline uptake inhibitor desipramine to protect noradrenergic neurones caused a reduction of ICSS rates to some 20 % of control. However, recovery was seen to better than 50 % of normal rates by 6 days post-operative. Experiments such as these serve to highlight the role of dopamine, rather than that of noradrenaline, in ICSS. They are, however, still confounded by the reward versus performance question.

Unilateral depletions. Christie, Ljungberg & Ungerstedt (1973) reported that unilateral lesion of the ascending nigrostriatal dopaminergic bundle with 6-OHDA (8 μg injected stereotaxically directly into that brain area) caused a failure to find ICSS from electrodes subsequently implanted in the lateral hypothalamus on the same side of the brain as the lesion, while ICSS could still be found from contralateral placements. Clearly, a lesion-induced motor impairment could not explain this set of results since a motor impairment should have interfered with ICSS from any electrode location, not merely those ipsilateral to the lesion.

Phillips, Carter & Fibiger (1976) found that ICSS could be obtained from the caudate–putamen, particularly from the anterior portion. ICSS from these electrodes decreased to less than 10 % of control values following injection of 8 μg of 6-OHDA dissolved in 4 μl of ascorbate–saline directly into the fibres of the ascending dopaminergic nigrostriatal bundle on one side of the brain. Animals with this lesion, which depleted striatal dopamine to 5 % of control with only minor effects on hypothalamic noradrenaline, ipsilateral to the ICSS electrode remained depressed in their ICSS rates for the whole of the 19 post-operative test days (see Fig. 3.24). Animals with a contralateral lesion recovered ICSS behaviour to better than 50 % within 10 days and at the end of 21 test days were lever-pressing at 72 % of normal rate. This clearly implicates brain dopamine, rather than noradrenaline. Further, the authors argue that although both ipsilateral and contralateral electrode ICSS was depressed immediately after the lesion, the fact that contralateral ICSS recovered, whereas ipsilateral ICSS did not, rules out a motor impairment. Thus, evidence for a role for dopamine in the reward element of caudate–putamen ICSS is offered.

A further example of the successful use of the ipsilateral versus contralateral 6-OHDA technique is reported by Clavier & Gerfen (1979). They showed with ICSS electrodes in the sulcal prefrontal cortex that 6-OHDA lesion to the ascending nigrostriatal bundle, which includes the dopaminergic innervation of the sulcal prefrontal cortex, that ICSS rates fell to less than 10 % of normal. If the lesion was contralateral to the

ICSS electrode rapid recovery was seen, with ICSS rates up to 50 % of normal by day 2 post-operative and fully recovered by day 11 post-operative. If, however, the lesion were ipsilateral to the electrode no recovery at all occurred, and rates remained depressed at less than 10 % for the full 14 day test period. This is powerful evidence that ICSS from sulcal prefrontal cortex depends on the integrity of dopaminergic systems. The use of the unilateral lesion also avoids the possible motoric confound.

Koob, Fray & Iversen (1978) conducted a similar experiment with electrodes in the lateral hypothalamus or LC. When animals had learned to lever-press for ICSS they received injection of 8 μg 6-OHDA into A9 and a second injection into the more medial A10, both injections being either ipsilateral or contralateral to the stimulating electrode. Pre-treatment with the noradrenaline uptake inhibitor desipramine (25 mg kg^{-1}) was used to protect noradrenergic neurones. In both cases (lateral hypothalamic or LC ICSS), a profound suppression of ICSS was seen with the ipsilateral injections to less than 30–40 % of control rates. This suppression lasted for the entire course of the 21 day test period. However, the animals with injection contralateral to the stimulating electrode, although showing a temporary suppression of 1 to 7 days post-operative in all cases, had recovered to 100 % by day 10 post-operative. Rates remained as high as, or higher than, pre-operative baselines for the subsequent 11 days of

Fig. 3.24. Effects of unilateral 6-hydroxydopamine (6-OHDA) lesions on ICSS from caudate–putamen electrodes. Lesions were either contralateral (Contra) or ipsilateral (Ipsi) to the stimulating electrode. (Redrawn from Phillips *et al*., 1976.)

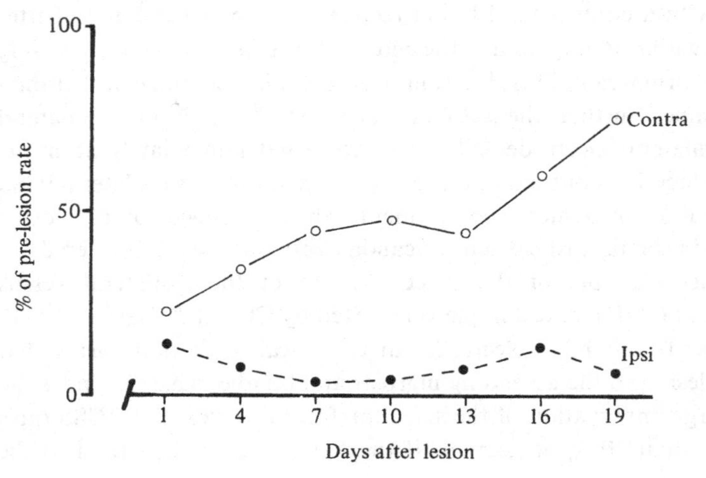

additional testing. Thus, lesions which reduce striatal dopamine to 1 % of normal, and nucleus accumbens olfactory tubercle dopamine to 5 %, permanently suppress ICSS if administered ipsilateral to the stimulating electrode, but have only a temporary effect of about 5 days duration if contralateral to it. This pattern is most consistent with a 6-OHDA attenuation of the reward element of the ICSS rather than a motor impairment. Of particular interest is the finding that unilateral depletion of dopamine permanently suppresses ICSS from a so-called noradrenergic site, the LC. This suggests that even if the stimulating electrode is not directly activating dopaminergic neurones at its tip (as in the LC), nonetheless its effects are ultimately funnelled through a dopaminergic reward pathway.

However, against this generality must be set the finding of Clavier & Fibiger (1977), who reported that with ICSS electrodes in the substantia nigra *both* ipsilateral *and* contralateral injections of 6-OHDA into the nigrostriatal dopaminergic bundle (depleting striatal dopamine to less than 3 % of normal) caused identical patterns of disruption of ICSS. A temporary decrease in lever-press rates down to a very small percentage of control was seen but this recovered to 50 % by day 5 and to 100 % post-operatively (see Fig. 3.25). This pattern is hence consistent with a motor impairment, not a reward deficit. ICSS from the substantia nigra

Fig. 3.25. Effects of 6-hydroxydopamine (6-OHDA) lesions on ICSS from substantia nigra electrodes. Lesions were either contralateral or ipsilateral to the stimulating electrode. 'C' indicates control, pre-lesion response rates. (Redrawn from Clavier & Fibiger, 1977.)

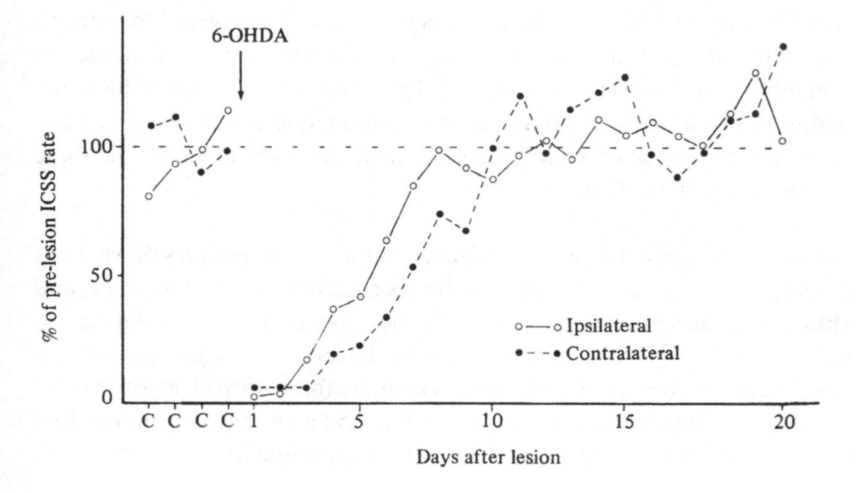

occurred at normal rates as from day 10 post-operative despite the virtual absence of dopamine in the ascending dopaminergic fibre system at the tip of the electrode. This suggests that some other, noncatecholaminergic, system is capable of mediating ICSS from the substantia nigra. Clavier & Corcoran (1976) have described one such descending system, originating in the sulcal prefrontal cortex and running through the substantia nigra, which might offer such a possibility.

A further demonstration that not all ICSS sites necessarily require intact dopaminergic systems came from Phillips & Fibiger (1978). They showed that unilateral destruction of the nigrostriatal dopaminergic bundle by injection of 6-OHDA into one side of the brain (which reduced striatal dopamine to less than 1 % of control values and accumbens plus frontal cortical dopamine to 4 % of normal) severely attenuated ICSS from A10 electrodes without affecting that from nucleus accumbens electrodes. The effect on prefrontal cortical ICSS was intermediate, with a severe suppression soon after the lesion recovering to 50 % of normal by 21 days post-operative. They conclude that although A10 ICSS requires an intact dopaminergic system, ICSS from the nucleus accumbens can proceed in its absence. Thus, a second brain region in addition to the substantia nigra appears to contain noncatecholaminergic elements which can mediate ICSS in their own right.

Further support for this idea was obtained by Simon *et al.* (1979), who found that ICSS from medial prefrontal cortical electrodes survived electrolytic lesions of the ascending dopaminergic fibres which depleted cortical dopamine to less than 20 % of control. In confirmation of the findings of Phillips & Fibiger (1978), ICSS from accumbens electrodes actually increased after this lesion, which reduced accumbens dopamine to less than 10 % of normal. This served to emphasise for the nucleus accumbens, and possibly for the medial prefrontal cortex, that ICSS in the region does not depend on an intact dopaminergic system. In that case, other noncatecholaminergic systems are presumably available in these regions to mediate ICSS.

Summary. It must be concluded that, although techniques have been developed to deplete one amine in the absence of effect on the other, and although it has proved possible with the unilateral lesion technique to separate out performance from reward effects, no single answer has emerged as to whether dopamine does mediate the reward element of ICSS or not. It would appear from the above evidence that dopamine does mediate if the stimulating electrode is in the LC, lateral hypothalamus, A10 area, sulcal prefrontal cortex and possibly the medial prefrontal cortex, but *not* if the electrode is in the substantia nigra, nucleus accumbens and possibly the medial prefrontal cortex. Attention has now moved on to possible noncatecholaminergic substrates of ICSS in these areas.

Conclusion

Briefly summarised, the preceding sections have suggested the following:

(1) Mapping studies have found ICSS-positive sites in many brain areas which also contain dopaminergic cell bodies, fibres or terminals. In contrast to the story with noradrenaline, micromapping has confirmed the association.
(2) Release studies, whether by radioactive tracers, depletion of amine levels, fluorescence or prior synthesis inhibition, have shown that dopaminergic fibres are indeed activated by electrical stimulation through ICSS electrodes in their neighbourhood.
(3) Drug studies using dopamine receptor blockers have convincingly shown an attenuation of ICSS. The problem of reward versus performance has, however, been raised by these studies.
(4) Drug studies using dopamine receptor stimulants, such as apomorphine, have yielded highly contradictory results.
(5) Interactions with other transmitter systems, such as acetylcholine and noradrenaline have been suggested and unusual effects of chronic treatments with drugs shown.
(6) Studies on the intracerebral localisation of the site of action of dopamine receptor blockers have highlighted the role of the nucleus accumbens and ventral anterior striatum.
(7) Studies of the performance versus reward question of the effects of dopamine receptor blockers on ICSS have strongly suggested, without ever being really convincing, that they act by blocking reward at doses lower than those which indisputably cause motor impairments.
(8) Studies with 6-OHDA lesions have convincingly demonstrated that it is dopamine rather than noradrenaline which is the amine of importance for ICSS. Further, use of the unilateral lesion technique has shown that this acts by a block of reward rather than by performance deficits. However, it has also shown that some sites in the brain can support ICSS in the absence of *either* catecholamine. Attention is now directed to the possible noncatecholaminergic substates of ICSS.

References

Amaral, D. G. & Routtenberg, A. (1975). Locus coeruleus and intracranial self-stimulation: a cautionary note. *Behavioral Biology*, **13**, 331–8.

Anlezark, G. M., Arbuthnott, G. W., Christie, J. E., Crow, T. J. & Spear, P. J. (1972). Electrical self-stimulation with brain stem electrodes. *Journal of Physiology*, **227**, 6P–7P.

Anlezark, G. M., Arbuthnott, G. W., Christie, J. E., Crow, T. J. & Spear, P. J. (1973). Electrical self-stimulation with electrodes in the region of the interpeduncular nucleus. *Journal of Physiology*, **234**, 103P–104P.

Anlezark, G. M., Arbuthnott, G. W., Christie, J. E., Crow, T. J. & Spear, P. J. (1974). Electrical self-stimulation in relation to cells of origin of catecholamine-containing neural systems ascending from the brain stem. *Journal of Physiology*, **237**, 31P–32P.

Antelman, S. M. & Caggiula, A. R. (1977). Norepinephrine–dopamine interactions and behavior. *Science*, **195**, 646–53.

Antelman, S. M., Canfield, D. R. & Fisher, A. E. (1976). Dopamine and self-stimulation. In *Brain stimulation reward*, eds. A. Wauquier & E. T. Rolls, p. 187. North-Holland: Amsterdam.

Antelman, S. M., Lippa, A. S. & Fisher, A. E. (1972). 6-Hydroxydopamine, noradrenergic reward and schizophrenia. *Science*, **175**, 919–20.

Arbuthnott, G. M., Crow, T. J., Fuxe, K., Olson, L. & Ungerstedt, U. (1970). Depletion of catecholamines *in vivo* induced by electrical stimulation of central monoamine pathways. *Brain Research*, **24**, 47–83.

Arbuthnott, C., Fuxe, K. & Ungerstedt, U. (1971). Central catecholamine turnover and self-stimulation behaviour. *Brain Research*, **27**, 406–13.

Arbuthnott, G. W., Mitchell, M. J., Nicolaou, N. M. & Yates, C. M. (1977). Changes in dopamine metabolism after intracranial self-stimulation in area ventralis tegmenti. *Journal of Physiology*, **267**, 36P–37P.

Atrens, D. M., Becker, F. T. & Hunt, C. E. (1980). Apomorphine: selective inhibition of the aversive component of lateral hypothalamic self-stimulation. *Psychopharmacology*, **71**, 97–9.

Atrens, D. M., Ljungberg, T. & Ungerstedt, U. (1976). Modulation of reward and aversion processes in the rat diencephalon by neuroleptics: differential effects of clozapine and haloperidol. *Psychopharmacology*, **49**, 97–100.

Atrens, D. M., Ungerstedt, U. & Ljungberg, T. (1977). Specific inhibition of hypothalamic self-stimulation by selective reuptake blockade of either 5-hydroxytryptamine or noradrenaline. *Psychopharmacology*, **52**, 177–80.

Audigier, Y., Virion, A. & Schwartz, J. C. (1976). Stimulation of cerebral histamine H_2 receptors by clonidine. *Nature*, **262**, 307–8.

Belluzzi, J. D., Ritter, S., Wise, C. D. & Stein, L. (1975). Substantia nigra self-stimulation: dependence on noradrenergic reward pathways. *Behavioral Biology*, **13**, 103–11.

Bolme, P., Corrodi, H., Hokfeldt, T., Lidbrink, P. & Goldstein, M. (1974). Possible involvement of central adrenaline neurons in vasomotor and respiratory control. Studies with clonidine and its interaction with yohimbine. *European Journal of Pharmacology*, **28**, 89–94.

Bowers, M. B. & van Woert, M. H. (1972). 6-Hydroxydopamine, noradrenergic reward and schizophrenia. *Science*, **175**, 920–1.

Breese, G. R., Howard, J. L. & Leahy, J. P. (1971). Effect of 6-hydroxydopamine on electrical self-stimulation of the brain. *British Journal of Pharmacology*, **43**, 255–7.

Broekkamp, C. L. E., Pijnenburg, A. J., Cools, A. R. & van Rossum, J. M. (1975). The effect of microinjections of amphetamine into the neostriatum and the nucleus accumbens on self-stimulation behaviour. *Psychopharmacologia*, **42**, 179–83.

Broekkamp, C. L. E. & van Rossum, J. M. (1974). Effects of apomorphine on self-stimulation behaviour. *Psychopharmacologia*, **34**, 71–80.

Broekkamp, C. L. E. & van Rossum, J. M. (1975). The effect of microinjection of morphine and haloperidol into the neostriatum and the nucleus accumbens on self-stimulation behavior. *Archives internationales de pharmacodynamie et de thérapie*, **217**, 110–17.

Cedarbaum, J. M. & Aghajanian, G. K. (1976). Noradrenergic neurons of the locus coeruleus: inhibition by epinephrine and activation by the α-antagonist piperoxane. *Brain Research*, **112**, 413–19.

Chiuch, C. C. & Moore, K. E. (1974). Relative potencies of *d*- and *l*-amphetamine in the release of dopamine from cat brain *in vivo*. *Research Communications in Chemistry, Pathology and Pharmacology*, **7**, 189–99.

Christie, J. E., Ljungberg, T. & Ungerstedt, U. (1973). Dopamine neurones and electrical self-stimulation in the lateral hypothalamus. *Journal of Physiology*, **234**, 80P–81P.

Clavier, R. M. & Corcoran, M. E. (1976). Attenuation of self-stimulation from substantia nigra but not dorsal tegmental noradrenergic bundle by lesions of sulcal prefrontal cortex. *Brain Research*, **113**, 59–69.

Clavier, R. M. & Fibiger, H. C. (1977). On the role of ascending catecholaminergic projections in intracranial self-stimulation of the substantia nigra. *Brain Research*, **131**, 271–86.

Clavier, R. M., Fibiger, H. C. & Phillips, A. C. (1976). Evidence that self-stimulation of the region of the locus coeruleus in rats does not depend upon noradrenergic projections to telencephalon. *Brain Research*, **113**, 71–81.

Clavier, R. M. & Gerfen, C. R. (1979). Self-stimulation of the sulcal prefrontal cortex in the rat: direct evidence for ascending dopaminergic mediation. *Brain Research Bulletin*, **12**, 183–7.

Clavier, R. M. & Routtenberg, A. (1974). Ascending monamine-containing fibre pathways related to intra-cranial self-stimulation: histochemical fluorescence study. *Brain Research*, **72**, 25–40.

Clavier, R. M. & Routtenberg, A. (1976*a*). Fibres associated with brain stem self-stimulation: Fink–Heimer study. *Brain Research*, **105**, 325–32.

Clavier, R. M. & Routtenberg, A. (1976*b*). Brain stem self-stimulation attenuated by lesions of medial forebrain bundle but not by lesions of locus coeruleus or the caudal norepinephrine bundle. *Brain Research*, **101**, 251–71.

Cooper, B. R., Black, W. C. & Paolino, R. M. (1971). Decreased septal-forebrain and lateral hypothalamic reward after alpha-methyl-*p*-tyrosine. *Physiology and Behavior*, **6**, 425–9.

Cooper, B. R., Cott, J. M. & Breese, G. R. (1974). Effects of catecholamine-depleting drugs and amphetamine on self-stimulation of brain following various 6-hydroxydopamine treatments. *Psychopharmacologia*, **37**, 235–48.

Cooper, B. R., Konkol, R. J. & Breese, C. R. (1978). Effects of catecholamine depleting drugs and *d*-amphetamine on self-stimulation of the substantia nigra and locus coeruleus. *Journal of Pharmacology and Experimental Therapeutics*, **204**, 592–605.

Corbett, D. (1980). Long term potentiation of lateral hypothalamic self-stimulation following parabrachial lesions in the rat. *Brain Research Bulletin*, **5**, 637–42.

Corbett, D., Skelton, R. W. & Wise, R. A. (1977). Dorsal noradrenergic bundle lesions fail to disrupt self-stimulation from region of locus coeruleus. *Brain Research*, **133**, 37–44.

Corbett, D. & Wise, R. A. (1979). Intracranial self-stimulation in relation to the ascending noradrenergic fibre systems of the pontine tegmentum and caudal midbrain: a moveable electrode mapping study. *Brain Research*, **177**, 324–36.

Corbett, D. & Wise, R. A. (1980). Intracranial self-stimulation in relation to the ascending dopaminergic systems of the midbrain: a moveable electrode mapping study. *Brain Research*, **185**, 1–15.

Coyle, J. T. & Snyder, S. H. (1969). Antiparkinsonian drugs: inhibition of dopamine uptake in the corpus striatum as a possible mechanism of action. *Science*, **166**, 899–901.

Coyle, J. T. & Snyder, S. H. (1970). Catecholamine uptake by synaptosomes in homogenates of rat brain: stereospecificity in different areas. *Journal of Pharmacology and Experimental Therapeutics*, **170**, 221–31.

Crow, T. J. (1971). The relation between electrical self-stimulation sites and catecholamine-containing neurones in the rat mesencephalon. *Experientia*, **27**, 662.

Crow, T. J. (1972). A map of the rat mesencephalon for electrical self-stimulation. *Brain Research*, **36**, 265–73.

Crow, T. J., Spear, P. J. & Arbuthnott, G. W. (1972). Intra-cranial self-stimulation with electrodes in the region of the locus coeruleus. *Brain Research*, **36**, 275–87.

Dahlstrom, A. & Fuxe, K. (1964). Evidence for the existence of monoamine-containing neurons in the central nervous system. I. Demonstration of monoamines in the cell bodies of brain stem neurons. *Acta physiologica scandinavica, suppl.*, **62**, 1–55.

Eichler, A. J. & Antelman, S. M. (1979). Sensitization to amphetamine and stress may involve nucleus accumbens and medial frontal cortex. *Brain Research*, **176**, 412–16.

Ellman, S. J., Ackermann, R. F., Bodnar, R. J., Jackler, F. & Steiner, S. S. (1976). *d*- and *l*-amphetamine differentially mediate self-stimulation in rat dorsal mid-brain area. *Physiology and Behavior*, **16**, 1–7.

Ettenberg, A. & Milner, P. M. (1977). Effects of dopamine supersensitivity on lateral hypothalamic self-stimulation in rats. *Pharmacology, Biochemistry & Behavior*, **7**, 507–14.

Ferris, R. M., Tang, F. L. M. & Maxwell, R. A. (1972). A comparison of the capacities of isomers of amphetamine, deoxypipradol and methylphenidate to inhibit the uptake of tritiated catecholamines into rat cerebral cortex, hypothalamus and striatum and adrenergic nerves of rabbit aorta. *Journal of Pharmacology and Experimental Therapeutics*, **181**, 407–17.

Fibiger, H. C., Carter, D. A. & Phillips, A. G. (1976). Decreased intracranial self-stimulation after neuroleptics or 6-hydroxydopamine: evidence for mediation by motor deficits rather than by reduced reward. *Psychopharmacology*, **47**, 21–7.

Fouriezos, G., Hansen, P. & Wise, R. A. (1978). Neuroleptic-induced attenuation of brain stimulation reward in rats. *Journal of Comparative and Physiological Psychology*, **92**, 661–71.

Fouriezos, G. & Wise, R. A. (1976). Pimozide-induced extinction of intracranial self-stimulation: response patterns rule out motor or performance deficits. *Brain Research*, **103**, 377–80.

Franklin, K. B. (1978). Catecholamines and self-stimulation: reward and performance effects dissociated. *Pharmacology, Biochemistry & Behavior*, **9**, 813–20.

Franklin, K. B. & Herberg, L. J. (1975). Self-stimulation: and noradrenaline: evidence that inhibition of synthesis abolishes responding only if the 'reserve' pool is dispersed first. *Brain Research*, **97**, 127–32.

Franklin, K. B. & McCoy, S. N. (1979). Pimozide-induced extinction in rats: stimulus control of responding rules out motor deficit. *Pharmacology, Biochemistry & Behavior*, **11,** 71–5.

German, D. C. & Bowden, D. M. (1974). Catecholamine systems as the neural substrate for intracranial self-stimulation: a hypothesis. *Brain Research*, **73,** 381–419.

Harris, J. E. & Baldessarini, R. J. (1973). Uptake of [^{3}H]catecholamine by homogenates of rat corpus striatum and cerebral cortex: effects of amphetamine analogues. *Neuropharmacology*, **12,** 669–79.

Hastings, L. & Stutz, R. M. (1973). The effect of alpha- and beta-antagonists on the self-stimulation phenomenon. *Life Sciences*, **13,** 1253–9.

Herberg, L. J., Stephens, D. N. & Franklin, K. B. (1976). Catecholamines and self-stimulation: evidence suggesting a reinforcing role for noradrenaline and a motivating role for dopamine. *Pharmacology, Biochemistry & Behavior*, **4,** 575–82.

Holloway, J. A. (1975). Norepinephrine and serotonin specificity of release with rewarding electrical stimulation of the brain. *Psychopharmacologia*, **42,** 127–34.

Horn, A. S., Cuello, A. C. & Miller, R. J. (1974). Dopamine in the mesolimbic system of the rat brain: endogenous levels and the effects of drugs on the uptake mechanisms and stimulation of adenylate cylase activity. *Journal of Neurochemistry*, **22,** 265–70.

Hunt, G. E., Atrens, D. M., Chesher, G. E. & Becker, F. T. (1976). Alpha-noradrenergic modulation of hypothalamic self-stimulation: studies employing clonidine, *l*-phenylephrine and alpha-methyl-para-tyrosine. *European Journal of Pharmacology*, **37,** 105–11.

Iversen, L. L. (1967). *The uptake and storage of noradrenaline in sympathetic nerves.* Cambridge University Press: Cambridge.

Kadzielawa, K. (1974). Dopamine receptor in the reward system of the rat. *Archives internationales de pharmacodynamie et de thérapie*, **209,** 214–26.

Koob, G. F., Balcolm, G. J. & Meyerhoff, J. L. (1976). Increases in intracranial self-stimulation in the posterior hypothalamus following lesions in the locus coeruleus. *Brain Research*, **101,** 554–60.

Koob, G. F., Fray, P. J. & Iversen, S. D. (1978). Self-stimulation at the lateral hypothalamus and locus coeruleus after specific unilateral lesions of the dopamine system. *Brain Research*, **146,** 123–40.

Langer, S. Z. (1980). Presynaptic receptors and modulation of neurotransmission: pharmacological implications and therapeutic relevance. *Trends in Neuroscience*, **3,** 110–12.

Liebman, J. M. & Butcher, L. L. (1973). Effects on self-stimulation behaviour of drugs influencing dopaminergic neurotransmission mechanisms. *Naunyn Schmiedeberg's Archives of Pharmacology*, **277,** 305–18.

Liebman, J. M. & Butcher, L. L. (1974). Comparative involvement of dopamine and noradrenaline in rate-free self-stimulation in substantia nigra, lateral hypothalamus and mesencephalic gray. *Naunyn Schmiedeberg's Archives of Pharmacology*, **284,** 167–94.

Liebman, J. M. & Segal, D. (1976). Lack of tolerance or sensitization to the effects of chronic *d*-amphetamine on substantia nigra self-stimulation. *Behavioral Biology*, **16,** 211–20.

Lindvall, O. & Bjorklund, A. (1974). The organization of the ascending catecholamine neuron systems in the rat brain. *Acta physiologica scandinavica, suppl.*, **412,** 1–48.

Lippa, A. S., Antelman, S. M., Fisher, A. E. & Canfield, D. R. (1973). Neurochemical

mediation of reward: a significant role for dopamine? *Pharmacology, Biochemistry & Behavior*, **1,** 23–38.
Miliaressis, E., Thoa, N. B., Tizabi, Y. & Jacobowitz, D. M. (1975). Catecholamine concentration of discrete brain areas following self-stimulation in the ventral tegmentum of the rat. *Brain Research*, **100,** 192–7.
Milner, P. M. (1970). *Physiological psychology*. Holt, Rinehart & Winston: New York.
Monasmith, B., Plotsky, P., Blank, C. L. & Adams, R. N. (1976). The effect of 6-aminodopamine on electrical self-stimulation in rats. *Pharmacology, Biochemistry & Behavior*, **5,** 19–21.
Mora, F. & Myers, R. D. (1977). Brain self-stimulation: direct evidence for the involvement of dopamine in the prefrontal cortex. *Science*, **197,** 1387–9.
Mora, F., Rolls, E. T., Burton, M. J. & Shaw, G. S. (1976). Effects of dopamine-receptor blockade on self-stimulation in the monkey. *Pharmacology, Biochemistry & Behavior*, **4,** 211–16.
Myers, R. D. & Mora, F. (1977). *In vivo* neurochemical analysis, by push–pull perfusion, of the mesocortical dopaminergic system of the rat during self-stimulation. *Brain Research Bulletin*, **2,** 105–12.
Neill, D. B., Parker, S. D & Gold, M. S. (1975). Striatal dopaminergic modulation of lateral hypothalamic self-stimulation. *Pharmacology, Biochemistry & Behavior*, **3,** 485–91.
Neill, D. B., Peay, L. A. & Gold, M. S. (1978). Identification of a subregion within rat neostriatum for the dopaminergic modulation of lateral hypothalamic self-stimulation. *Brain Research*, **153,** 515–28.
Nimitkitpaisan, Y., Bose, S., Kumar, R. & Pradhan, S. E. (1977). Effects of L-dopa on self-stimulation and brain biogenic amines in rats. *Neuropharmacology*, **16,** 657–61.
Olds, J. & Milner, P. (1954). Positive reinforcement produced by electrical stimulation of septal areas and other regions of rat brain. *Journal of Comparative and Physiological Psychology*, **47,** 419–27.
Olds, J., Yuwiler, A., Olds, M. E. & Yun, C. (1964). Neurohumors in hypothalamic substrates of reward. *American Journal of Physiology*, **207,** 242–54.
Olds, M. E. (1974). Effect of intraventricular norepinephrine on neuron activity in the medial forebrain bundle during self-stimulation behaviour. *Brain Research*, **80,** 461–77.
Olds, M. E. (1975). Effects of intraventricular 6-hydroxydopamine and replacement therapy with norepinephrine, dopamine and serotonin on self-stimulation in diencephalic and mesencephalic regions of the rat. *Brain Research*, **98,** 327–42.
Olds, M. E. & Yuwiler, A. (1972). Effect of brain stimulation in positive and negative reinforcing regions in the rat on content of catecholamines in hypothalamus and brain. *Brain Research*, **36,** 385–98.
Phillips, A. C., Carter, D. A. & Fibiger, H. C. (1976). Dopaminergic substrates of intracranial self-stimulation in the caudate-putamen. *Brain Research*, **104,** 221–32.
Phillips, A. G. & Fibiger, H. C. (1973). Dopaminergic and noradrenergic substrates of positive reinforcement: differential effects of *d*- and *l*-amphetamine. *Science*, **179,** 575–7.
Phillips, A. G. & Fibiger, H. C. (1976). Long-term deficits in stimulation-induced behaviors and self-stimulation after 6-hydroxydopamine administration in rats. *Behavioral Biology*, **16,** 127–43.
Phillips, A. G. & Fibiger, H. C. (1978). The role of dopamine in maintaining intracranial self-stimulation in the ventral tegmentum, nucleus accumbens and medial prefrontal cortex. *Canadian Journal of Psychology*, **32,** 58–66.
Phillips, A. G., van der Kooy, D. & Fibiger, H. C. (1977). Maintenance of intracranial

self-stimulation in hippocampus and olfactory bulb following regional depletion of noradrenaline. *Neuroscience Letters*, **4,** 77–84.

Redgrave, P. (1978). Modulation of intracranial self-stimulation behaviour by local perfusions of dopamine, noradrenaline and serotonin within the caudate nucleus and nucleus accumbens. *Brain Research*, **155,** 277–95.

Ritter, S. & Stein, L. (1973). Self-stimulation of noradrenergic cell group (A6) in locus coeruleus of rats. *Journal of Comparative and Physiological Psychology*, **85,** 443–52.

Ritter, S. & Stein, L. (1974). Self-stimulation in the mesencephalic trajectory of the ventral noradrenergic bundle. *Brain Research*, **81,** 145–57.

Robertson, A. & Mogenson, G. J. (1978). Evidence for a role of dopamine in self-stimulation of the nucleus accumbens of the rat. *Canadian Journal of Psychology*, **32,** 67–76.

Robertson, A. & Mogenson, G. J. (1979). Facilitation of self-stimulation of the prefrontal cortex in rats following chronic administration of spiroperidol or amphetamine. *Psychopharmacology*, **65,** 149–54.

Roll, S. K. (1970). Intracranial self-stimulation and wakefulness: effect of manipulating ambient brain catecholamines. *Science*, **168,** 1370–2.

Rolls, E. T., Kelly, P. H. & Shaw, S. G. (1974*a*). Noradrenaline, dopamine and brain-stimulation reward. *Pharmacology Biochemistry & Behavior*, **2,** 735–40.

Rolls, E. T., Rolls, B. J., Kelly, P. H., Shaw, P. G., Wood, R. J. & Dale, R. (1974*b*). The relative attenuation of self-stimulation, eating and drinking produced by dopamine-receptor blockade. *Psychopharmacologia*, **38,** 219–30.

St Laurent, J., Leclerc, R. R., Mitchell, M. L. & Milliaressis, T. E. (1973). Effects of apomorphine on self-stimulation. *Pharmacology, Biochemistry & Behavior*, **1,** 581–5.

St Laurent, J., Roizen, M. F., Miliaressis, E. & Jacobowitz, D. M. (1975). The effects of self-stimulation on the catecholamine concentration of discrete areas of the rat brain. *Brain Research*, **99,** 194–200.

Sastry, R. S. R. & Phillis, J. W. (1977). Evidence that clonidine can activate histamine H_2-receptors in rat cerebral cortex. *Neuropharmacology*, **16,** 223–5.

Schaefer, C. J. & Michael, R. P. (1980). Acute effects of neuroleptics on brain self-stimulation thresholds in rats. *Psychopharmacology*, **67,** 9–15.

Segal, M. & Bloom, F. E. (1976). The action of norepinephrine in the rat hippocampus. III. Hippocampal cellular responses to locus coeruleus stimulation in the awake rat. *Brain Research*, **107,** 499–511.

Shaw, S. G. & Rolls, E. T. (1976). Is the release of noradrenaline necessary for self-stimulation of the brain? *Pharmacology, Biochemistry & Behavior*, **4,** 375–9.

Simon, H., Stinus, L., Tassin, J. P., Lavielle, S., Blanc, G., Thierry, A. M., Glowinski, J. & Le Moal, M. (1979). Is the dopaminergic mesocorticolimbic system necessary for intracranial self-stimulation? Biochemical and behavioral studies from A10 cell bodies and terminals. *Behavioral and Neural Biology*, **27,** 125–45.

Stein, L. & Wise, C. D. (1969). Release of norepinephrine from hypothalamus and amygdala by rewarding medial forebrain bundle stimulation and amphetamine. *Journal of Comparative and Physiological Psychology*, **67,** 189–98.

Stein, L. & Wise, C. D. (1971). Possible etiology of schizophrenia: progressive damage to the noradrenergic reward system by 6-hydroxydopamine. *Science*, **171,** 1032–6.

Stephens, D. N. & Herberg, L.J. (1975). Catecholamines and self-stimulation: pharmacological differences between near- and far-lateral hypothalamic sites. *Brain Research*, **90,** 348–51.

Stephens, D. N. & Herberg, L. J. (1977). Effects on hypothalamic self-stimulation of drugs influencing dopaminergic neurotransmission injected into nucleus accumbens and corpus striatum of rats. *Psychopharmacology*, **54,** 81–5.

Stephens, D. N. & Herberg, L. J. (1979). Dopamine–acetylcholine 'balance' in nucleus accumbens and corpus striatum and its effects on hypothalamic self-stimulation. *European Journal of Pharmacology*, **54,** 331–9.

Stiglick, A. & White, N. (1977). Effects of lesions of various medial forebrain bundle components on lateral hypothalamic self-stimulation. *Brain Research*, **133,** 45–63.

Stinus, L. & Thierry, A. M. (1973). Self-stimulation and catecholamines. II. Blockade of self-stimulation by treatment with alpha-methyl-*para*-tyrosine and the reinstatement by catecholamine precursor administration. *Brain Research*, **64,** 189–98.

Stinus, L., Thierry, A.M., Blanc, G., Glowinski, J. & Cardo, B. (1973). Self-stimulation and catecholamines. III. Effect of imposed or self-stimulation in the area ventralis tegmenti on catecholamine utilization in the rat brain. *Brain Research*, **64,** 199–210.

Stinus, L., Thierry, A.M. & Cardo, B. (1976). Effects of various inhibitors of tyrosine hydroxylase and dopamine-beta-hydroxylase on rat self-stimulation after reserpine treatment. *Psychopharmacology*, **45,** 287–94.

Svensson, T. H. (1971). Functional and biochemical effects of *d*- and *l*-amphetamine on central catecholamine neurons. *Naunyn-Schmiedeberg's Archiv für experimentelle Pathologie und Pharmakologie*, **271,** 170–80.

Taylor, K. M. & Snyder, S. H. (1970). Differential effects of *d*- and *l*-amphetamine on behavior and on catecholamine disposition in dopamine and norepinephrine-containing neurons of rat brain. *Brain Research*, **28,** 295–309.

Ungerstedt, U. (1971). Stereotaxic mapping of monoamine pathways in the rat brain. *Acta physiologica scandinavica, suppl.*, **367,** 1–49.

U'Pritchard, D. C., Reisine, T. D., Mason, S. T., Fibiger, H. C. & Yamamura, H. I. (1980). Modulation of rat brain alpha- and beta-adrenergic receptor populations by lesion of the dorsal noradrenergic bundle. *Brain Research*, **187,** 143–54.

van der Kooy, D., Fibiger, H. C. & Phillips, A. G. (1977). Monoamine involvement in hippocampal self-stimulation. *Brain Research*, **136,** 119–30.

Vetulani, J., Leith, N. J., Stawarz, R. J. & Sulser, F. (1977). Effect of clonidine on the noradrenergic cyclic AMP generating system in the limbic forebrain and on medial forebrain bundle self-stimulation behavior. *Experientia*, **33,** 1490–1.

Wauquier, A. (1978). Differential action of dopamine agonists on brain self-stimulation behaviour in rats. *Archives internationales de pharmacodynamie et de thérapie*, **236,** 325–8.

Wauquier, A. & Niemegeers, C. J. E. (1972). Intracranial self-stimulation in rats as a function of various stimulus parameters. II. Influence of haloperidol, pimozide and pipamperone on medial forebrain bundle stimulation with monopolar electrodes. *Psychopharmacologia*, **27,** 191–202.

Wauquier, A. & Niemegeers, C. J. E. (1973). Intracranial self-stimulation in rats as a function of various stimulus parameters. III. Influence of apomorphine on medial forebrain bundle stimulation with monopolar electrodes. *Psychopharmacologia*, **30,** 163–72.

Wauquier, A. & Niemegeers, C. J. E. (1974*a*) Intracranial self-stimulation in rats as a function of various stimulus parameters. V. Influence of cocaine on medial forebrain bundle stimulation with monopolar electrodes. *Psychopharmacologia*, **38,** 201–10.

Wauquier, A. & Niemegeers, C. J. E. (1974*b*). Intracranial self-stimulation in rats as a function of various stimulation parameters. IV. Influence of amphetamine on medial forebrain bundle stimulation with monopolar electrodes. *Psychopharmacologia*, **34**, 265–74.

Wauquier, A. & Niemegeers, C. J. E. (1976). Restoration of self-stimulation inhibited by neuroleptics. *European Journal of Pharmacology*, **40**, 191–4.

Wise, C. D., Berger, B. D. & Stein, L. (1973). Evidence of alpha-adrenergic reward receptors and serotonergic punishment receptors in the rat brain. *Biological Psychiatry*, **6**, 3–21.

Wise, C. D. & Stein, L. (1969). Facilitation of brain self-stimulation by central administration of norepinephrine. *Science*, **163**, 299–301.

Yuwiler, A. & Olds, M. E. (1973). Catecholamines and self-stimulation behaviour: effects on brain levels after stimulation, and pretreatment with *d, l*-alpha-methyl-*p*-tyrosine. *Brain Research*, **50**, 331–340.

Zarevics, P. & Setler, P. E. (1979). Simultaneous rate-independent and rate-dependent assessment of intracranial self-stimulation: evidence for the direct involvement of dopamine in brain reinforcement mechanisms. *Brain Research*, **169**, 499–512.

4

Catecholamines and motor behaviour

Introduction

Motor behaviour is obviously of great importance to the organism. It is vital for the capture of food, the consummation of mating and the escape from predators. The involvement of the catecholamines in this process seems to lie not in the fine, detailed control of a single muscle contraction, as does, say, acetylcholine at the neuromuscular junction. Rather, catecholamines are involved in the so-called extrapyramidal motor system whose function appears to regulate the level of motor output to whole muscle groups and thus locomotor activity of the whole animal. We shall be concerned here with the amount of forward locomotion, and the amount of stereotyped movements of the mouth and whole head, rather than with the contraction of, say, the gastrocnemius; and even less with the contraction of a single muscle fibre within this muscle group.

The paradigm

Locomotor activity is often measured in photocell cages. These are circular or square arenas, typically measuring 25–50 cm across, with anything from one to six light or infrared beams projecting from one side to the other, about 3–5 cm above the floor. Thus, the rat running about will interrupt these light beams and this is sensed by photocells at the opposite end of the beam, which then increment a counter. Typically, counts are cumulated for 5 min and then printed and the counter reset to zero (see Fig. 4.1). This is repeated for 1 or 2 h until the drug has ceased to act and locomotor activity has fallen to low levels. With this paradigm large movements such as forward running are detected by the photocell assembly while small ones such as grooming or rearing are not. The weakness of the photocell beam technique is that an animal which repeatedly grooms or rears by chance in a light beam *itself* will generate a huge count which will be mistaken for forward running. Visual inspection of the rat during automated recording is highly recommended. Other locomotor activity apparatus includes those operating on the Doppler principle of reflected sound to detect movement, and even some radar-based devices. Some more advanced models can discriminate between small, grooming-type movements and larger, whole body running movements.

Another type of apparatus uses the slight displacement of the floor of the cage as the rat runs from one side to another to trigger microswitches mounted underneath it or to activate tension transducers of various kinds. These are generally termed jiggle cages or stabilimeters. A further, and rather different piece of apparatus, is the running wheel. Here the rat can run inside a wheel held with the axle stationary and so turn the wheel. The number of completed turns is usually recorded as the measure of activity. It

Fig. 4.1. (*a*) Locomotor activity cage. Plan view. (*b*) Locomotor activity arena or open field. Plan view. (*c*) Rotometer for quantifying rotational behaviour. (After Ungerstedt & Arbuthnott, 1970.)

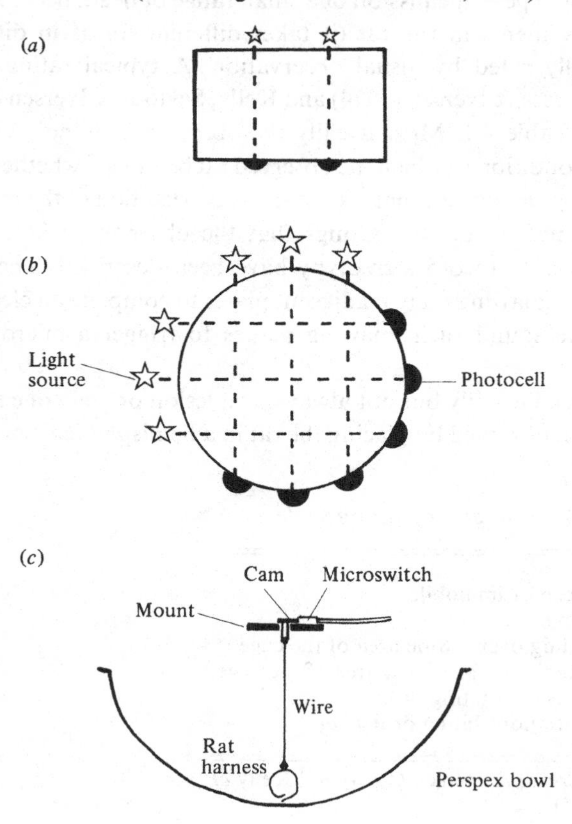

is a general, if possibly somewhat superstitious, feeling among most researchers that running wheels seem to yield results very different in terms of drug action and effects from almost any other type of activity measuring device!

When a rat is initially put into a new apparatus it will run around exploring the environment and hence generate numerous locomotor counts. With time and familiarity to the apparatus, the rat will cease to be active. Many drugs will modify this process of exploration *per se*, independently of any effect they may have on simple locomotor activity. Thus, to obtain a steady and nondecreasing baseline of activity and also to ensure that one is observing simple locomotor stimulation, rather than exploration, a period of habituation is given. This consists of allowing the rat from 30 min to 1 h to explore the apparatus each time it is introduced into it before injecting the drug you wish to test. Thus, in almost all modern experiments, the rat will have been habituated to the apparatus prior to drug injection and subsequent activity measurement.

Stereotypy, the repeated emission of a small range of head, neck, mouth and tongue movements in the rat (it takes different forms in different species) is usually rated by visual observation. A typical rating scale, adopted from Creese & Iversen (1974) and Kelly, Seviour & Iversen (1975) is illustrated in Table 4.1. Most usually the observer is 'blind'. That is, unaware of the condition to which the observed rat belongs – whether it is a control or an experimental animal – so that any expectation of the result to be obtained cannot distort the ratings that the observer makes. Some automated devices to record stereotypy have been described. They rely either on the rat's gnawing on two adjacent plates to complete an electrical circuit or on the sound such gnawing makes to trigger a microphone detector.

Rotation, caused usually but not always by a lesion on only one side of the brain, is often observed by placing the rat in a hemispherical container

Table 4.1. *Stereotypy rating scale*

0	Asleep or immobile
1	Active
2	Sniffing over a wide area of the cage
3	Sniffing in a restricted area of the cage
4	Occasional biting or licking
5	Continuous biting or licking

From Creese & Iversen (1974) and Kelly *et al.* (1975).

about 30 cm across and counting completed turns by eye. Some automated devices (see Fig. 4.1*c*) which affix a saddle to the rat attached to a leash which pulls round a rotating lever in time with the animal, have also been described (Ungerstedt & Arbuthnott, 1970). Rotation is usually obtained more easily in circular rather than square enclosures, unlike stereotypy which is best seen in the corners of a square box, particularly one with bars making up the wall on which the rat can chew.

Early evidence for catecholamine involvement

Precursor administration

A typical early technique used to investigate the role of catecholamines in locomotor activity was that of precursor administration. The idea being to increase the concentration of neurotransmitter in the brain and hence increase its normal role in the expression of locomotor activity. L-Dihydroxyphenylalanine (L-DOPA) was a favourite tool since it crosses the blood–brain barrier and hence can be administered intraperitoneally. Once in the brain it is converted into both dopamine and noradrenaline by decarboxylation. As such it immediately has the drawback of failing to dissociate the two amines one from the other. Further, L-DOPA is also active in the periphery when injected intraperitoneally and early experiments yielded conflicting results for just this reason. To ensure that the effects of L-DOPA that are seen are the result of its action on the central, rather than the peripheral, nervous system, the rat must be pre-treated with a decarboxylase inhibitor, such as Ro 4-4602 which does not cross the blood–brain barrier. Thus, decarboxylation of L-DOPA to physiologically active dopamine is prevented in the periphery but still occurs in the central nervous system. An early experiment using this technique was reported by Butcher, Engel & Fuxe (1972). They found that rats treated with a peripheral decarboxylase inhibitor and given 200 mg kg^{-1} L-DOPA showed a pronounced increase in locomotor activity about 60 min after injection (see Fig. 4.2). They also performed biochemical measurement of the levels of dopamine, noradrenaline and serotonin on other animals at different times after L-DOPA administration. They found a 700 % increase in whole brain dopamine, no great change in noradrenaline and a marked decrease in brain serotonin. They further used a histofluorescence technique to visualise the amount of amine in different brain areas after L-DOPA administration and observed marked increases in the fluorescence of dopamine cell bodies of the A8, A9, A12 and to a lesser extent A10 groups. No great change in fluorescence in noradrenaline cell bodies was seen. Increases were also noted in the dopaminergic terminal areas such as the neostriatum, nucleus accumbens, olfactory tubercle and amygdala. All of this would seem to suggest that the increase in locomotor activity caused by 200 mg kg^{-1} L-DOPA might be the result of increased activity in brain dopaminergic systems. We cannot from this experiment say which particular ones (A9 versus A10, with A8 and A12 also possible). A further caution is the marked decrease in serotonin levels also found. The lack of effect of L-DOPA on either biochemically measured noradrenaline or fluorescence in noradrenergic cell bodies suggests noradrenaline is not the critical amine.

Fig. 4.2. Locomotor activity in rats treated with the peripheral decarboxylase inhibitor Ro 4-4602, L-DOPA or a combination of the two. Ro 4-4602 was administered at the start of the session and L-DOPA was given 30 min later. (Redrawn from Butcher, Engel & Fuxe, 1972.)

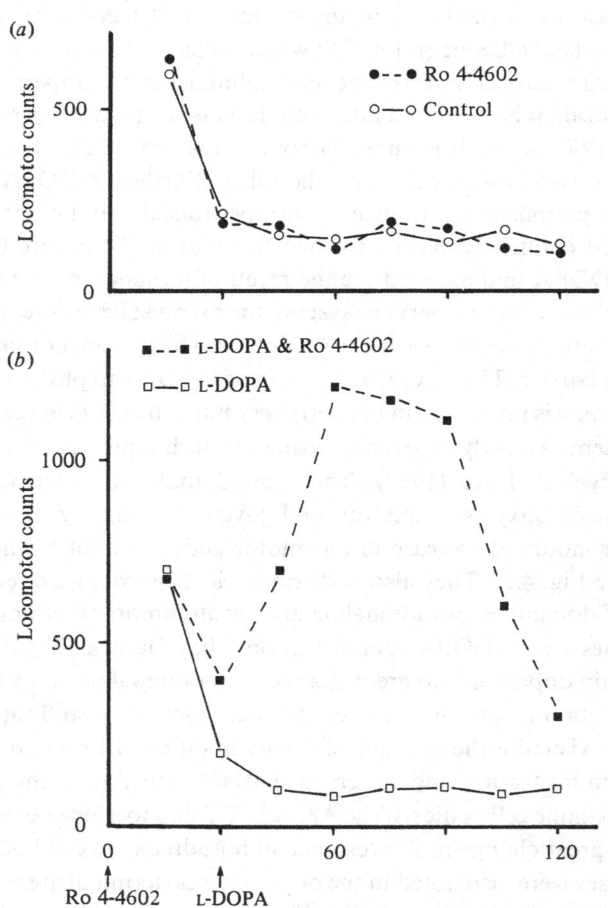

Consistent with the view that it is the increased dopamine synthesised from L-DOPA which brings about the observed locomotor stimulation is the report from Przegalinski & Kleinrok (1972). They found that the effect of L-DOPA and peripheral decarboxylase inhibitor (Ro 4-4602) in causing locomotor activity could be blocked by the neuroleptic drugs haloperidol or chlorpromazine: at the doses given (1 mg kg^{-1} and 0.25 mg kg^{-1}, respectively) these drugs are believed to block dopamine, rather than noradrenaline, receptors.

Another, more acute preparation which has been used to demonstrate the effect of L-DOPA in releasing motor output is that of the decerebrate rabbit, in which the motor nerves to the tibialis anterior and gastrocnemius medialis muscles were dissected free under anaesthesia and electrical activity recorded from them when the animal had recovered from anaesthesia. Pressure points and cut portions were infused with local anaesthetic and, of course, being decerebrate the animal was unable to perceive pain. Administration of L-DOPA led to a marked increase in the firing of these motor nerves (see Fig. 4.3) and even bigger effects were seen after peripheral decarboxylase inhibition with Ro 4-4602, indicating the central nature of the effect. Again this fails to dissociate effects on noradrenergic from those on dopaminergic systems and further, the severe surgery (total decerebration) renders the relevance to normal locomotion in the intact animal questionable.

Fig. 4.3. Effect of injection of L-DOPA alone, or following pre-treatment with Ro 4-4602, on spontaneous locomotor discharge from central end of nerve trunk to tibialis anterior or gastrocnemius medialis muscles in the rabbit. L-DOPA (100 mg kg^{-1} i.v.) was injected at time zero. Ro 4-4602 (50 mg kg^{-1}) was given 2 h previously. Activity in the hour baseline prior to L-DOPA administration was low and is omitted from the graphs. (Redrawn from Viala & Buser, 1974.)

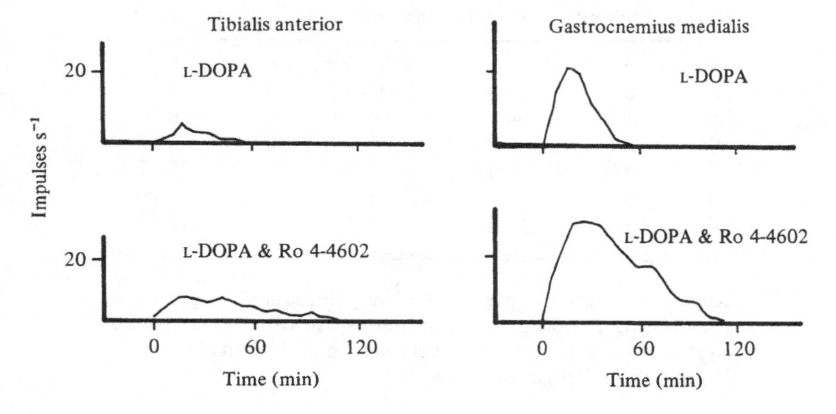

L-DOPA administered to human patients, usually for the treatment of Parkinson's disease, is also reported to induce excitation and arousal, leading sometimes to restlessness and aggression (Murphy, 1973).

Agents which serve to release endogenous stores of amines, such as amphetamine and tetrabenazine also induce intense locomotor activity (Scheel-Kruger & Jonas, 1973). An increase in both rearing and locomotion is seen with doses of amphetamine ranging from 1.0 to 5.0 mg kg^{-1}. At 5 mg kg^{-1} stereotyped behaviour starts to be seen, and at and above 10 mg kg^{-1} stereotypy predominates with forward locomotion and rearing declining (see Table 4.2). Again, tetrabenazine and amphetamine release transmitter from both noradrenaline and dopamine terminals, and so fail to dissociate the relative action of the two amines.

Inhibition of breakdown

As well as inactivation by active reuptake from the presynaptic cleft, noradrenaline and dopamine are broken down by the enzyme monoamine oxidase (MAO), which exists in two forms, type A and type B. Compounds exist which inhibit this enzyme and so act to increase the level of amine available in the synaptic cleft following release. Braestrup, Andersen & Randrup (1975) reported that the monoamine oxidase inhibitor (MAOI) *l*-deprenyl at 100 mg kg^{-1} induced mild locomotor activation while the isomeric mixture *d,l*-deprenyl induced an amphetamine-like stereotypy at 100 mg kg^{-1}. *l*-Deprenyl also markedly potentiated the stereotypy induced by phenylethylamine. While *l*-deprenyl blocks the type B monoamine oxidase, the type A MAOI clorgyline was ineffective in

Table 4.2. *Effects of amphetamine on rat behaviour*

No. of rats	Dose (mg kg^{-1})	Locomotion	Rearing	Sniffing	Licking/ biting
12	0.25	(+)	(+)	0	0
8	0.50	(+)	(+)	0	0
25	1.0	+	+	0	0
40	2.5	+++	+++	0	0
52	5.0	+++	+++	25/52	27/52
20	10	++	+++	—	20/20
10	20	++	++	—	10/10

Key: 0 behaviour not present; (+) very infrequent; + present in shortlasting periods; ++ continuous but of moderate intensity; +++ continuous and very strong in intensity.
From Scheel-Kruger & Jonas, 1973.

potentiating phenylethylamine stereotypy or in causing locomotor stimulation on its own. Again, since both noradrenaline and dopamine are substrates for type B monoamine oxidase no distinction between these two amines can be made.

False transmitters

α-Methyldopa is converted in the brain into α-methyldopamine and α-methylnoradrenaline. These are released from presynaptic stores but fail to stimulate the postsynaptic receptor. Since they take up storage sites usually used for the natural neurotransmitters, and also do not activate the postsynaptic membrane, they serve to reduce transmission across the synapse and are referred to as 'false transmitters' (Carlsson & Lindquist, 1962). Dominic & Moore (1971) reported that the behavioural action of α-methyldopa was to cause a profound drop in the locomotor activity of mice which remained depressed for over 6 h. Parallel biochemical measurements on other mice showed that both brain dopamine and brain noradrenaline fell after α-methyldopa. The time course of recovery of locomotor activity correlated more with the recovery of normal levels of brain dopamine than that of levels of noradrenaline. Here, at least, a small hint appeared of a separation in the roles of the two amines, with dopamine being of more importance.

Synthesis inhibitors and reserpine

Another technique to deplete brain amines is to use drugs such as reserpine or prenylamine which render the presynaptic storage mechanisms defective.

Depletion of brain noradrenaline by the drug prenylamine was found by Broitman & Donoso (1971) to reduce spontaneous motor activity measured in an open field. Effects were particularly marked 4 to 6 h after drug administration. Depletion of noradrenaline occurred throughout the brain (cortex, hypothalamus) but the decrease in activity correlated best with the decrease in hypothalamic noradrenaline. Prenylamine, like the widely used drug reserpine, suffers from the drawback of reducing serotonin as much as noradrenaline and dopamine. Very little which is definitive may thus be concluded from this study.

Ahlenius, Anden & Engel (1973) describe an experiment in which mice were administered with either 150 mg kg^{-1} of the tyrosine hydroxylase inhibitor α-methyl-*para*-tyrosine (AMPT) or with the storage granule depleting agent reserpine (0.6 mg kg^{-1}). Locomotor activity was markedly depressed. Since inhibition of tyrosine hydroxylase will prevent the synthesis of both noradrenaline and dopamine, and since reserpine depletes not only these catecholamines but also serotonin, this by itself tells us little of the critical amine involved. However, Ahlenius *et al.* next administered L-DOPA to the animals (10–25 mg kg^{-1}) and found that it restored normal activity in the AMPT-treated, but not in the reserpine pre-treated rats. Levels of noradrenaline in the brain were restored to normal and brain dopamine in fact rose to some 200 % of normal. Only much larger doses of L-DOPA (around 200 mg kg^{-1}) would restore normal activity after reserpine. The L-DOPA reversal of AMPT hypoactivity certainly seems to implicate catecholamines, but again fails to dissociate noradrenaline from dopamine. A further experiment sought to rectify that shortcoming. Ahlenius (1974) repeated the AMPT-induced suppression of locomotor activity and its restoration by L-DOPA. However, he also included a group which, as well as L-DOPA, received the dopamine-β-hydroxylase (DBH) inhibitor FLA 63. Thus, this group of rats cannot use the L-DOPA to synthesise new noradrenaline, whereas replenishment of brain dopamine occurs as normal. This was confirmed by biochemical studies in which L-DOPA, after AMPT, elevated brain dopamine levels to some 200 % of normal whether given on its own or in combination with FLA 63 to prevent noradrenaline synthesis. In animals receiving L-DOPA on its own, brain noradrenaline returned to 80–100 % of normal, but in rats also receiving FLA 63 no accumulation of noradrenaline occurred. The behavioural results were that L-DOPA restored locomotor activity after AMPT whether given on its own or in combination with FLA 63. Thus, it does not seem that the noradrenaline synthesised from L-DOPA was necessary for the reversal of AMPT-induced hypoactivity. It is suggested that AMPT reduced activity by preventing the synthesis of dopamine specifically. So, an indication of a critical role for dopamine in locomotor activity was starting to emerge from some of these early experiments.

Noradrenaline and motor behaviour

Agonist drugs

One possible solution to the inability of precursor administration to distinguish the amine of importance for the increase in locomotor activity might lie in the direct intraventricular administration of the amines themselves. Since these are rapidly inactivated, and may fail to distinguish between the different subtypes and classes of receptors (see p. 14), direct administration of artificial agonist drugs might also be of use.

Herman (1970) found that noradrenaline injected into the lateral ventricle of the rat brain while the animal was under light anaesthesia resulted in a small increase in locomotor activity and grooming at 10 μg for those animals which were spontaneously inactive. At 50 and 100 μg, this increase in activity was seen in all rats. However, at 200 μg a very profound sedation occurred which lasted for more than 2 h of total immobility.

This immediately demonstrates the need for a full dose–response curve when examining the effects of drugs on behaviour, since a marked excitation seen at lower doses of noradrenaline was converted at the highest dose into a profound sedation. It is also difficult to conclude whether noradrenaline is facilitatory or inhibitory to locomotor output, since both excitation and sedation were seen, but at different doses. Similar sedatory effects have been reported incidental to studies of ICSS behaviour by Wise & Stein (1969) (see p. 37). Herman's study also failed to rule out a role for dopamine, since injections of this amine were not examined on locomotor activity.

This last point assumed greater significance with the report of Geyer, Segal & Mandell (1972) that infusions of dopamine through chronically implanted ventricular cannulae *also* caused an increase in locomotor activity in an open field (similar to an activity cage, except that it tends to be larger, have smooth rather than barred walls and be divided into a number of squares to permit finer analysis of the pattern and extent of movement). However, infusion of noradrenaline was more effective than that of dopamine. Thus, 6 μg dopamine was needed to produce the same magnitude of increase in locomotor activity seen after 1–2 μg noradrenaline (see Fig. 4.4). These authors further suggested that dopamine might be

Fig. 4.4. Locomotor activity following intraventricular injections of noradrenaline (NA) or dopamine (DA) in saline (Sal), either with or without pre-treatment with the uptake inhibitor imipramine (IMI) at 5 mg kg^{-1}. (Redrawn from Geyer, Segal & Mandell, 1972.)

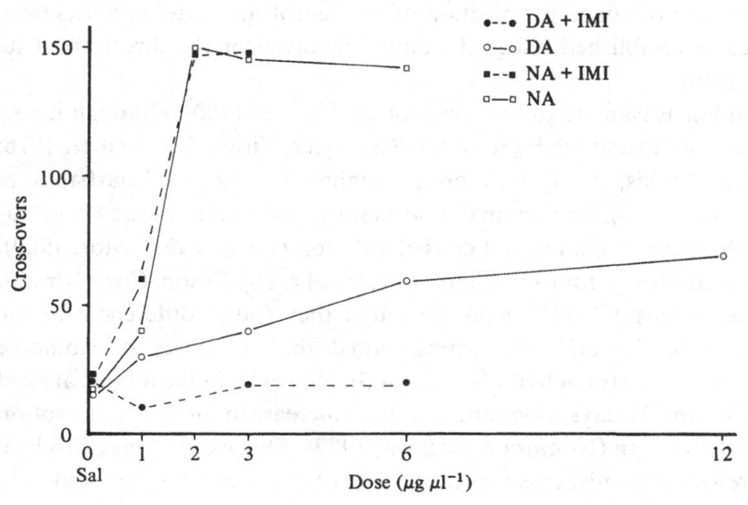

effective only after it had been converted into noradrenaline, since it is of course its natural precursor in the brain. Thus, the dopamine might be taken up into noradrenergic neurones and used to synthesise noradrenaline, which was then released and had the direct effect on locomotor activity. To support this, the authors treated the rats with imipramine, which blocks uptake into catecholaminergic neurones. Now, dopamine no longer caused locomotor stimulation. Infusions of noradrenaline were as effective as before, suggesting a direct receptor activation. Dopamine, however, seemed to be unable to activate this receptor until it had been taken up by noradrenergic neurones and converted into noradrenaline. The only fly in the ointment was the further finding of these authors that the noradrenaline-induced locomotor activity could be blocked by the dopamine receptor antagonist haloperidol. The doses used (0.1–0.5 mg kg^{-1}) are specific for the dopamine receptor and additional, fortuitous blockade of the noradrenergic receptor cannot be claimed.

In a further study it was found that chronic administration of AMPT would lead to an apparent supersensitivity of the noradrenergic receptor to noradrenaline, but not to dopamine, since the behavioural locomotor stimulation was enhanced as a result of 8 days of injection of 125 mg kg^{-1} AMPT per day. Since the enzyme DBH would still be able to convert injected dopamine into noradrenaline in order to stimulate the noradrenaline receptor it is unclear why a supersensitive response to injected dopamine was also not obtained.

The fact that both dopamine and noradrenaline will increase locomotor activity if injected intraventricularly, coupled with the evidence that higher doses of noradrenaline will yield an actual sedation in behaviour, means that the data obtained by injection of the neurotransmitter agonists have not clearly established either the amine involved or the direction of its involvement.

Clonidine is claimed to be an α-agonist (Langer, 1980), although it also has effects on histamine H_2 receptors (Audigier, Virion & Schwartz, 1976; Sastry & Phillis, 1977) and on adrenaline receptors (Cedarbaum & Aghajanian, 1976). The original reports suggested that it lowered exploratory behaviour in the rat and caused sedation (Laverty & Taylor, 1969). This was confirmed for four different strains of rat by Tilson, Chamberlain, Gylis & Buyinski (1977), who, although they found differences in the sensitivity of the different strains, noted that all doses of clonidine decreased locomotor activity (see Fig. 4.5). However, in the infant rat aged between 1 and 11 days, clonidine causes an increase in forward locomotion and wall climbing (Nomura & Segawa, 1979). This would appear to be a noradrenergic agonist effect since it could be prevented by the α-blockers phentolamine, phenoxybenzamine, yohimbine and piperoxan, but not by the histamine H_2 blocker metiamide (Nomura & Segawa, 1979). The evidence gleaned for the role of noradrenergic systems in locomotor behaviour from studies on clonidine is unclear, suggesting at the very least that a switch in functional sign occurs with maturation.

Stimulation of central β-receptors by intraventricular injection of isoprenaline also leads to behaviourally suppressive effects on locomotor activity and open field behaviour, according to Ksiazek & Kleinrok

Fig. 4.5. Locomotor activity in four strains of rat following injection with clonidine. Vertical bars show standard error of the mean. SD, Sprague-Dawley; LE, Long-Evans; KWN, Kyoto Wistar normotensive; KWH, spontaneously hypertensives; Sal, saline. Values are cumulative counts over a 2 h session. Clonidine was injected 15 min prior to testing. (Redrawn from Tilson, Chamberlain, Gylis & Buyinski, 1977.)

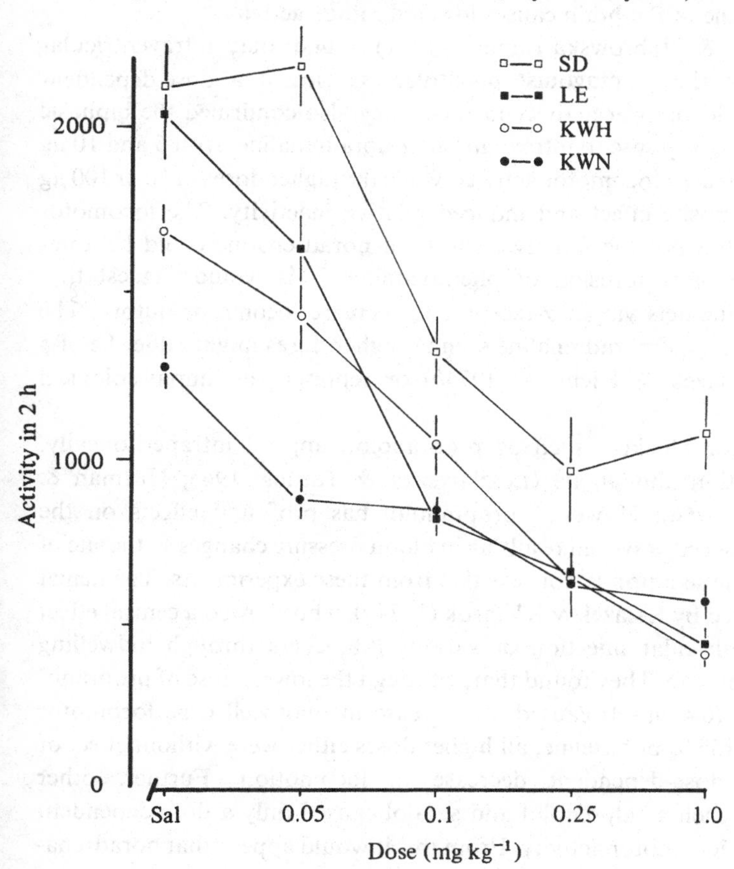

(1974*b*). The effects on locomotion were dose-dependent and, typically, after 0.2 μmol isoprenaline, lasted some 45 min.

Antagonist drugs and synthesis inhibitors

The opposite strategy might be that if release of noradrenaline were believed to be necessary to maintain locomotor activity, then administration of noradrenaline-receptor blocking agents or agents which reduced noradrenaline biosynthesis should cause sedation and a reduction in locomotor output.

Both Roll (1970) and Wise & Stein (1969) report that rats administered disulfiram or diethyldithiocarbamate (both DBH inhibitors) during an experiment on ICSS behaviour appeared sedated and inactive (see p. 43).

Rolls, Kelly & Shaw (1974) examined the dose–response curve of disulfiram on rearing and on locomotor behaviour and found a dose-dependent decrease in locomotor activity (see p. 46). Thus, lowered noradrenaline in the brain causes lowered motor activity.

Kleinrok & Zebrowska-Lupina (1971) found that intraventricular injection of the α-antagonist phentolamine caused a dose-dependent decrease in locomotor activity in rats. They also confirmed the biphasic nature of the response to intraventricular noradrenaline; with 5 and 10 μg causing increased locomotor activity, while the higher doses of 50 or 100 μg had the opposite effect and induced relative inactivity. The locomotor stimulant, but not the sedative, effects of noradrenaline could be antagonised by prior infusion of phentolamine. This would suggest that noradrenaline acts via an α-receptor to increase locomotor output. The sedative effects of noradrenaline seen at higher doses might either be of a β type (Ksiazek & Kleinrok, 1974*b*) or represent an unphysiological effect.

β-Receptor blockers such as propranolol, applied intraperitoneally, cause sedation and ataxia (Leszkovszky & Tardos, 1965; Hurman & Almirante, 1966). However, propranolol has profound effects on the peripheral nervous system resulting in blood pressure changes so the site of action of the sedation is not revealed from these experiments. The deficit was remedied by Ksiazek & Kleinrok (1974*a*), who showed a central effect by intraventricular injection of various β-blockers through indwelling chronic cannulae. They found that, although the lowest dose of propranolol studied (0.4 μmol) caused an increase in photocell cage locomotor activity to 150 % of baseline, all higher doses either were without effect or caused a dose-dependent decrease in locomotion. Further, other β-blockers such as alprenolol and sotalol caused only a dose-dependent decrease in locomotor activity. From this it would appear that noradrenaline is involved in maintaining normal locomotor output via a β-receptor, central blockade of which causes sedation.

Paradoxically, however, the same group of workers report that administration of a β-agonist drug, isoprenaline, via intraventricular cannulae also caused a dose-dependent depressive effect on locomotor activity (Ksiazek & Kleinrok, 1974*b*). Thus, it would appear that either blocking or stimulating a central β-receptor can lead to decreases in locomotor activity. Such findings hardly clarify the role of noradrenaline in normal motor output.

6-OHDA lesions

The most selective technique currently available for manipulation of central noradrenergic systems, without concurrent disruption of dopamine pathways, is the intraventricular or preferably intracerebral injection of small amounts of the neurotoxin 6-OHDA. When injected intraventricularly the neurotoxin can cause a severe and permanent destruction of both noradrenergic and dopaminergic neurones. When injected into the fibres of the ascending noradrenergic dorsal bundle it will deplete forebrain noradrenaline with no effect on brain dopaminergic systems. If injected into the nigrostriatal dopaminergic bundle after pre-treatment with the noradrenaline uptake inhibitor desipramine (DMI) 6-OHDA can cause a similarly profound and selective destruction of dopamine brain systems. I will examine the effects of the later lesions in the second half of this chapter, but here concentrate on the effects of lesions which deplete brain noradrenaline and their effect on locomotor activity.

Early experiments. Many of the earlier uses of 6-OHDA depleted both noradrenergic and dopaminergic systems because of the intraventricular route of administration of the neurotoxin. They thus failed to discriminate between the two amines in the profound locomotor effects often observed subsequent to lesion.

Laverty & Taylor (1970), for example, using a dose of 6-OHDA which would deplete both amines found a reduction in nocturnal activity in rats to 30 % of normal which lasted for at least 8 days after injection. Evetts, Uretsky, Iversen & Iversen (1970) report a reduction in spontaneous locomotor activity immediately after lesion, but rats tested 10 days post-operatively showed as high an activity as controls. Jalfre & Haefely (1971) found a reduction in activity and rearing in an open field after two intraventricular 6-OHDA injections of 500 μg each. On the other hand there are reports in the early literature of *increased* activity after 6-OHDA treatment. Thus, Fibiger, Lonsbury & Cooper (1972) and Vetulani,

Reichenberger & Wiszniowska (1972) claimed a hyperactivity following 250 μg of 6-OHDA intraventricularly in combination with an MAOI given intraperitoneally. This effect of increased activity lasted only 1 or 2 days post-lesion. The variable results of early administration of 6-OHDA on locomotor activity, and the frequent recovery to pre-operative performance, probably reflect the relatively poor depletions of amines obtained in the early experiments. Subsequent experimenters have obtained more severe depletions and have seen more consistent and longer lasting effects.

Selective depletions. Smith, Cooper & Breese (1973) used intracisternal administration of 6-OHDA between days 5 and 7 after birth to cause a depletion of brain dopamine to less than 5 % of normal and hypothalamic noradrenaline to 8 % of normal. These rats were allowed to grow to maturity (which, incidentally involved a slower increase in body weight in the lesioned rats than seen with the normal control growth curve) and were then tested in circular photocell activity cages. They were less active than controls but showed a similar habituation curve, that is, decline in activity over time after introduction to the cages (see Fig. 4.6). Rats, depleted only

Fig. 4.6. Effects of 6-hydroxydopamine (6-OHDA) treatments on locomotor activity in rats. See text for details of treatments which deplete brain dopamine (DA), brain (NA) or both DA and NA together. (Redrawn from Smith, Cooper & Breese, 1973.)

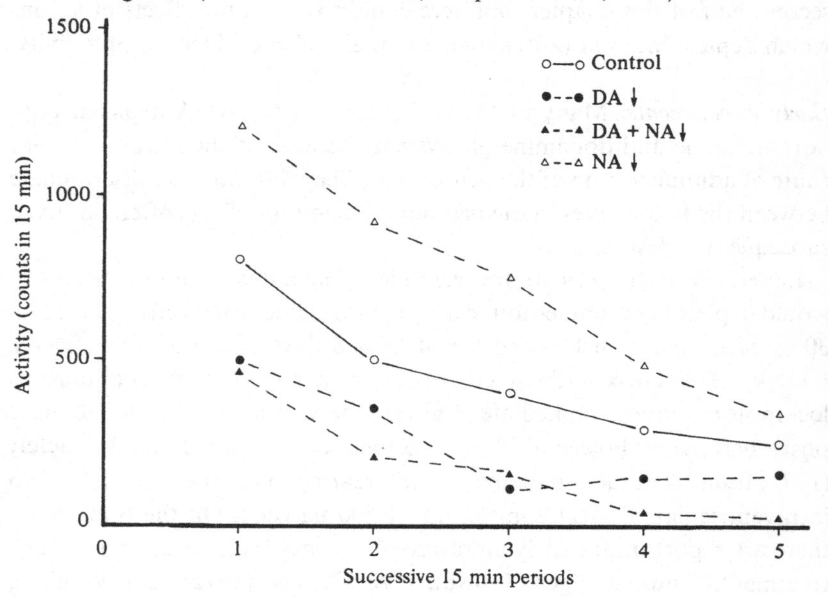

of brain dopamine by intracerebral 6-OHDA (100 μg on day 7 postnatal, with DMI pre-treatment to protect noradrenaline neurones) were similarly less active. The later treatment was successful in depleting brain dopamine to 10 % of control with complete sparing of brain noradrenaline. Rats in which brain noradrenaline was depleted to 46 % by two small doses of 6-OHDA (15 μg on day 5 and 25 μg on day 7) with only an 18 % loss of brain dopamine showed an actual *increase* in locomotor activity compared to controls when tested in adulthood (see Fig. 4.6). This, then seems to assign the activity-decreasing effects of 6-OHDA treatments to their dopamine-depleting effects, but leaves in question the role of noradrenaline in locomotion.

An opposite effect on activity in an open field was reported by Ellison & Bresler (1974), following selective noradrenaline depletion caused by three daily doses of 6-OHDA (25 μg each) per week for two weeks. This reduced forebrain noradrenaline by only 47 % but did not affect dopamine and caused decreased rearing in an open field during the first few minutes (Fig. 4.7*b*) but was without any effect on forward locomotion (Fig. 4.7*a*). Such small effects were probably the result of the extremely modest depletion of brain noradrenaline. It was, however, no less than that which, in the Smith *et al.* (1973) study, led to a marked *increase* in open field activity. It might be relevant at this stage to draw a possible distinction between the effects of amine depletions in the adult (acute) and those effected in the neonate rat which is then allowed to grow to maturity before testing (chronic). Although the actual levels of brain catecholamines may be the same at the point of testing, the behavioural experience of the two groups is very different. One has had a long period of its life in the absence of catecholamines during which learning experiences or brain plasticity changes can have occurred.

Pappas *et al.* (1974) used intraperitoneal administration of 50 mg kg^{-1} 6-OHDA on days 1 to 7 after birth to deplete noradrenaline to 41 % of normal without effect on brain dopamine. Such animals, when tested in adulthood, showed unaltered activity in an open field and the same patterning of activity in running wheels as controls across the light–dark periods of the day. Again, while achieving selectivity of amine depletion (i.e. sparing brain dopamine) the degree of noradrenaline depletion was not very impressive and negative conclusions made on such poor depletions are risky.

Other authors were able to achieve more profound destruction of the noradrenergic system and still found unaltered locomotion. Thus, Creese & Iversen (1975) injected 8 μg of 6-OHDA in 2 μl saline bilaterally into either the ascending fibres of the dorsal bundle or those of the ventral bundle. The dorsal bundle lesions reduced cortical noradrenaline to around 20 % of normal, while the ventral bundle 6-OHDA injections depleted hypothalamic noradrenaline to 18 % of normal and cortical noradrenaline to some 12 %. Neither of these severe lesions markedly altered either the habituation of locomotion when newly introduced to the photocell cage, or the baseline activity for the 2 h following subsequent intraperitoneal injection of saline.

Similarly, Mason & Iversen (1977) found that 6-OHDA dorsal bundle lesions in the adult failed to affect locomotor activity in photocell cages despite reducing cortical noradrenaline by more than 80 %. Further, a second noradrenaline depletion technique, that of intraperitoneal injection of 100 mg kg^{-1} 6-OHDA on days 1, 3, 5, 7, 9, 11 and 13 after birth, failed to alter either photocell cage activity or open field locomotion despite reducing cortical noradrenaline to less than 5 % of normal. A similar failure of adult dorsal bundle 6-OHDA lesions to affect locomotor activity in

Fig. 4.7. Effect of 6-hydroxydopamine (6-OHDA) treatment on open field behaviour. (Redrawn from Ellison & Bresler, 1974.)

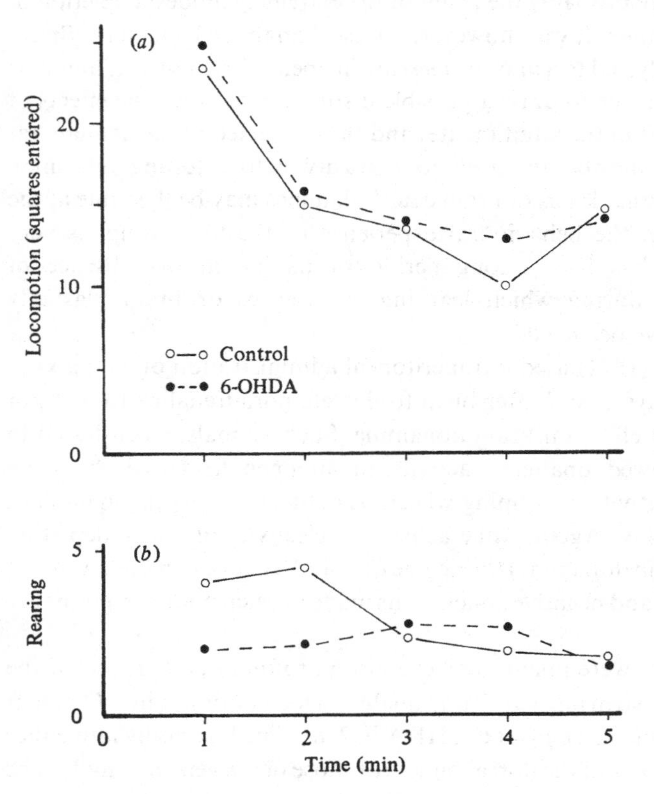

photocell cages despite reduction in brain noradrenaline to less than 5 % of control was reported by Roberts, Zis & Fibiger (1975).

Finally, electrolytic lesions of the source nucleus of the dorsal noradrenergic system, the locus coeruleus (LC), which depleted cortical noradrenaline to 14 % of control, failed to affect locomotor activity either in photocell cages or in the open field (Koob, Kelley & Mason, 1978).

Even in an acute physiological preparation, in which locomotor-like movements are induced by electrical stimulation of the so-called mesencephalic locomotor region of the cat, no effect of noradrenaline loss was seen. In a decerebrate cat, previous depletion of the noradrenergic innervation to the spinal cord by intraspinal injection of between 8 and 24 μg of 6-OHDA into each side of the L_1 cord failed to affect electrically induced limb movements (Steeves, Schmidt, Skovgaard & Jordan, 1980).

Summary. From the above it can be seen that 6-OHDA lesion of brain catecholamines can have a profound effect in reducing locomotor activity. However, subsequent studies with 6-OHDA techniques selective for noradrenaline depletion have failed to reveal any effect on simple locomotor activity in photocell cages or unstructured open fields. Noradrenaline seems not to play any role in simple forward locomotion *per se*. However, it may be involved in the response to more complex environments and subsequent habituation (see Chapt. 6).

Noradrenaline and rotation

When animals, either intact or with an asymmetrical unilateral lesion of the brain, are injected with the psychomotor stimulant drug amphetamine, as well as showing a marked increase in forward locomotor activity, they may also demonstrate reliable rotational behaviour. That is, they will turn in one direction for a protracted period of time. This rotational model has been used extensively to investigate functioning of the dopaminergic nigrostriatal pathway (see p. 148), but may also involve a noradrenergic component. Thus, Kokkinidis & Anisman (1979) reported that pre-treatment of intact mice with either the noradrenaline synthesis inhibiting drug FLA 63 or with the β-receptor blocker propranolol, but not with the α-receptor blocker phenoxybenzamine, could prevent the usual circling behaviour induced by amphetamine. This suggests a β-noradrenergic component for such rotation following amphetamine administration to normal, unlesioned mice. However, the dopamine receptor blocker haloperidol also blocked amphetamine-induced turning, suggesting an additional dopaminergic component; this point is abundantly amplified in the section on Dopamine and rotation behaviour (see pp. 148–54). Of interest was the finding that neither FLA 63, nor the β-blocker propranolol, affected the locomotor stimulant effects of amphetamine, which were, however, blocked by haloperidol. The whole question of how amphetamine exerts its locomotor stimulant and stereotypy-inducing effects, and whether these are through either aminergic system, is the subject of one of the following sections (see pp. 134–47).

The involvement of noradrenergic systems in rotational behaviour is confirmed by a series of lesion studies. Pycock, Donaldson & Marsden (1975*a*) found that unilateral electrolytic lesion to the LC caused rats to turn away from the lesioned side (contraversively) when administered amphetamine or apomorphine. A 50 % increase in the striatal dopamine content was seen on the same side as the lesion (ipsilateral). Similar results were obtained with 6-OHDA lesion of the LC (Donaldson *et al.*, 1976*b*). Donaldson *et al.* suggest a pathway from the LC to the nigrostriatal axis which modulates the activity of the dopaminergic systems. Thus, unilateral lesion of noradrenaline results in a functional imbalance of dopaminergic activity and rotation subsequent to amphetamine or apomorphine administration. At least two points of interaction of the LC system with the dopaminergic nigrostriatal system have been demonstrated. Collingridge, James & McLeod (1979) have shown that electrical stimulation of the LC causes a long-latency inhibition of firing of cells in the dopaminergic pars compacta of the substantia nigra. Further, Mason & Fibiger (1979*a*) have shown that HRP (horse-radish peroxidase) injected into the caudate–putamen of the rat is retrogradely transported back to cells in the LC. This demonstrates a LC projection to the terminal area of the nigrostriatal axis, as well as one to the dopaminergic cell bodies which project via the nigrostriatal system itself.

Donaldson *et al.* (1976*a*) have shown that the noradrenergic pathway involved in rotation runs in the ventral bundle, since lesion to the ventral bundle at a point where it is anatomically separated from the dorsal bundle also yields contraversive turning in response to amphetamine or apomorphine. Thus, lesion to the ventral bundle gives the same rotational effect as lesion to the LC itself. These authors found that electrolytic lesion to the dorsal bundle also caused rotation in response to amphetamine or apomorphine *but in the opposite direction*. That is, ipsiversive turning, rotation towards the side of the lesion, occurred after dorsal bundle lesions.

With electrolytic ventral bundle lesions the 60 % increase in striatal dopamine content seen after electrolytic LC lesions was also observed, whereas no change in dopamine content was seen after dorsal bundle lesions (Donaldson *et al.*, 1977). The authors question whether the rotational behaviour ipsiversive to the electrolytic dorsal bundle lesion was caused by dorsal noradrenergic fibre destruction (Donaldson *et al.*, 1976*b*, p. 330). Certainly, 6-OHDA lesions of the dorsal bundle, which affect only noradrenergic fibres, and not other possible types such as might be damaged by electrolytic lesions, fail to cause any sort of turning in response to amphetamine (G. F. Koob, personal communication).

Thus, it may be concluded that noradrenergic systems appear to have an effect on rotational behaviour induced by amphetamine or apomorphine in the rat. The system of importance would seem to arise in the LC but project via the ventral, rather than the dorsal, noradrenergic bundle. It may involve a direct modulation of dopaminergic systems either in the substantia nigra or in the caudate–putamen, but this is less clear.

Interactions with other systems

Acetylcholine systems. Cholinergic agonist drugs cause a behavioural state of catalepsy (see p. 126). That is, stimulation of cholinergic receptors in the brain, usually assumed to be in the caudate-putamen, prevents the rat from emitting a voluntary movement. Thus, a cataleptic animal will remain in an abnormal imposed posture for many minutes, whereas a normal rat would quickly correct the abnormal position. Catalepsy may be measured by placing the forepaws of the animal on a horizontal bar some 7 cm above the ground and timing how long it takes for the rat to remove its front paws from the bar. Normal, active rats step down from the bar within 10 to 15 s. Cataleptic rats may remain immobile for many *minutes*. Mason & Fibiger (1979*b*) reported that noradrenaline depletion caused by 6-OHDA injection into the fibres of the dorsal bundle would almost completely block the usual catalepsy caused by the muscarinic cholinergic agonist arecoline (see Fig. 4.8). Another cholinergic agonist drug of the muscarinic type, pilocarpine, also induced catalepsy and this was also blocked by prior depletion of forebrain noradrenaline (Mason, 1978). This is specific for the catalepsy induced by cholinergic drugs, since haloperidol also causes catalepsy by blocking dopamine receptors, but is not affected by noradrenaline depletion. This effect was muscarinic in nature since catalepsy induced by nicotine was unaffected by noradrenaline destruction (Mason, 1979) (see Fig. 4.8).

Cholinergic antagonist drugs, such as scopolamine and atropine, block cholinergic receptors (assumedly in the caudate–putamen) and cause locomotor stimulation. This locomotor stimulation of anticholinergics is potentiated by prior noradrenaline depletion (Mason & Fibiger, 1979*c*). This is shown in Fig. 4.9. Again, this effect is muscarinic since the mild locomotor stimulation caused by the nicotinic antagonist mecamylamine is unaffected by noradrenaline loss (Mason, 1979). That these are central, rather than peripheral, cholinergic effects is shown in Fig. 4.9 by the ineffectiveness of the centrally inactive anticholinergic drugs methylscopolamine and hexamethonium, which do not cross the blood–brain barrier and so are respectively limited to *peripheral* muscarinic and nicotinic blockade.

The anatomical site of the interaction of noradrenergic systems with cholinergic drugs is unknown but a few possibilities are suggested in Fig. 4.10.

Opiate systems. Opiate agonists such as morphine induce, in low doses, a motor stimulation, while in higher doses a behavioural depression and catalepsy prevail. Mason, Roberts & Fibiger (1978) found that noradrenaline depletion by dorsal bundle 6-OHDA lesion greatly potentiated the catalepsy induced by morphine. Further, the depression of locomotor activity caused by high doses (10–20 mg kg^{-1}) of morphine was also increased by noradrenaline lesion (Roberts, Mason & Fibiger, 1978) while the locomotor stimulation caused by lower doses of morphine (5 mg kg^{-1}) was unaltered by the noradrenaline depletion. This is summarised in Fig. 4.11. Again, the site of interaction of noradrenergic with opiate systems is unknown, although probably the same models indicated for cholinergic systems in Fig. 4.10 could also be invoked for the opiate systems.

Fig. 4.8. Catalepsy in response to doses of nicotine, pilocarpine, arecoline and haloperidol in control and noradrenaline-depleted (NA↓) rats. Catalepsy scores are time to remove front paws from a horizontal bar 7 cm above the floor. Sal = saline. (Compiled from Mason, 1978, 1979; and Mason & Fibiger, 1979*b*.)

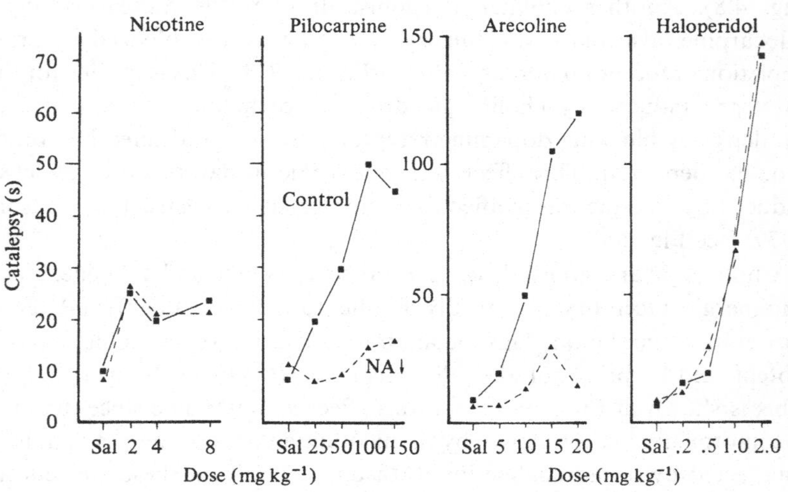

Ethanol. Injections of ethanol intraperitoneally also alter the locomotor activity of the rat. Again, a biphasic response is usually seen, with high doses of ethanol causing considerable sedation, although this could not now be termed catalepsy. Lower doses of ethanol may instead bring about a state of locomotor stimulation. Mason, Corcoran & Fibiger (1979) showed an interaction with noradrenergic systems in that depletion of noradrenaline by 6-OHDA injection into the dorsal bundle abolished the

Fig. 4.9. Locomotor activity in response to scopolamine (0.5 mg kg^{-1}), methylscopolamine (1 mg kg^{-1}), mecamylamine (4 mg kg^{-1}) and hexamethonium (5 mg kg^{-1}) in control and noradrenaline-depleted (NA↓) rats. (Compiled from Mason, 1979; and Mason & Fibiger, 1979*c*.)

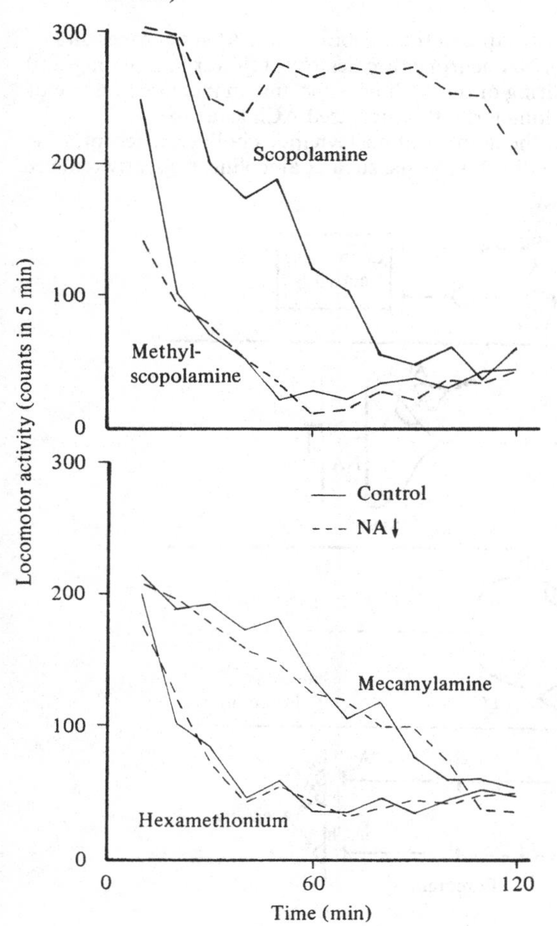

Fig. 4.10. Possible models of noradrenaline–acetylcholine (NA–ACh) interaction in brain. (*a*) NA and ACh neurones do not interact directly but merely converge a number of synapses later onto a mechanism controlling locomotor activity.

(*b*) ACh neurones regulate the release of noradrenaline from terminals in the forebrain by a presynaptic mechanism.

(*c*) Three alternative interactions of ACh neurones in the locus coeruleus (LC). *At top*: ACh neurones synapse with LC cell bodies – binding studies seeking to find a decrease in ACh receptors after 6-OHDA destruction of LC cell bodies suggest this does not occur. *In middle*: ACh neurones synapse not with LC cells but with interneurones within the LC which then synapse with LC NA cells – anatomical and ultrastructural studies suggest that interneurones do not occur within the LC itself, making this alternative unlikely. *At bottom*: ACh neurones synapse presynaptically with afferent fibres to the LC and these then synapse with LC NA cells – a few axo-axonic synapses have been identified ultrastructurally within the LC so this alternative remains a possibility.

(*d*) NA neurones synapse in the forebrain onto ACh neurones and inhibit them. When NA neurones are destroyed (lower half of diagram) an increase in the firing of the ACh neurone and an increased release of ACh occur. In the long term, this increased ACh leads to down-regulation of the number of postsynaptic cholinergic receptors so that the response to direct agonists, such as arecoline, is greatly reduced.

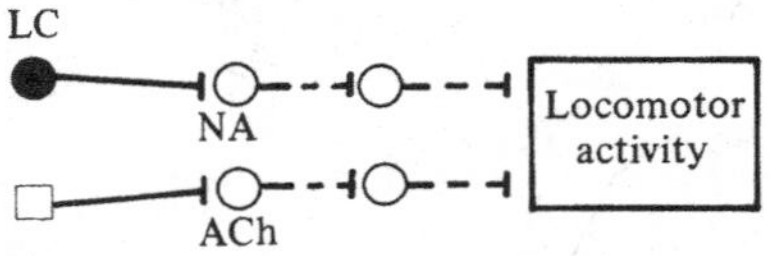

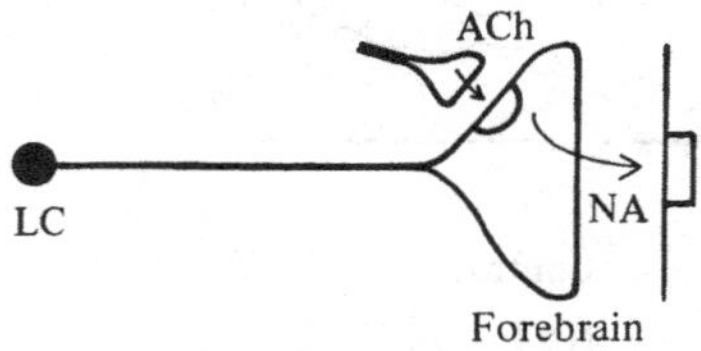

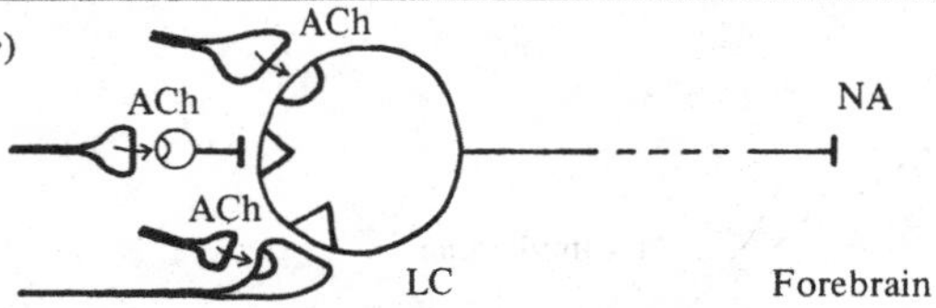

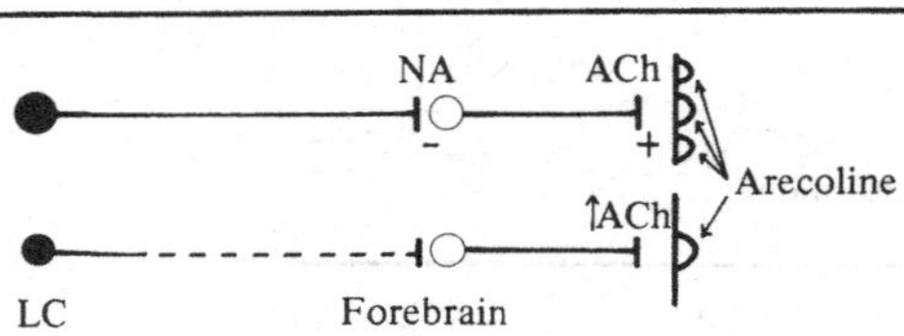

Fig. 4.11. Locomotor activity induced by morphine in control and noradrenaline-depleted (NA↓) rats. Values are total locomotor counts per hour over a 3 h session for saline (Sal), 5, 10 or 20 mg kg^{-1} of morphine injected intraperitoneally at the start of the session. (Redrawn from Roberts, Mason & Fibiger, 1978.)

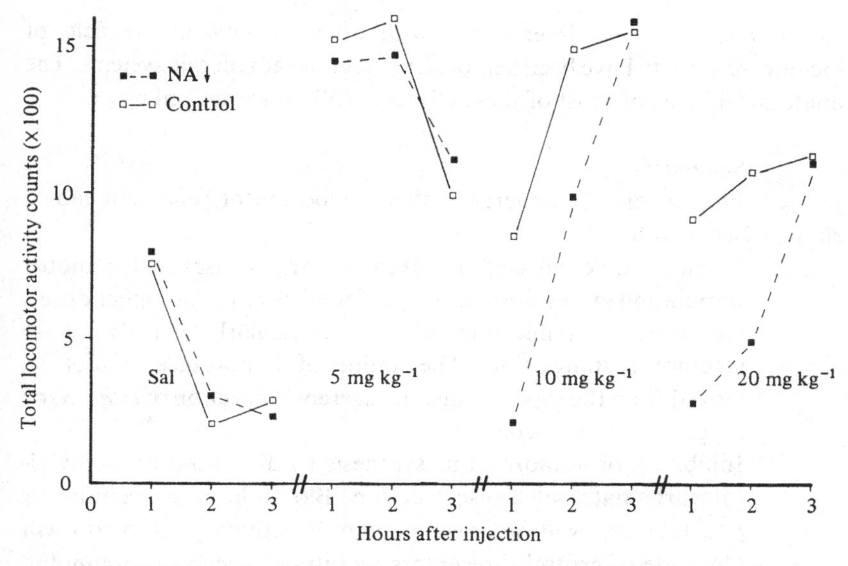

Fig. 4.12. Locomotor activity in response to ethanol in control and noradrenaline-depleted (NA↓) rats. Responses to saline (Sal), 1 and 0.1 g kg^{-1} ethanol are shown. Note that in normal rats 0.1 g kg^{-1} causes a pure hyperactivity. (Compiled from Mason, Corcoran & Fibiger, 1979.)

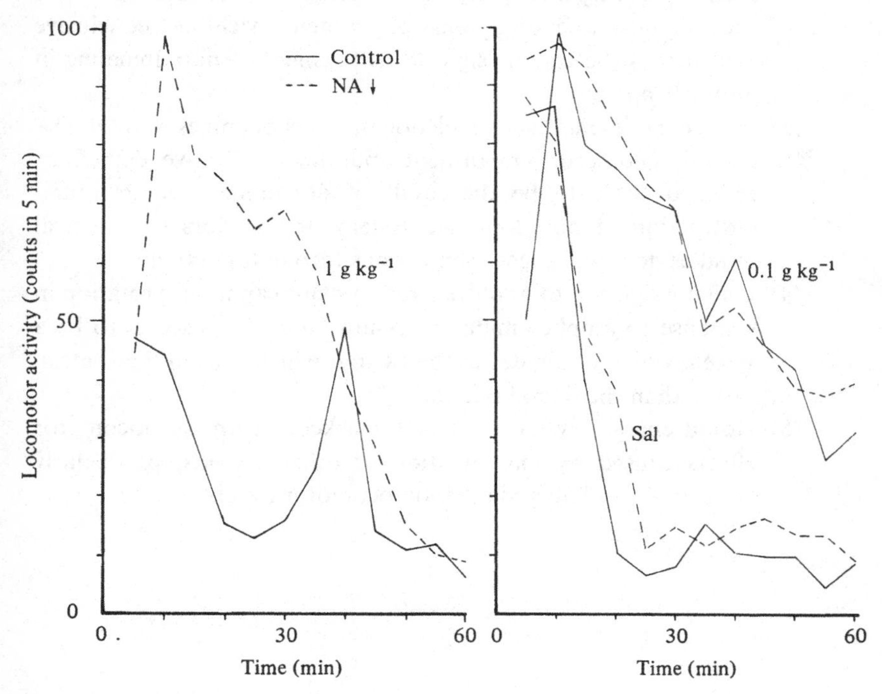

sedation seen after high doses of ethanol, but still permitted the locomotor stimulatory effects of ethanol to occur (see Fig. 4.12).

Conclusions. Complex interactions with other systems in the field of locomotor activity have been demonstrated for noradrenergic systems. The anatomical locus of most of these effects is still, however, unknown.

Summary

The role of noradrenergic systems in locomotor behaviour can be summarised as follows;

(1) Noradrenaline injected intraventricularly causes a locomotor stimulation at low doses but a profound sedation at higher doses. However, dopamine injected intraventricularly will also cause locomotor stimulation. The amine of importance cannot be judged from the experiments. Intracerebral injection of *either* α- or β-agonists causes sedation.
(2) Inhibition of noradrenergic synthesis by disulfiram or diethyldithiocarbamate will cause sedation. Blockade of α-receptors by phentolamine will decrease locomotor activity, but so too will blockade of central β-receptors by intraventricular propranolol. It appears that stimulating or blocking either α- or β-receptors gives exactly the same result in all cases – viz. sedation. Current research is desperately trying to explain this contradiction. As such agonist and antagonist experiments yield no conclusive evidence of the role or sign of involvement of noradrenaline in motor behaviour.
(3) Early experiments using depletion of catecholamines by 6-OHDA caused locomotor impairments but more selective depletions revealed these to be the result of dopamine loss. Profound destruction of noradrenergic systems in the dorsal or ventral bundles does not affect spontaneous locomotor activity.
(4) Unilateral lesion to noradrenergic systems can induce rotation in response to amphetamine or apomorphine. This seems to be a system which originates in the LC but which runs in the ventral, rather than the dorsal bundle.
(5) Noradrenergic systems interact markedly with the locomotor effects caused by manipulation of other systems, particularly cholinergic, opiate and ethanol locomotor effects.

Dopamine and motor behaviour

Introduction

Many of the early experiments mentioned in the first section of this chapter clearly implicated the catecholamines, noradrenaline or dopamine, in locomotor behaviour, but they failed to decide which of the two was critically involved.

The conclusion of the last section on Noradrenaline and motor behaviour is that noradrenaline is not primarily involved in that behaviour, although a complex modulatory action does seem likely. Thus, by logic alone, it would seem that it has to be dopamine which is primarily involved in motor function. Nonetheless, specific experiments are needed to prove this in any scientific sense and these will be covered below.

Dopamine agonists

One of the most powerful indications of the function of dopaminergic systems in the brain in motor behaviour has come from the effects on behaviour of dopamine agonist drugs. Either from drugs such as apomorphine, which directly stimulate the dopamine receptor, and which will be considered in the present section, or from indirectly-acting releasing agents, such as amphetamine, which will receive a whole subsequent section on the question of the noradrenergic or dopaminergic mechanism of its action.

Apomorphine. Perhaps the most classic dopamine receptor agonist drug is apomorphine. Administration of between 1 and 10 mg kg^{-1} of the drug intraperitoneally to the rat yields a syndrome called stereotypy. This, in the rat, consists of vigorous sniffing, usually with the head down near the floor of the cage, together with repeated head and neck movements which have a very repetitious, almost mechanical, nature. This mild form of stereotypy may be followed, in the rat, by a more intense behaviour consisting of biting, licking or gnawing movements of the mouth, often directed toward the cage bars, occasionally even involving the animal's own body or tail. In primates, including man, the stereotypies induced (in the latter case by amphetamine, rather than apomorphine which also has a violent emetic action) focus more on the arms and fingers and may involve the repetitious performance of quite complex motor sequences (assembling and repeatedly disassembling a motor car engine, tidying up a handbag or chest of drawers over and over again). Such behaviours have been extensively documented for human amphetamine users (Randrup & Munkvad, 1967).

Not only higher, or mammalian, species show stereotypy. Pigeons show a repeated pecking at the same, vacant spot after apomorphine injection (Dhawan, Saxena & Gupta, 1961). Similar pecking can be elicited by apomorphine in parrots, chickens, hens and quail (Deshpande, Sharma, Kherdikar & Grewal, 1961; Osuide & Adejoh, 1973). Anderson, Braestrup & Randrup (1975) have reported stereotyped biting from tortoises given 2 mg kg^{-1} apomorphine. However, lower vertebrates such as frogs and toads do not apparently show stereotyped responses to apomorphine (Nilakantan & Randrup, 1968).

In mice, unlike rats, apomorphine does not induce gnawing behaviour. However, such behaviour can be induced following 'potentiation' of apomorphine by another drug (Pedersen, 1967). Such potentiating drugs include imipramine (Dadkar, Dohadwella & Bhattacharya, 1976*a*) and cocaine (Dadkar *et al.*, 1976*b*, 1977). The usual behaviour seen in mice in response to apomorphine is a peculiar motor behaviour involving rearing or climbing along the walls of the cage (Protais, Costentin & Schwartz, 1976). A similar response in the mouse is seen after administration of amphetamine and L-DOPA in combination (Lal, Colpaert & Laduron, 1975).

In the guinea pig stereotypy takes the oral form seen at higher doses of apomorphine in the rat. Sniffing and repetitive head and neck movements are rarely seen (Costall & Naylor, 1976). In this species so-called 'dyskinetic movements' may be induced by injection of dopamine into the corpus striatum (Costall, Naylor & Pinder, 1975*d*). Stereotypy and dyskinesia, which are apparently similar behavioural syndromes, may turn out to be mediated by different mechanisms since they are differentially antagonised by a range of neuroleptic drugs (for example haloperidol antagonises stereotyped oral behaviour but not the dyskinesias, whereas oxiperomide (1 mg kg^{-1}) gave evidence of preferential antagonism of the dyskinesias; Costall & Naylor, 1976).

Other dopaminergic agonists. Piribedil [ET 495 or 1-(3,4-methylenedioxybenzyl-4-(2-pyrimidyl)-piperazine)] appears to be a direct dopamine receptor agonist drug. Thus, Costall & Naylor (1973) reported that it induced stereotypy in rats. Biochemical studies suggest that it may be a metabolite of piribedil called S584 which is actually the active agent (Miller & Iversen, 1974). That it acts postsynaptically was shown by Creese (1974), who destroyed the dopaminergic terminals in the forebrain by neonatal intraventricular injections of 100 μg of 6-OHDA. Both piribedil and its metabolite S584 induced marked stereotypy and locomotor stimulation in the lesioned rats. This argues for a direct receptor stimulant effect.

Costall, Kelly & Naylor (1975*a*) showed that the drug nomifensine also induced stereotypy. These effects could be blocked by the dopamine receptor antagonist haloperidol, as were those of apomorphine. Further, lesions of the dopamine terminals in the forebrain by electrolytic lesion of the nigrostriatal pathway in the substantia nigra, partially prevented the stereotypy-inducing effect. Again, it was suggested that a metabolite of the parent drug was responsible for the effects and injection of the metabolite into the striatum of the rat caused stereotyped behaviour, as it did on intraperitoneal injection. Similar results for the dependence of nomifensine-induced stereotypy on the integrity of the nigrostriatal pathway were shown by Price & Fibiger (1976), using 6-OHDA lesion.

An interesting hint emerged from the work of Costall, Naylor & Neumeyer (1975*b*) that the derivative of apomorphine (−)-*N*,*n*-propylnorapomorphine ((−)-NPA) might act differently from the parent drug. Thus, with increasing doses of apomorphine progressive stages of stereotypy are seen. The progression goes from sniffing and head movement at low doses of apomorphine to biting, licking and gnawing at higher doses. With (−)-NPA, the biting component was present right from the initial appearance of stereotypy. Doses of (−)-NPA lower than about 0.0063 mg kg^{-1} simply failed to induce any manifestation of stereotypy. Doses above this always caused biting and licking, albeit episodically, as well as sniffing and head movements. The significance of this remains to be elucidated.

The duration of action of apomorphine can be increased from its usual 60 min or so to 300 min or more by administration in the form of the so-called diester 'prodrug' diisobutyrylapomorphine. This is markedly lypophylic and may be stored in fat depots and only released slowly over time (Baldessarini, Kula & Walton, 1977). The time course of the stereotyped climbing behaviour seen in the mouse is compared in Fig. 4.13 for apomorphine and for diisobutyrylapomorphine. This prolonged action may make apomorphine derivatives of use in the treatment of Parkinson's disease, for which they have always been too short-acting in the past.

Other derivatives of apomorphine may be more potent than the parent compound in causing stereotypy. Menon, Clark & Neumeyer (1978) report that (−)-NPA is two to three times more effective than apomorphine in inducing stereotypy in mice. (−)-NPA was also some two to six times more potent in causing locomotor activity in mice previously treated with reserpine. Table 4.3 summarises some of the many comparisons of the efficiencies of the two compounds. It is in fact the (−)-isomer of apomorphine that has the greater activity. Thus, Schoenfeld, Neumeyer,

Dafeldecker & Roffler-Tarlov (1975) find an ED_{50} for the induction of stereotyped behaviour in rats of 0.54 mg kg^{-1} for (−)-apomorphine but a value of 1.1 mg kg^{-1} for a (±)-racemic mixture of the drug. In their experiments, they found (−)NPA to be some 35 times more potent than (−)-apomorphine in this test. Other substituted apomorphines were generally less active than the parent molecule (see Table 4.4).

Fuxe *et al.* (1978*a*) reported that the drug bromocryptine will induce stereotyped sniffing behaviour in rats. However, it seems to require the integrity of presynaptic dopaminergic terminals, suggesting a possible releasing action. Its stereotypical action is also weaker than that of apomorphine.

Bromocryptine is a member of the ergot family of drugs and other agents in this family, such as lergotrile, agrolavine and elymoclavine also cause mild stereotypy and increased locomotion (Fuxe *et al.*, 1978*b*).

Presynaptic receptors. Very low doses of direct dopamine agonist drugs cause not a locomotion stimulation, but a sedation. This is suggested to occur via dopamine receptors on dopaminergic terminals themselves, which inhibit the release of endogenous dopamine. Doses of apomorphine in the range 0.05 to 0.1 mg kg^{-1} cause a *decrease* in locomotor activity in the first hour after administration. Other ergot drugs have similar effects

Fig. 4. 13. Time course of stereotyped cage-climbing in mice compared to the time course of the brain levels of apomorphine following either apomorphine itself (APO) or the diisobutyryl ester (DIB). Note the much more prolonged time course of the behavioural and biochemical responses to DIB.

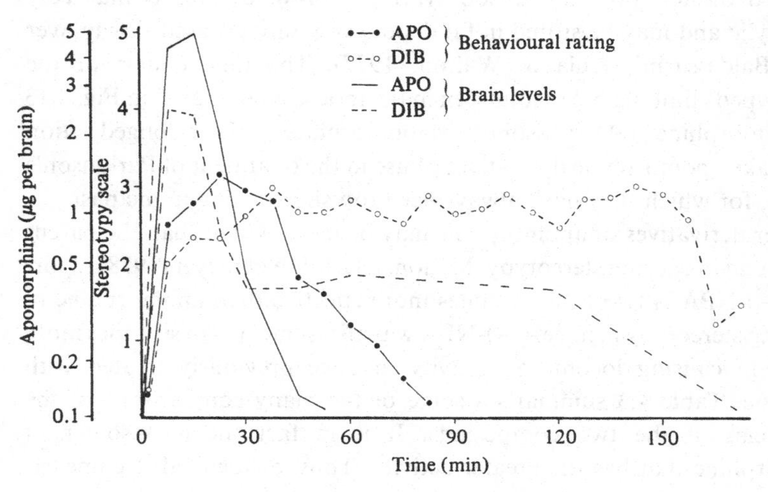

with one of them, Cf 25-397, causing *only* reduced locomotor activity no matter what the dose (Fuxe *et al.*, 1978*b*). Other dopamine agonists also, at low doses, inhibit locomotor activity. Thus, piribedil has been reported to reduce activity in doses in the range of 2.5 to 5.0 mg kg^{-1} (Dourish & Cooper, 1981). Indeed, Corsini *et al.* (1977) have reported sedation and sleep from low doses of apomorphine in humans.

The dopaminergic nature of this presynaptic, sedative effect of low doses of apomorphine is shown by the fact that it can be blocked by very low doses of neuroleptics such as pimozide, haloperidol, droperidol and sulpiride (Di Chiara *et al.*, 1976).

Conclusion. As has been seen, a wide range of drugs which activate the dopamine receptor lead to a behavioural state of increased locomotor

Table 4.3. *Comparison of NPA with apomorphine*

Species	Experimental paradigm	Relative potency of NPA (apomorphine = 1)
Man	Antiparkinson effect	10–15
Monkey	Overt behaviour	67
Rat	Stereotypy	10–35
	Injection into accumbens	12–16
	Rotation in nigra-lesioned rat	12–20
	Injection into 6-OHDA lesioned accumbens	40
	Rotation in raphe-lesioned rat	10
	Stimulation of adenylate cyclase	1
	Respiratory stimulation	3
Mouse	Gnawing	1.2
	Behavioural changes	13
	Postural asymmetries in 6-OHDA lesioned mouse	1–7
	Postural asymmetries in mechanically lesioned mouse	16–44
	Locomotor effects	2
	Stimulation of adenylate cyclase	2
	Hypothermia	85
Pigeon	Pecking	3
Cat	Behavioural stereotypies	200
	Hypotension	13
Dog	Emesis	2–40

Taken from Menon, Clark & Neumeyer, 1978.

activity and, eventually, stereotypy. This then seems very direct evidence for a role of dopamine in the control of motor output. Of interest, also, is the sedatory, activity-reducing effects of very low doses of dopamine receptor agonists. Further evidence for a role of dopaminergic systems in motor behaviour has come from the effects of dopamine receptor blockers on behaviour.

Dopamine antagonists

Introduction. The classic dopamine receptor blocking agents are the neuroleptic drugs. These are tertiary, or exceptionally, secondary amines (Janssen, 1970). They can be subdivided into four major classes: the butyrophenones, the phenothiazines, the rauwolfia alkaloids and the 3-phenoxypropylamines (such as spiramide). They characteristically inhibit exploratory and locomotor behaviour (Janssen, 1970) and cause a behavioural state called catalepsy. This is a profound inability to initiate a voluntary movement and may be demonstrated experimentally by placing the rat in an abnormal posture such as hanging from the top of a vertical wall by its forepaws or with one hind leg raised onto a 3–5 cm high platform. Normal rats will correct this abnormal posture within 5 to 15 s.

Table 4.4. *Comparison of apomorphine isomers*

Compound	R_1	R_2	R_3	ED_{50}
Apomorphine	OH	OH	CH_3	1.1 ± 0.13
10-hydroxyapomorphine	H	OH	CH_3	greater than 30
10-hydroxy-*N,n*-propyl-norapomorphine	H	OH	C_3H_7	11.0 ± 3.1
11-hydroxy-*N,n*-propyl-norapomorphine	OH	H	C_3H_7	8.0 ± 2.8
N,n-propyl-norapomorphine	OH	OH	C_3H_7	0.04 ± 0.01

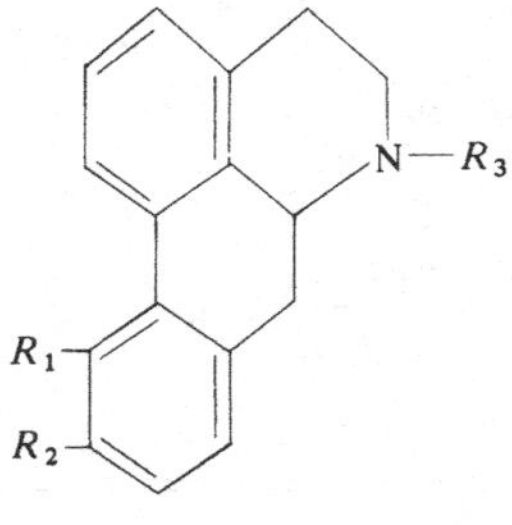

N—R_3

R_1

R_2

Taken from Schoenfeld *et al.*, 1975.

Rats displaying catalepsy will remain immobile in these imposed positions for many minutes. They are clearly awake (indeed it is not possible to demonstrate catalepsy in an anaesthetised or unconscious rat since the loss of muscle tone causes it to fall passively from the imposed posture), but unable to initiate the movement necessary to escape from the imposed posture. Figure 4.14 shows the dose of several different neuroleptics needed to induce a cataleptic state in rats. The route of administration by mouth makes the time course of catalepsy induction longer than is typically seen after intraperitoneal injection of the drug (Janssen *et al.*, 1975). Catalepsy may also be measured by placing the rats forepaws on a horizontal bar 7 cm above the floor, and timing how long it takes the rat to step down from this abnormal imposed position. Normal rats injected only with saline typically take 10 to 15 s to accomplish this. Cataleptic animals may take many minutes. Figure 4.15 shows this for an increasing series of doses of the neuroleptic haloperidol (Mason *et al.*, 1978).

Chronic administration of neuroleptic drugs may, upon termination of the treatment, cause a state of supersensitivity. Behaviourally, this reveals itself as a marked increase in the stereotypy and locomotor activity induced by direct dopamine receptor agonists such as apomorphine. A similar

Fig. 4.14. Estimated ED_{50} to cause catalepsy in rats following oral administration of four neuroleptics. The slow times for maximum effectiveness are due in part to the oral route of administration. (Redrawn from Janssen *et al.*, 1975.)

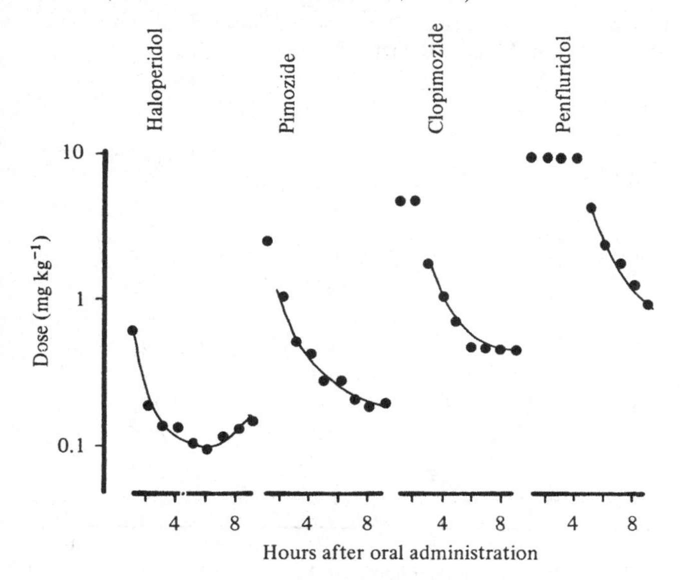

supersensitivity may be induced by chronic blockade of dopamine synthesis and release by administration of the synthesis inhibitor AMPT for 14 days (Gudelsky, Thornburg & Moore, 1975).

Smith & Davis (1976) in a typical experiment showed that administration of neuroleptics for six to seven weeks in the rat would markedly enhance apomorphine-induced stereotypy upon subsequent withdrawal of neuroleptics. This effect lasted for some 7 days after withdrawal for the atypical neuroleptics, clozapine and thioridazine and for some 14 or more days after chronic haloperidol administration and withdrawal. Thioridazine and clozapine also increased the apomorphine-induced locomotor activity on day 8 after termination of treatment and the rats showed enhanced spontaneous locomotor activity when tested 5 or 10 days after withdrawal.

Conclusion. In summary, the profound effects that dopamine receptor blockers have in decreasing spontaneous locomotor activity and in inducing a state of catalepsy complement the evidence from dopaminergic agonist drugs, such as apomorphine, in increasing locomotor activity and causing the expression of behavioural stereotypies.

Intracerebral localisation

While the agonist and antagonist data serve to implicate dopamine in the control of motor output they fail to solve the following problems.

Fig. 4.15. Cataleptic response to four doses of haloperidol in normal rats. (Compiled from Mason, Roberts & Fibiger, 1978.)

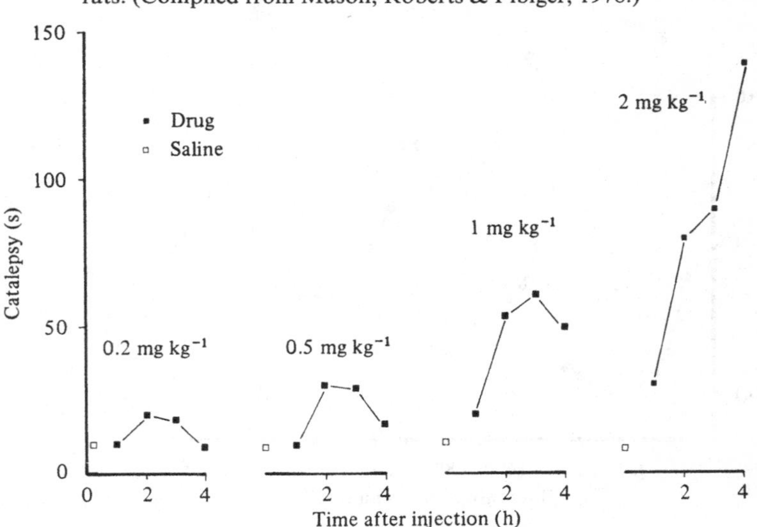

Which dopamine synapse, in which brain area, is the one which increases motor output when stimulated with apomorphine and decreases it when blocked by neuroleptics? Is this the same synapse which causes stereotyped behaviour to emerge? Or could there be two dopamine synapses, one involved in the occurrence of stereotypy and one in the forward locomotion? Intracerebral injection of agonists and antagonists may answer these questions. Further answers are provided by the localised 6-OHDA lesions covered in a subsequent section.

Pijnenburg, Honig & van Rossum (1975*a*) found that bilateral injection of 2.5 or 5 μg haloperidol into the nucleus accumbens via chronically indwelling cannulae caused a marked block of the locomotor activity elicited by injection of *d*-amphetamine intraperitoneally. Similar injection into the neighbouring caudate nucleus was without effect. Injection of either of the noradrenaline receptor blockers phentolamine or propranolol into the accumbens failed to block the amphetamine-induced locomotor activity. This suggests that one dopamine synapse important for locomotor activity may lie in the nucleus accumbens. Since haloperidol is considered to be a reasonably selective dopamine receptor blocker and since neither the α- nor the β-noradrenergic blockers (phentolamine and propranolol, respectively) had any effect, this would appear to be strong evidence for a role of dopamine, specifically in the nucleus accumbens, in locomotor activity. One problem with the present study, however, was that intracerebral injection of saline also slightly reduced the amphetamine locomotor activity (see Fig. 4.16). This suggests that some of the effect was an artefact of the mere injection of fluid into the accumbens, and as such probably unrelated to dopamine receptor blockade. However, the saline effect was relatively small and the comparison of importance was made between haloperidol injection and saline injection (as opposed to normal amphetamine locomotor activity without intracerebral injection of anything).

Pijnenburg & van Rossum (1973) found that injection of the natural neurotransmitter dopamine into the nucleus accumbens could elicit increased locomotor activity, as could similar injection of the dopamine agonist ergometrine (Pijnenburg, Woodruff & van Rossum, 1973). Pijnenburg *et al.* (1975*b*) further sought to dissociate the effects of the two catecholamines upon locomotor activity, by comparing injections of dopamine into the nucleus accumbens with similar injections of noradrenaline. Dopamine caused a stronger stimulation of locomotor activity than did noradrenaline. Injections of serotonin were virtually without effect. The stimulant effects could be reversed by injection of haloperidol into the nucleus accumbens 15 min after the initial catecholamine injection. Since haloperidol is regarded as a dopamine receptor blocker, this, together with the best locomotion being obtained by dopamine rather than noradrenaline, is further evidence of dopamine mediation. However, injection of simple saline after the initial catecholamine injection also partially reduced the locomotor stimulation. This serves to highlight the problems associated with intracerebral injections of drug during behaviour.

Stereotyped behaviour induced by *d*-amphetamine, on the other hand, seems to be antagonised by haloperidol injections into the caudate nucleus, not into the accumbens. Thus, Fog, Randrup & Pakkenberg (1971) and Costall, Naylor & Olley (1972) report blockade or reduction of amphetamine stereotypy after caudate injections of haloperidol. It should be noted, however, that the doses required (20–100 μg bilaterally) were rather larger than those mentioned above which were used to block the amphetamine locomotor activity after injection into the accumbens. This may simply be because of the much larger volume of the caudate compared to the accumbens.

However, the dopamine receptor agonists apomorphine and NPA caused much weaker increases in locomotor activity when injected into the accumbens than did injection of dopamine itself (Costall *et al.*, 1975*c*).

Fig. 4.16. Locomotor activity caused by intraperitoneal injection of 1 mg kg^{-1} amphetamine (A) and its modification by intra-accumbens injection of saline (AS) or 5 μg haloperidol (AH). For comparison the activity in response to intraperitoneal saline (S) is shown. (Redrawn from Pijnenburg, Honig & van Rossum, 1975*a*.)

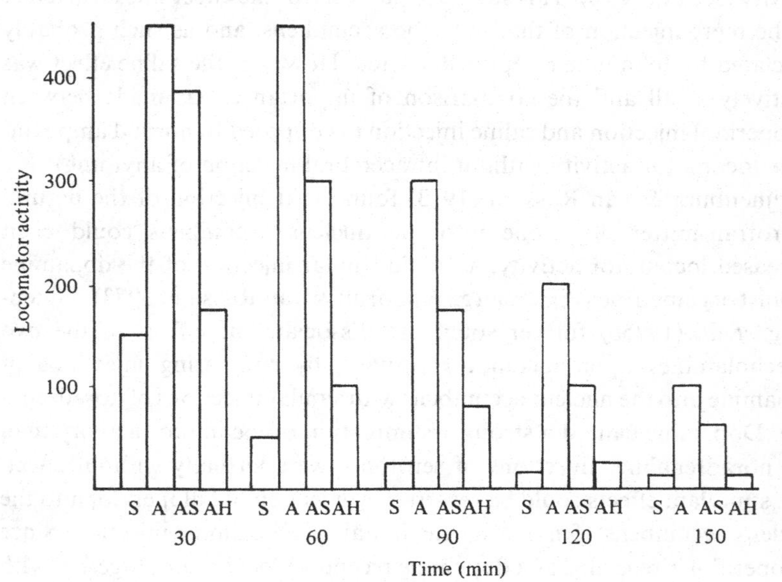

They did, however, induce a stereotyped biting response. Dopamine itself when injected into the accumbens produced no more intense stereotypy than mild sniffing. Further evidence for two separate dopaminergic mechanisms in the accumbens controlling respectively stereotypy and locomotor activity was produced by Costall, Naylor, Cannon & Lee (1977). The compound 2-(*N*,*N*-dipropyl)amino-5,6-dihydroxytetralin when injected into the nucleus accumbens caused marked stereotypy but no locomotor activity. 2-(*N*-propyl)amino-5,6-dihydroxytetralin induced both locomotor activity and stereotypy. 2-Amino-6,7-dihydroxytetralin appeared to induce locomotor activity and only very mild stereotypy (see Fig. 4.17). Thus, there might be two populations of receptors in the accumbens which can be differentially stimulated by the above compounds and which control locomotor activity and stereotypy independently.

Jones, Mogenson & Wu (1981) injected dopamine, together with other neurotransmitters and blocking agents, into the nucleus accumbens. They found that co-injection with the cholinergic agonist carbachol enhanced locomotor activity over and beyond that caused by dopamine on its own. This effect could be blocked by addition of atropine to the injected fluid (a veritable cocktail!). However, atropine *per se* was without effect on dopamine elicited locomotor activity. Serotonin co-injection reduced dopamine elicited locomotor activity, but confusingly the supposed serotonin antagonist methysergide *also* reduced dopamine-elicited locomotor activity. Co-injection of γ-aminobutyric acid (GABA) increased locomotion at low doses (2.25 μg) while decreasing it at higher doses (10 or 33 μg). The GABA antagonist picrotoxin increased locomotion at all doses. A wiring diagram similar to that shown in Fig. 4.18 is proposed to explain these sundry effects.

From the above, strong evidence can be drawn to support a role of dopamine in the nucleus accumbens in locomotor activity. Some hint of a second set of receptors in this nucleus which might additionally control stereotypy has emerged. Manipulation of dopaminergic receptors in the caudate seems to affect stereotypy in a clear fashion. Interaction with other receptor populations and neurotransmitter systems in the control of locomotor activity seems certain.

Other manipulations of locomotor behaviour

It has been suggested that some of the actions of dopamine as a neurotransmitter may be mediated through activation of the enzyme adenylcyclase (Kebabian, Petzold & Greengard, 1972). This enzyme acts to form cyclic adenosine monophosphate (cAMP) which has been suggested to be the 'second messenger' which carries out the effects of dopamine on the postsynaptic cell. cAMP is then broken down by the enzyme phosphodiesterase. This enzyme can be inhibited by compounds such as papaverine or caffeine. If the stereotypy-inducing and locomotor stimulant effects of dopamine receptor activation were mediated via the cAMP second-messenger system it might be expected that such phosphodiesterase inhibitors would affect the behaviours elicited by amphetamine or apomorphine. Kostowski, Gajewska, Bidinski & Hauptman (1976) report

Fig. 4.17. Hyperactivity and stereotypy induced by 2-(*N*-propyl)-amino-5,6-dihydroxytetralin (*top*), 2-amino-6,7-dihydroxytetralin (*middle*) and 2-(*N, N*-propyl)amino-5,6-dihydroxytetralin (*bottom*) in rats. Compounds were injected directly into the accumbens 2 h after 100 mg kg^{-1} nialamide i.p. Doses to right of each graph are in $\mu g\ \mu l^{-1}$. (Redrawn from Costall, Naylor, Cannon & Lee, 1977.)

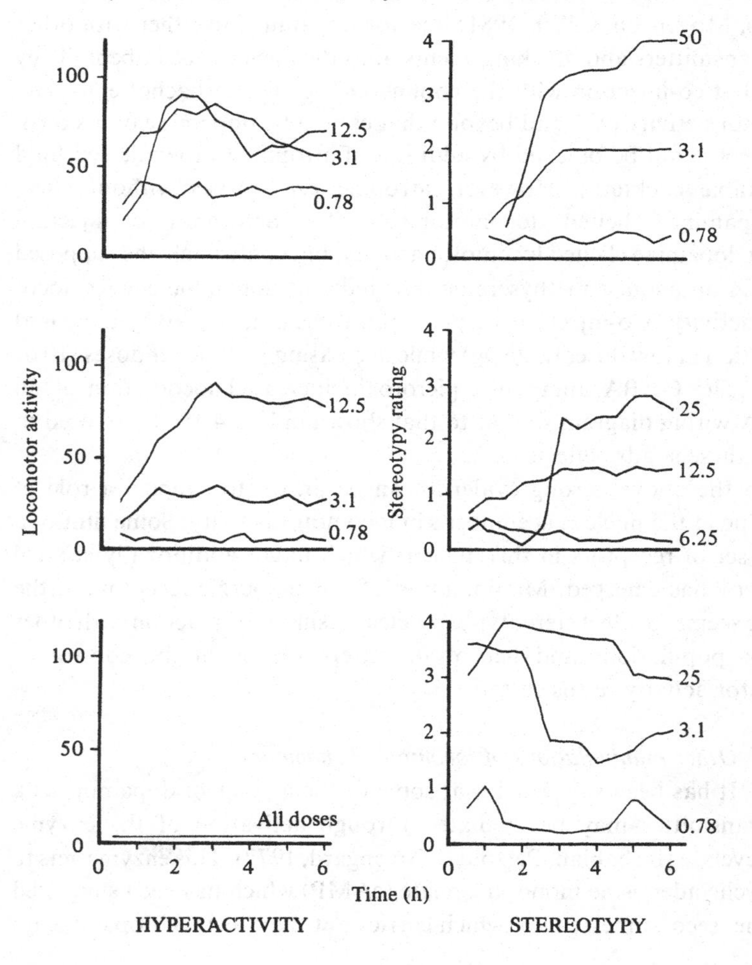

such an experiment. They found papaverine caused a slight decrease in the stereotypy induced by apomorphine and a marked increase in the catalepsy induced by haloperidol. Thus, although phosphodiesterase inhibition affects these dopaminergic behaviours it appears to be in the wrong direction for the most simple of the second messenger hypotheses. On the other hand, another inhibitor of phosphodiesterase, caffeine, does have effects in the predicted direction. Stromberg & Waldeck (1973) found that caffeine caused a 100 % increase in locomotor activity in mice, consistent with a potentiation of postsynaptic dopaminergic events. Further, caffeine markedly potentiated the locomotor stimulant effects of L-DOPA. In the latter cases more than a 400 % increase in locomotor activity was recorded. However, it has recently been suggested (Daly, Bruns & Snyder, 1981) that the locomotor effects of caffeine owe more to an interaction with adenosine receptors than to an inhibition of phosphodiesterase.

Gordon & Shellenberger (1974) reported sex differences in the levels of spontaneous activity recorded in running wheels and activity cages. Female rats were slightly, but significantly, more active on both measures than males. Miller & Sahakian (1974) also observed an age effect in that the atypical neuroleptic drugs, clozapine and thioridazine blocked amphetamine-induced locomotor activity in 11-day-old rats but failed to do this in adult rats. Further, Sahakian, Robbins, Morgan & Iversen (1975) reported an effect of rearing experiences on amphetamine responses. They found

Fig. 4.18. Simplified block diagram of putative neurotransmitters of the nucleus accumbens. ACh, acetylcholine; 5HT, serotonin; DA, dopamine, GABA, γ-aminobutyric acid; VTA, ventral tegmental area; GP, globus pallidus. (Redrawn from Jones, Mogenson & Wu, 1981.)

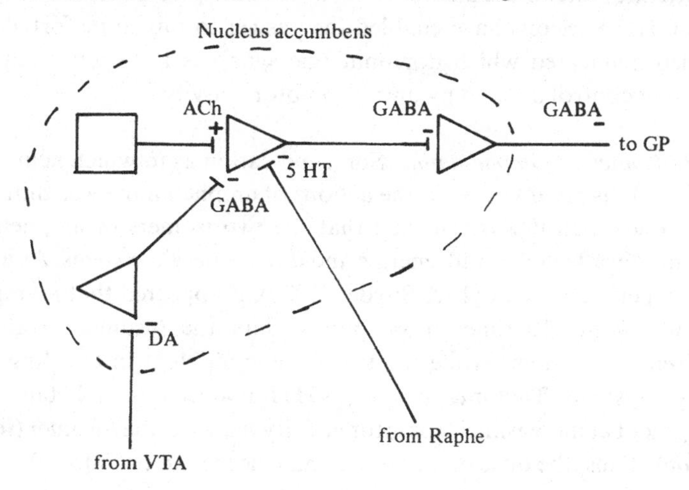

that rats raised in deprived environments (isolated from other rats at 16 days postnatal) showed more intense stereotypy with amphetamine than did rats raised in control environments (group housed four per cage). Of interest was the observation that no such potentiation of the amphetamine *locomotor* response was seen. This again suggests that stereotypy and locomotor behaviour may be controlled by different dopaminergic systems.

Smith & Leelavathi (1978) found that the stereotypical effects of apomorphine and *d*-amphetamine increase with age. Thus, 2-month-old rats showed the weakest stereotypy to either 1 mg kg^{-1} apomorphine or 3.5 mg kg^{-1} amphetamine, while 10-month-old rats showed a greater response and 20-month-old rats exhibited the largest response of the three groups. the authors speculate that changes in either receptor mechanisms or drug metabolism may occur with age.

6-OHDA lesions and the action of amphetamine

Introduction. Some of the most powerful evidence for a role of dopamine in the control of motor behaviour has come from the effects of 6-OHDA lesion of these systems. Some experiments have merely implicated dopaminergic systems as a whole, while a number of others have indicated specific brain areas containing dopaminergic synapses to be of importance. A related question, which has also proved susceptible to answer by the use of 6-OHDA lesion techniques, is that of the neurochemical basis of amphetamine effects. Through what neurotransmitter systems does amphetamine induce stereotypy and increase locomotor activity? Since amphetamine has effects *in vitro* on noradrenergic, dopaminergic and serotonergic systems the answer is not immediately obvious. It will be seen that 6-OHDA lesions have enabled the answer to this to be forthcoming and also delineated which dopaminergic synapses in which brain areas separately control stereotypy and locomotor activity.

d- and l-isomers of amphetamine. Some indication as to which neurotransmitter systems are involved in the actions of amphetamine was thought to be possible when it was reported that the two isomers of amphetamine differentially affected noradrenergic and dopaminergic systems. As a result of the experiment of Coyle & Snyder (1970), it appeared that *d*-amphetamine was some 10 times more potent than the *l*-isomer in affecting noradrenergic systems while the two were equipotent in affecting dopaminergic systems. Taylor & Snyder (1971) found that *d*-amphetamine was more potent at increasing locomotor activity than was the *l*-isomer (see Fig. 4.19 *top*). Thus, the dose of *d*-amphetamine needed to produce 50 % of the maximum increase in locomotor activity was some 0.9 mg kg^{-1} while that of *l*-amphetamine was 8.8 mg kg^{-1} – approximately the 10 to 1 ratio which suggested, at that time, an involvement of noradrenergic systems. The relative potencies of the two isomers were much closer together (some 2:1) for stereotyped gnawing (Fig. 4.19 *bottom*) and this lead the authors to suggest a dopaminergic mechanism for this type of motor output.

Dandiya & Kulkarni (1974) examined the effects of the two isomers on open field behaviour in the rat. They found that *d*-amphetamine induced perhaps twice as much locomotor stimulation in the horizontal direction as did the *l*-isomer. This they interpreted, on the basis of the Coyle & Snyder (1970) findings, as indicating a noradrenergic basis for forward locomotion. It is interesting that they found the two isomers to be roughly equipotent in enhancing vertical rearing. At the very least this suggests a separate system involved in the control of horizontal and vertical movement in the open field.

With the re-evaluation of the neurochemical effects of the two isomers (see p. 42) it now appears that the two are equipotent on noradrenergic systems, but the *d*-isomer is 7–10 times more effective on dopaminergic systems than the *l*-isomer. This then suggests, as do the above behavioural data, that dopamine is involved clearly in the locomotor effects of amphetamine (with a 10:1 isomer ratio). The 2:1 superiority of the *d*-isomer in eliciting stereotypy might implicate dopamine, or it might in reality be closer to equipotence, which implicates noradrenergic systems.

It would appear that it is actually the former interpretation which is correct. Costall & Naylor (1974*a*) showed that electrolytic lesion of the dopaminergic pathways' projection in the anterior hypothalamus blocked both the stereotypic and locomotor activity enhancing effects of *both* the *d*- and the *l*-isomers of amphetamine. Lesions of dopaminergic terminal areas such as the globus pallidus and the olfactory tubercle also abolished or markedly reduced the stereotypy seen after either *d*- or *l*-amphetamine. This suggests that the 2:1 ratio of potency in inducing stereotypy seen by Taylor & Snyder (1971) reflects a real difference in the activity of the two isomers and implicates a dopamine substrate. Additionally, in normal rats Costall & Naylor (1974*a*) observed an approximately 8–10 to 1 ratio of the *d*- and *l*-isomers on their four point rating scale for the intensity of stereotypy induced.

In summary, the evidence to date would suggest a dopaminergic role for both stereotypy and locomotor activity on the basis of the *current* knowledge of the neurochemical effects of the *d*- and *l*-isomers of amphetamine.

Lesions of the ascending dopaminergic systems. An early experiment by Iversen (1971) used electrolytic lesions of the substantia nigra to deplete striatal dopamine. She found some rats which showed a reduction in the locomotor stimulant effects of *d,l*-amphetamine, while other rats seemed to show a potentiation of the locomotor stimulant effects. Dopamine

Fig. 4.19. Dose–response curves for the locomotor effect (*top*) and the stereotypy effect (*bottom*) of the *d*- and *l*-isomers of amphetamine. Amphetamine was administered subcutaneously and the locomotor counts over 30 min totalled or the number of rats (out of six) displaying stereotypy expressed as a percentage. (Redrawn from Taylor & Snyder, 1971.)

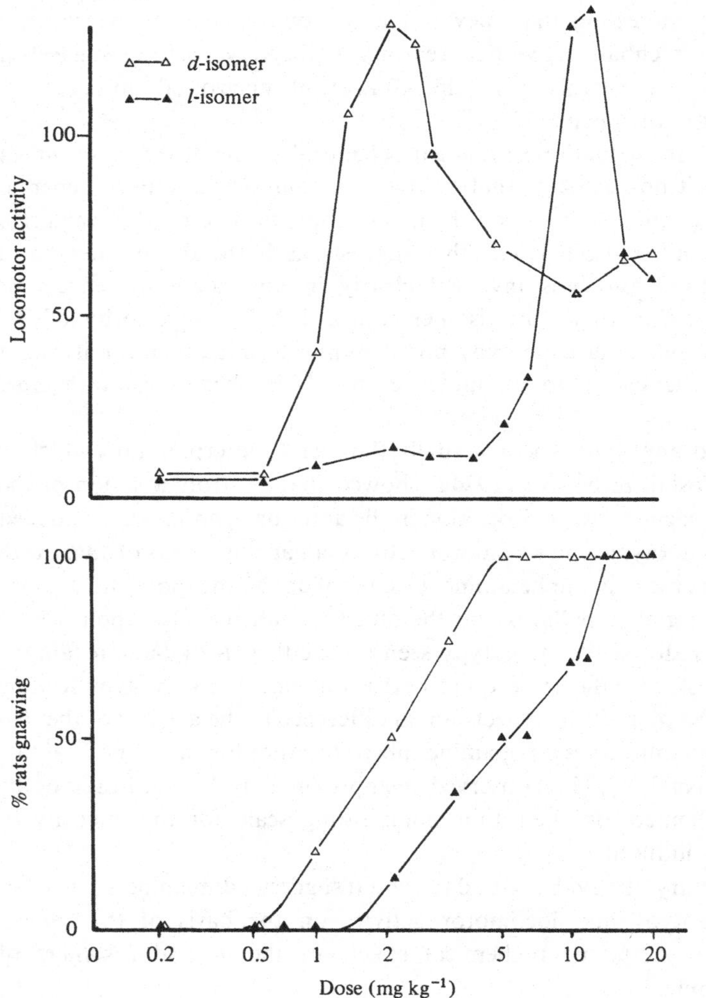

depletions in the striatum rarely exceeded 50 % and this may explain the varied behavioural responses.

Later experiments by Creese & Iversen used 6-OHDA to destroy the ascending dopaminergic systems with greater success. Thus, Creese & Iversen (1973) administered 100 μg 6-OHDA intraventricularly to neonatal rat pups on days 5, 7 and 9 after birth. This depleted striatal dopamine to unmeasurable levels, but also affected hypothalamic noradrenaline by reducing it to 35 % of normal. These rats, when tested in adulthood, showed a total block of both the stereotypic and the locomotor stimulant effects of *d*-amphetamine. Further, they showed evidence of postsynaptic supersensitivity as a result of chronic understimulation of the dopamine receptors because of the loss of the normal dopamine releasing presynaptic terminals. When given apomorphine in doses ineffective in normal rats (0.5–1.5 mg kg^{-1}) they showed marked locomotor stimulation and profound stereotypy. A similar pattern of results was obtained if dopamine was deleted in the adult by intraventricular injection of 150 μg 6-OHDA following tranylcypromine pre-treatment (5 mg kg^{-1}). Again, the locomotor response to apomorphine was slightly enhanced whilst that to amphetamine was reduced (Iversen & Creese, 1975). The stereotypy in response to apomorphine was greatly enhanced and that to amphetamine reduced. Further, Creese & Iversen (1974) showed the same results with local injections of 6-OHDA into the substantia nigra itself. Injection bilaterally of 8 μg 6-OHDA in 2 μl ascorbate–saline caused a depletion of better than 99 % in striatal tyrosine hydroxylase activity. It also, however, appeared to deplete hypothalamic noradrenaline.

So far, it has been shown that depletion of forebrain catecholamines blocks the amphetamine response, both locomotor and stereotypy. Apomorphine locomotor, and particularly stereotypy, responses were greatly increased in these rats. However, in the experiments described so far depletion of noradrenaline has accompanied dopamine loss. This was remedied by 6-OHDA lesion of either the ventral or the dorsal bundle which did not affect amphetamine locomotor or stereotypy (Creese & Iversen, 1975). Further, Price & Fibiger (1974) lesioned the ascending dopaminergic pathways in the adult rat by bilateral injection of 8 μg 6-OHDA into the substantia nigra. Again, this blocked the stereotypy response to 5 mg kg^{-1} *d*-amphetamine but greatly potentiated that seen in response to 0.2 mg kg^{-1} apomorphine. This lesion also reduced hypothalamic noradrenaline to 30 % of normal but again destruction of the ventral bundle by 6-OHDA failed to alter either response, ruling out the loss of noradrenaline as the causative agent. Similar failure of either dorsal or ventral bundle 6-OHDA lesions to affect either the locomotor or the stereotyped response to amphetamine has been reported by Roberts *et al.* (1975).

The nature of the enhanced response to apomorphine seen after 6-OHDA destruction of ascending dopaminergic systems was questioned by Costall, Naylor & Pycock (1976*b*), who found that responses to noradrenaline and serotonin injected directly into the striatum increased markedly after 6-OHDA destruction of the afferent dopamine input. While the increase in response to dopamine injected into the striatum which was also observed may be explained by a supersensitivity of the dopamine receptors as a result of denervation, the much greater increase in the responses to noradrenaline and serotonin suggest a generalised loss of selectivity of the receptor as well as supersensitivity. On the other hand, Schultz & Ungerstedt (1978) found that the enhanced sensitivity of striatal cells following 6-OHDA denervation as measured electrophysiologically by their inhibition by apomorphine correlated well with the rotational behaviour shown by this unilaterally lesioned rat. The time course of the two events was virtually identical (inhibition of striatal cell and initiation of rotation occurred at similar times after apomorphine injection). Further, bigger doses of apomorphine caused larger inhibition of striatal single unit activity as well as a greater magnitude of rotation. However, a slight discrepancy was noticed in that the electrophysiological data suggested a 10–80-fold increase in sensitivity to apomorphine whereas behavioural studies have sometimes reported only a 2–10-fold increase in apomorphine stereotypy or rotation.

Costall, Hui & Naylor (1980) have reported that the baseline response of an animal to a dopaminergic agonist may also determine whether this animal shows marked or meagre supersensitivity to that agonist following 6-OHDA-induced dopamine denervation. Thus, rats which showed much hyperactivity to peripheral injection of (−)-NPA before the lesion developed much more pronounced supersensitivity as a result of 6-OHDA injection into the nucleus accumbens than did rats which showed only low or moderate response to (−)-NPA pre-operatively; and all of this despite similar dopamine depletions by 6-OHDA in the three groups.

Heikkila, Shapiro & Duvoisin (1981) directly measured the increase in sensitivity of postsynaptic dopamine receptors after 6-OHDA denervation using receptor binding techniques (Creese & Snyder, 1979). Rats which rotated in response to L-DOPA showed an increase in [^{3}H]spiroperidol binding compared either to control or to nonrotators. The dopamine uptake (an index of the number of dopaminergic terminals) in the animals which turned was some 5% of control, whereas injected rats which failed to turn (nonrotators) showed a reduction to only 42% of normal. This would appear to support the idea of the enhanced effects of dopamine agonists being caused by receptor supersensitivity.

However, Staunton, Wolfe, Groves & Molinoff (1981), although obtaining similar increases in [^{3}H]spiperone binding after dopamine denervation, find that the behavioural effects to apomorphine occur earlier in time after the lesion than the receptor changes. Thus, the rotational response to apomorphine peaked 3 days post-operatively and remained high throughout subsequent testing (see Fig. 4.20). The increase in dopamine receptors, as measured by the increase in [^{3}H]spiperone binding, however, did not increase until 14 days post-operatively. Animals which were vigorously rotating 7 days post-lesion did not show any increase in the number of dopamine receptors (see Fig. 4.20). This suggests that the mechanism of the increased response to direct dopamine agonist drugs after dopamine denervation may be more complex than had been thought.

The preceding results strongly suggest that both the locomotor stimulant

Fig. 4.20. Time course of the increase in [^{3}H]spiperone binding to the caudate–putamen of rats following 6-OHDA destruction of the nigrostriatal dopaminergic pathway compared to the time-course of rotational behaviour in response to 0.24 mg kg^{-1} apomorphine averaged over 5 min. (Redrawn from Staunton, Wolfe, Groves & Molinoff, 1981.)

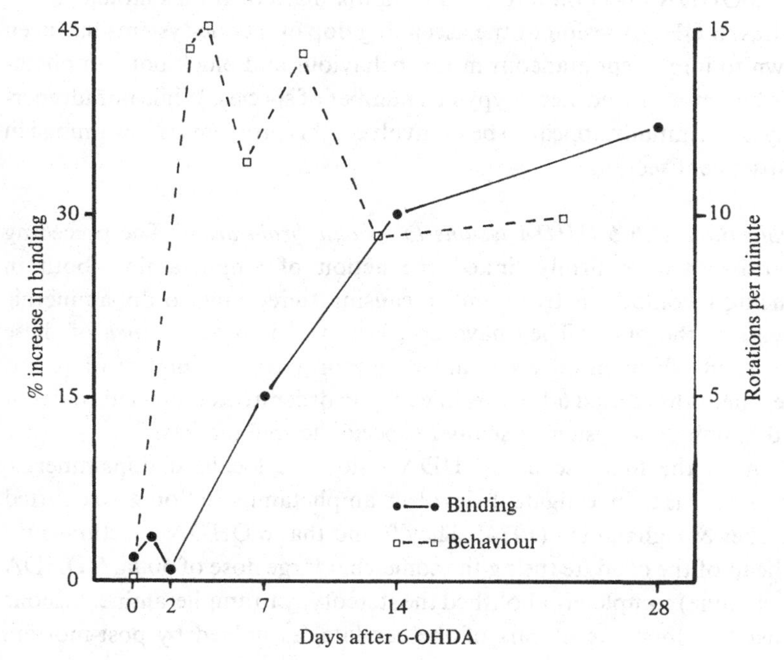

and the stereotypy-inducing effects of amphetamine depend on the integrity of ascending dopamine systems. They also suggest that, in the absence of such systems, supersensitivity of the denervated dopamine receptors occurs which increases the locomotor and stereotypy response to the direct agonist apomorphine. Little evidence emerged for a direct role of noradrenergic systems in the action of amphetamine on motor behaviour.

Similar evidence of the dopaminergically mediated nature of amphetamine action has been obtained in species other than the rat. Thus, Cheng, Bhatnagar & Long (1975) report that amphetamine-induced pecking in the pigeon is blocked by neuroleptic drugs such as haloperidol. Prevention of synthesis of both dopamine and noradrenaline by AMPT blocked the amphetamine pecking, whereas prevention of noradrenaline synthesis alone with U 14,624 did not. This is consistent with the observation that the direct dopamine receptor agonist apomorphine also causes stereotyped pecking in pigeons (Cheng & Long, 1974).

Wolfarth *et al.* (1979) found that amphetamine and apomorphine caused stereotyped sniffing and head-moving in the cat but did not cause licking or gnawing as seen in the rat. Both drugs resulted in an increase in locomotor activity in the cat. Lesion to the substantia nigra, with 6-OHDA (10 μg in 4 μl saline) reduced the stereotyped head-nodding caused by amphetamine.

Kraemer *et al.* (1981) report motor and social behaviour disturbances after 6-OHDA injection into the substantia nigra of rhesus monkeys.

Thus, 6-OHDA lesion of the ascending dopaminergic systems has been shown to impair spontaneous motor behaviour and block both amphetamine locomotion and stereotypy in a number of species. While noradrenergic systems initially appear to be uninvolved, this problem is re-examined in a subsequent section.

Localisation with 6-OHDA lesions to specific brain areas. The preceding experiments have firmly linked the action of amphetamine, both in inducing locomotor activity and in causing stereotypy, to dopaminergic systems in the brain. They have not, however, indicated *which* of these systems in which brain areas are of importance. A number of recent experiments have used 6-OHDA injection in order to lesion specific parts of the dopaminergic systems usually in specific terminal areas.

Perhaps the first use of 6-OHDA lesions of localised dopaminergic terminal areas to investigate their role in amphetamine action was reported by Asher & Aghajanian (1974). They found that 6-OHDA injections into the head of the caudate (using the somewhat large dose of 40 μg 6-OHDA in 8 μl saline) completely abolished the stereotypy to amphetamine. Lesions of just the dorsolateral part of the caudate, as judged by post-mortem histofluorescence, were also effective (animal 96-08). 6-OHDA lesions of the nucleus accumbens and olfactory tubercle were without effect on the occurrence of amphetamine stereotypy, although they did induce the emergence of a 'staring' behaviour with amphetamine.

Creese & Iversen (1974) found a similar result. 6-OHDA lesion of the caudate which reduced its dopamine content to less than 10 % of normal virtually abolished the stereotypy response to 5 mg kg^{-1} *d*-amphetamine. The locomotor stimulation to 1.5 mg kg^{-1} *d*-amphetamine occurred as normal in these 6-OHDA caudate-lesioned rats. Stereotypy in response to the direct receptor agonist apomorphine was enhanced by the caudate lesions. 6-OHDA lesions to the olfactory tubercle were without effect on either stereotypy or locomotion. It is unclear from the post-mortem amine assays whether or not the nucleus accumbens was affected by these later lesions. This assumes importance in view of the study by Kelly *et al.* (1975) to be described below. From the lateral coordinate of stereotaxic injection into the olfactory tubercle (3.0 mm from midline) it is possible that the accumbens was spared.

Kelly *et al.* (1975) used 6-OHDA to deplete dopamine from either the caudate or from both the nucleus accumbens and olfactory tubercle. Caudate dopamine was reduced to 50 % of normal by 6-OHDA injection here with no effect on dopamine in the accumbens or the olfactory tubercle. 6-OHDA injection into the later structures depleted them both to around 20 % of normal with only a small effect on caudate dopamine. Stereotypy to *d*-amphetamine (5 mg kg^{-1}) was reduced by the caudate lesion but not by the accumbens–tubercle depletion (see Fig. 4.21*a*). Stereotypy induced by apomorphine was markedly enhanced by caudate dopamine loss. The locomotor response to a lower dose of amphetamine (1.5 mg kg^{-1}) was virtually absent in accumbens–tubercle lesioned rats but unaffected by caudate 6-OHDA (see Fig. 4.21*b*). The locomotor activity induced by apomorphine (1 mg kg^{-1}) was markedly potentiated in accumbens–tubercle lesioned rats (Fig. 4.21*c*) but unaffected by caudate 6-OHDA. Thus, a clear double dissociation was demonstrated between the caudate which seemed involved in the stereotypy aspect of amphetamine action, and the accumbens–tubercle involved with the locomotor release seen after lower doses of amphetamine. The potentiation of apomorphine stereotypy and locomotion suggests that, following denervation, supersensitive receptors develop in, respectively, the caudate and the accumbens–tubercle.

Costall *et al.* (1975*b*) found that electrolytic lesions of the accumbens abolished the sniffing response associated with forward locomotion and seen in response to low doses of apomorphine or (−)-NPA. Similar effects were seen with lesion of the tuberculum olfactorium, with this latter lesion

Fig. 4.21. Locomotor and stereotypy response to *d*-amphetamine or apomorphine following 6-hydroxydopamine (6-OHDA) lesion of either the striatum or the accumbens–tubercle in rats.
(*a*) Stereotypy in response to 5 mg kg^{-1} *d*-amphetamine. (*b*) Locomotor activity in response to 1.5 mg kg^{-1} *d*-amphetamine. (*c*) Locomotor activity in response to 1.0 mg kg^{-1} apomorphine. (Redrawn from Kelly, Seviour & Iversen, 1975.)

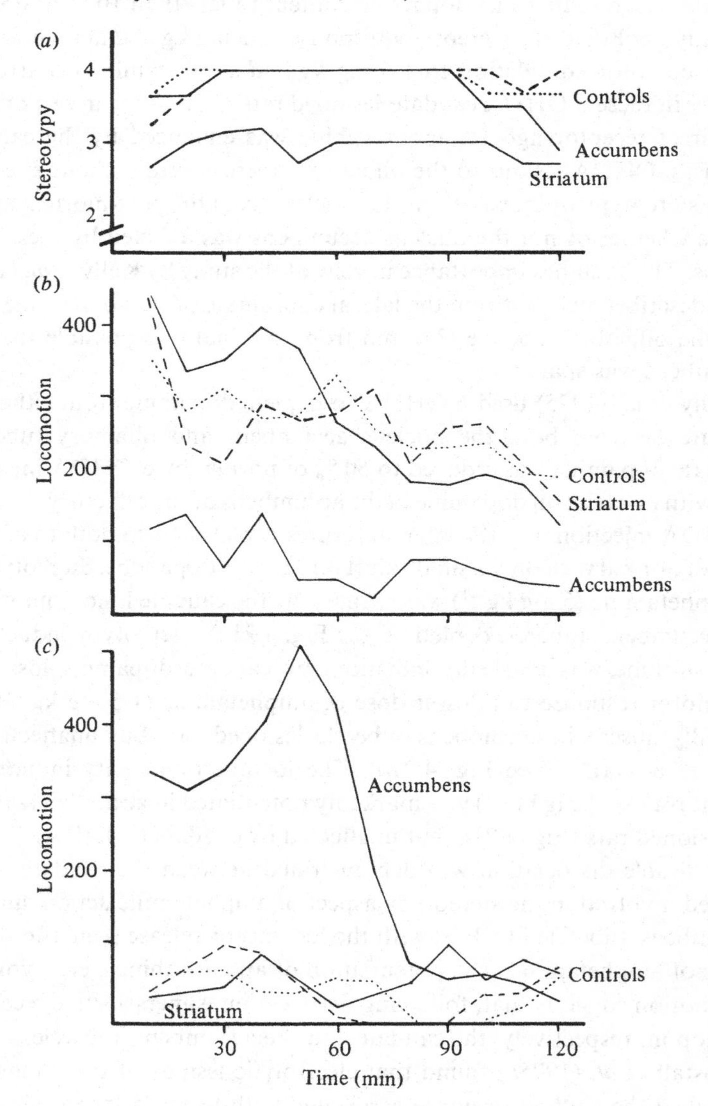

also having a small effect on the stereotypical biting and gnawing induced by apomorphine. Surprisingly, electrolytic lesion of the caudate failed to affect either the sniffing or the biting in response to apomorphine or (−)-NPA. Microinjection of either apomorphine of (−)-NPA into the olfactory tubercle caused a dose-dependent increase in sniffing and at high doses stereotyped biting occurred (possibly because of spread of the drug to other structures). A generally similar picture emerged with injection into the accumbens. Injection of apomorphine or (−)-NPA into the caudate failed to induce stereotyped biting at any dose. It is interesting to note that these injections, and presumably the electrolytic lesions, were more in the medial part of the caudate (Costall *et al.*, 1975*b*, Fig. 1) while the work of Asher & Aghajanian (1974) suggested that it was the dorsolateral caudate that was of importance.

This is supported by a later report from Costall *et al.* (1977) that injection of dopamine into the caudate induced predominantly stereotyped biting while if injected into the accumbens locomotor activity was seen. Other agonists gave different shades of effect. Apomorphine failed to enhance locomotor activity when injected into the accumbens, but did induce stereotyped biting from the caudate. (−)-NPA, however, failed to induce stereotypy when injected into the caudate but gave both stereotypy and locomotor stimulation from the accumbens. Other spectra of activity were shown by the derivatives 2-amino-5,6-dihydroxy-1,2,3,4-tetrahydronaphthalene, and related compounds. These findings suggest that, as well as stereotypy being induced by stimulation of receptors in the caudate, it may also be triggered from a population of receptors in the accumbens which differs from that set of accumbens receptors which controls the output of locomotor activity.

Other brain areas also appear to be involved in the expression of amphetamine stereotypy and locomotion. Thus, Glick & Greenstein (1973) showed that lesion to the frontal cortex on one side of the brain would induce a rotational response to amphetamine suggesting a modulatory influence of the frontal cortex on nigrostriatal dopaminergic transmission. Carter & Pycock (1980*a*) found that 6-OHDA injection into the medial prefrontal cortex, following DMI pre-treatment to protect noradrenergic neurones, caused a 70 % depletion of dopamine in this area and increased spontaneous motor activity, starting some 3 days post-operatively. A similar increase in both amphetamine-induced stereotypy and locomotor activity was also seen. However, apomorphine-induced stereotypy was significantly reduced by these frontal cortical dopamine depletions. This is, of course, exactly the opposite pattern seen with dopamine depletion in subcortical regions and serves to confirm Glick & Greenstein's (1973) suggestion of a behaviourally inhibitory effect of the frontal cortex on nigrostriatal dopamine systems.

Another brain area involved in the expression of amphetamine effects, probably by being on the output side of the nigrostriatal system, is the superior colliculus. Redgrave, Dean, Donohoe & Pope (1980) found that lesions of the superior colliculus attenuated apomorphine induced stereotypy while enhancing the locomotor activity induced by the drug. Similar results were obtained for amphetamine stereotypy by Pope, Dean & Redgrave (1979), who also saw a similar increase in the locomotor effects of the drug. This suggests that the output of the nigrostriatal system mediating stereotypy, but not that of the mesolimbic accumbens–tubercle system mediating locomotion, may be funnelled through the superior colliculus at one stage in its downward passage towards the spinal cord. This may come about since the pars reticulata of the substantia nigra, which receives a so-called feedback projection from the caudate, projects directly to the superior colliculus (Vincent, Hattori & McGeer, 1978).

In summary, the localised 6-OHDA lesion experiments have revealed control of (i) stereotypy by the caudate and (ii) locomotor activity induced by amphetamine by the accumbens–tubercle. However, a second population of receptors in the accumbens–tubercle may also be involved in the elicitation of apomorphine derivatives. The dopaminergic projection to the frontal cortex appears to have an opposite, behaviourally inhibitory effect on amphetamine responses to that of the nigrostriatal systems. The output from the latter, but not from the accumbens–tubercle, seems to involve the nigrotectal projection to the superior colliculus.

Noradrenaline and amphetamine action. Although the stereotypical and locomotor effects of amphetamine have been clearly shown to depend on the integrity of brain dopaminergic systems, this does not mean that they may not also, in addition, be modulated by noradrenergic pathways.

Evidence from agonists. The putative α-agonist clonidine has been found to intensify the locomotor response to apomorphine. Thus, Maj, Mogilnicka & Palider (1975) found that, although 1 mg kg^{-1} clonidine was without effect on spontaneous locomotor activity, it served to increase the locomotor stimulation caused by 1 mg kg^{-1} of apomorphine. In some cases (60, 90 and 120 min after injection), clonidine more than doubled the apomorphine locomotor effect. A higher dose of clonidine reduced spontaneous locomotor activity (see p. 106) and failed to enhance the apomorphine effect (Maj *et al.*, 1975). The stereotypy induced by apomorphine is also enhanced by clonidine, which does not induce stereotypy on its own. However, Pycock, Jenner & Marsden (1977) failed to find any effect of clonidine on stereotypy induced by apomorphine, although confirming its enhancing effect on the locomotor stimulation of apomorphine. This may be due to the lower systemic dose of clonidine used (0.5 mg kg^{-1} compared to 1 mg kg^{-1} in Maj *et al.*, 1975), since injection of clonidine directly into the nucleus accumbens did indeed potentiate the stereotypical response to systemically administered apomorphine (Pycock *et al.*, 1977).

Other α-agonists such as xylometazoline and naphazoline also potentiate apomorphine or amphetamine stereotypy (Zebrowska-Lupina, Kleinrok, Kozyrska & Wielosz, 1978), causing a transition from sniffing to overt licking behaviour. These may have their effects by presynaptic α-receptors on noradrenergic neurones (α_2) which inhibit the overall release of noradrenaline (Langer, 1980), since the postsynaptic α-agonist drugs methoxamine, phenylephrine and to a lesser extent noradrenaline itself acted to *inhibit* stereotyped behaviours (Zebrowska-Lupina *et al.*, 1978). These are believed to act via a postsynaptic α-receptor.

The β-agonist isoprenaline has been reported (Ksiazek & Kleinrok, 1974*a*) to inhibit the stereotypy induced by amphetamine when the former is injected intraventricularly. The very lowest dose (0.002 mol) was, however, reported to enhance amphetamine hypermotility very slightly.

Similar enhancement of the locomotor effects of amphetamine has been reported after combination with DMI which acts to prevent noradrenaline reuptake and hence increase the concentration of noradrenaline in the synaptic cleft (Dingell, Owens, Norvich & Sulser, 1967).

The conclusion from agonist studies on the action of amphetamine is unclear. Some enhance the effects of amphetamine, some inhibit these same effects. A possible, but rather speculative scheme at the present would be that clonidine actually acts by turning off noradrenaline release via α_2-receptors. Such reduced noradrenaline activity enhances the effects of apomorphine and amphetamine. Other α_1 drugs such as phenylephrine act by stimulating postsynaptic noradrenergic receptors and, together with postsynaptic β-stimulation by isoprenaline, act to inhibit amphetamine and apomorphine effects.

Antagonists. Weinstock & Speiser (1974) found that several β-blocking drugs (propranolol, oxprenolol and practolol) acted to decrease the locomotor response to amphetamine without affecting the stereotypy induced. The α-blocker thymoxamine also reduced amphetamine locomotion without effect on stereotypy. This is supported by a report from Zebrowska-Lupina (1977) that α-blockers phenoxybenzamine, phentolamine and aceperone inhibited amphetamine hypermobility. Strangely, only aceperone inhibited apomorphine hypermotility. This author also found that these α-blockers potentiated amphetamine stereotypy, contrary to the report of Weinstock & Speiser (1974).

In summary, the results from agonists and antagonists certainly suggest that amphetamine and apomorphine effects may be modulated by noradrenergic systems but leave the sign and direction of the interaction very unclear.

Synthesis inhibition and 6-OHDA lesion. Mogilnicka & Braestrup (1976) showed that inhibition of noradrenaline synthesis with either DDC or FLA 63 would markedly potentiate the stereotypy seen after amphetamine, apomorphine or phenethylamine administration. Similarly, Grabowska-Anden (1977) found that inhibition of noradrenaline synthesis with FLA 136 also enhanced amphetamine-induced stereotypy.

Braestrup (1977) reported that 6-OHDA injection into what appeared to be both ventral and dorsal noradrenergic bundles increased the frequency of licking stereotypy to both *d*-amphetamine and phenethylamine. A similar effect was also seen on apomorphine-induced sniffing stereotypy.

So far, the noradrenaline manipulations have affected *both* the dorsal and the ventral systems. Pycock (1977) destroyed specifically the dorsal pathway either by electrolytic lesions to the LC or by peripheral administration of 6-OHDA to neonatal rats on days 1 and 6 after birth. Despite depletion of forebrain noradrenaline to about 50 % of normal, no effect on either amphetamine or apomorphine stereotypy was seen. These are not very impressive depletions so caution is advised. However, Creese & Iversen (1974) and Roberts *et al.* (1975) used 6-OHDA injection into fibres of the dorsal bundle to achieve depletions better than 85–95 % and still saw no effect on amphetamine stereotypy or locomotor responses.

Kostowski, Jerlicz, Bidinski & Hauptmann (1978) examined the effects of ventral bundle lesions on apomorphine stereotypy and found them to be without effect. No regional amine assays were carried out so it is impossible to assess how effective these lesions were. This was rectified by Jerlicz, Kostowski, Bidinski & Hauptmann (1978), who showed a less than 50 % fall in noradrenaline content in hypothalamic regions after ventral bundle lesions. Neither apomorphine nor amphetamine stereotypy was affected. Thus, there appears to be little evidence for a role of ventral bundle noradrenaline in stereotypy induced by amphetamine or apomorphine. This contrasts with its role in rotation (see p. 113) caused by these drugs. These same authors (Kostowski, Jerlicz, Bidinski & Hauptmann, 1977) found that electrolytic lesion to the LC, which decreased forebrain noradrenaline by 50 % enhanced apomorphine stereotypy in rats, with a rather smaller enhancement in amphetamine stereotypy. Since these lesions were no better than those of Pycock (1977), which were clearly without effect on either amphetamine or apomorphine stereotypy, it is hard to see what the resolution is.

In summary, it is clear from synthesis inhibition studies that a reduction in the activity in noradrenaline systems *as a whole* enhances amphetamine and apomorphine stereotypy. It is also clear from 6-OHDA lesion of the dorsal bundle that this is not the system involved. Two possibilities remain: the effect might be mediated by the ventral bundle, much as are the noradrenaline effects on rotation (see p. 114). The lesions of Jerlicz *et al.* (1978) argue against this, although their depletions were only 50 %. The effect may be mediated by a projection of the LC which does not run in the dorsal bundle. This would reconcile the absence of effect after dorsal bundle 6-OHDA with the positive results of synthesis inhibitors. This might be a projection to the raphe, cerebellum or descending to the spinal cord. However, at least one study (Pycock, 1977) failed to find an effect even after electrolytic lesion of the LC. At the moment the involvement of noradrenaline in stereotypy seems clear, but the details remain murky.

Conclusions. The neurochemical basis of the stereotypical and the locomotor stimulant effects of amphetamine has been investigated in a number of ways. The two isomers of amphetamine, based on the original evidence of Coyle & Snyder (1970), were employed because they have different neurochemical effects. Although the two isomers yield different potencies on different behaviours the subsequent confusion concerning their actual neurochemical effects rather vitiates any suggestions from this direction.

6-OHDA lesions of the ascending dopaminergic systems have convincingly implicated dopamine in both the stereotypical and the locomotor stimulant effects of the drug. More selective terminal area lesions have suggested that the stereotypy may be mediated via dopamine release in the caudate nucleus while locomotor stimulation and exploratory sniffing involves the accumbens–olfactory tubercle. The dopamine innervation to the medial prefrontal cortex appears to have an effect opposite to that to the subcortical dopaminergic terminal zones. Thus, an *increase* in the effects of amphetamine is seen after 6-OHDA lesion of the frontal cortex dopamine input.

Noradrenaline seems to have a modulatory influence on the effects of amphetamine and apomorphine. It would seem to inhibit these effects when normally present as judged by the increases in stereotypy reported after either synthesis inhibition or 6-OHDA lesion. It would appear not to be the dorsal bundle system that is involved, nor does the ventral bundle *per se* seem involved. It may or may not originate from the LC and run in the central tegmental tracts rather than in the dorsal bundle. This problem is in need of more detailed investigation.

Chronic amphetamine administration

Repeated administration of amphetamine in rats leads to a potentiation of some types of elicited behaviour. Thus, for example, Segal, Weinberger, Cahill & McCunney (1980) reported that administration of 2.5 mg kg^{-1} amphetamine every 4 h resulted in a more rapid onset of stereotypy after the first few days on this regime. The magnitude of the stereotypy was also increased by this treatment. Parallel neurochemical measures revealed a depletion of both noradrenaline and dopamine during the chronic amphetamine administration. It is not presently known whether this is related to the observed behavioural potentiation in any causative fashion. Similar potentiation of the locomotor and stereotypical effects of amphetamine has been seen in other species, such as monkeys (Garver, Schlemm, Maas & Davis, 1975) and cats (Ellingwood & Kilbey, 1976).

Not all components of the stereotyped behaviour show similar potentiation. Eichler, Antelman & Black (1980) report that sniffing showed a potentiation over days with repeated administration of doses of amphetamine ranging from 2 to 4 mg kg^{-1}, whereas licking behaviour dropped in intensity over the first 15 or so days. This is, of course, consistent with the suggestions reviewed previously that licking and sniffing/locomotion may be mediated by different neurochemical substrates.

Dopamine and rotation behaviour

Introduction. When dopaminergic systems on one side of the brain are lesioned, the rat will subsequently show rotational behaviour. That is, instead of walking in a straight line, its forward locomotion carries it round in a circle to one side or the other. The direction of rotation varies depending on the brain lesion and stimulating drug. An early description of rotation after dopamine lesion was given by Ungerstedt & Arbuthnott (1970), who quantified this behaviour with a so-called rotometer. This is a perspex hemispherical bowl (see p. 97) in which the rat is placed. A harness is fitted to the rat's body and attached by wire to a cam placed above the centre of the bowl. Rotations in either direction cause small increments on a counter. Turns per minute can be cumulated and then printed out with the counter being reset to zero. Ungerstedt & Arbuthnott (1970) describe intense rotation in response to 5 mg kg^{-1} amphetamine after unilateral lesion to the ascending nigrostriatal dopaminergic pathway by injection of 6-OHDA (8 μg in 4 μl saline). Animals so lesioned rotated ipsiversively.

That is *towards* the lesioned side. Figure 4.22*a* shows the time course of this rotational behaviour, which lasted 1 to 2 h, depending on the dose of amphetamine. Higher doses induced more rotation (Fig. 2.22*b*). The maximum rate of rotation was some 18 turns per minute. The authors suggested that the rotation was caused by an asymmetry in the release of dopamine in the brain caused by the drug. On the intact side a considerable release of dopamine will occur from terminals as a result of amphetamine administration. On the lesioned side very few dopaminergic terminals remain to release dopamine in response to amphetamine. It is supposed

Fig. 4.22. Rotational behaviour to amphetamine in rats with a unilateral 6-OHDA lesion of the nigrostriatal dopaminergic pathway. The dose–response curve is shown at the top while the time-course of three doses of the drug is shown at the bottom. (Redrawn from Ungerstedt, 1971*a*.)

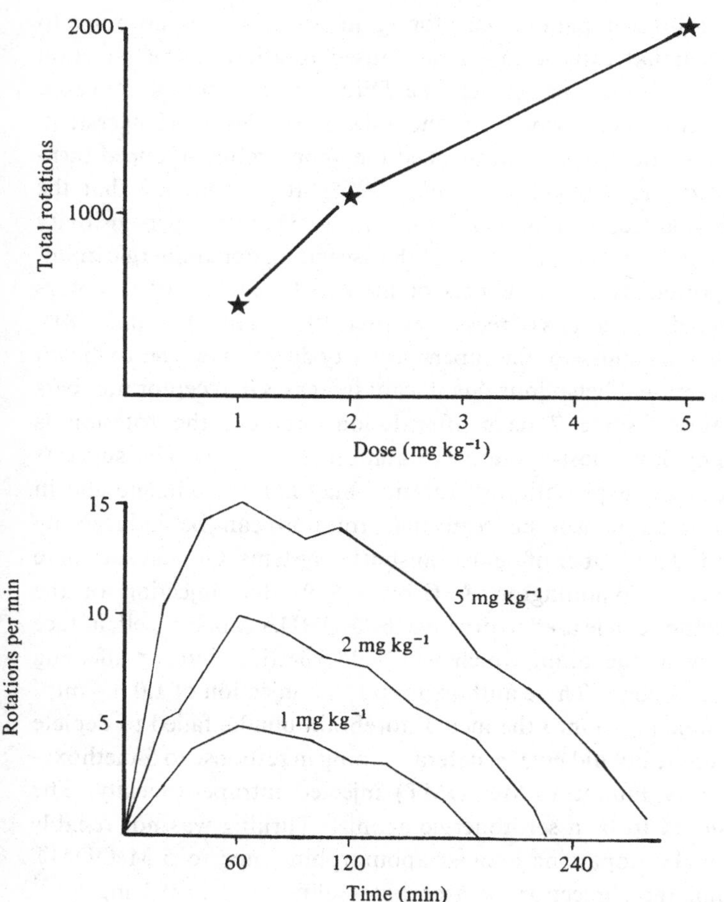

that the release of dopamine facilitates locomotion, so that a greater amount of locomotion occurs in the intact than on the lesioned side of the body and thus the animal rotates towards the lesioned side.

A number of drugs which interfere with dopaminergic neurotransmission act to block the amphetamine-induced rotation of the unilaterally lesioned rat. Synthesis inhibition with AMPT, depletion of storage granules with reserpine or receptor blockade with haloperidol or spiroperidol all prevent amphetamine-induced rotation (Ungerstedt, 1971*a*). Other neuroleptic drugs such as pimozide, chlorpromazine, metoclopramide or clozapine were also effective, but the α-blocker phenoxybenzamine was not (Pycock, Tarsy & Marsden, 1975*b*). Inhibition of noradrenaline synthesis by FLA 63 has been reported either to have no effect on amphetamine-induced rotation in the 6-OHDA lesioned rat (Pycock *et al.*, 1975*b*) or to potentiate it (Ungerstedt, 1971*a*).

Injection of direct dopamine receptor agonists such as apomorphine to the 6-OHDA unilaterally lesioned rat caused rotation in the direction opposite to that caused by amphetamine. Direct receptor stimulants cause *contraversive* rotation, away from the side of the lesion (Ungerstedt, 1971*b*). Neuroleptic drugs again blocked the apomorphine-induced turning (Ungerstedt, 1971*b*; Pycock *et al.*, 1975*b*). It is suggested that the apomorphine-induced turning is a result of postsynaptic supersensitivity caused by the unilateral denervation of the ascending dopaminergic input. This is supported by many reports of increased numbers of receptors following such denervations (Creese & Snyder, 1979; Heikkila *et al.*, 1981). However, the time course of the supersensitivity does not appear to match that of the rotational behaviour. Significant increases in receptor numbers were only found some 7 days after lesion, whereas the rotation is pronounced by day 3 post-operative (Staunton *et al.*, 1981). This suggests that the dopamine explanation of rotation may not be complete and in following sections it will be seen that rotation can be induced by manipulation of a number of neurotransmitter systems. Of relevance here is the finding of Waddington & Crow (1979) that injection of the ascorbate–saline vehicle used to dissolve the 6-OHDA can, by itself, induce an asymmetry in the brain which will yield rotation *without* affecting dopaminergic systems. These authors found that injection of 1.0 mg ml^{-1} ascorbate–saline (4 μl) into the medial forebrain bundle failed to deplete striatal dopamine but did give ipsilateral turning in response to 5-methoxy-*N*,*N*-dimethyltryptamine (5-MeODMT) injected intraperitoneally. The latter is believed to be a serotonergic agonist. Turning was not reliably attributed to the dopamine agonist apomorphine, nor to 5-MeODMT following unilateral injection of ascorbate–saline of only 0.2 mg ml^{-1}

concentration. Thus, the dopaminergic systems are not the only ones involved in rotational behaviour. This is made plain in subsequent sections (see p. 154).

Other drugs injected intraperitoneally in the unilaterally dopamine-depleted rat will also elicit rotation. Costall & Naylor (1975) report that in addition to *d*-amphetamine and apomorphine, the drugs methylphenidate, nomifensine, ET 495 (piribedil), L-DOPA and amantadine also caused rotation, as did cocaine analogues (Heikkila, Cabbat, Manzino & Duvoisin, 1979). A further indication of the role of nondopaminergic systems in rotation is the fact that after 6-OHDA lesions L-DOPA, ET 495 and apomorphine induce contralateral rotation (Ungerstedt, 1971*b*), whereas following *electrolytic* lesion of the substantia nigra, which as well as depleting striatal dopamine may also additionally affect other systems, *ipsiversive* rotation is described in response to these agents (Costall & Naylor, 1975). Contralateral rotation was also seen with bromocryptine, and (−)-NPA, while D145, nomifensine and its metabolite induced ipsilateral circling (Costall *et al*., 1975*a*).

The role of nondopaminergic systems in turning is partially explained by work of Marshall & Ungerstedt (1977). The efferent output pathways from the striatum descend in close proximity to the ascending dopaminergic nigrostriatal afferent fibres. Thus, animals given electrolytic lesions along the course of the efferent fibres in the internal capsule rotated towards the side of the lesion when given apomorphine. This was true even if the animals had previously been given a 6-OHDA lesion in the *opposite* nigrostriatal dopaminergic pathway. This is illustrated in Fig. 4.23 and indicates that the dopaminergic systems do not single-handedly control the direction of rotation, but are only one link in a multi-synapsed pathway controlling the motor output of that side of the body.

Rotation may also be caused, not in response to intraperitoneal administration of drugs, following 6-OHDA unilateral lesion, but in response to injection of drugs directly into the caudate nucleus. Thus, Setler, Malesky, McDevitt & Turner (1978) showed that dopamine injected into the denervated caudate of the rat caused vigorous turning away from the side of the injection. Although dopamine was the most potent substance, a cautionary note is added by the finding that injection of noradrenaline or adrenaline but not isoproterenol also caused contraversive circling.

Rotation in normal rats. Rotational behaviour has also been reported in normal, intact rats in response to apomorphine and amphetamine. Jerussi & Glick (1975) reported that apomorphine would cause rotation in normal rats. This increased with dose up to 10 mg kg^{-1} and the direction varied from rat to rat but was consistent for a given individual from test to test. Similar results were reported for *d*-amphetamine by Jerussi & Glick (1974). This is interpreted by these authors as a natural imbalance in the nigrostriatal pathways in the two halves of the brain.

It has also been reported that normal, intact rats will rotate even without drugs. Glick & Cox (1978) claim that normal rats rotate at night, with the preferred direction being consistent for a given rat across hours and across tests. It was also in the same direction as that rotation induced in these normal rats by *d*-amphetamine. It has also been claimed that the direction of rotation in intact, unlesioned rats correlates with neurochemical indices of dopaminergic activity in the two nigrostriatal pathways (Glick, Jerussi, Waters & Green, 1974). Jerussi, Glick & Johnson (1977) suggested that the levels of the dopamine metabolites dihydroxyphenylacetic acid and homovanillic acid (HVA) were higher in the striatum ipsilateral to the direction of rotation. On the other hand, activity of adenylate cyclase was higher on the side contralateral to the direction of rotation. These studies remain rather controversial.

Uptake inhibitors. Consistent with the view that rotation is dependent on dopamine release is the finding (Pycock, Milson, Tarsy & Marsden, 1976) that the dopamine uptake inhibitor benztropine potentiated the rotation induced by amphetamine in unilateral 6-OHDA rats. This can be explained by a prolongation and potentiation of the rotational effects of the

Fig. 4.23. Schematic representation of the two surgical preparations used by Marshall & Ungerstedt (1977). (*a*) Intact nigrostriatal dopaminergic (DA) fibres and efferent fibres in left hemisphere, with electrolytically lesioned nigrostriatal and efferent fibres in right hemisphere. (*b*) 6-OHDA lesioned nigrostriatal DA fibres but *intact* efferent fibres in left hemisphere, with electrolytically lesioned nigrostriatal and efferent fibres in right.

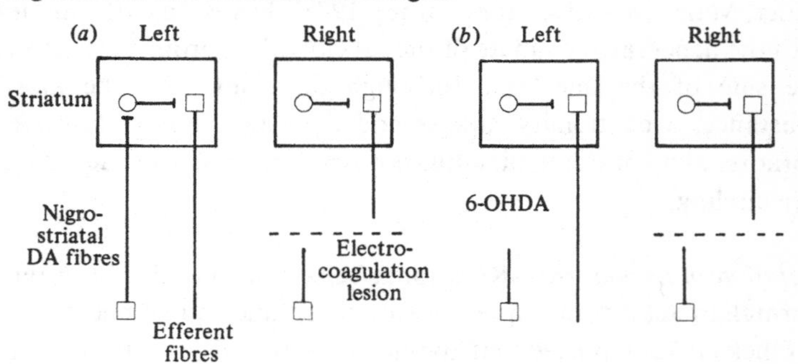

dopamine released by amphetamine as a result of blockage of the normal inactivation mechanism, presynaptic reuptake. Noradrenaline uptake inhibitors amitriptyline or DMI failed to potentiate amphetamine turning. Similar potentiation was seen with another dopamine uptake inhibitor nomifensine, although this drug also induced rotation in the absence of amphetamine, suggesting that it might have a releasing action in its own right.

Other presynaptic uptake blockers, such as cocaine and its analogues, Win 35,428 and Win 35,065 also cause ipsiversive rotational behaviour in unilaterally 6-OHDA lesioned rats (Heikkila, Cabbat, Manzino & Duvoisin, 1980). These effects could be blocked by the dopamine receptor antagonist haloperidol.

Complications of the rotational model. If the rat is tested 24 h after 6-OHDA lesion, rather than the usual two weeks, a different behavioural pattern emerges. Now, amphetamine caused contralateral turning. This is opposite to its effect in chronically denervated animals. It is suggested that a brief increase in dopamine synthesis may occur shortly after the lesion in an attempt to restore normal function and amphetamine may interact with this newly synthesised pool of dopamine to cause contralateral turning (Oberlander, Euvrard, Dumont & Boissier, 1979). On the other hand, methylphenidate, pipradrol and nomifensine still induce ipsiversive rotation at 24 h post-lesion. These drugs may interact with the reserve, rather than the newly synthesised pool of dopamine.

Another complication embodies the two-receptor idea of Cools & van Rossum (1976). Tye *et al.* (1977) report that both the analogues of apomorphine, di-isobutylapomorphine and lergotrile, will cause contraversive turning in unilaterally 6-OHDA lesioned rats. However, haloperidol blocked the turning caused by di-isobutylapomorphine but not that caused by lergotrile. This suggests two populations of receptors involved in turning, only one of which is blocked by haloperidol. The finding that the neuroleptic clozapine would block the lergotrile turning without blocking that caused by di-isobutylapomorphine suggests that the second receptor may also be dopaminergic. However, it should be noted that clozapine also has anticholinergic activity (Miller & Hiley, 1975).

Along similar lines is the finding of Kelly & Moore (1976) that, although unilateral 6-OHDA lesions to the accumbens–olfactory tubercle do not of themselves cause rotation in response to amphetamine or apomorphine, bilateral lesions do greatly reduce the magnitude of that turning caused by unilateral 6-OHDA lesion of the caudate. It thus appears that intact dopaminergic systems in the accumbens–tubercle are needed for the expression of caudate lesion-induced turning, as perhaps a form of amplifier (Kelly & Moore, 1976), even though an asymmetry in the accumbens–tubercle system fails to *cause* turning.

The role of nondopaminergic systems in rotation (which will be taken up again in a later section, see below) is shown by an experiment of Reavill, Jenner, Leigh & Marsden (1981). They investigated the role of the nigrothalamic output pathway in expression of circling behaviour caused by unilateral 6-OHDA lesion of the ascending dopaminergic nigrostriatal pathway. Lesion of the ventromedial thalamic nucleus ipsilateral to the lesioned nigrostriatal pathway reduced apomorphine rotation but not that in response to amphetamine. Conversely, lesion of the ventromedial thalamic nucleus contralateral to the lesioned nigrostriatal pathway decreased amphetamine rotation but not that in response to apomorphine. Lesion of the neighbouring parafascicular thalamic nucleus was without effect. Again, lesion to the ventromedial thalamic nucleus did not *cause* rotation, either spontaneously or in response to amphetamine or apomorphine.

In summary, the asymmetry of nigrostriatal dopaminergic pathways which induces rotational behaviour can be traced to a number of synapses further on where it seems to interact with several different neurochemical systems. I shall now look at the interaction of dopaminergic systems with other neurotransmitters.

Dopamine interaction with other systems

Acetylcholine systems. Anticholinergic drugs such as scopolamine or atropine potentiate amphetamine-induced stereotyped behaviour (Arnfred & Randrup, 1968; Scheel-Kruger, 1970). In the rotational model with unilateral 6-OHDA lesion of the nigrostriatal pathway, scopolamine by itself induces ipsiversive turning (Pycock *et al.*, 1978) and potentiates the ipsiversive turning caused by amphetamine. Scopolamine did not affect apomorphine-induced turning. The cholinergic agonists, arecoline and pilocarpine, and the esterase inhibitor, physostygmine, although not actually causing turning, did depress the rate of amphetamine-induced rotation (Pycock *et al.*, 1978). These were central effects, since the peripherally active, but centrally inactive, drugs methylscopolamine or neostygmine failed to affect apomorphine- or amphetamine-induced rotation.

In addition to the muscarinic cholinergic interaction evidenced above there also appears to be a nicotinic interaction. Kaakkola (1981) found that, in the unilaterally 6-OHDA lesioned rat, nicotine increased amphetamine-induced rotation. The nicotinic antagonists mecamylamine and pempidine slightly reduced amphetamine-induced rotation. In confirmation of the above work by Pycock *et al.*, (1978), Kaakkola (1981) found that the muscarinic antagonist atropine potentiated amphetamine-induced rotation and pilocarpine, a muscarinic agonist, decreased it. Again, the nicotinic effects were central in origin since the peripherally active hexamethonium failed to alter amphetamine-induced rotation. Thus, there appears to be a nicotinic cholinergic effect on dopaminergic systems which is opposite in sign to that of the muscarinic effect. These effects all seem to be presynaptic, since none of the nicotinic agents affected apomorphine-induced rotation. This was further confirmed by Georgiev, Markovska & Petkova (1978), who found that neither intrastriatal application of muscarinic agents (atropine) nor nicotinic ones (nicotine or mecamylamine) altered apomorphine-induced stereotyped behaviour in the mouse. However, one report (Grabowska, 1975) did find that atropine injected intraperitoneally potentiated apomorphine stereotypy in rats, raising the possibility of an extrastriatal polysynaptic interaction. This effect, though, was weak and not dose-dependent.

The rotation induced by other drugs in the unilaterally lesioned rat is also affected by cholinergic manipulations. Finnegan, Kanner & Meltzer (1976) report that phencyclidine-induced rotation was also inhibited by arecoline and potentiated by the anticholinergic agent trihexylphenidyl.

Opiate systems. Large doses of morphine may induce a stereotyped form of behaviour (Fog, 1970; Charness, Amit & Taylor, 1975). The morphine stereotyped behaviour consisted of bursts of motor activity, stereotyped biting of the cage bars or of the rat's own body. An increase in grooming behaviour is also seen (Ayhan & Randrup, 1973). Haloperidol would reduce but not completely prevent morphine-induced stereotypies (Charness *et al.*, 1975). Ayhan & Randrup (1973) found that morphine-induced locomotor excitation could be blocked by synthesis inhibition with AMPT. Two dopamine receptor blockers, spiramide and pimozide, also prevented this behaviour. However, the behaviour could also be antagonised by noradrenaline synthesis inhibition with FLA 63 or receptor blockade with the α-antagonists phenoxybenzamine or aceperone. It would appear that both dopamine *and* noradrenaline may be involved in this type of behaviour.

Injection of morphine or heroin directly into the midbrain reticular formation induced extremely rapid rotation ipsiversive to the injection (Jacquet, Carol & Russell, 1976). Dopaminergic antagonists, such as pimozide, failed to block this effect and strangely the opiate antagonist naloxone actually potentiated it.

Morphine injected intraperitoneally may also induce rotation in intact normal mice. The direction of rotation varying from mouse to mouse but being consistent from test to test for a given individual (Glick & Morihisa, 1976). This effect can also be obtained in rats and is blocked by naloxone (Morihisa & Glick, 1977).

In rats with unilateral 6-OHDA lesion of the nigrostriatal dopamine systems intraperitoneal morphine, methadone, levorphanol, nalorphine and pentazocine all cause ipsiversive rotation. The opiate antagonist naloxone did not itself induce circling but it did block the circling induced by morphine (Iwamoto, Loh & Way, 1976).

Cools, Janssen & Broekkamp (1974) reported that morphine injected intraperitoneally in cats in doses of 5 to 20 mg kg^{-1} gave stereotyped head and neck movements occurring in a repeated and ritualised sequence. Subsequently, postural changes involving placing of the limbs occurred and these too were stereotyped. Towards the end of drug action, violent locomotor movements were seen, such as backwards jumping. Haloperidol injected directly into the caudate nucleus of the cat completely blocked the morphine-induced stereotypies involving postural changes or locomotion but did not affect the stereotyped head movements (Cools *et al.*, 1974). Opiate drugs of the partial-agonist type such as cyclazocine and levallorphan induce a bizarre behavioural syndrome of head movements, backwards walking and hind paw pivoting. Paradoxically, these effects are blocked by dopaminergic agonists, apomorphine, piribedil, amphetamine or L-DOPA (Buckett & Shaw, 1975).

Holtzman & Jewett (1973) report that the opiate antagonist naloxone reduced the effects of *d*-amphetamine on avoidance responding and Moon, Feigenbaum, Carson & Klawans (1980) found that naloxone partially antagonised apomorphine-induced stereotypy. Further, low doses of naloxone potentiated low doses of haloperidol in disrupting apomorphine-induced stereotyped behaviours (see Fig. 4.24).

Similarly, naloxone blockade of apomorphine stereotypy has also been reported in guinea pigs (Margolin & Moon, 1979).

Thus, in summary it seems clear that opiate systems can influence dopaminergic ones, eliciting overt stereotyped behaviours and causing rotation either in normal mice and rats or in those with unilateral 6-OHDA lesion to the dopaminergic systems. The opiate antagonist naloxone can reduce dopaminergic behaviours such as amphetamine- or apomorphine-induced stereotypy.

Ethanol. There is ample biochemical evidence for an interaction between ethanol and dopaminergic systems. Griffiths, Littleton & Ortiz (1974) reported that acute administration of ethanol (0 to 3 h by inhalation) caused an increase in locomotor activity in mice and a drop in brain levels of noradrenaline, dopamine and serotonin. The greatest drop was seen in serotonin, with dopamine following second. Chronic ethanol administration (24 h to 10 days) caused locomotor depression and ataxia and both dopamine and noradrenaline levels rose to be some 40–50 % higher than normal after 10 days of administration. Less effect was seen on serotonin. Administration of AMPT can prevent the locomotor excitation caused by ethanol in mice (Carlsson, Engel & Svensson, 1972). Further, the locomotor stimulation induced by ethanol, but not the behavioural depression, was antagonised by the dopamine agonist apomorphine (Carlsson *et al.*, 1974).

This dopaminergic basis for the locomotor stimulant effects of ethanol is further supported by the finding that caffeine, which prolongs the effect of dopamine receptor stimulation by preventing the breakdown of cAMP second messenger by its phosphodiesterase-inhibiting effects, potentiates the ethanol-induced locomotor stimulation. Thus, 25 mg kg^{-1} caffeine

Fig. 4.24. Effect of the opiate antagonist naloxone (NX) on apomorphine (APO) stereotypy. (*a*) Stereotypy in response to 0.5 mg kg^{-1} apomorphine together with different doses of naloxone. (*b*) Effect of 0.4 mg kg^{-1} naloxone on stereotypy induced by different doses of apomorphine. (Redrawn from Moon, Feigenbaum, Carson & Klawans, 1980.)

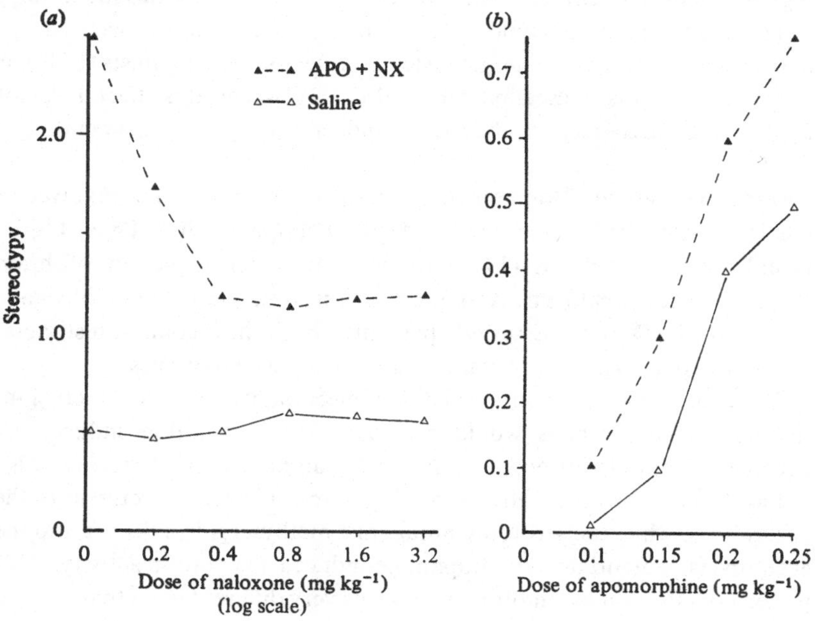

administered with 1 g kg^{-1} ethanol induced twice as much activity as that recorded with 1 g kg^{-1} ethanol on its own (Waldeck, 1974).

Hallucinogens. Both locomotor excitation and aberrant stereotyped behaviours are induced in the rat by high doses of the hallucinogen lysergic acid diethylamide (LSD) (Dixon, 1968). The neuroleptic drugs chlorpromazine and perphenazine could block both the locomotor excitation and the stereotyped biting and gnawing, as could the synthesis inhibitor AMPT. On the other hand, noradrenergic manipulations, such as the α-antagonist phenoxybenzamine and the β-blocker propranolol, together with the synthesis inhibitor disulfiram, all reduced locomotor excitation and stereotypy, albeit to slightly lesser extents (Dixon, 1968). Blockade of noradrenaline reuptake by DMI, acted to enhance both the locomotor and stereotypical effects of LSD.

In the unilaterally 6-OHDA nigrostriatal lesioned rat, LSD is a potent dopamine agonist, since it induces strong contraversive turning, as would apomorphine for example (Trulson, Stark & Jacobs, 1977). The nonhallucinogenic analogue BOL (2-bromo-*d*-lysergic acid diethylamide) had a much weaker effect. On the other hand the hallucinogens mescaline and STP (2,5-dimethoxy-4-methylamphetamine caused ipsiversive turning, suggesting a releasing effect, as did DMT (*N*,*N*-dimethyltryptamine) and 5-MeODMT (5-methoxy-*N*,*N*-dimethyltryptamine). Additionally, the highly potent hallucinogen psilocybin failed to cause any turning at all.

Thus, in summary, it would appear that some of the hallucinogenic drugs may interact with dopaminergic systems, either as direct agonists (LSD) or as releasing agents (mescaline, STP, DMT). However, this effect does not appear to be necessary for their hallucinogenic action (e.g. psilocin).

Serotonergic systems. Brain serotonin has been suggested to be involved in the locomotor and stereotypical effects of amphetamine. Thus, Lapin, Oxenkrug & Azbekyan (1972) found that either depletion of brain serotonin with *p*-chlorphenylalanine, or serotonergic receptor blockade with BOL, LSD or methysergide prevented both the locomotion stimulation with 10 mg kg^{-1} amphetamine and also the stereotypies.

Carter & Pycock (1978) reported that injection of serotonin directly into the nucleus accumbens would attenuate the locomotion induced by peripheral administration of 1 mg kg^{-1} *d*-amphetamine. Serotonin also reduced the locomotor effects of dopamine injected directly into the accumbens, while the serotonin antagonist methysergide, injected into the accumbens, potentiated the dopamine-induced locomotor activity. This pattern of effect on locomotor activity was very different from that seen on stereotyped behaviour. Here, apomorphine-induced stereotypy was potentiated by peripheral injection of the serotonin agonist quipazine or the reuptake inhibitor ORG 6582. Injection of the antagonists metergoline or cyproheptadine, reduced the intensity of apomorphine stereotypy. Carter & Pycock (1980*b*) found that destruction of the serotonin terminals in the amygdala by injection of the neurotoxin 5,7-dihydroxytryptamine (5,7-DHT) resulted in a reduction of the stereotyped responses to both amphetamine and apomorphine. Electrolytic lesion of the source nuclei, either the medial or the dorsal raphe, reduced the stereotypy seen in response to amphetamine or apomorphine (Costall & Naylor, 1974*b*) or to intracaudate injection of dopamine itself. However, 5,7-DHT lesion of the nucleus accumbens, striatum and substantia nigra potentiated amphetamine stereotypy (Carter & Pycock, 1979).

Alterations to the serotonergic systems will cause rotational behaviour. Thus, Costall & Naylor (1974*b*) and Costall, Naylor, Marsden & Pycock (1976*a*) report that asymmetrical lesion in the medial raphe nucleus caused rats to circle away from the side of the lesion spontaneously during the first few days post-operative. Administration of either amphetamine or apomorphine greatly enhanced this contraversive rotation. Depletions of cortical, striatal and mesolimbic serotonin were found, but not of cortical noradrenaline, thus ruling out the effect being caused by damage to the noradrenergic fibres of the dorsal or ventral bundles which run in the vicinity of the medial raphe nucleus (slightly more laterally by about 0.5 mm or so).

Similar results were found by Giambalvo & Snodgrass (1978), who injected the neurotoxins 5,7-DHT or *p*-chloroamphetamine, into one side of the medial raphe and found circling contraversive to the injected side. The effect seemed to depend on dopaminergic systems since rotation could be blocked by the dopamine antagonist, haloperidol. Amphetamine or apomorphine both enhanced this contraversive turning. Injection of 5,7-DHT into the substantia nigra elicited a similar syndrome.

Serotonergic agents may modify the turning seen after unilateral 6-OHDA lesion to the ascending dopamine systems in the nigrostriatal bundle. Thus, Milson & Pycock (1976) found that increasing serotonin by administration of L-tryptophan or 5-hydroxytryptophan decreased amphetamine- or apomorphine-induced rotation. Reduction of serotonin by the synthesis inhibitor *p*-chlorphenylalanine increased amphetamine- and apomorphine-induced circling. Surprisingly, the receptor antagonists methysergide, LSD or cyproheptadine failed to alter amphetamine- or apomorphine-induced turning.

In summary, a clear interaction of the dopaminergic systems with serotonin has been demonstrated. The overall effect of serotonin is to facilitate dopamine transmission, possibly through terminals in the amygdala. Other terminals in the striatum and mesolimbic areas seem to have an opposing effect and serotonin here inhibits dopamine-induced locomotion.

GABAminergic systems. Muscimol is a highly potent γ-aminobutyric acid (GABA) agonist and if injected intraperitoneally causes a marked potentiation of stereotypy induced by apomorphine, methylphenidate or cocaine (Scheel-Kruger, Christensen & Arnt, 1978). By itself, muscimol does not induce stereotypy and serves to reduce the locomotor stimulation seen with lower doses of the above drugs. Further, muscimol injected into the substantia nigra will elicit strong contraversive turning away from the side of the injection (Martin *et al*., 1978). Similar, but shorter-lasting, responses were seen after injection of GABA itself, or of another GABA agonist – imidazole acetic acid. Further, the GABA transaminase inhibitor ethanolamine-*o*-sulphate, which increases GABA levels by inhibiting the enzyme responsible for its breakdown, also induced contraversive turning. GABA antagonists, such as picrotoxin or bicuculline, did not of themselves induce turning, but picrotoxin did block the muscimol effect. Haloperidol only partially reduced muscimol turning, raising the possibility of a nondopaminergic system arising in the substantia nigra (such as reticulata projections to thalamus or colliculus).

Arnt & Scheel-Kruger (1979) additionally reported blocking muscimol-induced turning by the GABA antagonist bicuculline, whereas strychnine (a glycine antagonist) and naloxone (an opiate antagonist) were without effect. Again, muscimol turning could not be blocked by dopamine receptor blockade with haloperidol. It seemed to depend on cell bodies in the substantia nigra, pars reticulata since the effect could be abolished by kainic acid destruction of cell bodies in that region.

Injection of GABA directly into the striatum of mouse markedly reduced the stereotypy caused by apomorphine injected subcutaneously (Georgiev *et al*., 1978), as did the weaker GABA agonist amino-oxyacetic acid.

Another GABA link in the extrapyramidal pathway seems to connect the nucleus accumbens with the globus pallidus. Jones & Mogenson (1980) found that injection of the GABA antagonist picrotoxin into the globus pallidus increased locomotor activity, while GABA injected here decreased activity. This is consistent with the work of Pycock & Horton (1976) that infusions of ethanolamine-*o*-sulphate, which increases GABA levels, into the globus pallidus would block the increase in locomotor activity caused by injection of dopamine into the accumbens. Jones & Mogenson (1980) report that injection of GABA itself into the globus pallidus would also attenuate the locomotor activity caused by intra-accumbens dopamine injection.

Thus, in summary, two points of interaction of dopaminergic systems with GABA are to be seen. One involves the output pathway from the substantia nigra, pars reticulata to the thalamus and colliculus. This is in turn influenced by dopaminergic systems through the descending striatonigral pathway from the striatum. A second interaction is at the level of the GABAminergic projection from the accumbens to the globus pallidus. Both the striatum and the accumbens, of course, receive dopaminergic innervation.

Anticonvulsants. These drugs are sometimes thought to act via changes in the brain levels of GABA and are included here for this reason.

Elliott *et al.* (1977) found that the anticonvulsant diphenylhydantoin inhibited both apomorphine and amphetamine-induced circling behaviour in mice with unilateral 6-OHDA nigrostriatal lesions. It failed, however, to affect either apomorphine stereotypy or locomotor activation. When injected into one striatum following peripheral injection of apomorphine it induced contraversive turning, suggesting a dopamine receptor blocking action.

The situation is consequently unclear.

Benzodiazepines. Arnt, Christensen & Scheel-Kruger (1979) reported that the benzodiazepines diazepam, chlordiazepoxide, clonazepam and nitrazepam would potentiate the locomotor activity induced by methylphenidate in mice. A potentiation of some 40–50 % was also seen in the methylphenidate-induced stereotyped gnawing. It was suggested that this effect might be mediated via GABA mechanisms and some support for this was gained by the finding that when benzodiazepines were administered to animals which had received both methylphenidate and the GABA agonist muscimol a much larger potentiation of the stereotypy was seen than after methylphenidate without muscimol.

In contrast to this, Weiner, Goetz, Nausleda & Klawans (1977) found that one benzodiazepine, clonazepam, reduced L-DOPA or amphetamine-induced stereotypy in guinea-pigs. However, it was without effect on apomorphine-induced stereotypy and the authors suggest an inhibitory interaction of the drug with presynaptic dopaminergic mechanisms.

The effects of benzodiazepines, and their possible mediation via GABA systems, remain unclear.

Glycine. Mendez, Finn & Dahl (1976) report that injection of glycine into the substantia nigra of the rat induced contraversive turning away from the side of the injection. Similar injection into the caudate nucleus was ineffective.

Arnt & Scheel-Kruger (1979) found that glycine injection into the substantia nigra also induced contraversive turning and showed that this effect could be blocked by the glycine antagonist strychnine, but not by antagonists of GABA (picrotoxin or bicuculline). Such turning, like that induced by GABA, was *not* blocked by haloperidol, suggesting a route independent of the nigrostriatal system. The current assumption is that glycine is contained within interneurones in the substantia nigra (Dray *et al.*, 1977), but no further details of the anatomical organisation and interconnections are known.

Substance P. Injection of substance P into the substantia nigra induced contralateral turning in rats if the injection site was in the pars reticulata (James & Starr, 1977), while ipsiversive rotation was seen with sites dorsal or anterior to the reticulata. This effect may be mediated by dopaminergic systems since post-mortem biochemical measurements revealed that contralateral turning was accompanied by a marked increase in the dopamine metabolite HVA on the side of the injection, while ipsiversive response caused a decrease in this metabolite. It seems that substance P may excite neurones in the substantia nigra, probably dopaminergic dendrites in the reticulata, to elicit the contralateral turning. The mode of action of the ipsiversive turning remains unknown.

Other peptides. The octapeptide cholecystokinin enhanced apomorphine-induced cage-climbing (see p. 122) in mice. This effect was only seen with the nonsulphated octapeptide and not with the sulphated form (Kovacs, Szabo, Penke & Telegdy, 1981).

Summary

The evidence for a role of brain dopaminergic systems in motor behaviour can be summarised as follows;

(1) Dopamine agonist drugs (such as apomorphine, piribedil, nomifensine, bromocryptine) and apomorphine derivatives (such as (−)-NPA) or 'prodrugs' (such as di-isobutyrylapomorphine) all induce locomotor activation in low doses and bizarre, stereotyped behaviours at higher doses. These stereotypies consist of repeated motor movements, and in the rat oral stereotypies such as licking or gnawing. At very low doses these dopamine agonists may have

the reverse effect, i.e. sedation, which is believed to be mediated by presynaptic inhibitory receptors on dopaminergic terminals.

(2) Dopaminergic antagonists, such as the neuroleptic drugs, cause a reduction in exploratory and locomotor activity at low doses, and at higher doses a state of complete immobility, termed catalepsy, is seen. Chronic treatment with neuroleptic drugs may, upon withdrawal, reveal a supersensitivity because of increased receptor number. This may show itself as the complication of tardive dyskinesia during the long-term treatment of chronic schizophrenia with drugs.

(3) Local injection of drugs into discrete brain areas highlights the role of the caudate in stereotypy and the role of the accumbens–tubercle in locomotor activity. A second population of receptors in the accumbens–tubercle may also be involved in stereotypy.

(4) The neurochemical basis of the action of amphetamine in causing stereotypy and locomotor activation has been investigated. The use of *d*- and *l*-isomers failed to be conclusive, but 6-OHDA lesion studies showed an involvement of the ascending dopaminergic systems and more localised 6-OHDA lesion implicated the caudate in stereotypy and the accumbens–tubercle in the locomotor activation of the drug. Other dopaminergic areas such as the frontal cortex seem to have an effect opposing that of the subcortical dopaminergic systems, being inhibitory to the elicitation of stereotypy and locomotor activation. Other nondopaminergic areas, such as the superior colliculus and the ventromedial thalamus, may be involved in the nondopaminergic output side of the extrapyramidal motor system. A modulatory role for noradrenaline has been found in the action of amphetamine in causing stereotypy, although the evidence is contradictory as to which specific noradrenaline system is involved. It is clearer in the case of rotational behaviour (see p. 113), where a noradrenergic system arising in the LC and running in the ventral (*not* the dorsal) bundle seems to modulate dopaminergic function.

(5) Chronic administration of amphetamine results in a potentiation of some of the aspects of stereotypy (sniffing) whereas other aspects (licking) may fail to show potentiation. The neurochemical basis of these effects is unknown.

(6) Rotation is induced by unilateral 6-OHDA lesion to the nigrostriatal dopaminergic system. Dopamine receptor agonists, such as apomorphine, cause contraversive turning while presynaptic dopamine releasing agents, such as amphetamine, cause ipsiver-

sive turning. Rotation may also be seen in normal, intact rats but it is much weaker and it is questionable whether it correlates with any underlying asymmetry of the nigrostriatal dopaminergic systems.

(7) The accumbens–tubercle dopaminergic system, while not *causing* asymmetrical rotation, does modulate its vigour. Thus, bilateral 6-OHDA lesion of the accumbens–tubercle can abolish rotational behaviour previously established by unilateral 6-OHDA caudate lesion.

(8) Considerable interaction with other neurotransmitter systems has been demonstrated for the dopamine involvement in motor behaviour. Thus, cholinergic, opiate, ethanol, hallucinogen, serotonergic, GABAminergic, glycinergic, anticonvulsant, benzodiazepine and substance P systems all interact in one way or another with dopaminergic systems in the control of extrapyramidal motor output.

References

Ahlenius, S. (1974). Reversal by L-DOPA of the suppression of locomotor activity induced by inhibition of tyrosine-hydroxylase and DA-beta-hydroxylase in mice. *Brain Research*, **69**, 57–65.

Ahlenius, S., Anden, N. E. & Engel, J. (1973). Restoration of locomotor activity in mice by low L-DOPA doses after suppression by alpha-methyltyrosine but not by reserpine. *Brain Research*, **62**, 189–99.

Anderson, H., Braestrup, C. & Randrup, A. (1975). Apomorphine-induced stereotyped biting in the tortoise in relation to dopaminergic mechanisms. *Brain, Behaviour and Evolution*, **11**, 365–73.

Arnfred, T. & Randrup, A. (1968). Cholinergic mechanisms in brain inhibiting amphetamine-induced stereotyped behaviour. *Acta pharmacologica et toxicologica*, **26**, 384–94.

Arnt, J., Christensen, A.V. & Scheel-Kruger, J. (1979). Benzodiazepenes potentiate GABA–dopamine dependent stereotyped gnawing in mice. *Journal of Pharmacology and Pharmacy*, **31**, 56–8.

Arnt, J. & Scheel-Kruger, J. (1979). GABAergic and glycinergic mechanisms within the substantia nigra: pharmacological specificity of dopamine-independent contralateral turning behaviour and interactions with other neurotransmitters. *Psychopharmacology*, **62**, 267–77.

Asher, I. M. & Aghajanian, G. K. (1974). 6-Hydroxydopamine lesions of olfactory tubercles and caudate nuclei: effect on amphetamine stereotyped behaviour in rats. *Brain Research*, **82**, 1–12.

Audigier, Y., Virion, A. & Schwartz, J. C. (1976). Stimulation of cerebral histamine H_2 receptors by clonidine. *Nature*, **262**, 307–8.

Ayhan, I. H. & Randrup, A. (1973) Behavioural and pharmacological studies on

morphine-induced excitation of rats. Possible relationships to brain catecholamines. *Psychopharmacologia*, **29**, 317–28.
Baldessarini, R. J., Kula, N. S. & Walton, K. G. (1977). Behavioural effects of apomorphine and diisobutyrylapomorphine in the mouse. *Psychopharmacology*, **53**, 45–53.
Braestrup, C. (1977). Changes in drug-induced stereotyped behaviour after 6-OHDA lesions in noradrenergic neurons. *Psychopharmacology*, **51**, 199–204.
Braestrup, C., Anderson, H. & Randrup, A. (1975). The monoamine oxidase inhibitor deprenyl potentiates phenylethylamine behavior in rats without inhibition of catecholamine metabolite formation. *European Journal of Pharmacology*, **34**, 181–7.
Broitman, S. T. & Donoso, A. O. (1971). Locomotor activity and regional brain noradrenaline levels in rats treated with prenylamine. *Experientia*, **27**, 1308–9.
Buckett, W. R. & Shaw, J. S. (1975). Dopaminergic drugs antagonize the psychotomimetic effects of partial-agonist analgesics. *Psychopharmacologia*, **42**, 293–7.
Butcher, L. L., Engel, J. & Fuxe, K. (1972). Behavioural, biochemical and histological analyses of the central effects of monoamine precursors after peripheral decarboxylase inhibition. *Brain Research*, **41**, 387–411.
Carlsson, A., Engel, J., Strombom, U., Svensson, T. H. & Waldeck, B. (1974). Suppression by dopamine-agonists of the ethanol induced stimulation of locomotor activity and brain dopamine synthesis. *Naunyn-Schmiedeberg's Archives of Pharmacology*, **283**, 117–22.
Carlsson, A., Engel, J. & Svensson, T. H. (1972). Inhibition of ethanol-induced excitation in mice and rats by alpha-methyl-para-tyrosine. *Psychopharmacologia*, **26**, 307–12.
Carlsson, A. & Lindquist, M. (1962). *In vivo* decarboxylation of alpha-methyl-DOPA and alpha-methyl-metatyrosine. *Acta physiologica scandinavica*, **54**, 87–94.
Carter, C. J. & Pycock, C. J. (1978). Differential effects of central serotonin manipulation on hyperactive and stereotyped behavior. *Life Sciences*, **23**, 953–60.
Carter, C. J. & Pycock, C. J. (1979). The effects of 5,7-dihydroxytryptamine lesions of extrapyramidal and mesolimbic sites on spontaneous motor behaviour and amphetamine-induced stereotypy. *Naunyn-Schmiedeberg's Archives of Pharmacology*, **308**, 51–4.
Carter, C. J. & Pycock, C. J. (1980*a*). Behavioural and biochemical effects of dopamine and noradrenaline depletion within the medial prefrontal cortex of the rat. *Brain Research*, **192**, 163–76.
Carter, C. J. & Pycock, C. J. (1980*b*). 5,7-Dihydroxtryptamine lesions of the amygdala reduce amphetamine- and apomorphine-induced stereotyped behaviour in the rat. *Naunyn-Schmiedeberg's Archives of Pharmacology*, **312**, 235–8.
Cedarbaum, J. M. & Aghajanian, G. K. (1976). Noradrenergic neurons of the locus coeruleus: inhibition by epinephrine and activation by the alpha-antagonist piperoxan. *Brain Research*, **112**, 413–19.
Charness, M. E., Amit, Z. & Taylor, M. (1975). Morphine-induced stereotypic behaviour in rat. *Behavioral Biology*, **13**, 71–80.
Cheng, H. C., Bhatnagar, R. K. & Long, J. P. (1975). Dopaminergic nature of amphetamine-induced pecking in pigeon. *European Journal of Pharmacology*, **33**, 319–24.
Cheng, H. C. & Long, J. P. (1974). Dopaminergic nature of apomorphine induced pecking in pigeons. *European Journal of Pharmacology*, **26**, 313–20.
Collingridge, G. L., James, T. A. & McLeod, N. K. (1979). Neurochemical and electrophysiological evidence for a projection from the locus coeruleus to the substantia nigra. *Journal of Physiology*, **290**, 44P.

Cools, A. R., Janssen, H. J. & Broekkamp, C. L. E. (1974). The differential role of the caudate nucleus and the linear raphe nucleus in the initiation and the maintenance of morphine-induced behaviour in cats. *Archives internationales de pharmacodynamie et thérapie*, **210,** 163–74.

Cools, A. R. & van Rossum, J. M. (1976). Excitation-mediating and inhibition-mediating dopamine receptors: a new concept towards a better understanding of electrophysiological, biochemical, pharmacological, functional and clinical data. *Psychopharmacology*, **45,** 243–54.

Corsini, G. U., Del Zompo, M., Manconi, S., Piccardi, M. P., Onali, P. L. & Mangoni, A. (1977). Evidence for dopamine receptors in the human brain mediating sedation and sleep. *Life Sciences*, **20,** 1613–18.

Costall, B., Hui, S-C. G. & Naylor, R. J. (1980). Denervation in the dopaminergic mesolimbic system: functional changes followed using (−)*N-n*-propylnorapomorphine depend on the basal activity level of rats. *Neuropharmacology*, **19,** 1039–48.

Costall, B., Kelly, D. M. & Naylor, R. J. (1975*a*). Nomifensine: a potent dopaminergic agonist of antiparkinson potential. *Psychopharmacologia*, **41,** 153–64.

Costall, B. & Naylor, R. J. (1973). The site and mode of action of ET495 for the mediation of stereotyped behaviour in the rat. *Naunyn-Schmiedeberg's Archives of Pharmacology*, **278,** 117–20.

Costall, B. & Naylor, R. J. (1974*a*). Extrapyramidal and mesolimbic involvement with the stereotypic activity of D- and L-amphetamine. *European Journal of Pharmacology*, **25,** 121–9.

Costall, B. & Naylor, R. J. (1974*b*). Stereotyped and circling behaviour induced by dopaminergic agonists after lesions of the midbrain raphe nuclei. *European Journal of Pharmacology*, **29,** 206–22.

Costall, B. & Naylor, R. J. (1975). A comparison of circling models for the detection of antiparkinson activity. *Psychopharmacologia*, **41,** 57–64.

Costall, B. & Naylor, R. J. (1976). Dissociation of stereotyped biting responses and oro-bucco-lingual dyskinesias. *European Journal of Pharmacology*, **36,** 423–9.

Costall, B., Naylor, R. J., Cannon, J. G. & Lee, T. (1977). Differential activation by some 2-aminotetralin derivatives of the receptor mechanisms in the nucleus accumbens of rats which mediate hyperactivity and stereotyped biting. *European Journal of Pharmacology*, **41,** 307–19.

Costall, B., Naylor, R. J., Marsden, C. D. & Pycock, C. J. (1976*a*). Circling behaviour produced by asymmetric medial raphe nuclei lesions in rats. *Journal of Pharmacology and Pharmacy*, **28,** 248–9.

Costall, B., Naylor, R. J. & Neumeyer, J. L. (1975*b*). Differences in the nature of the stereotyped behaviour induced by apomorphine derivatives in the rat and in their actions in extrapyramidal and mesolimbic brain areas. *European Journal of Pharmacology*, **31,** 1–16.

Costall, B., Naylor, R. J. & Neumeyer, J. L. (1975*c*). Dissociation by the apomorphine derivatives of the stereotypic and hyperactivity responses resulting from injections into the nucleus accumbens septi. *Journal of Pharmacy and Pharmacology*, **27,** 875–7.

Costall, B., Naylor, R. J. & Olley, J. E. (1972). Stereotypic and anticataleptic activities of amphetamine after intracerebral injections. *European Journal of Pharmacology*, **18,** 83–94.

Costall, B., Naylor, R. J. & Pinder, R. M. (1975*d*). Dyskinetic phenomena caused by the intrastriatal injection of phenylethylamine, phenylpiperazine, tetrahydroisoquinoline and tetrahydronaphthalese derivatives in the guinea pig. *European Journal of Pharmacology*, **31**, 94–9.

Costall, B., Naylor, R. J. & Pycock, C. (1975*e*). The 6-hydroxydopamine rotational model for the detection of dopamine agonist activity: reliability of effect from different locations of 6-hydroxydopamine. *Journal of Pharmacology and Pharmacy*, **27**, 943–6.

Costall, B., Naylor, R. J. & Pycock, C. (1976*b*). Non-specific supersensitivity of striatal dopamine receptors after 6-hydroxydopamine lesion of the nigrostriatal pathway. *European Journal of Pharmacology*, **35**, 276–83.

Coyle, J. T. & Snyder, S. H. (1970). Catecholamine uptake by synaptosomes in homogenates of rat brain: stereospecificity in different areas. *Journal of Pharmacology and Experimental Therapeutics*, **170**, 221–31.

Creese, I. (1974). Behavioural evidence of dopamine receptor stimulation by piribedil (ET 495) and its metabolite S584. *European Journal of Pharmacology*, **28**, 55–8.

Creese, I. & Iversen, S. D. (1973). Blockage of amphetamine induced motor stimulation and stereotypy in the adult rat following neonatal treatment with 6-hydroxydopamine. *Brain Research*, **55**, 369–82.

Creese, I. & Iversen, S. D. (1974). The role of forebrain dopamine systems in amphetamine induced stereotyped behaviour in the rat. *Psychopharmacologia*, **39**, 345–57.

Creese, I. & Iversen, S. D. (1975). The pharmacological and anatomical substrates of the amphetamine response in the rat. *Brain Research*, **83**, 419–36.

Creese, I. & Snyder, S.H. (1979). Nigrostriatal lesions enhance ^{3}H-spiroperidol binding. *European Journal of Pharmacology*, **56**, 277–81.

Dadkar, N. K., Dohadwella, A. N. & Bhattacharya, B. K. (1976*a*). The involvement of serotonergic and noradrenergic systems in the compulsive gnawing in mice induced by imipramine and apomorphine. *Journal of Pharmacy and Pharmacology*, **28**, 68–9.

Dadkar, N. K., Dohadwella, A. N. & Bhattacharya, B. K. (1976*b*) Influence of pheniramine and chlorpheniramine on apomorphine induced compulsive gnawing in mice. *Psychopharmacology*, **48**, 7–10.

Dadkar, N. K., Dohadwella, A. N. & Bhattacharya, B. K. (1977). Possible role of dopamine in central effects of cocaine as measured by apomorphine gnawing test in mice. *Psychopharmacology*, **52**, 115–17.

Daly, J. W., Bruns, R. F. & Snyder, S. H. (1981). Adenosine receptors in the central nervous system: relationship to the central actions of methylxanthines. *Life Sciences*, **28**, 2083–99.

Dandiya, P. C. & Kulkarni, S. K. (1974). A comparative study of *d*- and *l*-amphetamine on the open field performance of rats. *Psychopharmacologia*, **39**, 67–70.

Deshpande, V., Sharma, M., Kherdikar, P. & Grewal, R. (1961). Some observations on pecking in pigeons. *British Journal of Pharmacology and Chemotherapeutics*, **17**, 7–11.

Dhawan, B., Saxena, N. & Gupta, G. (1961). Antagonism of apomorphine-induced pecking in pigeons. *British Journal of Pharmacology and Chemotherapy*, **16**, 137–45.

Di Chiara, G., Porceddu, M. L., Vargiu, L., Argiolas, A. & Gessa, G. L. (1976). Evidence for dopamine receptor mediating sedation in the mouse brain. *Nature*, **264**, 564–7.

Dingell, J. V., Owens, M. L., Norvich, M. R. & Sulser, F. (1967). On the role of norepinephrine biosynthesis in the central action of amphetamine. *Life Sciences*, **6**, 1155–62.

Dixon, A. K. (1968). Evidence of catecholamine mediation in the 'aberrant' behaviour induced by lysergic acid diethylamide (LSD) in the rat. *Experientia*, **24**, 743–7.

Dominic, J. A. & Moore, K. E. (1971). Depression of behaviour and the brain content of alpha-methylnorepinephrine and alpha-methyldopamine following the administration of alpha-methyldopa. *Neuropharmacology*, **10,** 33–44.

Donaldson, I. M., Dolphin, A., Jenner, P., Marsden, C. D. & Pycock, C. (1976*a*). Contraversive circling behaviour produced by unilateral electrolytic lesions of the ventral noradrenergic bundle mimicking the changes seen with unilateral electrolytic lesions of the locus coeruleus. *Journal of Pharmacy and Pharmacology*, **28,** 329–31.

Donaldson, I. M., Dolphin, A., Jenner, P., Marsden, C. D. & Pycock, C. (1976*b*). The involvement of noradrenaline in motor activity as shown by rotational behaviour after unilateral lesions of the locus coeruleus. *Brain*, **99,** 427–46.

Donaldson, I. M., Dolphin, A., Jenner, P., Pycock, C. & Marsden, C. D. (1977). Rotational behaviour produced in rats by unilateral electrolytic lesions of the ascending noradrenergic bundles. *Brain Research*, **138,** 487–509.

Dourish, C. T. & Cooper, S. J. (1981). Effects of acute and chronic administration of low doses of a dopamine agonist on drinking and locomotor activity in the rat. *Psychopharmacology*, **72,** 197–202.

Dray, A., Fowler, L. J., Oakley, N. R., Simmonds, M. A. & Tanner, T. (1977). Regulation of nigro-striatal dopaminergic neurotransmission in the rat. *Neuropharmacology*, **16,** 511–18.

Eichler, A. J., Antelman, S. M. & Black, C. A. (1980). Amphetamine stereotypy is not a homogeneous phenomenon: sniffing and licking show distinct profiles of sensitization and tolerance. *Psychopharmacology*, **68,** 287–90.

Ellingwood, E. H. & Kilbey, M. M. (1976). Stimulant abuse in man: the use of animal models to assess and predict behavioural toxicity. In *Predicting dependence liability of stimulant and depressant drugs*, ed. T. Thompson & K. Unna, pp. 114–19. University Park Press: Baltimore.

Elliott, P. N., Jenner, P., Chadwick, D., Reynolds, E. & Marsden, C. D. (1977). The effect of diphenylhydantoin on central catecholamine containing neuronal systems. *Journal of Pharmacology and Pharmacy*, **29,** 41–3.

Ellison, G. D. & Bresler, D. E. (1974). Tests of emotional behaviour in rats following depletion of norepinephrine, of serotonin, or of both. *Psychopharmacologia*, **34,** 275–88.

Evetts, K. D., Uretsky, N. J., Iversen, L. L. & Iversen, S. D. (1970). Effects of 6-hydroxydopamine on CNS catecholamines, spontaneous motor activity and amphetamine induced hyperactivity in rats. *Nature*, **225,** 961–2.

Fibiger, H. C., Lonsbury, B. & Cooper, H. P. (1972). Early behavioural effects of intraventricular administration of 6-hydroxydopamine in rat. *Nature*, **236,** 209–11.

Finnegan, K. T., Kanner, M. I. & Meltzer, H. Y. (1976). Phencyclidine-induced rotational behaviour in rats with nigrostriatal lesions and its modulation by dopaminergic and cholinergic agents. *Pharmacology, Biochemistry & Behavior*, **5,** 651–60.

Fog, R. (1970). Behavioural effects in rats of morphine and amphetamine and of a combination of the two drugs. *Psychopharmacologia*, **16,** 305–12.

Fog, R., Randrup, A. & Pakkenberg, H. (1971). Intrastriatal injection of quaternary butyrophenones and oxypertine – neuroleptic effect in rats. *Psychopharmacologia*, **19,** 224–34.

Fuxe, K., Fredholm, B. B., Ogren, S. O., Agnati, L. F., Hokfeldt, T. & Gustafsson, J. A. (1978*a*). Pharmacological and biochemical evidence for the dopamine agonistic effect of bromocryptine. *Acta endocrinologica (Kbh)*, **88,** *suppl.* 216, 27–56.

Fuxe, K., Fredholme, B. B., Ogren, S. O., Agnati, L. F., Hokfeldt, T. & Gustafsson, J. A.

(1978*b*). Ergot drugs and central monoaminergic mechanisms: a histochemical, biochemical and behavioural analysis. *Federation Proceedings*, **37**, 2181–91.

Garver, D. L., Schlemm, F., Mass, J. W. & Davis, J. M. (1975). Schizophreniform behavioural psychosis mediated by dopamine. *American Journal of Psychiatry*, **132**, 33–9.

Georgiev, V. P., Markovska, V. L. & Petkova, B. P. (1978). Apomorphine stereotypies and transmitter mechanisms in the striatum. I. Changes in the apomorphine stereotypies caused by drugs acting on the GABA-ergic, dopaminergic and cholinergic transmission. *Acta physiologica et pharmacologica bulgarica*, **4**, 56–64.

Geyer, M. A., Segal, D. S. & Mandell, A. J. (1972). Effect of intraventricular infusion of dopamine and norepinephrine on motor activity. *Physiology and Behavior*, **8**, 653–8.

Giambalvo, C. T. & Snodgrass, S. R. (1978). Effect of *p*-chloroamphetamine and 5,7-dihydroxytryptamine on rotation and dopamine turnover. *Brain Research*, **149**, 453–67.

Glick, S. D. & Cox, R. D. (1978). Nocturnal rotation in normal rats: correlation with amphetamine-induced rotation and effects of nigrostriatal lesions. *Brain Research*, **150**, 149–61.

Glick, S. D. & Greenstein, S. (1973). Possible modulating influence of frontal cortex on nigro-striatal function. *British Journal of Pharmacology*, **49**, 316–21.

Glick, S. D., Jerussi, T. P., Waters, D. H. & Green, J. P. (1974). Amphetamine-induced changes in striatal dopamine and acetyl-choline levels and relationship to rotation (circling behaviour) in rats. *Biochemical Pharmacology*, **23**, 3223–5.

Glick, S. D. & Morihisa, J. M. (1976). Changes in sensitivity of morphine-induced circling behaviour after chronic treatment and persistence after withdrawal in rats. *Nature*, **260**, 159–61.

Gordon, J. H. & Shellenberger, K. (1974). Regional catecholamine content in the rat brain: sex differences and correlation with motor activity. *Neuropharmacology*, **13**, 129–37.

Grabowska, M. (1975). Influence of dopamine-like compounds on stereotypy and locomotor activity in atropinized rats. *Archivum immunologiae et therapiae experimentalis*, **23**, 753–61.

Grabowska-Anden, M. (1977). Modification of the amphetamine-induced stereotypy in rats following inhibition of noradrenaline release by FLA 136. *Journal of Pharmacology and Pharmacy*, **29**, 566–7.

Griffiths, P. J., Littleton, J. M. & Ortiz, A. (1974). Changes in monoamine concentrations in mouse brain associated with ethanol dependence and withdrawal. *British Journal of Pharmacology*, **50**, 489–98.

Gudelsky, G. A., Thornburg, J. E. & Moore, K. E. (1975). Blockade of alpha-methyltyrosine-induced supersensitivity to apomorphine by chronic administration of L-DOPA. *Life Sciences*, **16**, 1331–8.

Heikkila, R. E., Cabbat, F. S., Manzino, L. & Duvoisin, R. C. (1979). Rotational behaviour induced by cocaine analogs in rats with unilateral 6-hydroxydopamine lesions of the substantia nigra: dependence upon dopamine uptake inhibition. *Journal of Pharmacology and Experimental Therapeutics*, **211**, 189–94.

Heikkila, R. E., Shapiro, B. S. & Duvoisin, R. C. (1981). The relationship between loss of dopamine nerve terminals, striatal [^{3}H]spiroperidol binding and rotational behaviour in unilaterally 6-hydroxydopamine-lesioned rats. *Brain Research*, **211**, 285–92.

Herman, Z. S. (1970). The effects of noradrenaline on rats behaviour. *Psychopharmacologia*, **16**, 369–74.

Holtzman, S. G. & Jewett, R. E. (1973). Stimulation of behaviour in the rat by cyclazocine:

effects of naloxone. *Journal of Pharmacology and Experimental Therapeutics*, **187,** 380–90.

Hurman, W. & Almirante, L. (1966). Central nervous system effects of four beta-adrenergic receptor blocking drugs. *Journal of Pharmacy and Pharmacology*, **18,** 317–18.

Iversen, S. D. (1971). The effect of surgical lesions to frontal cortex and substantia nigra on amphetamine responses in rats. *Brain Research*, **31,** 295–311.

Iversen, S. D. & Creese, I. (1975). Behavioural correlates of dopaminergic supersensitivity. In *Advances in neurology*, vol. 9, ed. D. B. Calne, T. N. Chase & A. Barbeau, pp. 81–92. Raven Press: New York.

Iwamoto, E. T., Loh, H. H. & Way, E. L. (1976). Circling behaviour after narcotic drugs and during naloxone-precipitated abstinence in rats with unilateral nigral lesions. *Journal of Pharmacology and Experimental Therapeutics*, **197,** 503–16.

Jacquet, Y. F., Carol, M. & Russell, I. S. (1976). Morphine-induced rotation in naive, nonlesioned rats. *Science*, **192,** 261–3.

Jalfre, M. & Haefely, W. (1972). Effect of some centrally acting agents in rats after intraventricular injections of 6-hydroxydopamine. In *6-Hydroxydopamine and catecholamine neurons*, ed. T. Malmfors & H. Thoenen, pp. 333–46. North-Holland: Amsterdam.

James, T. A. & Starr, M. S. (1977). Behavioural and biochemical effects of substance P injected into the substantia nigra of the rat. *Journal of Pharmacology and Pharmacy*, **29,** 181–2.

Janssen, P. A. J. (1970). Chemical and pharmacological classification of neuroleptics. *Modern Problems in Pharmacopsychiatry*, **5,** 33–44.

Janssen, P. A. J., Niemegeers, C. J. E., Schellekens, K. H. L., Lenaerts, F. M. & Wauquier, A. (1975). Clopimozide (R 29,764), a new highly potent and orally long-lasting neuroleptic of the diphenylbutylpiperidine series. *Arzneimittel-Forschung*, **25,** 1287–94.

Jerlicz, M., Kostowski, W., Bidinski, A. & Hauptmann, M. (1978). Effects of lesions in the ventral noradrenergic bundle on behavior and response to psychotropic drugs. *Pharmacology, Biochemistry & Behavior*, **9,** 721–4.

Jerussi, T. P. & Glick, S. D. (1974). Amphetamine-induced rotation in rats without lesions. *Neuropharmacology*, **13,** 283–6.

Jerussi, T. P. & Glick, S. D. (1975). Apomorphine-induced rotation in normal rats and interaction with unilateral caudate lesions. *Psychopharmacologia*, **40,** 329–34.

Jerussi, T. P., Glick, S. D. & Johnson, C. D. (1977). Reciprocity of pre- and post-synaptic mechanisms involved in rotation as revealed by dopamine metabolism and adenylate cyclase stimulation. *Brain Research*, **129,** 385–8.

Jones, D. L. & Mogenson, G. J. (1980). Nucleus accumbens to globus pallidus GABA projection subserving ambulatory activity. *American Journal of Physiology*, **238,** R65–9.

Jones, D. L., Mogenson, G. J. & Wu, M. (1981). Injections of dopaminergic, cholinergic, serotonergic and GABAergic drugs into the nucleus accumbens: effects on locomotor activity of the rat. *Neuropharmacology*, **20,** 29–38.

Kaakkola, S. (1981). Effect of nicotinic and muscarinic drugs on amphetamine- and apomorphine-induced circling behaviour in rats. *Acta pharmacologica et toxicologica*, **48,** 162–7.

Kebabian, J. W., Petzold, G. L. & Greengard, P. (1972). Dopamine-sensitive adenylate cyclase in caudate nucleus of rat brain and its similarity to the dopamine receptor. *Proceedings of the National Academy of Sciences, USA*, **69,** 2145–9.

Kelly, P. H. & Moore, K. E. (1976). Mesolimbic dopaminergic neurones in the rotational model of nigrostriatal function. *Nature*, **263**, 695–6.
Kelly, P.H., Seviour, P. W. & Iversen, S. D. (1975). Amphetamine and apomorphine responses in the rat following 6-OHDA lesions of the nucleus accumbens septi and corpus striatum. *Brain Research*, **94**, 507–22.
Kleinrok, Z. & Zebrowska-Lupina, I. (1971). Central action of phentolamine administered intraventricularly in the rat. *Psychopharmacologia*, **20**, 348–54.
Kokkinidis, L. & Anisman, H. (1979). Circling behavior following systemic *d*-amphetamine administration: potential noradrenergic and dopaminergic involvement. *Psychopharmacology*, **64**, 45–54.
Koob, G. F., Kelley, A. F. & Mason, S. T. (1978). Locus coeruleus lesions: learning and extinction. *Physiology and Behavior*, **20**, 709–16.
Kostowski, W., Gajewska, S., Bidinski, A. & Hauptmann, M. (1976). Papaverine, drug-induced stereotypy and catalepsy and biogenic amines in the brain of the rat. *Pharmacology, Biochemistry & Behavior*, **5**, 15–17.
Kostowski, W., Jerlicz, M., Bidinski, A. & Hauptmann, M. (1977). Behavioural effects of neuroleptics, apomorphine and amphetamine after bilateral lesion of the locus coeruleus. *Pharmacology, Biochemistry & Behavior*, **7**, 289–93.
Kostowski, W., Jerlicz, M., Bidinski, A. & Hauptmann, M. (1978). Evidence for existence of two opposite noradrenergic brain systems controlling behavior. *Psychopharmacology*, **59**, 311–12.
Kovacs, G. L., Szabo, G., Penke, B. & Telegdy, G. (1981). Effects of cholecystokinin octapeptide on striatal dopamine metabolism and on apomorphine-induced stereotyped cage-climbing in mice. *European Journal of Pharmacology*, **69**, 313–19.
Kraemer, G. W., Breese, G. R., Prange, A. J., Moran, E. C., Lewis, J. K., Kemnitz, J. W., Bushnell, P. J., Howard, J. L. & McKinney, W. T. (1981). Use of 6-hydroxydopamine to deplete brain catecholamines in the rhesus monkey: effects on urinary catecholamine metabolites and behavior. *Psychopharmacology*, **73**, 1–11.
Ksiazek, A. & Kleinrok, Z. (1974*a*). Central action of drugs affecting beta-adrenergic receptor. *Polish Journal of Pharmacology and Pharmacy*, **26**, 297–304.
Ksiazek, A. & Kleinrok, Z. (1974*b*). The central action of beta-adrenergic receptor blocking agents. I. The central action of intraventricularly administered isoprenaline in the rat. *Polish Journal of Pharmacology and Pharmacy*, **26**, 287–95.
Lal, H., Colpaert, F. C. & Laduron, P. (1975). Narcotic withdrawal-like mouse jumping produced by amphetamine and L-DOPA. *European Journal of Pharmacology*, **30**, 113–16.
Langer, S. Z. (1980). Presynaptic receptors and modulation of neuro-transmission: pharmacological implications and therapeutic relevance. *Trends in Neuroscience*, **3**, 110–12.
Lapin, I. P., Oxenkrug, G. F. & Azbekyan, S. G. (1972). Involvement of brain serotonin in the stimulant action of amphetamine and of cholinolytics. *Archives internationales de pharmacodynamie et thérapie*, **197**, 350–61.
Laverty, R. & Taylor, K. M. (1969). Behavioural and biochemical effects of 2-(2,6-dichlorophenylamino)-2-imidazoline hydrochloride (St 155) on the central nervous system. *British Journal of Pharmacology*, **35**, 253–64.
Laverty, R. & Taylor, K. M. (1970). Effects of intraventricular 2,4,5-trihydroxyphenylethylamine (6-hydroxydopamine) on rat behaviour and brain catecholamine metabolism. *British Journal of Pharmacology*, **40**, 836–46.

Leszkovszky, G. & Tardos, L. (1965). Some effects of propranolol on the central nervous system. *Journal of Pharmacy and Pharmacology*, **17**, 518–20.
Maj, J., Mogilnicka, E. & Palider, W. (1975). Serotonergic mechanism of clonidine and apomorphine interaction. *Polish Journal of Pharmacology and Pharmacy*, **27**, 27–35.
Margolin, D. I. & Moon, B. H. (1979). Naloxone blockade of apomorphine-induced stereotyped behaviour: interaction of endogenous opiates with dopamine. *Journal of the Neurological Sciences*, **43**, 13–17.
Marshall, J. F. & Ungerstedt, U. (1977). Supersensitivity to apomorphine following destruction of the ascending dopamine neurons: quantification using the rotational model. *European Journal of Pharmacology*, **41**, 361–7.
Martin, G. E., Papp, N. L. & Bacino, C. B. (1978). Contralateral turning evoked by the intranigral microinjection of muscimol and other GABA agonists. *Brain Research*, **155**, 297–312.
Mason, S. T. (1978). Pilocarpine: noradrenergic mechanism of a cholinergic drug. *Neuropharmacology*, **17**, 1015–21.
Mason, S. T. (1979). Central noradrenergic–cholinergic interaction and locomotor behaviour. *European Journal of Pharmacology*, **56**, 131–7.
Mason, S. T., Corcoran, M. E. & Fibiger, H. C. (1979). Noradrenergic processes involved in the locomotor effects of ethanol. *European Journal of Pharmacology*, **54**, 383–7.
Mason, S. T. & Fibiger, H. C. (1979*a*). Regional topography within noradrenergic locus coeruleus as revealed by retrograde transport of horse-radish peroxidase. *Journal of Comparative Neurology*, **187**, 703–24.
Mason, S. T. & Fibiger, H. C. (1979*b*). Noradrenaline–acetylcholine interactions in brain: possible behavioural function in locomotor activity. *Nature*, **277**, 396–7.
Mason, S. T. & Fibiger, H. C. (1979*c*). Noradrenaline–acetylcholine interactions in brain: behavioural functions in locomotor activity. *Neuroscience*, **4**, 517–26.
Mason, S. T. & Iversen, S. D. (1977). Behavioural basis of the dorsal bundle extinction effect. *Pharmacology, Biochemistry & Behavior*, **7**, 373–9.
Mason, S. T., Roberts, D. C. S. & Fibiger, H. C. (1978). Noradrenergic influences on catalepsy. *Psychopharmacology*, **60**, 53–7.
Mendez, J. S., Finn, B. W. & Dahl, K. E. (1976). Rotatory behaviour induced by glycine injected into the substantia nigra of the rat. *Experimental Neurology*, **50**, 174–9.
Menon, M. K., Clark, W. G. & Neumeyer, J. L. (1978). Comparison of the dopaminergic effects of apomorphine and (−)-*N*,*n*-propyl-norapomorphine. *European Journal of Pharmacology*, **52**, 1–9.
Miller, R. J. & Hiley, R. (1975). Antimuscarinic actions of neuroleptic drugs. *Advances in Neurology*, **9**, 141–54.
Miller, R. J. & Iversen, L. L. (1974). Stimulation of a dopamine-sensitive adenylate cyclase in homogenates of rat brain by a metabolite of ET 495. *Naunyn–Schmiedeberg's Archives of Pharmacology*, **282**, 213–8.
Miller, R. J. & Sahakian, B. (1974). Differential effects of neuroleptic drugs on amphetamine-induced stimulation of locomotor activity in 11 day-old and adult rats. *Brain Research*, **81**, 387–92.
Milson, J. A. & Pycock, C. J. (1976). Effects of drugs acting on cerebral 5-hydroxytryptamine mechanisms on dopamine-dependent turning behaviour in mice. *British Journal of Pharmacology*, **56**, 77–85.
Mogilnicka, E. & Braestrup, C. (1976). Noradrenergic influences on the stereotyped

behavior induced by amphetamine, phenethylamine and apomorphine. *Journal of Pharmacology and Pharmacy*, **28,** 253–5.

Moon, B. H., Feigenbaum, J. J., Carson, P. E. & Klawans, H. L. (1980). The role of dopaminergic mechanisms in naloxone-induced inhibition of apomorphine-induced stereotyped behaviour. *European Journal of Pharmacology*, **61,** 71–8.

Morihisa, J. M. & Glick, S. K. (1977). Morphine-induced rotation (circling behaviour) in rats and mice: species differences, persistence of withdrawal-induced rotation and antagonism by naloxone. *Brain Research*, **123,** 180–7.

Murphy, D. L. (1973). Mental effects of L-dopa. *Annual Review of Medicine*, **24,** 209–16.

Nilakantan, B. & Randrup, A. (1968). Phylogenetic approach to the study of brain mechanisms involved in the action of amphetamine and other drugs. In *The present status of psychotropic drugs*, ed. C. Carletti & J. M. Bove, pp. 262–5. International congress series, No. 180. Excerpta Medica: Amsterdam.

Nomura, Y. & Segawa, T. (1979). The effect of alpha-adrenoreceptor antagonists and metiamide on clonidine-induced locomotor stimulation in the infant rat. *British Journal of Pharmacology*, **66,** 531–5.

Oberlander, C., Euvrard, C., Dumont, C. & Boissier, J. R. (1979). Circling behaviour induced by dopamine releasers and/or uptake inhibitors during degeneration of the nigrostriatal pathway. *European Journal of Pharmacology*, **60,** 163–70.

Osuide, G. & Adejoh, P. (1973). Effect of apomorphine and its interaction with other drugs in the domestic fowl. *European Journal of Pharmacology*, **23,** 56–66.

Pappas, B. A., Peters, D. A., Saari, M., Sobrian, S. K. & Minch, E. (1974). Neonatal 6-hydroxydopamine sympathectomy in normotensive and spontaneously hypertensive rat. *Pharmacology, Biochemistry & Behavior*, **2,** 381–6.

Pedersen, V. (1967). Potentiation of apomorphine effect (compulsive gnawing behaviour) in mice. *Acta pharmacologica et toxicologia* **25,** *suppl.*, 2, 23–9.

Pijnenburg, A. J. J., Honig, W. M. M. & van Rossum, J. M. (1975*a*). Inhibition of *d*-amphetamine-induced locomotor activity by injection of haloperidol into the nucleus accumbens of the rat. *Psychopharmacologia*, **41,** 87–95.

Pijnenburg, A. J. J., Honig, W. M. M. & van Rossum, J. M. (1975*b*). Effects of antagonists upon locomotor stimulation induced by injection of dopamine and noradrenaline into the nucleus accumbens of nialamide-pretreated rats. *Psychopharmacologia*, **41,** 175–80.

Pijnenburg, A. J. J. & van Rossum, J. M. (1973). Stimulation of locomotor activity following injection of dopamine into the nucleus accumbens. *Journal of Pharmacy and Pharmacology*, **25,** 1003–5.

Pijnenburg, A. J. J., Woodruff, G. N. & van Rossum, J. M. (1973). Ergometrine induced locomotor activity following intracerebral injection into the nucleus accumbens. *Brain Research*, **59,** 289–302.

Pope, S. G., Dean, P. & Redgrave, P. (1979). Action of amphetamine on hyperactivity produced by lesions of the superior colliculus in rats. *Neuroscience Letters, suppl.*, 3, S247.

Price, M. T. C. & Fibiger, H. C. (1974). Apomorphine and amphetamine stereotypy after 6-hydroxydopamine lesions of the substantia nigra. *European Journal of Pharmacology*, **29,** 249–52.

Price, M. T. C. & Fibiger, H. C. (1976). Abolition of nomifensine-induced stereotypy after 6-hydroxydopamine lesions of ascending dopaminergic projections. *Pharmacology, Biochemistry & Behavior*, **5,** 107–9.

Protais, P., Costentin, J. & Schwartz, J. C. (1976). Climbing behaviour induced by apomorphine in mice: a simple test for the study of dopamine receptors in striatum. *Psychopharmacology*, **50,** 1–6.

Przegalinski, E. & Kleinrok, Z. (1972). An analysis of the DOPA-induced locomotion stimulation in mice with inhibited extra-cerebral decarboxylase. *Psychopharmacologia*, **23,** 278–88.

Pycock, C. (1977). Noradrenergic involvement in dopamine-dependent stereotyped and cataleptic responses in the rat. *Naunyn-Schmiedeberg's Archives of Pharmacology*, **298,** 15–22.

Pycock, C., Donaldson, I. M. & Marsden, C. D. (1975*a*). Circling behaviour produced by unilateral lesions in the region of the locus coeruleus in rats. *Brain Research*, **97,** 317–29.

Pycock, C. & Horton, R. (1976). Evidence for an accumbens–pallidal pathway in the rat and its possible gabaminergic control. *Brain Research*, **110,** 629–34.

Pycock, C. J., Jenner, P. G. & Marsden, C. D. (1977). The interaction of clonidine with dopamine-dependent behaviour in rodents. *Naunyn-Schmiedeberg's Archives of Pharmacology*, **297,** 133–41.

Pycock, C., Milson, J. A., Tarsy, D. & Marsden, C. D. (1976). The effects of blocking catecholamine uptake on amphetamine-induced circling behaviour in mice with unilateral destruction of striatal dopaminergic nerve terminals. *Journal of Pharmacology and Pharmacy*, **28,** 530–2.

Pycock, C., Milson, J., Tarsy, D. & Marsden, C. D. (1978). The effect of manipulation of cholinergic mechanisms on turning behaviour in mice with unilateral destruction of the nigro-striatal dopaminergic systems. *Neuropharmacology*, **17,** 175–83.

Pycock, C., Tarsy, D. & Marsden, C. D. (1975*b*). Inhibition of circling behaviour by neuroleptic drugs in mice with unilateral 6-hydroxydopamine lesions of the striatum. *Psychopharmacologia*, **45,** 211–19.

Randrup, A. & Munkvad, I. (1967). Stereotyped activities produced by amphetamine in several animal species and man. *Psychopharmacologia*, **11,** 300–10.

Reavill, C., Jenner, P., Leigh, N. & Marsden, C. D. (1981). The role of nigral projections to the thalamus in drug-induced circling behavior in the rat. *Life Sciences*, **28,** 1457–66.

Redgrave, P., Dean, P., Donohoe, T. P. & Pope, S. G. (1980). Superior colliculus lesions selectively attenuate apomorphine-induced oral stereotypy: a possible role for the nigrotectal pathway. *Brain Research*, **196,** 541–6.

Roberts, D. C. S., Mason, S. T. & Fibiger, H. C. (1978). 6-OHDA lesions to forebrain noradrenaline innervation alters the locomotor response and cataleptic response to morphine. *European Journal of Pharmacology*, **52,** 209–14.

Roberts, D. C. S., Zis, A. P. & Fibiger, H. C. (1975). Ascending catecholamine pathways and amphetamine induced locomotor activity: importance of dopamine and apparent non-involvement of norepinephrine. *Brain Research*, **93,** 441–54.

Roll, S. K. (1970). Intracranial self-stimulation and wakefulness: effect of manipulating ambient brain catecholamines. *Science*, **168,** 1370–2.

Rolls, E. T., Kelly, P. H. & Shaw, S. G. (1974). Noradrenaline, dopamine and brain-stimulation reward. *Pharmacology, Biochemistry and Behavior*, **2,** 735–40.

Sahakian, B. J., Robbins, T. W., Morgan, M. J. & Iversen, S. D. (1975). The effects of psychomotor stimulants on stereotypy and locomotor activity in socially-deprived and control rats. *Brain Research*, **84,** 195–205.

Sastry, B. S. R. & Phillis, J. W. (1977). Evidence that clonidine can activate histamine H_2-receptors in rat cerebral cortex. *Neuropharmacology*, **16,** 223–5.

Scheel-Kruger, J. (1970). Central effects of anticholinergic drugs measured by the apomorphine gnawing test in mice. *Acta pharmacologica et toxicologica*, **28,** 1–16.

Scheel-Kruger, J., Christensen, A. V. & Arnt, J. (1978). Muscimol differentially facilitates stereotypy but antagonises motility induced by dopaminergic drugs: a complex GABA–dopamine interaction. *Life Sciences*, **22,** 75–84.

Scheel-Kruger, J. & Jonas, W. (1973). Pharmacological studies on tetrabenazine-induced excited behaviour of rats pretreated with amphetamine or nialamid. *Archives internationales de pharmacodynamie et thérapie*, **206,** 47–65.

Schoenfeld, R. I., Neumeyer, J. L., Dafeldecker, W. & Roffler-Tarlov, S. (1975). Comparison of structural and stereoisomers of apomorphine on stereotyped behaviour of the rat. *European Journal of Pharmacology*, **30,** 63–8.

Schultz, W. & Ungerstedt, U. (1978). Striatal cell supersensitivity to apomorphine in dopamine-lesioned rats correlated to behaviour. *Neuropharmacology*, **17,** 349–53.

Segal, D. S., Weinberger, S., Cahill, J. & McCunney, S. J. (1980). Multiple daily amphetamine administration: behavioral and neurochemical alterations. *Science*, **207,** 905–7.

Setler, P. E., Malesky, M., McDevitt, J. & Turner, K. (1978). Rotation produced by administration of dopamine and related substances into the supersensitive caudate nucleus. *Life Sciences*, **23,** 1277–84.

Smith, R. C. & Davis, J. M. (1976). Behavioural evidence for supersensitivity after chronic administration of haloperidol, clozapine and thioridazine. *Life Sciences*, **19,** 725–31.

Smith, R. C. & Leelavathi, D. E. (1978). Behavioral effects of dopamine agonists increase with age. *Communications in Psychopharmacology*, **2,** 39–43.

Smith, R. D., Cooper, B. R. & Breese, G. R. (1973). Growth and behavioural changes in developing rats treated intracisternally with 6-hydroxydopamine: evidence for involvement of brain dopamine. *Journal of Pharmacology and Experimental Therapeutics*, **185,** 609–19.

Staunton, D. A., Wolfe, B. B., Groves, P. M. & Molinoff, P. B. (1981). Dopamine receptor changes following destruction of the nigrostriatal pathway: lack of relationship to rotational behavior. *Brain Research*, **211,** 315–29.

Steeves, J. D., Schmidt, B. J., Skovgaard, B. J. & Jordan, L. M. (1980). Effect of noradrenaline and 5-hydroxytryptamine depletion on locomotion in the cat. *Brain Research*, **185,** 349–62.

Stromberg, U. & Waldeck, B. (1973). Further studies on the behavioral and biochemical interaction between caffeine and L-DOPA. *Journal of Neural Transmission*, **34,** 241–52.

Taylor, K. M. & Snyder, S. H. (1971). Differential effects of *d*- and *l*-amphetamine on behavior and on catecholamine disposition in dopamine and norepinephrine containing neurons of rat brain. *Brain Research*, **28,** 295–309.

Tilson, H. A., Chamberlain, J. H., Gylis, J. A. & Buyinski, J. P. (1977). Behavioural suppressant effects of clonidine in strains of normotensive and hypertensive rats. *European Journal of Pharmacology*, **43,** 99–105.

Trulson, M. E., Stark, A. D. & Jacobs, B. L. (1977). Comparative effects of hallucinogenic drugs on rotational behavior in rats with unilateral 6-hydroxydopamine lesions. *European Journal of Pharmacology*, **44,** 113–19.

Tye, N. C., Horsman, L., Wright, F. C., Large, B. T. & Pullar, I. A. (1977). Two dopamine receptors: supportive evidence with the rat rotational model. *European Journal of Pharmacology*, **45,** 87–90.

Ungerstedt, U. (1971*a*). Striatal dopamine release after amphetamine or nerve degeneration revealed by rotational behaviour. *Acta physiologica scandinavica, suppl.*, **367,** 49–68.

Ungerstedt, U. (1971*b*). Postsynaptic supersensitivity after 6-hydroxy-dopamine induced degeneration of the nigro-striatal dopamine system. *Acta physiologica scandinavica, suppl.*, **367,** 69–93.

Ungerstedt, U. & Arbuthnott, G. W. (1970). Quantitative recording of rotational behavior in rats after 6-hydroxy-dopamine lesions of the nigrostriatal dopamine system. *Brain Research*, **24,** 485–93.

Vetulani, J., Reichenberger, K. & Wiszniowska, G. (1972). Asymmetric behavioral and biochemical effects of unilateral injections of 6-hydroxydopamine into the lateral brain ventricle of the rat. *European Journal of Pharmacology*, **19,** 231–8.

Viala, D. & Buser, P. (1974). Effects of a decarboxylase inhibitor on the DOPA and 5-HTP induced changes in the locomotor-like discharge of rabbit hind limb nerves. *Psychopharmacologia*, **40,** 225–33.

Vincent, S. R., Hattori, T. & McGeer, E. G. (1978). The nigrotectal projection: a biochemical and ultrastructural characterization. *Brain Research*, **151,** 159–64.

Waddington, J. L. & Crow, T. J. (1979). Drug-induced rotational behaviour following unilateral intracerebral injection of saline–ascorbate solution: neurotoxicity of ascorbic acid and monoamine-independent circling. *Brain Research*, **161,** 371–6.

Waldeck, B. (1974). Ethanol and caffeine: a complex interaction with respect to locomotor activity and central catecholamines. *Psychopharmacologia*, **36,** 209–20.

Weiner, W. J., Goetz, C., Nausleda, P. A. & Klawans, H. L. (1977). Clonazepam and dopamine-related stereotyped behaviour. *Life Sciences*, **21,** 901–6.

Weinstock, M. & Speiser, Z. (1974). Modification by propranolol and related compounds of motor activity and stereotype behaviour induced in the rat by amphetamine. *European Journal of Pharmacology*, **25,** 29–35.

Wise, C. D. & Stein, L. (1969). Facilitation of brain self-stimulation by central administration of norepinephrine. *Science*, **163,** 299–301.

Wolfarth, S., Coelle, E. F., Osborne, N. N., Sontag, K. H. & Wand, P. (1979). Drug-induced stereotypes and asymmetric behaviour after substantia nigra pars posterior (SNPP) lesions in cats. *Brain Research*, **178,** 545–54.

Zebrowska-Lupina, I. (1977). The effect of alpha-adrenolytics on central action of agonists and antagonists of dopaminergic system. *Polish Journal of Pharmacology and Pharmacy*, **29,** 393–404.

Zebrowska-Lupina, I., Kleinrok, Z., Kozyrska, C. & Wielosz, M. (1978). The effect of alpha-adrenergic receptor stimulant drugs on amphetamine or apomorphine-induced stereotypy in rats. *Polish Journal of Pharmacology and Pharmacy*, **30,** 459–67.

5

Catecholamines and learning behaviour

Introduction

The role of catecholamines, and especially noradrenaline, in learning behaviour has been a major subject of catecholamine research since the early 1970s. The initial suggestion of such a role came from Kety (1970, 1972) and was more clearly defined for noradrenaline by Crow (1972, 1973). It is a natural development of the putative role of catecholamines in ICSS (see Chapter 3). If catecholamine neurotransmitters in the brain are involved in the reinforcing element of electrical brain stimulation then perhaps they are also involved in the rewarding element of natural reinforcers. If that is so, then behaviours such as learning, which require the integrity of an endogenous reward system, should be severely disrupted in its absence. A prediction of the catecholamine theory of reinforcement is therefore that modification of behaviour caused by natural reinforcers (that is, learning) should be impaired after manipulations which deplete catecholamines or otherwise interfere with their neurotransmitter function in the brain. This prediction, and its subsequent testing for both noradrenergic and dopaminergic manipulations, will be the central theme of the present chapter.

The paradigm

Unlike descriptions in previous chapters, no single paradigm or experimental apparatus has been exclusively used in the study of catecholamines and learning. A bewildering plethora of learning situations has been invoked in the pursuit of this hypothesis. Rather than define and describe each such behavioural situation in the text as it occurs, they have been gathered together into an Appendix. This has the virtue of enabling those already familiar with a paradigm to skip a superfluous description of it and also avoiding numerous repetitions of exactly the same description when, as often happens, the paradigm is reused a few pages on by another experimenter. Thus, for all the common, and a few of the less common, behavioural paradigms used in this chapter the reader is referred to the Appendix.

Early evidence for catecholamine involvement

Much of the early evidence suffers from a two-fold weakness. Not only were manipulations used which altered both noradrenaline and dopamine together, but attention was focussed on the continued performance of a previously learned response under the drug, rather than on the acquisition *de novo* of a response. The effects of a drug on performance of a response may be very different from those on the initial acquisition of that response. In as much that the behaviour being studied was being maintained by the constant presence of a reinforcer and that the removal of the reinforcer (either by changing the schedule to one of extinction or by blocking the central rewarding effects of that reinforcer while continuing to present it) would cause such behaviour to cease eventually, then a decrease in performance of a previously learned response could be taken as evidence for catecholamines in the central mediation of reinforcing events. However, many other effects, such as a motor impairment, would also cause a decrease in performance of a previously learned response. Thus, the early evidence of catecholaminergic involvement in reward and learning must remain ambiguous, not only on a neurochemical but also on the behavioural level. It is only when experiments were carried out on the initial acquisition of a response, with appropriate control procedures to rule out a motor impairment, that such evidence gained in strength.

With these caveats in mind I will now examine the early studies on the role of catecholamines in learning behaviour.

L-DOPA and MAOIs

One approach, discussed in previous chapters, has been to increase the availability of the given neurotransmitter in the brain and look for enhancement of the putative behavioural function. Catecholamine function can be increased by peripheral administration of L-dihydroxyphenylalanine (L-DOPA), preferably together with a peripherally active decarboxylase inhibitor. This precursor is then used to synthesise additional amounts of noradrenaline and dopamine in the brain. A second approach aimed at increasing the availability of neurotransmitter is to prevent its breakdown. Hence, administration of inhibitors of the enzyme monoamine oxidase (MAOIs) was used.

Kitsikis, Roberge & Frenette (1972*a*) found that 30 mg kg^{-1} L-DOPA administered orally to cats led to a significant improvement in their performance of a delayed response task. Here, one of two wells set into horizontal board was baited with food in sight of the animal; a screen was used to prevent selection of a well unit until a delay period of 5 s had elapsed; and then the cat was allowed to select one of the two wells with its paw and retrieve the food. L-DOPA was particularly effective in low-performance animals. Chronic administration of L-DOPA to these animals over 6 or 7 days led to a continued improvement in performance and progressive elimination of errors. This would suggest an effect on learning as well as on performance of the delayed response task were it not for the fact that the authors omitted a similar chronic group receiving merely placebo treatment. This so called 'enhanced learning' could be due simply to the passage of time and repeated training, not necessarily related to administration of L-DOPA. Similar acute treatment with L-DOPA of cats performing a visual discrimination task, in which the well containing food was signalled by being covered with a red card and the empty well by a blue card, was ineffective (see Fig. 5.1). Biochemical measurements revealed that a marked increase in dopamine levels in the striatum occurred about 4–6 h after L-DOPA administration, with no great effect on the levels of noradrenaline or serotonin. From this, the transmitter of importance would seem to be dopamine, but the behavioural paradigm has demonstrated an effect on performance of a previously learned task, not improved learning of a new task.

However, L-DOPA does not always have a facilitatory effect on previously learned behaviour. Thut (1977) reported that L-DOPA in doses from 100 mg kg^{-1} to 320 mg kg^{-1} depressed the performance of mice previously trained on a Sidman avoidance schedule (see Appendix). The number of lever-presses emitted in order to avoid receiving electric footshock dropped as a linear function of increasing dose of L-DOPA. This was particularly marked between 30 and 45 min after injection of the drug.

Use of MAOIs such as tranylcypromine to increase brain levels of dopamine and noradrenaline has been found to cause a transient decrease in two-way active avoidance responding (see Appendix) in rats, followed some 5 h later by a marked improvement in performance (Bucci, 1974). On the other hand, a related MAOI, iproniazid, failed to show the delayed increase in performance and only the immediate decrement was seen (Bucci & Bovet, 1974). Since tranylcypromine but not iproniazid caused a marked increase in brain noradrenaline some 5 h after injection, the authors suggest that noradrenaline might underlie the observed improvement in performance.

It can be seen that, because of the impurity of the neurochemical changes caused (alterations in both noradrenaline and dopamine) and the failure to examine the acquisition *per se* of a response, very little can be claimed from these studies of L-DOPA and MAOIs in so far as a role of catecholamines in learning is concerned. More evidence has been obtained from studies on synthesis inhibitors such as α-methyl-*para*-tyrosine (AMPT).

Synthesis inhibition with AMPT

Hanson (1965) found that 150–200 mg kg^{-1} AMPT disrupted an already learned conditioned avoidance reaction (two-way active avoidance) in cats. From 100 % performance before injection, impairments in performance to worse than 80 % were seen. Quite often the cats showed a progressively slower avoidance response throughout the session until they failed to avoid and would only move when the electric shock actually came on (escape). Some cats eventually failed even to escape. The severity of the drug is shown by the fact that four out of the nine cats died, one soon after

Fig. 5.1. Number of errors on a delayed response task (DR) or on a visual discrimination task. L-DOPA columns indicate animals treated with L-DOPA some 4–8 h previously. 'DR low' are low performing cats, 'DR naive' are untrained cats. (Redrawn from Kitsikis *et al.*, 1972*a*.)

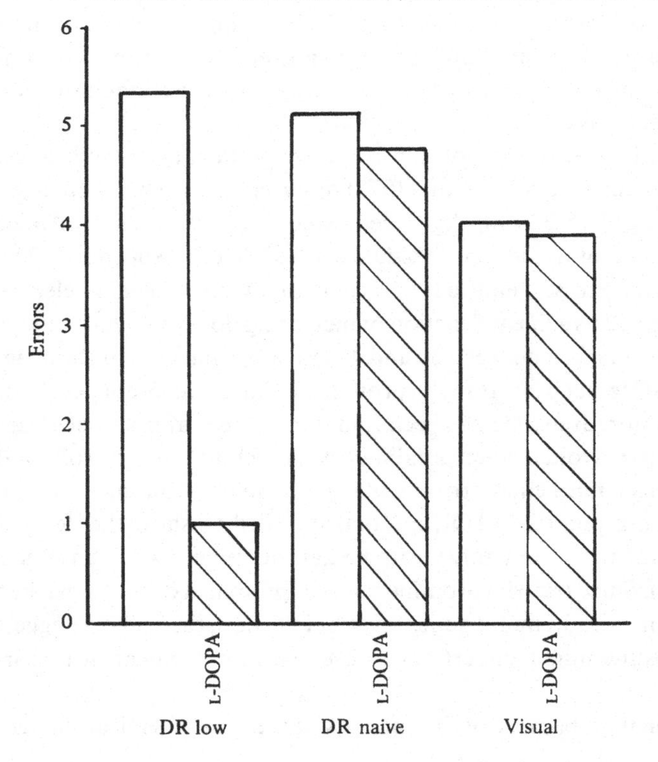

injection, and three others some days later despite glucose injections and external temperature maintenance. Similar results were obtained in rats trained on a signalled shock avoidance schedule in a Skinner box (see Appendix). Lever-pressing to avoid electric shock dropped markedly, particularly 5–10 h after 150 mg kg^{-1} AMPT. Three out of the six rats tested died 19 h to 2 days after injection. On another group of 150 mg kg^{-1} AMPT was found to deplete brain dopamine to 20 % of normal, brain noradrenaline to 25 % with no effect on serotonin levels. Similar dopamine depletions were seen in cats, but with smaller apparent effects on brain noradrenaline.

This experiment is typical of those using AMPT in the early literature. First, reductions in both dopamine and noradrenaline are caused so that no differentiation can be made between the two amines for the behavioural effects. Secondly, performance on an already learned task is examined. The learning of the task is not tested. Finally, the deaths of so many of the experimental subjects must raise doubts as to the specificity of the drug treatment.

Weissman & Koe (1965) showed similar disruption of a task in which the rat had to jump from the grid floor of the box onto a shelf in order to avoid electric footshock. AMPT (178 mg kg^{-1}) disrupted this response in two out of ten rats. This experiment was also instructive in that the compound α-methyl-*meta*-tyrosine was without this effect. AMMT is in fact an amine *releasing* agent and so the two compounds are not to be confused.

Hanson (1967) used another inhibitor of tyrosine hydroxylase, H 59/64, and found disruption of two-way active avoidance behaviour in rats, especially after a dose of 500 mg kg^{-1}. Dopamine levels in the brain fell to 50 % while those of noradrenaline fell to 35 % of normal. On positively reinforced schedules such as fixed-ratio (FR) lever-pressing (see Appendix) AMPT also caused a 36 % decrease in response rate in the first 4 min of the session with even more dramatic effects as the session progressed (Schoenfeld & Seiden, 1967). Again brain dopamine fell to about 30 % and brain noradrenaline to 40 % of normal. Lever-pressing on fixed interval schedules (FI, see Appendix) also declined after AMPT treatment, but less severely than did responding on fixed ratio (Schoenfeld & Seiden, 1969).

Thus, in summary, whatever the response or reinforcer used, AMPT treatment seems to cause a decline in the emission of an already learned behaviour. However, no differentiation between the roles of noradrenaline and dopamine was made.

The deficit in responding after AMPT can be corrected by the administration of DOPA, which replenishes *both* dopamine and the noradrenaline levels in the brain (Moore & Rech, 1967; Ahlenius, 1974). This does not help us to determine the amine of importance but Fuxe & Hanson (1967) sought to determine which aminergic system was activated during avoidance behaviour by blocking new synthesis with AMPT and examining the depletion of fluorescence caused by a 4 h avoidance session. Marked depletion of fluorescence was seen upon sacrifice immediately after the avoidance session in noradrenergic areas of the brain. Lesser depletion of fluorescence in dopaminergic brain areas was seen although some increased dopamine release seemed to occur in caudate, accumbens and olfactory tubercle. This again confirms the role of catecholamine systems in the emission of already learned avoidance behaviour and suggests a greater importance for noradrenaline.

Other authors have examined the acquisition learning of avoidance behaviour after AMPT. Ahlenius (1973) tested the effects of AMPT and the dopamine-β-hydroxylase (DBH) inhibitor FLA 63 on the acquisition of a two-way active avoidance response in female mice. Despite reduction in brain dopamine to 40 % by AMPT and of brain noradrenaline to 50 % by both AMPT and FLA 63 no impairment of learning was seen. However, injection of AMPT when the avoidance task had been acquired resulted in a disruption of that behaviour. FLA 63 did not affect performance of the already learned task. Hall & Mayer (1975) also examined the effect of treatment with AMPT of the acquisition learning of a passive avoidance step-through task (see Appendix). Despite a reduction in brain dopamine of 40 % and noradrenaline of 50 % no impairment of learning was seen at a shock intensity of 0.8 mA. Although at a shock intensity of 1.6 mA mice treated with AMPT re-entered the compartment in which they had been shocked more quickly than controls, suggesting a deficit in learning, at 0.16 mA shock the reverse pattern was seen, with the AMPT-treated mice avoiding the shock compartment *more* than saline controls! AMPT thus appears to have no consistent effect on passive avoidance learning. Similar results were obtained by Zornetzer, Gold & Hendrickson (1974) using a passive avoidance step-through task. If mice trained under AMPT also received a similar injection of AMPT prior to re-test 24 h later, they showed learning and retention of the passive avoidance task as good as those of saline controls. Mice treated with AMPT prior to training and tested 30 min later showed unimpaired learning and retention even without a second, pre-test injection of AMPT.

In summary, it may be concluded that although catecholamines are involved in the expression of previously learned behaviour, the use of AMPT synthesis inhibition has failed to demonstrate which catecholamine is of importance and has also failed to show a marked involvement of catecholamines in the learning, as opposed to the performance, of avoidance tasks.

Early 6-OHDA lesions

Early treatments with 6-hydroxydopamine (6-OHDA) usually depleted both noradrenaline and dopamine together and so served to provide evidence for catecholamine involvement without separating the two amines. Some experiments examined the effect of catecholamine depletion on performance of an already learned task while others were more useful in delineating a role for catecholamines in learning by examining the acquisition of that task.

Lenard & Beer (1975*a*) found that treatment with 250 μg 6-OHDA injected intraventricularly after pargyline pre-treatment (to potentiate the depletion of both amines) reduced brain dopamine to about 15% of control and brain noradrenaline to about the same. This very severely reduced the avoidance performance of rats trained previously to jump onto a shelf in order to avoid electric footshock. Once the electric footshock came on, however, the rats' escape response appeared to be normal. A similar disruption of an approach response for food reward was also reported (Beer & Lenard, 1975) although there was considerable recovery over the 43 days of post-operative testing. Correlation of the degree of amine depletion with the severity of the avoidance deficit seemed to suggest that brain dopamine was of greater importance (Lenard & Beer, 1975*b*), although restoration of the avoidance response after 6-OHDA treatment by intraventricular injection of noradrenaline and dopamine revealed that *both* were effective (Lenard & Beer, 1975*c*). Intraperitoneal injections of either the dopaminergic agonist apomorphine or the α-noradrenergic agonist clonidine were also effective in reversing the behavioural deficit. Thus, a clear role for catecholamines in the performance of both a previously learned avoidance and an approach response was demonstrated by the use of 6-OHDA, but little convincing differentiation of the amines of importance was achieved.

Similar results were reported by Ruiz & Monti (1975), who used 200 μg 6-OHDA injected into the left ventricle. Performance on a previously learned pole-jump avoidance task (see Appendix) dropped in 11 out of 12 rats and remained depressed for up to 40 days. Intraventricular injection of noradrenaline, clonidine or DOPA were all effective in restoring the behaviour. Again, the results fail to differentiate between noradrenaline and dopamine as the basis for the 6-OHDA-induced drop in avoidance performance.

On the other hand, not all early experiments succeeded in finding an impairment in the performance of a previously learned task. Cooper, Grant & Breese (1973*b*) found that neither depletion of noradrenaline on its own by three injections of 25 μg 6-OHDA intracisternally nor depletion of dopamine on its own by intracisternal injection of 200 μg 6-OHDA after desipramine pre-treatment altered a continuously reinforced (CRF, see Appendix) lever-press for food. Even simultaneous depletion of *both* amines by intracisternal injection of 200 μg of 6-OHDA after pargyline pre-treatment had no effect on lever-pressing. Depletions of brain amines were not impressive, since about 50 % of noradrenaline survived the noradrenaline depletion technique while 25 % of brain dopamine survived the second technique. Better depletions were obtained after combined depletion (30 % noradrenaline and 16 % dopamine surviving), but still no deficit on an already learned lever-press was seen.

Even more paradoxical was the finding of Schoenfeld & Uretsky (1972) that two intraventricular injections of 250 μg of 6-OHDA led to a four-fold *increase* in the response rate on a variable interval (VI) schedule (see Appendix) for water reinforcement. This was seen not only on performance of a previously learned variable interval schedule but even on the acquisition *de novo*. Response rates on the continuously reinforced schedule which preceded variable interval training were not altered, in agreement with the negative findings of Cooper, Grant & Breese (1973*b*). Brain dopamine was reduced to 20 % of control and brain noradrenaline to 12 %.

Other authors focussed more intently on the acquisition process, rather than the performance of an already learned response. Thus Mason & Iversen (1974) examined the acquisition of a complex motor response involving pulling or pushing a ball out of a narrow tunnel in order to allow the rat to run through to reach food. Intraventricular injection of 6-OHDA which severely depleted both noradrenaline and dopamine resulted in a slower learning of this response. It is not known what effect such depletion would have on the performance of an already learned response of this nature. Cooper *et al.* (1974) found that depletion of brain dopamine by injection of 8 μg of 6-OHDA in 4 μl saline into either the caudate nucleus or the cell body region of A9/A10 slowed the acquisition learning of two-way active avoidance. Similar injection into the ascending noradrenergic pathways failed to impair such learning. Virtually identical results were obtained by this group using other techniques of 6-OHDA administration to deplete noradrenaline and dopamine. Cooper, Breese, Howard & Grant (1972) used three intracisternal injections of 25 μg 6-OHDA to deplete brain noradrenaline on its own, injection of 200 μg 6-OHDA to deplete brain dopamine predominately but not exclusively, and two doses of 200 μg of 6-OHDA with pargyline pre-treatment to deplete both amines. Only the rats with dopamine depletions (with or without noradrenaline loss) showed impairment of learning of two-way active avoidance. In fact, the noradrenaline and dopamine depleted group (pargyline pre-treatment) showed no sign of learning the task even after 100 trials. Rats with noradrenaline depletions on their own were significantly better than controls in learning the two-way active avoidance task.

Howard & Breese (1974) found that 6-OHDA injection in cats also impaired learning. This time a conditioned heart rate response was used. Administration of electric shock with an initially neutral stimulus, such as a light or a tone, eventually endows the stimulus with the power to slow the heart rate on its own in the absence of shock. 6-OHDA lesioned cats failed to show the learning of the association of shock with tone. However, depletions of both noradrenaline and dopamine were found, thus failing to indicate the amine of importance.

Finally, Howard, Grant & Breese (1974) examined the learning of spatial discrimination in a T-maze (see Appendix) for food reward. Depletion of both noradrenaline and dopamine by 6-OHDA and pargyline severely impaired the acquisition of this response. However, this was an effect on performance rather than learning, since similar depletions given after learning had occurred also led to a drop in responding. Correlation of the behavioural deficit with the levels of noradrenaline and dopamine in individual rats suggested that the amine of importance might be dopamine.

Thus, in summary, the early experiments using 6-OHDA have established a role for catecholamines in the performance of a previously learned response. Impairments in acquisition learning may also be seen, but these may be because of the inability of the animal to perform the response rather than a learning deficit. Such deficits as were seen related more to dopamine than to noradrenaline, but some of the evidence was correlational rather than direct. More pertinent evidence will be presented in subsequent sections on specific 6-OHDA lesions of noradrenergic and, independently, dopaminergic systems.

Catecholamine release/levels after learning

A technique not used extensively heretofore in areas such as ICSS and motor behaviour is that of the release of catecholamines as a result of learning behaviour. The logic of the technique is simple. Thus, if catecholaminergic systems in the brain are involved in learning, measureable changes in these systems should occur as a result of a learning experience. Some studies have sought to find a decrease in endogenous catecholamines as a result of release during learning. Other, more sophisticated studies have monitored such postulated release by the efflux of radioactively labelled catecholamine during learning. Often changes in both dopamine and noradrenaline have been detected. Thus, the results are often ambiguous on these biochemical grounds. A further problem is to ensure that the observed change is due solely to the learning part of the behavioural situation to which the animal has been exposed and not to its increased locomotor activity in the situation or to the stress it experiences or to any one of a hundred other simultaneously occurring psychological processes *other than* that of learning. It is this later difficulty which has perhaps most severely limited the application of this approach (Rose, 1981).

Weiss, Stone & Harrell (1970) found that whole brain noradrenaline levels were increased about 10 % after a 70 trial one-way active avoidance session (see Appendix). Other animals which received as many electric shocks but were not allowed an avoidance response were used as controls. The latter group would be expected to experience the same stress resulting from electric tailshock, but, since no avoidance response was available, would not invoke learning processes. Such an experiment is relatively sophisticated behaviourally but is simplistic biochemically since dopamine levels were not measured and whole brain, rather than regional, noradrenaline concentrations were determined.

Further ambiguity is added to the above experiment by the finding of DiCara & Stone (1970) that instrumental conditioning in rats of the heart rate response to initially neutral stimuli such as light and tone caused an increase in noradrenaline levels not only in the brain but also in the heart. Although changes in noradrenaline levels in the central nervous system might be attributed to learning, changes in cardiac noradrenaline content manifestly cannot be!

Gold & van Buskirk (1978*a*) found that training on step-through passive avoidance (see Appendix) at high footshock levels produced a decrease in whole brain noradrenaline some 10 min after training. Retention of this task some 24 h later was good. Similar training with low footshock intensity failed to alter whole brain noradrenaline and 24 h retention was very poor. A correlation existed between the degree of depletion of whole brain noradrenaline and the amount learned as shown by 24 h retention. This monotonic relationship was found down to a depletion of whole brain noradrenaline to 70 or 80 % of normal. However, rats with depletions to 60 % of normal showed poor retention. Thus, a U-shaped curve has to be invoked. Again brain dopamine was not measured. No regional assay of brain noradrenaline was carried out. The latter criticism was rectified by Gold & Murphy (1980), who measured noradrenaline in forebrain and hindbrain after step-through passive avoidance training. A 20 % decrease in noradrenaline levels was seen in *both* areas. Thus, these experiments fail to highlight the specific brain area involved in learning.

Some experiments fail to find any relation between whole brain noradrenaline levels and retention of passive avoidance step-through learning. Palfai, Brown & Walsh (1978) found that reserpine reduced 7 day retention of the task and also reduced brain noradrenaline and dopamine levels. However, doses of reserpine which caused the greatest noradrenaline depletion failed to cause retention impairments. Thus, no correlation between brain catecholamine levels and retention emerged. A similar conclusion was reached by Bauer (1973), who found that retention of a one-way active avoidance task was good 10 min after training, became worse at 3.5 h post-training (so called incubation or Kamin effect) and was again good some 24 h after training. However, brain noradrenaline levels were reduced 10 min *and* 3.5 h after training but were back to normal after 24 h. Thus, since brain noradrenaline (or serotonin for that matter) failed to show a U-shaped time course after training it did not correlate with the degree of retention.

Thus, there is some doubt as to whether whole brain noradrenaline levels are indeed related to step-through passive avoidance learning or not.

Vachon & Roberge (1981) found a lower level of noradrenaline in the amygdala and frontal cortex after 11 days of training cats on a delayed response task (see p. 178). Higher levels of dopamine were found in the amygdala with no change in the neostriatum. This the authors interpret as evidence of increased noradrenaline release (and therefore depletion of levels) during delayed response learning, together with reduced activity in amygdaloid dopaminergic systems (thus increasing transmitter level).

Levels of a neurotransmitter often fail to reflect the release of that substance. Usually, compensatory increases in synthesis occur together with any increase in release. Thus, methods other than the measurement of overall levels may be required to detect any change in the activity of that system. Lewy & Seiden (1972) used the technique of release of [^{3}H]noradrenaline to monitor functional activity in the system. They injected rats with 10 μl saline containing 1 μCi of [^{3}H]noradrenaline via an indwelling ventricular cannula and allowed them to lever-press for water reward on a VI 30 schedule for the next 2 h. The animals were then sacrificed and the amount of [^{3}H]noradrenaline remaining in their brains determined. Animals which had been lever-pressing showed 25 % less radioactivity remaining in the brain than rats which had been handled but not allowed to lever-press. This would suggest a greater release of noradrenaline as a consequence of greater activity in the system during lever-pressing. Again brain dopamine release was not measured and no localisation of the brain area responsible for the increased noradrenaline release can be obtained from the published data. The behavioural controls exclude the mere deprivation of water as being responsible for the noradrenaline release, but do not differentiate between the motor activity of lever-pressing and the central processes activated by the receipt of the water reward.

Coyle, Wender & Lipsky (1973) found a higher activity of the noradrenaline synthesising enzyme DBH in whole brain of the Roman Low avoider (RLA) strain of rat compared to controls or the Roman High avoider strain. The RLA rat learns a two-way active avoidance task more slowly than other rats. A higher incorporation of [^{3}H]tyrosine into dopamine was also seen in this rat strain. Hraschek, Pavlik & Endroczi (1977) found that rats which showed the best performance on two-way active avoidance also showed the greatest release of noradrenaline during the task as judged by the disappearance of [^{3}H]noradrenaline from the brain after prior intraventricular injection. The greater learning showed by these rats is here irretrievably confounded with the greater motor activity as a consequence of the greater number of responses emitted by these animals and as such is ambiguous.

Greater behavioural specificity was achieved in an experiment by Tilson, Rech & Sparber (1975). They trained rats on a one-way active avoidance task in which a light signalled the onset of electric footshock 40 s later. Prior to this training session, the animals received an injection of [^{14}C]noradrenaline via an intraventricular cannula and release of the label was determined by push–pull perfusion of the ventricles over the next 40 min. Control groups were presented with the light in the absence of the shock. A 24 % increase in the release of labelled noradrenaline occurred in the group in which avoidance learning was also occurring. To control for the motor activity which accompanied the learning of the avoidance task another group was trained on the task and only then presented with the light in the absence of the footshock. Now, neither motor activity (the avoidance response was unavailable) nor stress (the electric shock did not occur) could be invoked to cause the increase in release of labelled noradrenaline seen here. The release of brain dopamine was not examined and the authors indeed suggest that the [^{14}C]noradrenaline may well be taken up by the dopaminergic caudate and released into the ventricles from there (Tilson *et al.*, 1975, p. 391). Again, studies which achieve adequate behavioural controls often fail to achieve similar neurochemical precision and vice versa.

Rigorous control of both aspects was achieved by Albert, Emmett-Oglesby & Seiden (1977), who labelled noradrenaline stores with [^{3}H]noradrenaline injected intraventricularly, and sacrificed the animals after sessions of various schedules of reinforcement. The radioactivity remaining in the cortex and brainstem was measured. Other experiments used similar behavioural schedules and measured both noradrenaline and dopamine after synthesis inhibition with AMPT. The decrease in noradrenaline as a result of lever-pressing on various FR schedules (see Appendix) correlated with the number of reinforcers presented. Presentation of free reinforcements on a variable time schedule (VT, see Appendix) resulted in a similar decrease in noradrenaline which correlated with the number of presentations of the reward rather than the size of the reward (i.e. volume of water consumed). Caudate dopamine was lowered irrespective of the number of rewards or size of reward. Thus, the authors suggest that the presentation of reward is the factor in the highly complex behavioural situation which causes an increased release of noradrenaline. They were, however, unable to distinguish between the consumption of the reward (involving motor elements) and the mere presentation of it (involving only reward processes).

The latter limitation may be overcome by presenting not rewards as such but stimuli which have previously been paired with primary reward. Such secondary conditioned stimuli cannot be 'consumed' and so are free of the confound mentioned above. Cessens *et al.* (1980) report such an experiment in which presentation of stimuli associated with footshock (stimuli associated with the experimental chamber) *by themselves* caused an increased release of brain noradrenaline as judged by the increase in whole brain metabolites of noradrenaline (3-methoxy-4-hydroxyphenylglycol, MHPG). Animals which had not previously experienced footshock in the experimental chamber did not show an increase in MHPG when placed in the chamber. This seems to implicate brain noradrenaline in the presentation of reward rather than in the motor activity of its consumption.

A role for brain dopamine was evidenced by Heffner & Seiden (1980), who examined the synthesis of noradrenaline and dopamine from radioactively labelled precursor tyrosine. Rats which lever-pressed for water reward on a FR5 schedule showed greater synthesis of dopamine in the caudate than did controls which were not allowed to lever-press. Dopamine synthesis in mesolimbic and hypothalamic areas was not altered. Noradrenaline synthesis in the hypothalamus was 48 % higher than in controls but did not differ in other brain areas. Further, the increase in dopamine synthesis in the caudate of individual animals correlated with the number of lever-presses made on the FR5 schedule. No correlation with the increased noradrenaline synthesis in the hypothalamus was seen. This suggests that dopamine release and compensatory synthesis were related to the motor act of lever-pressing rather than to reward processes *per se*. It also suggests a high degree of localisation of function since changes in dopamine synthesis were not seen in mesolimbic or hypothalamic regions.

A further technique touched on in the above experiment is the use of synthesis inhibition to allow the decrease in noradrenaline levels as a result of increased release to be measured without its being compensated for by increased synthesis. Synthesis inhibitors such as AMPT and disulfiram have been administered to rats prior to learning experiences and then the catecholamine levels measured afterwards. The comparison is to controls which have not undergone the learning experience. If catecholamines are involved in learning the experimental animals should show greater decreases in catecholamine content than the controls as a result of the increased release during learning. Hurwitz, Robinson & Barofsky (1971*a*) used disulfiram to prevent the resynthesis of noradrenaline in rats previously trained on a signalled shock avoidance lever-press. Performance of the task while under disulfiram resulted in a 20 % greater depletion of brain noradrenaline than occurred in control rats placed in the chamber but not allowed to perform the response (i.e. controls depleted to 80 % of normal while experimentals showed a depletion to 60 %).

Kovacs, Versteeg, de Kloet & Bohus (1981) examined animals trained on a step-through passive avoidance task and divided them into two groups on the basis of their 24 h retention of this task. 'Good' avoiders showed a higher turnover of noradrenaline in the hippocampus and a higher dopamine turnover in the amygdala than controls whereas 'poor' learners showed a higher turnover of noradrenaline in the amygdala. Turnover was assessed by the rate of disappearance of noradrenaline and dopamine in these areas after synthesis inhibition with AMPT. No change in the total levels of noradrenaline and dopamine in the absence of AMPT was found, in agreement with Palfai *et al.* (1978). In that good avoiders showed totally different changes, rather than more of the same, from those shown by poor avoiders in comparison with untrained controls, the significance of these results is uncertain. *A priori* it would have been expected that the 'strength' of the biochemical change would be related to the 'intensity' of learning, not that altogether different areas would be involved (Rose, 1981).

Boarder, Feldon, Gray & Fillenz (1979) found that, compared to untrained but handled control rats, the activity of the catecholamine-synthesising enzyme tyrosine hydroxylase was increased by 50 % in rats trained to run down an alley-way to food reward. This effect was seen in samples from the hippocampus, but other brain areas were not reported so it is unknown if the effect is specific for this structure. It did not seem to be a motor effect, since animals trained on partial reinforcement (PRF, see Appendix), although emitting the same number of responses (and at the same vigour since the running times did not differ), failed to show an increase in tyrosine hydroxylase activity. It also does not seem to be due to presentation of reward since, of course, the partially reinforced animals also received food, albeit on only half of the trials. This experiment raises more questions than it solves.

Summary. From the above data on release of noradrenaline and dopamine during learning it may be concluded that a clear correlation exists between dopamine release, especially in the caudate and the emission of the motor behaviour required for the appropriate response. Release of noradrenaline seems to occur in both appetitive and aversive situations in response to presentation of the reinforcer and, as shown by secondary conditioned reinforcers, is unrelated to the actual consumption (i.e. motor movements of the jaw, etc.) of the reward. However, the fact that noradrenaline release occurs at the same time as reward presentation does not necessarily mean that one underlies the other. Correlations do not necessarily indicate causes. This point will be illustrated further by the sections on selective 6-OHDA lesions and learning.

Conclusions from early experiments

(1) The use of L-DOPA and MAOIs to increase catecholamine levels found some evidence for an improvement of performance of an already learned task but failed to examine the learning *de novo* of new tasks. The neurochemical changes caused by these treatments involve both noradrenaline and dopamine.

(2) Synthesis inhibition with AMPT, while still involving both noradrenaline and dopamine, confirmed the role of catecholamines in the performance of already learned tasks but failed to yield evidence of a further role in the learning *de novo* of the behaviours.

(3) Early 6-OHDA lesions which generally but not exclusively depleted both noradrenaline and dopamine caused impairment in the expression of already learned behaviours. Some impairment in the learning *de novo* of a behaviour was seen but could be the result of an inability to perform the required response rather than a learning deficit *per se*.

(4) The release of catecholamines as a result of learning showed a role for dopamine in the motor output of the response and that noradrenaline was released upon presentation of reward (either primary or secondary). That the release of noradrenaline served to

code for the central representation of reinforcement remained untested by these correlative approaches.

Noradrenaline and learning behaviour

Agonist drugs

The literature on the effects of noradrenaline agonist drugs on learning is rather patchy, with no systematic examination having been carried out. The effects of some agonists, such as amphetamine, have been studied in detail while others have barely been touched upon. This may well reflect the uncertainty as to what result would actually be expected from administration of agonist drugs even if noradrenergic systems were crucial for learning. A prediction would be that if noradrenaline is released during learning then more noradrenaline (as mimicked by an agonist) might give more learning. Closer examination shows this is not necessarily the case. If noradrenaline is released to code for reinforcement in a discrete fashion immediately after the occurrence of a reward presentation, it is essentially being released in an impulse-contingent fashion. An agonist acts continuously on the postsynaptic receptors, without regard for what is occurring in the environment. Rather than learn more readily a specific response which leads to reward, the animal under a noradrenaline agonist drug, since it is experiencing continuing and unfluctuating reward, might well learn nothing at all. Since it is experiencing continuing pharmacological reward, even in the absence of any act on its own part, there seems little reason why it should learn to do anything. This is a similar problem to that of impulse-contingent versus noncontingent reward which was examined earlier in relation to ICSS and agonist drugs (see p. 40). Thus, the noradrenaline theory of reward cannot predict what would be observed in rats treated with a noradrenaline agonist and placed into a learning situation. As we shall see the experimental picture is equally clouded. The first agonist to be reviewed will be noradrenaline itself.

Carley & Haynes (1969) reported that, in an approach-avoidance situation in a three-compartment linear maze (see Appendix), intraperitoneal injection of adrenaline caused an increase in avoidance behaviour in mice while similar injection of noradrenaline caused greater approach behaviour. The study is confounded by possible motor effects of the drugs, by the fact that emission of already learned behaviour was examined rather than learning *de novo* and by the peripheral route of administration of the drugs. Although there is evidence for a role of the peripheral sympathetic nervous system in avoidance behaviour – for example, Holz, Pendleton, Fry & Gill (1977) showed that removal of the medulla of the adrenal glands shortened the recovery of behaviour suppressed by punishment and that this could be reversed by adrenaline injection – this will not be pursued here (see p. 222).

Merlo & Izquierdo (1967) found that peripheral injection of adrenaline, or the β-agonist isoproterenol, would slow learning of a trace-avoidance conditioning task (see Appendix) in which the rat was required to lift one or both forepaws within 1 s of the termination of a 5 s warning buzzer. Noradrenaline was without effect on acquisition but when injected peripherally during extinction (see Appendix) served to speed up this process. The β-agonist isoproterenol was equally effective in speeding up extinction, but adrenaline was markedly less affective here. This suggests a role for adrenaline in fear-motivated acquisition learning but fails to differentiate central from peripheral systems (it is unclear how much, if any, adrenaline would penetrate into the central nervous system after a peripheral injection). Further, since both the α-antagonist phentolamine *and* the putative β-antagonist nethalide blocked this effect when co-injected with adrenaline the receptor type remains ambiguous. The α-agonist clonidine (0.2–0.4 mg kg^{-1}) also disrupts avoidance behaviour (Taboada, Souto, Hawkins & Monti, 1979).

The effect of noradrenaline in speeding up extinction is interesting since it is the mirror image of the effect that 6-OHDA lesion to the ascending central noradrenergic system has in slowing down extinction of both one-way (Fibiger & Mason, 1978) and two-way active avoidance behaviours (Mason & Fibiger 1979*a*). The role of central noradrenaline in extinction is examined in more detail in Chapter 6.

To get around the effects on the peripheral sympathetic nervous system use has been made of the intraventricular route of noradrenaline administration. Haycock, van Buskirk, Ryan & McGaugh (1977) administered drugs intraventricularly via indwelling cannulae and found that injection of 0.1 μg of dopamine improved the retention of a passive avoidance step-through task while noradrenaline at any of five doses was without effect. Higher doses of dopamine failed to improve retention, which is difficult to explain. When a suppression of a licking response as a result of a single footshock was examined, the opposite picture emerged. Now, intraventricular injection of 1 μg noradrenaline improved retention but a similar dose of dopamine was without effect. Although this would suggest a role for catecholamines in memory, it would further suggest that it is not generalised, but that one transmitter is involved in one task while the other catecholamine is involved in another behaviour. Since considerable spread of the injected amine was seen in control experiments in which ^{3}H-labelled catecholamine was injected and the brain dissected into serial anterior–

posterior sections, it was not possible to localise the site of action to a specific brain structure.

Such localisation was achieved by Margules (1971*a*), who found that injection of noradrenaline into the dorsal portion of the corticomedial part of the amygdala would release lever-pressing suppressed by electric shock. Other sites were less effective, and dopamine injections had a much weaker effect.

Margules (1968) found that the physiologically inactive isomer, *d*-noradrenaline, was virtually without effect when injected into the amygdala as was simple saline injection. The release of punished responding by *l*-noradrenaline could be blocked by the α-antagonist phentolamine but not by the β-blocker LB-46 (Margules, 1971*b*). Similarly, Komissarov & Talalaenko (1974) found that injection of 5 μg noradrenaline into the amygdala depressed conditioned avoidance behaviour, suggesting again that noradrenaline may antagonise the behavioural effects of punishment. However, since such injections also decreased simple motor activity, the interpretation is ambiguous. Dopamine, on the other hand, depressed conditioned avoidance while increasing motor activity. A lower dose of noradrenaline (2 μg) had the opposite effect of facilitating conditioned avoidance behaviour without altering locomotor activity. Such results are irretrievably confusing. Again the limitation on these experiments are that they examined the emission of an already learned behaviour rather than learning *de novo*.

Administration of amphetamine, which releases both noradrenaline and dopamine, has also been reported to improve retention of a step-through passive avoidance task. Thus, Martinez *et al.* (1980*a*) found that injection of amphetamine peripherally was successful in improving 72 h retention, whereas a wide range of doses of the drug (50, 100, and 500 μg) administered centrally via indwelling ventricular cannulae were not. While it is possible that amphetamine had a central action when injected peripherally and simply failed to reach the requisite brain area when administered intraventricularly, it is highly suggestive that much of the memory-enhancing effect of amphetamine is of peripheral origin. This is further indicated by the finding of Martinez *et al.* (1980*b*) that destruction of the medulla of the adrenal glands blocked the memory-enhancing effects of amphetamine injected peripherally. Further, the compound *d,l*-4-hydroxyamphetamine, which does not enter the brain, also enhanced retention of a step-through passive avoidance task. All of this suggests that the well-established effects of amphetamine in improving retention cannot be invoked as evidence for a role of *central* noradrenaline in learning and memory.

Other methods of increasing noradrenaline activity have been used. Cairncross, King & Schofield (1975) report that the deficit in avoidance learning caused by lesion to the olfactory bulbs in rats can be reversed if the animals are also treated with the noradrenaline uptake inhibitor amitriptyline. Acquisition learning of one-way active avoidance proceeded more quickly, and higher levels of performance were maintained, in the lesioned animals receiving amitriptyline than in lesioned rats under saline.

A further technique for activating noradrenergic systems, albeit with less pharmacological specificity, is to stimulate electrically the source nucleus of the ascending noradrenergic pathway, that is the locus coeruleus (LC). This runs the risk of activating systems other than just the noradrenergic ones, and for this reason has been little applied. However, Segal & Edelson (1978) found that priming stimulation of 5–100 μA intensity applied bilaterally to the LC through 62 μm wire electrodes served to speed up the reversal of a black–white discrimination task in a double runway. Such stimulation was without effect on the initial learning to run along the runway or in the acquisition of the discrimination *per se*. Thus, no evidence for a general role in learning is to be found from this experiment (that would require that both the initial running and the discrimination acquisition should occur more quickly in the stimulated rats, as well as the better reversal that was observed). This finding is, however, interesting in terms of the involvement of noradrenaline in cognitive behaviour (see Chapter 6, p. 277) and is the mirror image of the finding that 6-OHDA lesion to the ascending noradrenergic systems markedly impairs visual discrimination reversal (Mason & Iversen, 1977*a*).

Other, more indirect, evidence for a role of brain noradrenaline in memory comes from the findings that the amnesia induced by the protein synthesis inhibitor cycloheximide can be reversed by catecholaminergic manipulations. Thus, Gibbs & Ng (1979) used an unusual passive avoidance task in which day-old chicks are allowed to peck at a metal bead for 10 s. Typically, disgust reactions occur (since it is not edible) and upon a second presentation of the bead most animals refuse to peck at it. Intracranial administration of cycloheximide reduced the number of chicks refusing to peck after 3 h to only 20 %, compared to 90–95 % seen normally. Subcutaneous administration of the MAOI pargyline, or the indirect amine releasing agent metaraminol, reversed this amnesia in a more of less dose-dependent fashion. Noradrenaline and amphetamine were similarly effective (Gibbs, 1976). Amnesia induced in this paradigm by ouabain can also be reversed by noradrenaline or amphetamine administration (Gibbs & Ng, 1977). However, since it is a matter of debate exactly how cycloheximide or ouabain acts to cause amnesia, the relevance of its subsequent reversal by noradrenaline to memory processes occurring normally in the intact animal is unclear.

In summary, a clear role of the peripheral noradrenergic system in learning has been revealed by studies with noradrenaline agonists. Indeed amphetamine may act solely via the periphery. However, less evidence of a role for *central* noradrenaline in learning has been forthcoming from agonist studies. What central effects noradrenaline agonists have been reported to have are on emission of an already learned behaviour. I shall now turn to noradrenaline antagonists and learning.

Antagonist drugs

The evidence here is sparse indeed. Taboada *et al.* (1979) reported that the α-antagonist phenoxybenzamine produced a dose-dependent decrease in the performance of an already learned pole-jump avoidance task. Doses used ranged from 4 to 12 mg kg^{-1} which is in the range for selective α-receptor blockade. A similar finding was independently reported by Plech, Herman & Wierzbicki (1975) for 20 mg kg^{-1} phenoxybenzamine.

The β-blockers seem to have less effect upon performance of an already learned task than α-blockers, since both Laverty & Taylor (1968) and Singh, Bhandari & Mahawar (1971) find that propranolol fails to affect Y-maze performance in rats; however, more complex behaviour such as differential reinforcement of low rates of responding (DRL, see Appendix) has been reported to suffer disruption after propranolol treatment (Richardson, Stacey, Cerauskis & Musty, 1971). The latter effect required daily injection of the drug, raising the possibility of toxic accumulation and consequent nonspecificity.

Gold & van Buskirk (1978*b*), however, found that neither the β-blocker propranolol nor the α-blocker phenoxybenzamine affected the acquisition learning or 24 h retention of a step-through passive avoidance task. This contrasts with Cohen & Hamburg (1975), who reported that the rather large dose of 45 mg kg^{-1} propranolol (a β-blocker) injected immediately after step-down passive avoidance training would prevent recall of the task when tested 1, 3 or 7 days later. However, such a large dose may well be in the toxic region. Similar retention failure upon re-test 8 days later was reported by Dismukes & Rake (1972) in mice injected subcutaneously with the putative β-blocker dichloroisoproterenol (75 mg kg^{-1}) immediately after two-way active avoidance training.

Stephenson & Andrew (1981) found that all of the β-blockers sotalol, timolol and nadolol were effective in causing amnesia in chicks following peripheral administration using a pecking response (see p. 195). The

α-antagonist phenoxybenzamine and the α_2-antagonist piperoxan were ineffective, as was the β_2-antagonist atenolol, suggesting the possible involvement of a β_2-receptor in memory consolidation.

Hraschek & Endroczi (1979) found that intraperitoneal injection of either of the β-blockers GYKI 41099 or Trasicor failed to alter performance of rats on an already learned two-way active avoidance task. However, administration of either drug significantly slowed the acquisition learning of this task. Similar injections into the vicinity of the LC of 10 μg of the β-blockers also impaired acquisition learning. This is somewhat unexpected since there are no β-receptors on LC cells (Dahlof, Engberg & Svensson, 1981), only α_2-receptors (Cedarbaum & Aghajanian, 1976). Perhaps the drugs spread either to adjacent brain areas receiving noradrenergic innervation or into the ventricles themselves. Injection, either intraperitoneally or intracerebrally, of the α-blocker phenoxybenzamine failed to affect learning of two-way active avoidance. The authors also showed that intraperitoneal injection of the β-blockers slowed running during the learning of a new maze for food reward. However, since the drugs were not tested on the performance of an already learned maze task, the effects on acquisition might reflect a simple performance deficit rather than a deficit of learning *per se*.

A recent study by Ridley, Haystead, Baker & Crow (1981) using the α-blocker aceperone in marmoset monkeys (*Callithrix jacchus*) found that it failed to block the acquisition of a visual discrimination when this was the first task learned on that day, but a second discrimination learning session (with different positive and negative visual stimuli, but otherwise identical to the first) was impaired. This cannot be explained by a global deficit in learning as a result of α-blockade. Ridley *et al.* (1981) invoke a block of 'learning set' to explain these results but this must remain a guess until tested more directly.

In summary, most of the studies with noradrenaline antagonists suggest that α-receptors are not necessary for learning to occur while others demonstrate a role of α-receptors only in the performance of an already learned task, not necessarily in its learning *per se*. There is more evidence for a role of β-receptors in learning since β-blockers do not generally affect performance of an already learned task but may impair retention. A β_2 subtype may be of importance.

Synthesis inhibitors

A lot of the evidence implicating noradrenaline in memory and learning has come from the use of compounds which inhibit the synthesis of noradrenaline in the brain. Commonly used compounds include FLA 63, diethyldithiocarbamate (DDC) and disulfiram. The results obtained are only as good as the specificity of these inhibitors. That they inhibit the synthesis of noradrenaline is undeniable, the problem is what else do they inhibit or disrupt biochemically? If the compound not only inhibits the synthesis of noradrenaline but also, say, of transmitter *Y*, then the behavioural effect observed cannot uniquely be ascribed to noradrenergic systems. I shall return to this later after a review of the behavioural effects which are observed.

Some early experiments claimed to demonstrate a role for noradrenergic systems in reward by the use of AMPT. This compound, it will be remembered, inhibits tyrosine hydroxylase and so prevents the synthesis of *both* noradrenaline and dopamine. The evidence obtained is hence ambiguous and it is surprising that early authors were not aware of this. Thus, Beaton & Crow (1969) found that 100 mg kg^{-1} AMPT severely depressed lever-pressing for food reward while having a smaller effect on lever-pressing for water reward. The authors say (*ibid.* pp. 1134–5), 'However, the possibility that dopaminergic neurones are involved cannot be excluded, but seems unlikely . . . .' Subsequent interpretation in the light of the effects of dopaminergic lesions on ICSS and motor behaviour (see Chapters 3 and 4) would completely reverse this. The effects were almost certainly due to dopamine depletion, not that of noradrenaline.

Similarly, Miller, Cox & Maickel (1970) chose to interpret the well-known effects of AMPT in depressing Sidman avoidance responding (see section on Early experiments, above) as evidence for a role of noradrenaline in the control of this behaviour while again stating (*ibid.* p. 516), 'Finally, one cannot ignore the possibility that dopamine is in some manner involved in the results presented here . . . .' Present knowledge would suggest that indeed it is.

Let us now turn to the many experiments carried out using selective noradrenaline synthesis inhibitors, which, whatever else they may do, do not concomitantly affect dopamine.

There are three main paradigms used here. One examines the effect of synthesis inhibition of noradrenaline on already learned behaviours. Another administers the drug prior to training on the task and hence examines the acquisition of learning *per se*. The third paradigm concerns itself with the putative stages of consolidation of the memory trace after initial learning has taken place. It is supposed that initially labile and transient memories encoded in short-term memory stores are made permanent in a long-term memory store by undergoing some biochemical or structural change. Administration of a drug *after* training is complete but within the time period of the transfer from short- to long-term stores might leave short-term memory unimpaired but prevent the establishment of a long-term memory. The duration of this fragility of memory is controversial but may be assumed for present purposes to be of the order of hours after training.

Expression of already learned behaviours. Krantz & Seiden (1968) showed that doses of DDC of 250 or 500 mg kg^{-1} would depress the performance of an already learned two-way active avoidance task by 50 and 80 %, respectively. A smaller dose of 125 mg kg^{-1} was without effect. Already, at this early stage in the history of the use of DDC, the authors caution that as well as inhibiting DBH, the noradrenaline-synthesising enzyme DDC is a general copper chelator and will hence inhibit all other body and brain enzymes requiring the presence of copper (Krantz & Seiden, 1968, p. 167). Hamburg & Cohen (1973) found that DDC (250 mg kg^{-1}) impaired the performance of a previously learned passive avoidance step-down task (see Appendix) if administered 1, 3, 5 or 7 days after initial training. However, if the drug was injected 30 min prior to training no impairment was seen, whereas administration 6 h prior to training caused a significant impairment in the learning of the passive avoidance task. This is interpreted as suggesting that initial learning is unaffected but long-term retention beyond 1 or 2 h fails to occur in the absence of brain noradrenaline. This point is expanded in the subsequent section.

Ahlenius & Engel (1973) found that another noradrenaline synthesis inhibitor, FLA 63, impaired two-way active avoidance performance in mice and decreased brain noradrenaline to 30 % of normal. Replenishment of brain noradrenaline levels by 3,4-dihydroxy-*ortho*-phenylserine reinstated avoidance behaviour. Another experiment by Ahlenius & Engel (1972) examined the effect of DBH inhibition with FLA 63 on performance of brain noradrenaline levels by 3,4-dihydroxy-*ortho*-phenylserine (DOPS) reinstated avoidance behaviour. Another experiment by Ahlenius & Engel open to many interpretations from a failure of reward mechanisms to a disruption of the perception of time.

In summary, depletion of brain noradrenaline by synthesis inhibition will disrupt the expression of a number of already learned behaviours. This may come about in a number of ways ranging from motor impairment to changes in temporal perception, in addition to an impairment of reward mechanisms. More direct evidence for a role of noradrenaline in learning has come from the following sections on synthesis inhibition and the acquisition *de novo* and subsequent retention of learned behaviours.

Synthesis inhibition during learning. Osbourne & Kerkut (1972) report that injection of 250 mg kg^{-1} DDC 10 min prior to each of three acquisition sessions on a two-way active avoidance task severely impaired learning of this task compared to controls. Injection of the drug some 6 h *after* each session also produced a slight impairment, again suggesting interference with memory consolidation processes (see next section). Interestingly a very strong correlation between the degree of loss of brain noradrenaline and the severity of the acquisition impairment was found (Spearman $\rho = 0.85$, $P < 0.001$). This experiment examined the learning of a novel task under DDC rather than the performance of previously learned behaviours as reviewed in the above section.

Randt, Quartermain, Goldstein & Anagoste (1971) reported that injection of 250 mg kg^{-1} DDC 30 min prior to training mice on a passive avoidance step-through task resulted in impaired retention of this task when tested between 1 and 24 h later. Surprisingly, improved retention (i.e. slower re-entry into the shocked compartment) was seen compared to controls at 1 and 5 min. Again, the authors found that DDC injected after training (in this case immediately after the trial) also caused retention failure when tested 24 h later. Thus, this experiment again shows a failure of learning *de novo* under DDC of a novel task, and additionally suggests an action on the consolidation of memory which takes place in the hour or two *after* learning.

It would appear that at least two processes are being altered by such noradrenaline synthesis inhibition since Solanto & Hamburg (1979) found that administration of DDC some 2 to 4 h prior to training on a step-down passive avoidance task produced a profound impairment of retention when tested 24 h later. Measurement of the brain levels of noradrenaline immediately after training in other rats showed that these were at their lowest for this drug-test interval. Other intervals between drug and training resulted in less impairment and less noradrenaline depletion at the moment of training (see Fig. 5.2). This seems to suggest a role for noradrenaline in processes which occur more or less at the moment of training rather than consolidation of memory in the hour or two subsequent to that training. However, a similar study from Haycock, van Buskirk & McGaugh (1977) found that injections of DDC timed to yield the minimum levels of noradrenaline 30 min to 90 min *after* training were the most efficacious. This would indeed suggest a role for noradrenaline in the consolidation rather than the initial learning processes. The situation remains ambiguous at present.

Other learning situations show impairment. Yonkov & Roussinov (1979) found that 400 mg kg^{-1} DDC either 3 h prior to training, or immediately after training of rats on a maze task severely impaired retention 24 h or 14 days later. Other DBH inhibitors also show this effect.

Thus, Botwinick, Quartermain, Freedman & Hallock (1977) found that FLA 63 in doses of 15 mg kg^{-1} upwards would cause amnesia in mice if injected 4 h prior to training on a spatial discrimination task in a T-maze for food reward. Izquierdo, Beamish & Anisman (1979) found that FLA 63 in a dose of 40 mg kg^{-1} would impair 72 h retention of three out of four different avoidance tasks (step-through passive, step-down passive, step-up active but not freezing avoidance; see Appendix) if administered 2 h prior to training but if given immediately after training was ineffective.

However, Ahlenius (1973) failed to find any effect of 20 mg kg^{-1} FLA 63 on acquisition learning of two-way active avoidance. This dose of FLA 63

Fig. 5.2. Retention of passive avoidance learning after diethyldithiocarbamate (DDC) treatment, together with whole brain noradrenaline (NA) levels at various times after the drug. (Redrawn from Solanto & Hamburg, 1979.)

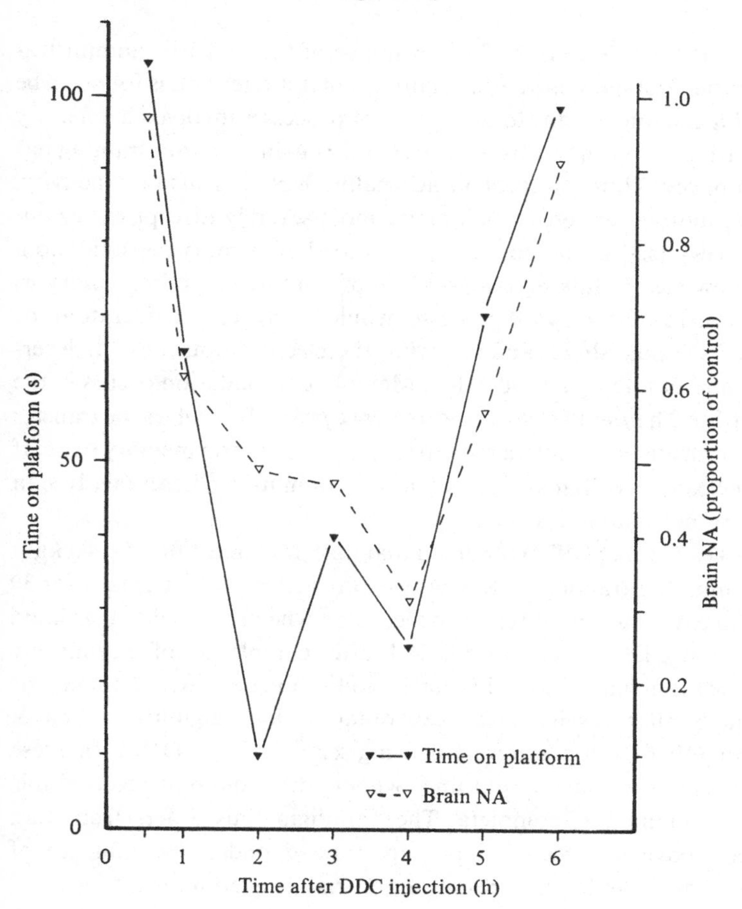

produced a 50 % decrease in whole brain noradrenaline and this may not have been profound enough to influence behaviour.

In summary, it is clear that pre-trial administration of DBH inhibitors such as DDC or FLA 63 will impair the acquisition and subsequent retention of many different tasks, both food rewarded and aversively motivated. However, impairment of acquisition may be caused by interference with processes other than those of reward, such as motor output, attention, etc. Training on a task under a drug cannot determine which of these processes has been impaired in order to produce the observed acquisition deficit. Thus, although suggestive evidence for a role of noradrenaline in learning has been provided by the effects of synthesis inhibitors on the learning *de novo* of a task, more convincing evidence has accumulated from the effects of post-trial administration of synthesis inhibitors.

Post-trial synthesis inhibition. The advantage of the post-trial administration of noradrenaline synthesis inhibitors is that if retention is found to be impaired it can only be due to disruption of processes involved in memory consolidation *per se* since this is the only thing going on *after* training has been completed. Thus, effects of noradrenaline depletion on motor activity, attention, motivation, etc. which might most severely disrupt the acquisition process itself would not be expected to alter memory consolidation. We are now clearly talking about a role of noradrenaline in memory as distinct from reward. Reward processes would be expected to operate at the instant of stepping down and receiving the electric footshock, or lever-pressing and obtaining a food pellet. Memory consolidation occurs in the period of 1 to 2 h *after* the learning experience *per se*. It involves the transfer of short-term memories into a more permanent long-term memory store. If the distinction is worth making, I am now examining processes involved in memory rather than in learning.

Dismukes & Rake (1972) observed that post-trial injection of 1.4 g kg^{-1} DDC 1 min after training mice on a two-way active avoidance task for 30 trials resulted in failure to retain the learning when re-tested 8 days later. Delaying the DDC injection until 24 h after completion of training, by which time the memory consolidation period would be over, did not impair retention. Similar results have been obtained by Fulginiti, Molina & Orsingher (1976) using a lower, 600 mg kg^{-1} dose of DDC. In these experiments the animals were trained *without* drug and only received this after the training was complete. The paradigm thus differs from that described above in which animals are trained under the influence of noradrenaline depletion as a result of pre-trial drug administration.

Hall (1976) reported similar effects on a different type of task, a discriminated (light v. dark) escape from electric footshock (see Appendix). In mice trained to a weak criterion of performance (five out of six responses correct) 1.2 g kg^{-1} DDC caused a deficit in retention when tested one week later. However, training the animals to a higher level of performance on the task prior to injection of DDC, or the use of spatial, rather than visual, discrimination resulted in no deficit in retention. Delay of DDC administration until 1 h after completion of training also abolished the deficit, as did the use of lower doses of DDC (900 and 600 mg kg^{-1}).

Other species. Amnesia as a result of DDC treatment has been reported in species other than rats and mice. Thus, Gilman & Fischer (1976) found that administration of DDC in doses from 50 to 250 mg kg^{-1} 30 min prior to training 4-day-old chicks (*Gallus domesticus*) resulted in amnesia of a runway passive avoidance task. Chicks were initially trained to traverse a runway in order to reach four other chicks in the end cage. Subsequently a section of the runway was electrified and the chick exposed to the apparatus until it had not crossed the electrified section for 3 min. A 24 h re-test revealed that the greater the dose of DDC administered prior to shock training the greater the number of drug-free trials required to re-learn the passive avoidance task.

Stahl, Zeller & Boshes (1971) found that the MAOI pargyline dissolved in the water of the aquarium at a concentration of 10^{-5} or 10^{-6} mol l^{-1} inhibited 24 and 48 h retention of a two-way active avoidance response in goldfish (*Carassius auratus*) and Stahl, Narotsky, Boshes & Zeller (1974) extended this to other MAOIs (tranylcypromine, iproniazid, Lilly 51641 and SU-11,739) and showed that significant changes in noradrenaline fluorescence in the hypothalamus accompanied these changes in memory.

Reversal of amnesia. It appears that the amnesia caused by post-trial and pre-trial injection of DDC can be reversed by intraventricular injections of noradrenaline. Thus, Stein, Belluzzi & Wise (1975) report that pre-trial DDC impaired retention of a passive avoidance step-down task when tested 3 days later. However, if noradrenaline (10 μg) was injected intraventricularly soon after training, the retention was normal despite DDC administration.

The non-physiological precursor of noradrenaline, DOPS, can also reverse the amnesia caused by DDC. Thus, Hamburg & Kerr (1976) injected DDC 30 min prior to training on a step-down passive avoidance task and found impaired retention when tested 24 h later. Injection of DOPS 1 h prior to DDC was effective in preventing this amnesia. DOPS is converted into noradrenaline by a general aromatic amino acid decarboxylase and so does not require DBH activity. Thus, DOPS can replenish brain noradrenaline even in the presence of DDC. Meligeni, Ledergerber & McGaugh (1978) confirmed the reversal of DDC amnesia by intraventricular noradrenaline but also showed the potentially disturbing fact that *peripheral* injections of noradrenaline could also reverse DDC amnesia. Since noradrenaline injected peripherally presumably failed to cross the blood–brain barrier it is difficult to see how it can be argued to be reinstating *central* noradrenaline transmission. This raises the possibility that the effects of DDC might be the result of its depletion of peripheral, rather than central noradrenaline. On the other hand, how then could central injections of noradrenaline such as Stein, Belluzzi & Wise administered be effective in reversing the amnesia, since these would not get to the peripheral noradrenergic system in significant quantities?

Walsh & Palfai (1981) confirmed that peripherally administered noradrenaline would prevent amnesia usually caused by DDC on a passive avoidance step-through task in mice. Further, they found that dopamine and adrenaline were also effective in this respect. Noradrenaline, dopamine or adrenaline injected 15 min prior to, immediately or 10 min after training on the passive avoidance task were all effective in preventing DDC amnesia but not if delayed until 90 min after training. Thus, although consistent with a role for catecholamines in memory consolidation, these latest findings raise questions as to *which* catecholamine is important and, more drastically, suggest that it may be peripheral rather than central in nature.

Problems with synthesis inhibition data. A further confound is that DDC administration is not always reported to give amnesia. In three studies a significant enhancement of retention has been reported. It was not merely that, for whatever reason, amnesia failed to be detected – such as too low a dose of drug – but an actual *improvement* in retention above control values was seen. Such are the results of the studies of Haycock, van Buskirk & McGaugh (1976), Hall (1977) and Danscher & Fjerdinstad (1975). This finding cannot be stressed too much. DDC does not always cause amnesia.

Additionally, a number of authors have found that noradrenaline synthesis inhibitors other than DDC may not cause amnesia despite inducing severe depletions of brain noradrenaline. Flexner & Flexner (1976) found that neither 13 mg kg^{-1} FLA-63 nor 200 mg kg^{-1} of another DBH inhibitor, U 14,624, induced amnesia on a maze learning task in mice. However, the low dose of FLA 63 and the fact that treatment was only started 24 h after training (although it continued for a subsequent 7 days) might explain this. More perplexing is the report of Haycock, van Buskirk

& McGaugh (1977) that, although DDC caused amnesia of a step-through passive avoidance task, other DBH inhibitors such as U 14,624, FLA 63 and fusaric acid failed. They produced as great or even greater depletions of brain noradrenaline as did DDC so their failure cannot be attributed to this.

Finally, the greatest problem with DDC amnesia is that since DDC is a copper chelator it will interfere not only with noradrenaline synthesis but with all other enzymatic processes requiring copper. These will include liver and brain aldehyde dehydrogenase, intestinal indoleamine-2,3-dioxygenase, liver D-glucuronolactone, mixed function oxygenase, superoxide dismutase, hepatic microsomal and plasma esterases and brain tyrosine metabolism (see Mason, 1979). DDC is also a diuretic and will induce a taste aversion in rats (Roberts & Fibiger, 1976). To ascribe its behavioural effects to only one out of a possible ten actions (i.e. noradrenaline synthesis inhibition) is, on the crudest level, to stand only one chance in ten of being right. This is especially emphasised, since inhibition of noradrenaline synthesis with other, possibly purer drugs such as FLA 63, fusaric acid and U14,624 in many cases does not cause amnesia (see above).

Summary. Although DDC injected either before or very shortly after training on many tasks will cause impaired retention when examined 24 h or more later this is not as strong evidence for a role of noradrenaline in learning and memory as might be supposed. DDC-induced impairments of performance on an already learned task are, of course, no evidence at all.

The internal inconsistencies include the following. DDC may sometimes improve retention rather than impair it. DDC inhibits many other enzymes than just that required for noradrenaline synthesis. Inhibition of noradrenaline synthesis by more selective agents may fail to yield amnesia. Reversal of DDC amnesia can be achieved by peripheral noradrenaline injections as well as by central ones.

I shall now examine evidence yielded by the more selective technique of lesion, usually with 6-OHDA, to central noradrenergic systems.

Lesions of noradrenergic systems

If noradrenergic systems are involved in learning it might be expected that a lesion to these would produce an animal either unable to learn or considerably slower in doing so. An early technique used to lesion central noradrenergic systems was that of electrolytic lesion to the source nucleus of the ascending noradrenergic innervation, the LC. This suffers from a number of drawbacks. By definition it is not pharmacologically specific, since electrolytic lesion will destroy all neurochemical systems in the vicinity of the lesion, not just those using noradrenaline as their neurotransmitter. A small degree of pharmacological specificity is conferred on this approach, however, by two further conditions. First, in the rat (but not necessarily in the monkey, and certainly not in the cat) the LC is made up exclusively of noradrenaline-containing cell bodies. Secondly, therefore, rigorous restriction of the electrolytic lesion to the confines of the LC without spread to *any* adjacent brainstem area will lesion only noradrenergic cell bodies. Thus ran the logic of the experimenters who used electrolytic LC lesion to examine the function of central noradrenergic systems in learning and memory.

However, the drawbacks of LC lesion probably outweigh its advantages. In some species, cell bodies containing transmitters other than noradrenaline are found in the LC, for example serotonergic perikarya reported in a species of monkey by Sladek & Walker (1977). Electrolytic lesion of the LC in this species would disturb serotonergic as well as noradrenergic systems. It would then not be possible to ascribe an observed behavioural effect unambiguously to one or other transmitter.

Further, it is virtually impossible with an electrolytic lesion to destroy all of the LC and nothing but the LC. This is because of the three-dimensional shape of the nucleus (see Fig. 5.3). In the anterior–posterior plane the LC has a long, narrow anterior pole broadening into a larger body near the main part of the nucleus and then more abruptly disappearing at the posterior extent of the nucleus. An electrolytic lesion is almost spherical in shape. Thus, to impose a spherical lesion such as to destroy the nucleus totally must in addition destroy adjacent, nonnoradrenergic parts of the brainstem. To restrict a spherical lesion only to the LC means that parts of the noradrenergic nucleus must consequently be spared. This is summarised in Fig. 5.3.

The first use of electrolytic LC lesion to test a role for brain noradrenaline in learning was reported by Anlezark, Crow & Greenway (1973). They found that animals with *large* bilateral LC lesions showed decreased noradrenaline content of the cortex to less than 30 % of normal. These animals were trained to run along an L-shaped runway to reach wet mash reward at the end of it. Figure 5.4 shows the learning over the first 4 or 5 days of control animals (either unoperated, with small lesion in the cerebellum dorsal to the LC or with lesions in the brainstem lateral and ventral to the LC). LC lesioned animals failed to show any learning in terms of the speed to traverse the runway despite training being continued for 16 days. This experiment then would seem directly to fit the predictions of the noradrenaline theory of learning.

However, other authors have failed to find learning impairments after

LC lesions. Amaral & Foss (1975) produced LC lesioned animals with *better* noradrenaline loss in the cortex than those in the study of Anlezark *et al.* (1973) but without finding a learning impairment in a T-maze. Amaral & Foss (1975) found a noradrenaline depletion to about 20 % of normal with some rats having a mere 3 % remaining. Although they report slower running along the T-maze, suggesting a possible motor deficit, they found that the two groups did not differ in the number of training days required to learn an olfactory discrimination for food reward (correct arm of the T-maze marked with 1 % amyl acetate solution).

Sessions, Kant & Koob (1976), although agreeing with Anlezark *et al.* (1973) that LC lesions which in this case depleted cortical noradrenaline to about 55 % of normal slowed the acquisition of a running response in an L-alley-way for one 90 mg food pellet, found that learning of other tasks was not affected. Acquisition of a taste aversion (see Appendix) to cyclophosphamide occurred identically in control and lesioned rats, as did learning of one-way active avoidance and CRF lever-pressing in an operant box for food reward (see Fig. 5.5).

Similar results were obtained by Koob, Kelley & Mason (1978), who again confirmed the slower running in an L-shaped runway by LC lesioned rats but found that these same animals could successfully learn a left–right discrimination in a T-maze, a CRF lever-pressing task and a successive visual discrimination task in an operant box (see Appendix). The conclusion of the last three studies then is that, although LC lesioned rats run more slowly to food reward in an alley-way, this is not a global learning deficit, since they can successfully learn a wide range of other tasks without difficulty.

Continuing the list of tasks showing unimpaired learning in LC lesioned

Fig. 5.3. Diagram of parasagittal view of locus coeruleus (LC) indicating in (*a*) that a small electrolytic lesion which does not damage nonnoradrenaline cells must inevitably spare much of the LC, and in (*b*) that an electrolytic lesion large enough to destroy all of the LC must inevitably damage nonnoradrenergic areas as well.

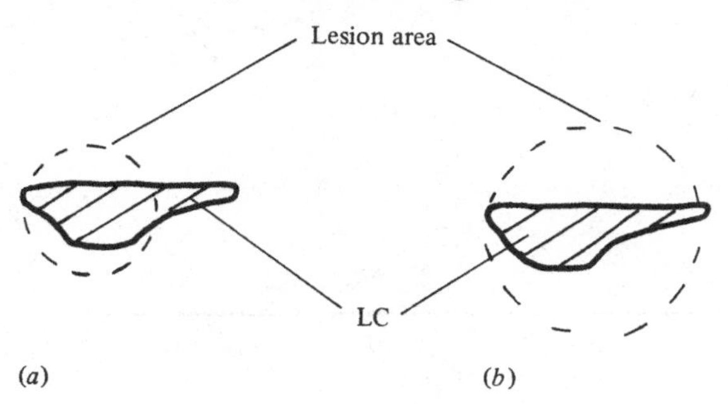

rats, Crow, Longden, Smith & Wendlandt (1977) found that LC lesions did not cause any alteration in the acquisition of a three-compartment linear maze task for food reward and only a marginally significant impairment in the acquisition of two-way active avoidance in a shuttle box (a one-tailed t-test was just significant at the 5% level for comparison of the most severe LC lesions with the least severe LC lesions). However, comparison of either LC lesioned group with more appropriate controls such as cerebellar or brainstem lesioned animals was *not* significant; see Fig. 5 and Table 2 of Crow *et al.* (1977).

A second approach to the role of the LC in learning and memory was adopted by Zornetzer and associates. They examined the period of time after learning for which the memory of the learning experience was susceptible to disruption by electroconvulsive shock (ECS). For a few

Fig. 5.4. Acquisition learning of a runway response for wet mash reward following large electrolytic lesions to the locus coeruleus (LC). Time to run down the alley-way is plotted against 16 days of training. Total LC is a group of three animals which had near total destruction of the LC. (Redrawn from Anlezark *et al.*, 1973.)

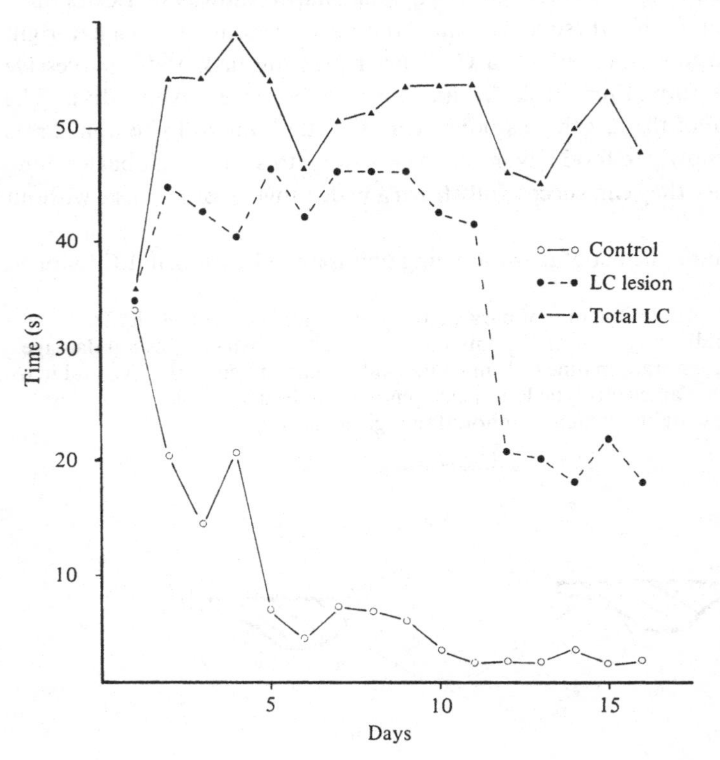

hours after learning ECS can induce retrograde amnesia, rather like DDC administration (see p. 202). Presumably during this time the memory is stored in a short-term memory store and is being consolidated into the more permanent long-term store.

Zornetzer & Gold (1976) made bilateral lesion of the LC through indwelling electrodes immediately after training mice on a step-through avoidance task. No impairment of retention when tested 48 h later was seen. If ECS was applied transcorneally immediately after training, normal

Fig. 5.5. Effects of locus coeruleus (LC) lesions on: (*a*) learning of a runway response for food reward. (*b*) learning and retention of a taste aversion. C = Conditioning day, cyclophosphamide injected after saccharin consumption on this day only. (*c*) learning of one-way active avoidance; (*d*) learning of a continuously reinforced lever-press response for food reward. (Redrawn from Sessions, Kant & Koob, 1976.)

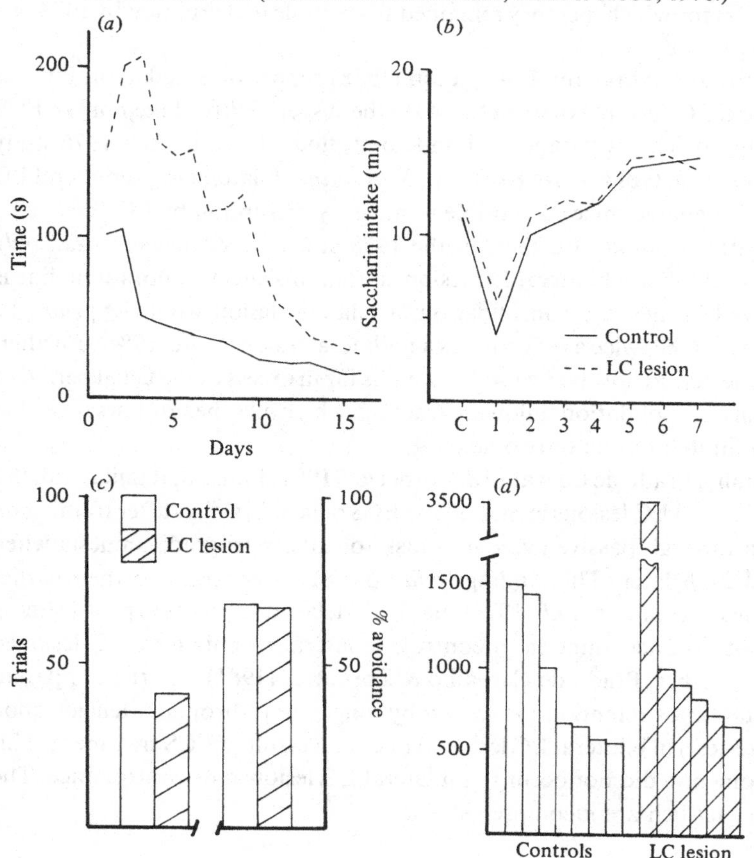

mice showed amnesia upon 8 or 24 h re-test. If the ECS was delayed until 40 h after training, normal mice did not show amnesia upon 8 or 24 h post-training. However, LC lesioned mice showed amnesia in response to ECS either administered immediately after training *or* delayed 40 h. Delay of the ECS to 7 days post-training resulted in its being ineffective in both the normal and the LC lesioned mice. This is interpreted by the authors to indicate that memory remains susceptible to disruption for longer in LC lesioned mice than in normals, and hence that the noradrenergic systems are involved in the process of memory consolidation in the intact animal. This argument is vitiated by the subsequent report that destruction of noradrenergic systems in the brain increased the intensity of ECS-induced seizures in rats (Mason & Corcoran, 1978). Thus, in effect the LC lesioned mice of Zornetzer & Gold (1976) were receiving a more severe amnestic treatment than their controls and it could be the effective magnitude of the disruption which was increased in the lesioned animals rather than the duration for which memory remained susceptible to disruption by ECS of a given intensity.

Zornetzer, Abraham & Appleton (1978) report that unilateral but not bilateral LC lesions extended to 7 days the susceptibility of memory to ECS disruption. This is paradoxical on two grounds. First, in their 1976 study (Zornetzer & Gold, 1976; Expt 3, p. 334) *neither* bilateral nor unilateral LC lesions rendered mice susceptible to memory disruption by ECS delayed 7 days after training. Further, in the 1978 study by Zornetzer *et al.*, *only* unilateral but *not* bilateral LC lesion caused this effect. If noradrenaline is involved in memory consolidation, a bilateral lesion would, *a priori*, be expected to be twice as effective as a unilateral lesion (Rose, 1981). Further, the time period now being used, 7 days, is far in excess of the usual period of memory consolidation following training, which all other authors regard as being limited to one to two *hours*.

Finally, Prado de Carvalho & Zornetzer (1981) found that unilateral, but not bilateral LC lesions in mice allow ECS delayed 14 days after training on a step-through passive avoidance task to cause retrograde amnesia when tested 24 h later. This finding (*ibid.*, p. 181) is contrary to their earlier findings (Zornetzer *et al.*, 1978) that ECS delayed for 14 days post-training did not produce amnesia in controls, bilateral or unilateral LC lesioned mice. Further, Prado de Carvalho & Zornetzer (1981) report an apparent facilitation of retention (as shown by longer step-through latencies upon 24 h re-test) in bilateral LC lesioned mice as a result of ECS treatment. This phenomenon did not occur in unilateral LC lesioned or control mice. The results as such are incomprehensible.

In summary, electrolytic lesions to the LC in rats appear to impede the running of an L-shaped runway for food reward but affect no other learning task examined (CRF lever-pressing, active avoidance, taste aversion, successive visual discrimination, spatial discrimination in a T-maze). Thus, the effects of LC lesions in rats do not support a role of noradrenaline in learning. Failure to find general learning impairments in such animals is in fact direct evidence against such a role. That is, animals can learn as well as normal even in the absence of over 80 % of the noradrenergic innervation to the cortex. This will be expanded further in the next section.

The data on the extended susceptibility of memory to ECS disruption by LC lesions cannot be taken as evidence for a role of noradrenaline in memory consolidation. First, ECS itself is potentiated by LC lesions, so a more intense amnestic treatment was being used in the lesioned compared to the control mice. Secondly, the apparent result that unilateral *but not bilateral* LC lesions give this effect has not been adequately explained. Thirdly, the internal inconsistencies as to how long after training ECS can disrupt memory in LC lesioned mice renders the results of dubious value.

It must be concluded that not only have the studies on LC lesions and learning failed to support a role for noradrenaline in learning and memory – they have actually provided some of the most direct evidence *against* such a role.

6-OHDA lesions and learning

Deletion of brain noradrenaline by the neurotoxin 6-OHDA overcomes some of the drawbacks of electrolytic lesion of the LC. Thus, no effect on other, nonnoradrenergic, systems will occur. A severe and permanent destruction of noradrenergic neurones can be achieved, some depletions exceed 95 % and all are routinely better than 85 %. The most used technique has been to inject 6-OHDA in microgram amounts into the dorsal bundle where the noradrenergic fibres are concentrated as they run from the LC to innervate the forebrain. It will be recalled that this is the pathway which was invoked by Crow (1973) as being the critical one for the noradrenaline theory of learning (see Chapter 3). Examination of a possible role of the ventral noradrenergic systems in learning will be postponed until a subsequent section in this chapter.

Table 5.1 shows the effect that bilateral injection of 4 μg 6-OHDA dissolved in 2 μl of saline with 0.2 mg $^{-1}$ ascorbic acid antioxidant on the concentration of noradrenaline, dopamine and serotonin and on the

Table 5.1. *Post-mortem assays on control and dorsal bundle lesion rats*

Dorsal bundle 6-OHDA	Control ($n=10$)	Lesion ($n=10$)	%	P
Noradrenaline				
Cortex	289±14	9±6	3	0.001
Hippocampus	305±18	15±3	5	0.001
Hypothalamus	2380±110	860±90	36	0.001
Amygdala	408±5	61±4	15	0.001
Septum	954±12	435±34	46	0.01
Cerebellum	219±12	271±8	124	NS
Spinal cord	255±6	307±12	120	NS
Dopamine				
Striatum	13570±660	1284±1230	95	NS
Amygdala	88±9	66±27	75	NS
Septum	650±70	500±20	77	NS
Hypothalamus	452±19	421±26	93	NS
Serotonin				
Cortex	377±21	382±19	101	NS
Hippocampus	493±38	485±41	98	NS
Hypothalamus	1206±110	1156±96	96	NS
Choline acetyltransferase				
Cortex	13.4±0.7	13.1±0.6	98	NS
Hippocampus	15.0±0.5	14.9±0.6	99	NS
Hypothalamus	11.9±1.68	10.2±1.14	86	NS
Septum	15.1±0.8	13.7±0.4	91	NS
Amygdala	18.4±1.5	19.6±1.2	106	NS
Glutamic acid decarboxylase				
Cortex	12.1±0.4	11.3±0.4	93	NS
Hippocampus	10.8±0.4	11.0±0.5	101	NS
Hypothalamus	25.0±1.2	23.6±1.5	91	NS
Septum	16.7±0.9	16.3±1.3	98	NS
Amygdala	10.2±0.2	11.0±0.8	108	NS

Values are means with standard error of the mean. Noradrenaline, dopamine and serotonin values are in nanograms of amine per gram wet weight of tissue and choline acetyltransferase and glutamic acid decarboxylase are micromol per 100 mg protein per hour. % column represents values of lesion groups expressed as percent of control values. *P* column is the two-tailed significance of the difference between control and lesion groups using Student's t-test. NS = not significant.

activities of cholinergic and GABAminergic enzymes in localised regions of the rat brain. It can be seen that severe and permanent loss of the noradrenergic innervation to the forebrain (cortex, hippocampus, and, due to the higher percentage of ventral bundle innervation, to a lesser extent of the amygdala and septum) occurs with no alteration in dopamine, serotonin, acetylcholine or GABAminergic systems in any brain area measured. The noradrenergic innervation of the cerebellum and spinal cord is also spared, perhaps even slightly elevated.

A second technique used to deplete dorsal bundle noradrenergic systems involves administration of 100 mg kg^{-1} 6-OHDA intraperitoneally to neonatal rats on days 1, 3, 5, 7, 9, 11 and 13 after birth. For reasons still not understood, this destroys the noradrenergic innervation of dorsal bundle terminal areas without affecting either the ventral noradrenergic system or the dopaminergic systems. This technique was first described by Clark, Laverty & Phelan (1972). The regional noradrenaline depletions typically obtained are shown in Table 5.2.

Positive reinforcement. The first learning task examined with these noradrenaline depleted rats was the L-shaped alley-way of Anlezark *et al.* (1973), in which LC lesioned rats are found to be impaired. Thus, Mason &

Table 5.2. *Neonatal peripheral treatment*

Regional concentrations of amines; values are means in ng g^{-1} tissue with standard error of the mean.

	Control ($n=10$)	Treated ($n=9$)	%	t-statistic
Noradrenaline				
Hippocampus	374±48	23±6	6.2	7.2**
Hypothalamus	756±97	843±146	112	0.49
Cortex	238±24	6±9	2.5	9.0**
Cerebellum	247±37	29±26	11.7	4.8**
Brainstem	185±53	342±29	184	2.6*
Dopamine				
Hypothalamus	361±152	339±87	94	0.12
Cortex	143±97	163±58	114	0.17
Striatum	4310±540	4639±496	107	0.65
Brainstem	40±15	58±18	145	0.76

* Significant at 5 % level.
** Significant at 0.1% level.
From Mason, 1976.

Iversen (1975) looked at the rate of learning of this approach response for food reward in both the dorsal bundle lesioned (Table 5.1) and the neonatally treated animal (Table 5.2). In neither group was any sign of slower learning observed (see Fig. 5.6). Both lesioned and control rats learned that food was present at the end of the runway, as shown by the speeding up of running times at an equal rate over days. Negative results are always to be interpreted with some caution. Perhaps the noradrenaline depletion technique had failed to work and one was really testing two groups of normals. To check this, the animals which were tested on the L-shaped alley-way were sacrificed and the noradrenaline levels in their brains assayed. This showed that the lesions were indeed effective in reducing noradrenaline levels to below the detection of a highly sensitive radio-enzymatic assay. Further, the lesions, although not altering acquisition learning in this alley-way, were effective in changing extinction behaviour (see Chapter 6). This must be interpreted as strong evidence against a role of noradrenaline in learning, despite the effects that LC lesions may have on this running behaviour.

Fig. 5.6. Effects of 6-hydroxydopamine (6-OHDA) lesions, either of the dorsal bundle when adult (DB), or via the intraperitoneal route when neonate (NPT), on learning of a runway response for food reward. (Redrawn from Mason & Iversen, 1975.)

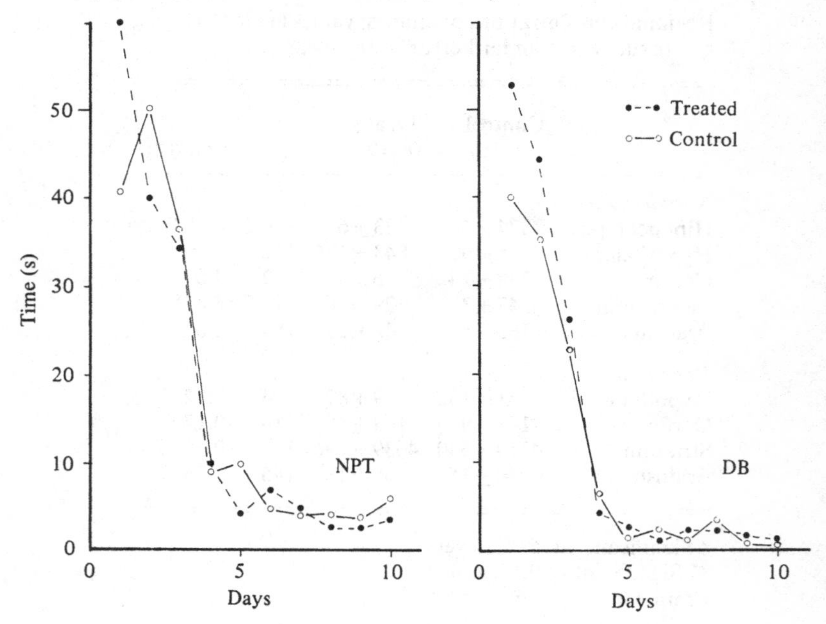

A virtually identical experiment carried out by Owens, Boarder, Gray & Fillenz (1982) using a straight alley-way showed the same result. Despite noradrenaline depletions better than 90 % the lesioned rats learned to run for food reward every bit as quickly as controls.

A third experiment to examine alley-way running was that of Roberts, Price & Fibiger (1976), who used either an L-shaped alleyway or a T-maze and found that, despite noradrenaline depletions better than 93 % after dorsal bundle 6-OHDA injection, both acquisition learning of the L- and a spatial discrimination in the T-maze proceeded as quickly as in controls. Reversal of the position habit in the T-maze was also unaffected by the noradrenaline depletion (see Fig. 5.7). Morris, Tremmel & Gebhart (1979) also reported unimpaired runway acquisition after noradrenaline depletions caused by electrolytic lesion of the dorsal bundle.

Various operant schedules have been examined after noradrenaline depletion. They range from simple lever-pressing on a CRF schedule to complex DRL tasks (see Appendix). Mason & Iversen (1977*a*, 1978*a*) examined both dorsal bundle and neonatally noradrenaline-depleted rats on the acquisition learning of a CRF operant lever-press. Animals were food deprived and allowed to explore the chamber to discover food pellets taped to the lever. During the course of trying to dislodge these pellets all the control animals learned that lever-pressing delivered a food pellet from

Fig. 5.7. Effects of 6-hydroxydopamine (6-OHDA) lesions of the dorsal bundle on learning and reversal of a spatial position habit in a T-maze. (Redrawn from Roberts *et al.*, 1976.)

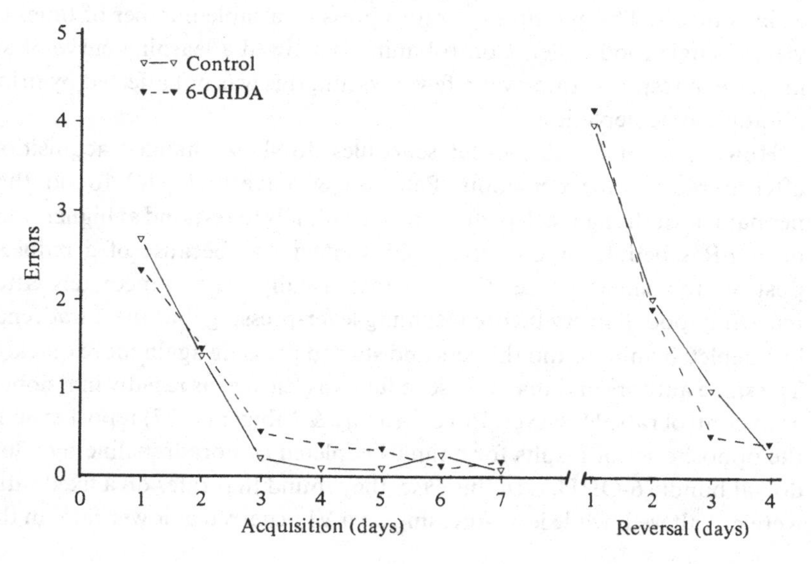

the automatic dispenser, and over the next 5 days their lever-pressing rates rose to a plateau. This occurred equally for either noradrenaline depleted group (dorsal bundle or neonatal). No impairment of the learning of a simple lever-press was seen despite noradrenaline loss exceeding 85 %.

Learning *de novo* of a FI was found to be unimpaired in dorsal bundle 6-OHDA rats by Mason & Iversen (1978*b*). This schedule is more complex than CRF, since it requires the animal to associate the passage of time with the point at which a lever-press becomes rewarded. Nonetheless, this more complex task was not affected by noradrenaline depletion. Nor were other tasks requiring timing or behavioural sequencing. Thus, DRL learning occurred as normal in either dorsal bundle or neonatally noradrenaline depleted rats (Mason & Iversen, 1977*b*) if the animals were placed in the operant box with a minimum of shaping. If they were trained extensively on CRF prior to the transfer to DRL, the lesion still failed to result in an impairment of DRL learning, in fact somewhat better learning was seen in the noradrenaline depleted groups (Mason & Iversen, 1977*b*). Noradrenaline depletions were in excess of 85 %. Similar successful learning of a DRL for water reinforcement was reported by Levine, McGuire, Heffner & Seiden (1980*b*) using the less sophisticated technique of intraventricular 6-OHDA administration to deplete forebrain noradrenaline to 20 % of normal.

Higher-order schedules, in which more than one lever-press has to be emitted per food pellet, have also been examined. Mason & Fibiger (1978*a*) found that acquisition learning of a variable ratio (VR, see Appendix) schedule occurred as rapidly in dorsal bundle noradrenaline depleted rats as in controls. This requires the rat to press a variable number of times to yield a single food pellet. Control animals showed a learning curve of an increase in response rate over a few days and this was not affected by prior noradrenaline depletion.

However, some higher-order schedules do show changed acquisition after noradrenaline depletion. Peterson & Laverty (1976) found that neonatal noradrenaline depletion caused animals to respond at higher rates on a FR schedule than controls. This effect was because of a reduced post-reinforcement pause. Control rats usually wait immediately after receiving food delivery before resuming lever-pressing, but the noradrenaline depleted animals did this less and started pressing again more quickly. The same authors find that a VI schedule was learned as rapidly in lesioned as in control rats. However, Price, Murray & Fibiger (1977) report exactly the opposite set of results for animals depleted of noradrenaline by adult dorsal bundle 6-OHDA. In this case, they found that rates on a fixed ratio were not altered, while lever-pressing on a VI occurred at lower rates in the noradrenaline depleted animals. They concur with Mason & Iversen (1978*a*) that response rates on CRF were unaltered. Use of an intraventricular 6-OHDA technique by Levine, Erinoff, Dregits & Seiden (1980*a*), which also depleted dopamine, resulted in a doubling of response rate on a VI schedule compared to controls. Similar intraventricular injections into neonates failed to yield a statistically significant elevation in VI responding (despite the hint to be seen in Fig. 1 in Levine *et al.*, 1980*a*). Thus, VI responding may decrease (Price *et al.*, 1977) or be unaffected by noradrenaline depletion (Peterson & Laverty, 1976; Levine *et al.*, 1980*a*). This suggests no great role of noradrenaline in the learning of the reinforcement contingencies involved in a VI operant task. Rates on a fixed ratio may either increase (Peterson & Laverty, 1976) or remain unaltered (Price *et al.*, 1977). Again no support for a global role of noradrenaline in learning is to be found herein.

Other complex tasks have been examined after noradrenaline depletion. In order to tax the system as much as possible a highly difficult motor manipulative task was examined by Mason & Iversen (1977*c*). This required the rat to pull a ball out of a tunnel by the handle attached to the ball, in order to clear the tunnel of the obstruction and allow the rat to run through to retrieve the food pellets present on the other side. So complex is this task that not all normal rats can manage to master it on occasion (T. J. Crow, personal communication). Nonetheless, all the dorsal bundle 6-OHDA and the neonatally noradrenaline depleted rats of Mason & Iversen (1977*c*) learned it successfully and as quickly as controls. Other subtle tests of association ability have been carried out. Mason & Robbins (1979) examined the ability of the rat to learn that two initially neutral stimuli (the click of the feeder and the flash of the light illuminating the food magazine) were associated with food. Every time rats pressing on a VR schedule received a food pellet the flash of light and click of the feeder were presented. The animals were then tested in the absence of food for their preference for a lever which produced one of the two initially neutral stimuli (light or click) over a second lever which did nothing. Normal rats markedly preferred the lever giving the stimuli, indicating that they had learned the association of stimulus with food. 6-OHDA bundle lesioned rats showed exactly the same degree of preference, indicating that the same stimulus–food associative learning had taken place despite noradrenaline depletions better than 90 %. Again, no evidence for a role of noradrenaline in learning can be drawn from this test.

However, not all acquisition tasks proceed as usual following noradrenaline depletion. Mason & Iversen (1977*a*, 1978*c*) found that acquisition learning of a successive visual discrimination task was markedly impaired by noradrenaline depletion. This task presented periods of time with a light stimulus (S+), indicating that lever-pressing would be rewarded on a CRF schedule, with other periods during which the light was absent (S−) and lever-pressing did not yield a food pellet. Noradrenaline depleted rats continued to press during the unrewarded periods while controls soon came to restrict their responding to the S+ periods only. Acquisition learning of a visual–auditory successive discrimination similar to the above was also impaired by noradrenaline depletion with adult dorsal bundle 6-OHDA injections (Mason & Fibiger, 1979*b*). Acquisition of a T-maze discrimination using different levels of overhead illumination as the sole cue was also slower in the dorsal bundle noradrenaline depleted rats (Mason & Lin, 1980) as was a discrimination task using three simultaneous visual dimensions (cited in Leconte & Hennevin, 1981, p. 592). It seems that any task requiring the use of stimuli, other than simple spatial ones, for a discrimination habit may be impaired by noradrenaline depletion. This will be expanded in Chapter 6 (see p. 277).

Leconte & Hennevin (1981) found that complex maze learning was impaired by noradrenaline depletion. Use of a multiple-choice point maze (essentially three T-mazes joined end to end) resulted in slowed learning in animals depleted of brain noradrenaline by localised 6-OHDA administration. Again this is covered in the section on noradrenaline and selective attention in the Chapter 6.

In summary, the vast majority of studies using positive reinforcement to cause learning in noradrenaline depleted rats have failed to find any impairment of that learning. Inconsistent changes have been reported on FR and VI schedules, but not in the fashion to support a role of noradrenaline in global learning processes. Clear deficits in acquisition of stimulus discriminations (other than simple spatial ones) have been found but seem to relate more to selective attention (see p. 277) than to a global block of learning. Thus, most of the evidence from 6-OHDA-induced noradrenaline depletion and positively reinforced tasks is directly against the noradrenaline theory of learning.

Negative reinforcement

Central noradrenaline. Crow & Wendlandt (1976) reported that 6-OHDA lesion to the dorsal bundle impaired 72 h retention of a passive avoidance step-down task. Immediate memory as measured 1 min after the single-shock learning trial was not affected (see Fig. 5.8). This would seem to support a direct parallel with the similar experiment carried out with DDC, a noradrenaline synthesis inhibitor (see p. 202). Rainbow, Adler & Flexner (1976) report a similar passive avoidance deficit after intraventricular injection of 25 μg of 6-OHDA in mice. However, against this must be set the results of Fibiger, Roberts & Price (1975*a*), who failed to find any effect of 6-OHDA lesions of the dorsal bundle on 72 h retention of passive avoidance learning. Similarly negative results were reported by Roberts & Fibiger (1977), by Mason & Fibiger (1979*c*), by Mason, Roberts & Fibiger (1979*b*), and by Mason & Fibiger (1978*b*) for retention of passive avoidance. On balance, it would appear that, despite depletions of forebrain noradrenaline in excess of 95%, retention of a passive avoidance task is not generally affected.

Fibiger & Mason (1978) examined the acquisition learning of a one-way active avoidance task after 6-OHDA dorsal bundle lesion. Despite large depletions of noradrenaline, which did in fact alter extinction behaviour

Fig. 5.8. Effects of 6-hydroxydopamine (6-OHDA) lesion of the locus coeruleus on passive avoidance learning and 72 h retention. Electric footshock was given upon stepping down during trial 3 (shock). Retention was tested 1 min and 72 h later. (Redrawn from Crow & Wendlandt, 1976.)

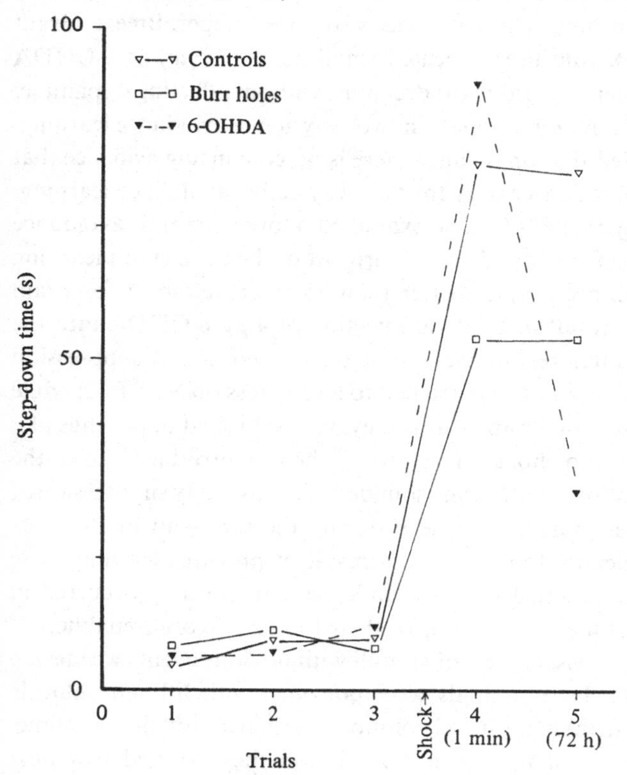

(see p. 265), no impairment of acquisition learning was seen. Using the novel neurotoxin of unknown mechanism DSP4 (*N*-(2-chloroethyl)-*N*-ethyl-2-bromobenzylamine), Ogren, Archer & Ross (1980) found an impairment in one active avoidance learning experiment but not in a second group of rats tested identically. On balance there is no convincing evidence that central noradrenaline depletion alters the acquisition learning of one-way active avoidance.

Mason & Fibiger (1979*a*) also examined the acquisition learning of a two-way active avoidance task in a shuttle box after 6-OHDA dorsal bundle lesion. Rather than an impairment, a small but significant improvement in learning was seen. Against this must be set the report of Ogren *et al.* (1980) that DSP4, which depleted brain noradrenaline to some 3–15 % of normal, severely impaired the acquisition learning in two groups of rats on two-way active avoidance. However, it must be remembered that peripheral administration of DSP4 will deplete the peripheral sympathetic, as well as the central noradrenergic nervous system. Since the peripheral noradrenergic system has a clear role in avoidance learning (see p. 222), DSP4 impairment cannot be accepted as evidence for a role of *central* noradrenaline in learning. This is especially so since Cooper, Breese, Grant & Howard (1973*a*), who used repeated small doses (25 μg) of 6-OHDA intracisternally to deplete brain noradrenaline without affecting dopamine, also found a significant *improvement* in two-way active avoidance learning. It must be concluded that on balance there is no convincing evidence that *central* noradrenaline is necessary for two-way active avoidance learning.

Mason & Fibiger (1979*d*) also examined more complex avoidance learning situations. They found no impairment of the acquisition learning of a Sidman avoidance despite better than 95 % depletion of forebrain noradrenaline as a result of bilateral injection of 4 μg 6-OHDA into the dorsal bundle. Similar results were seen on a conditioned suppression paradigm, in which animals were trained to lever-press on a CRF schedule for food reward. In another apparatus they were subjected to pairings of a tone with electric footshock. They were then reintroduced into the lever-pressing situation and the tone sounded. Considerably suppression of ongoing CRF lever-pressing was seen during the tone – an index of its association with electric footshock as a result of previous learning. The same degree of conditioned suppression as seen in controls occurred in animals depleted of more than 95 % of their forebrain noradrenaline.

A similar complex association of stimuli with noxious events was tested in a taste aversion. Here animals were poisoned with lithium chloride immediately after drinking a novel solution, saccharin, for the first time. On subsequent offers of the saccharin solution they rejected it almost completely as a result of associating its taste with the noxious consequences of lithium chloride injection. Such taste aversion learning was not impaired by 6-OHDA lesion to the dorsal noradrenergic bundle (Mason & Iversen, 1978*c*; Mason & Fibiger, 1979*e*).

Thus, none of the complex negatively reinforced learning situations (Sidman avoidance, conditioned suppression or taste aversion) is affected by severe noradrenaline loss.

Interaction with adrenal hormones. Although depletion of central noradrenaline in the dorsal bundle projection areas on its own does not affect a whole variety of negatively reinforced learning, additional removal of the adrenal glands does cause a profound deficit. The first report of this was from Ogren & Fuxe (1974), who found that neither dorsal bundle 6-OHDA lesion of central noradrenaline nor adrenalectomy on their own altered acquisition learning of a two-way active avoidance task. However, the two together produced animals which were unable to re-learn the two-way active avoidance response. Again, Ogren & Fuxe (1977) showed a similar effect on one-way active avoidance learning and performance. Roberts & Fibiger (1977) extended this pattern to passive avoidance step-down, and this was confirmed by Mason *et al.* (1979*b*).

In all of these experiments the pattern was identical. No learning or performance deficit was seen with 6-OHDA dorsal bundle lesion. No impairment was seen after adrenalectomy on its own. When dorsal bundle lesioned rats were adrenalectomised, then and only then, a severe impairment in the learning *de novo* and in the performance of a previously learned negatively reinforced task was seen. This applies for one-way, two-way and passive avoidance tasks. A neurochemical basis for this unexpected interaction is provided by the report of Roberts & Bloom (1981) that 6-OHDA lesion to the dorsal bundle raised β-binding in the hippocampus by 40 % while an additional adrenalectomy raised it to 70 %. Adrenalectomy *per se* did not affect β-receptor binding. It is, of course, unknown what would stimulate these β-receptors since the afferent noradrenergic innervation which normally does this has been almost totally removed by the lesion (greater than 95 %). It has been shown that the change in both binding and behaviour results from removal of corticosterone as a consequence of adrenalectomy, since replacement therapy prevents the acquisition impairments and the biochemical changes (Ogren & Fuxe, 1977; Roberts & Bloom, 1981). However, even this interaction with adrenal hormones is not evidence for a role of noradrenaline in general learning, since Mason *et al.* (1979*b*) showed that animals with dorsal bundle lesions and adrenalectomies could learn a CRF lever-press for food reward as well as did controls. These same animals were, however, impaired in 24 h retention when it came to a passive avoidance step-down task. This would imply that the deficit is one of fear motivation rather than a global block of learning such as would be required by the noradrenaline theory of learning.

In summary, an interaction of central noradrenergic systems with peripheral hormones is indicated, but only in fear motivation not in general learning.

Peripheral noradrenergic systems. A significant amount of evidence has accumulated concerning a role of the peripheral sympathetic nervous system in avoidance and fear learning. Although possibly 'peripheral' also to our main concern of a role for central noradrenaline in learning, it will be mentioned here, since much of the evidence has been derived from 6-OHDA lesions.

Di Guisto (1972) gave three injections of 6-OHDA (30 mg kg^{-1}) intraperitoneally separated by 24 h and then 6 h. This peripheral administration of 6-OHDA to adult animals results in temporary damage to the peripheral noradrenergic system without effect on central noradrenaline since 6-OHDA does not cross the blood–brain barrier in the adult, as opposed to the neonatal rat. After some two to four weeks, regeneration of the peripheral noradrenergic system occurs, something seen only to a much lesser extent in the central nervous system. Di Guisto (1972) found such peripheral noradrenaline depletion to slow learning of a step-down passive avoidance task compared either to saline controls or to rats in which adrenaline, but not noradrenaline, had been depleted peripherally by adrenal demedullation (see Fig. 5.9). Such impaired acquisition of a step-down passive avoidance task assumes especial importance when it is recalled (see above) that depletion of *both* central and peripheral noradrenaline by synthesis inhibition with DDC impairs this task whereas depletion of central noradrenaline only by 6-OHDA in general does not. It may be suggested, therefore, that the effects of DDC on memory, since they have almost always been studied in avoidance paradigms, might rather reflect a role of *peripheral* noradrenaline in fear-motivated learning, not a general role of central noradrenaline in global learning that some authors have espoused.

Di Guisto & King (1972) used a single peripheral injection of 30 mg kg^{-1} 6-OHDA 2 h prior to training on acquisition learning of two-way active, one-way active or trace one-way active avoidance tasks. Severe impairments were seen on the acquisition learning of two-way active and trace one-way avoidance but not on the simpler one-way active avoidance paradigm (see Fig. 5.10). The authors suggest that peripheral noradrenergic systems are particularly necessary for complex avoidance learning and the one-way active avoidance paradigm is too simple to reveal this. This is supported by the findings of Pappas & Sobrian (1972), who performed peripheral sympathectomy at birth with intraperitoneal 6-OHDA injections (at which time the destruction of the peripheral noradrenergic systems is permanent) and found unimpaired learning of one-way active avoidance when adult.

Similar deficits in two-way active avoidance learning were reported by Lord, King & Pfister (1976) for animals subjected to 6-OHDA sympathectomy when adult or at birth. Previous experiments generally failed to determine the degree of depletion caused in the peripheral nervous system by peripheral 6-OHDA injection but this was rectified by Oei (1978), who found that a single intraperitoneal injection of 50 mg kg^{-1} 6-OHDA caused a 90 % decrease in noradrenaline content of the heart with no effect on whole-brain noradrenaline. Such rats were again impaired in the acquisition learning of two-way active avoidance. Oei & King (1978) used a

Fig. 5.9. Effects of destruction of the peripheral noradrenergic nervous system by intraperitoneal 6-hydroxydopamine (6-OHDA) injection when adult on passive avoidance learning. A group of adrenalectomised rats (Adrenals) is also shown. (Redrawn from Di Guisto, 1972.)

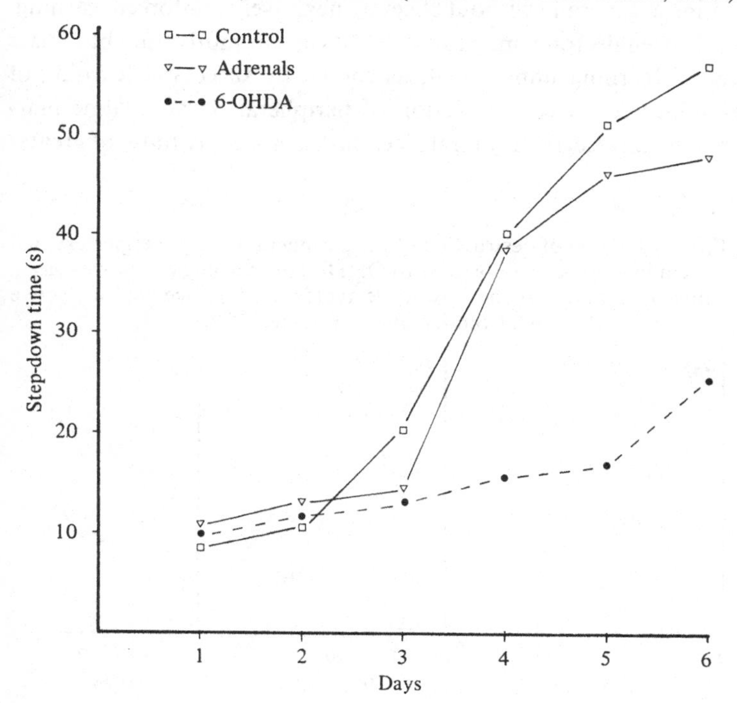

similar injection procedure to deplete heart noradrenaline to 17 % of normal and found impaired acquisition learning of a trace one-way active avoidance, thus confirming earlier results. Finally, Oei & Ng (1978) confirmed the impairment in learning of passive avoidance step-down as a result of a single intraperitoneal 6-OHDA injection of 50 mg kg^{-1}, which depleted heart noradrenaline to 28 % of normal.

It is thus manifestly clear that depletion of *peripheral* noradrenaline will severely retard learning of passive avoidance, two-way active or trace one-way active avoidance tasks but not the simpler one-way active paradigm. These results stand as a cautionary note to experimenters who administer noradrenaline synthesis inhibitors intraperitoneally and then assume that the results are due solely to the central noradrenaline loss caused.

It should be emphasised that the results described above are not brought about by a general learning impairment, but may be specific to fear motivation, since Mason & Iversen (1978*a*) showed that depletion of peripheral noradrenergic systems by two doses of 30 mg kg^{-1} 6-OHDA failed to impair acquisition learning of a positively reinforced CRF lever-press task for food.

Conclusion. The above sections indicate that depletion of central noradrenaline is generally without effect on negatively reinforced learning. Addition of adrenalectomy may cause deficits in fear motivation but this is not a general learning impairment, as shown by successful learning of positively reinforced tasks. Depletion of peripheral noradrenaline markedly impairs many negatively reinforced tasks, especially those of greater

Fig. 5.10. Effect of destruction of the peripheral noradrenergic nervous system by 6-hydroxydopamine (6-OHDA) on learning of (*a*) two-way active avoidance, (*b*) one-way active avoidance, (*c*) trace one-way active avoidance. (Redrawn from Di Guisto & King, 1972.)

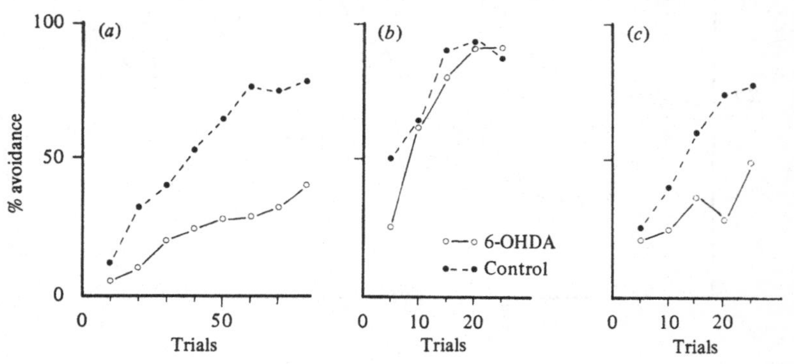

difficulty. Again, this is not a general learning impairment but perhaps a change in fear motivation, as shown again by the unimpaired learning of positively reinforced tasks.

Thus, 6-OHDA lesions and negatively reinforced learning provides direct evidence against a role of *central* noradrenaline in learning and memory. This, taken together with an equally strong negative conclusion from 6-OHDA lesions and positively reinforced learning, and that from LC electrolytic lesions, offers strong disproof of a role for central noradrenaline in learning; that is, as originally formulated by Crow (1972) to invoke the dorsal noradrenergic bundle system. I shall now examine the relatively sparse evidence concerning the role of ventral noradrenergic systems in learning.

Ventral bundle lesions

If manipulations which alter whole brain noradrenaline are found to affect learning and memory (such as synthesis inhibition by DDC) but manipulations of the dorsal system on its own are without effect (6-OHDA dorsal bundle lesion), it may be that the remaining noradrenergic systems (projections to the spinal cord or to the cerebellum or via the ventral systems) may be involved. Only a few studies have examined a role for ventral bundle noradrenaline in learning and memory. Kruk & Millar (1979) found that 6-OHDA lesion of the ventral bundle which depleted hypothalamic noradrenaline to 20 % of normal reduced the rate of lever-pressing for CRF food reward compared both to saline-injected controls and to pre-operative levels. When the schedule was increased to FR3 or FR10, the lesioned rats were reluctant to increase their response rates and considerable impairment relative to normal rats was seen. Since this was on the performance of an already learned task, no evidence for a role of ventral bundle noradrenaline in learning *de novo* is offered but it is suggestive of a motivational role in hunger or food intake. Relevant to this is the hyperphagia seen after ventral bundle lesions and the evidence for hypothalamic noradrenergic receptors in feeding behaviour (see p. 347).

Kostowski & Plaznik (1978) found that electrolytic lesion to the ventral noradrenergic bundle facilitated acquisition learning of two-way active avoidance. This came about primarily as an increased activity and reduced freezing in the apparatus (as shown by greater inter-trial interval crossings by the lesioned rats). Normal rats tend to freeze in the apparatus in the early stages of learning, and, since this response is incompatible with the active avoidance response, such freezing tends to retard learning. Thus, it is uncertain whether the altered learning after ventral bundle lesions reflects learning processes *per se* or merely a change in locomotor activity. No evidence is provided that the ventral noradrenergic bundle subserves central reinforcement, since it is manifestly not necessary for learning to occur.

Cerebellar projection

Gilbert (1975) suggested a variant of the noradrenergic theory of learning in which the noradrenergic input to the cerebellum acted as a reinforcement pathway for *motor* learning in much the same way as that to the cortex was suggested by Crow (1972) to be a reinforcement pathway for associative, cognitive learning. On Gilbert's model only tasks tapping into motor learning strategies would be impaired by noradrenaline depletion. Thus, a separate test of this theory may be needed. Such was presented by Mason & Iversen (1977*c*) in which depletion of cerebellar noradrenaline to less than 12 % was effected by neonatal peripheral administration of 6-OHDA on days 1, 3, 5, 7, 9, 11 and 13 after birth. Despite this severe loss of cerebellar noradrenaline, the animals when adult were able to learn a complex motor manipulative task of pulling a ball out of a tunnel (see above) as rapidly and as successfully as controls. This would seem to imply that cerebellar noradrenaline is not necessary for even complex motor learning to occur.

Self-administration

One additional line of evidence often quoted as relevant to a role of noradrenaline in central reward processes comes from the self-administration paradigm. This involves the voluntary administration by the rat of drug solutions through a chronic indwelling intravenous cannula, as a consequence of a lever-press. If the only consequence of a lever-press is that a shot of intravenous drug solution is delivered and if a naive rat placed into the apparatus eventually comes to lever-press at quite high rates, then the consequence of lever-pressing (viz. drug injection) must be operationally reinforcing.

Relevant to a role for noradrenaline in reward is the finding that rats will self-administer clonidine, an α_2-agonist (Davis & Smith, 1977). This effect occurred only at 15 $\mu g\ kg^{-1}$, since 5 $\mu g\ kg^{-1}$ was not self-administered and 30 $\mu g\ kg^{-1}$ was severely toxic. Clonidine has been shown to stimulate histamine H_2 receptors (Audigier, Virion & Schwartz, 1976; Sastry & Phillis, 1977) as well as α_2-adrenergic ones and may also interact with adrenaline ones (Bolme *et al.*, 1974). As such, the mere fact that clonidine is self-administered proves very little. However, Davis & Smith (1977) find that self-administration of clonidine is blocked by the α-blocker phenoxybenzamine. The latter did not, however, alter self-administration of the dopaminergic agonist drug apomorphine, suggesting at least two independent reinforcement systems, one noradrenergic and another dopaminergic.

On the other hand, phenoxybenzamine fails to block the self-administration of amphetamine, producing either a nonspecific depression of responding (Yokel & Wise, 1975) in rats and in monkeys (Goldberg & Gonzales, 1976) or no effect whatsoever (Risner & Jones, 1976) in dogs. Similarly, cocaine self-administration is unaffected by either the α-blocker phentolamine or the β-blocker propranolol (de Wit & Wise, 1977).

Noradrenaline synthesis inhibition with DBH inhibitors does not affect self-administration of morphine (Schwartz & Marchok, 1974) or clonidine (Davis & Smith, 1977). Lesion of the noradrenergic systems by 6-OHDA injection into the dorsal bundle similarly fails to alter cocaine self-administration (Roberts, Corcoran & Fibiger, 1977).

Thus, noradrenergic systems do not seem to be necessary for general mediation of reward as measured by the above studies on cocaine, morphine, amphetamine and apomorphine self-administration. However, a special case may exist for alcohol self-administration.

Amit, Brown, Levitan & Ogren (1977) found that synthesis inhibition of noradrenaline with FLA 57 would cause the cessation of oral drinking of ethanol solutions in rats, a finding confirmed by Brown *et al.* (1977). Davis, Smith & Werner (1978) reported that synthesis inhibition with U 14,624 would suppress lever responding for intragastric infusion of ethanol in rats, without affecting lever-pressing maintained by intragastric sweetened mild reward. Amit, Levitan & Lindros (1976) showed that inhibition of noradrenaline synthesis with FLA 63 or with disulfiram would also suppress oral intake of ethanol in rats. Intraventricular injection of 6-OHDA which depleted noradrenaline would similarly suppress ethanol intake (Kiianmaa, Fuxe, Jonsson & Ahtee, 1975; Myers & Melchior, 1975; Brown & Amit, 1977), and Mason, Corcoran & Fibiger (1979*a*) showed that lesion specifically to the ascending projections in the dorsal bundle would prevent alcohol consumption by mouth in the rat. The β-blocker propranolol has also been reported to reduce the euphoriant effects of alcohol in human patients (J. H. Mendelson, A. M. Rossi & J. Bernstein, 1972, unpublished; Tyrer, 1972).

All of the diverse lines of evidence most strongly suggest a role for central noradrenergic systems in alcohol intake. It must be emphasised that this is not evidence for a role of noradrenaline in general reward, since only ethanol self-administration and not that of cocaine, morphine, amphetamine or apomorphine responds to noradrenergic manipulations.

Other authors have reported that microinjections of noradrenaline directly into brain structures will act as reinforcers and cause the learning *de novo* of tasks for no other reward. Cytawa, Jurkowlaniec & Biatowas (1980) found that injection of 21 μg of *l*-noradrenaline via bilateral chronic cannulae aimed at the lateral hypothalamus would cause the learning of a position habit in a T-maze. Thus, animals were injected with noradrenaline if they turned against the spontaneous side preference but with an equal volume of saline if they chose the preferred side. Within 4 days, 15 out of 16 rats were now turning into the previously non-preferred arm (*ibid.*, Fig. 2, p. 617). Additionally the running speed in the T-maze, which in the 10 days prior to noradrenaline injections was getting progressively slower, now started to speed up and was quite rapid by the third day of noradrenaline injections. Jurkowlaniec & Biatowas (1981) found a similar effect with cannulae aimed at the amygdala, but only if the animals were food deprived when tested. Although no food was ever present in the T-maze itself, the animals had to be hungry for noradrenaline injections to cause learning. Satiated rats did not show such learning. This suggests that the two noradrenaline reward systems in the amygdala and in the lateral hypothalamus are functionally dissociable and may subserve different purposes in the normal rat.

In summary, intracerebral microinjection of noradrenaline seems to provide the most suggestive evidence of a role for central noradrenaline in learning yet found. It is still incomplete since it has not been demonstrated that the injected noradrenaline was actually stimulating a noradrenaline receptor. It is known that noradrenaline can, for example, activate dopamine receptors (Iversen, 1975). To make this evidence more conclusive it must be shown that injections of dopamine do not reinforce learning and that the effect of noradrenaline is blocked by α- or β-blocker, but not by dopamine receptor antagonists.

Interaction with other systems

Peptides. The peptide hormone vasopressin facilitates consolidation of memory, while another peptide, oxytocin, attenuates consolidation of memory (Bohus, Kovacs & de Wied, 1978). While oxytocin does not appear to affect catecholaminergic systems (D. H. Versteeg, unpublished data cited in Kovacs *et al.*, 1979*c*, p. 304), vasopressin has been reported to affect the turnover of catecholamine in limbic areas (Kovacs, Vecsei, Szabo & Telegdy, 1977). Kovacs *et al.* (1979*c*) found that immediate post-trial injection of 25 pg vasopressin into the dorsal septal nucleus or the hippocampal dentate gyrus facilitated 24 h and 48 h retention of a step-through passive avoidance task. Injection into the amygdala was ineffective, as was intraventricular administration or hippocampal administration outside of the dentate gyrus. The rate of disappearance of noradrenaline after synthesis inhibition with AMPT was markedly increased by vasopressin injection into the dentate gyrus whereas in the septal nuclei the rate of disappearance was reduced. This suggests an involvement of noradrenaline in the memory-facilitating effects of vasopressin. However, so far it is merely a correlation, this need not indicate causation. The latter step in the argument was filled by Kovacs, Bohus & Versteeg (1979*a*), who showed that prior lesion with 6-OHDA to the ascending noradrenergic dorsal bundle abolished the memory-facilitating effects of vasopressin on passive avoidance when injected into the dentate gyrus or septum. This appears to be an interaction with noradrenergic terminals, since injection of vasopressin directly into the cell body region of the LC failed to enhance passive avoidance retention (Kovacs *et al.*, 1979*b*, Table 1) despite activating the firing of LC cells (Olpe & Baltzer, 1981).

Hypoxia. Hurwitz, Robinson & Barofsky (1971*b*) found that hypobaric hypoxia (equivalent to a height of 7010 m above sea level) severely impaired performance on a signalled bar-press shock avoidance paradigm (see Appendix) and also reduced brain noradrenaline levels to 20 % of normal. The authors question the causal nature of the observed correlation, since administration of 0.5 mg kg^{-1} of the MAOI tranylcypromine restored brain noradrenaline levels to normal (or above) but failed to reinstate avoidance behaviour. Brown, Kehr & Carlsson (1975) found that rats exposed to 6 % O_2 showed a decreased activity of DBH, the noradrenaline synthesising enzyme, in brain and a decrement in two-way active avoidance to 28 % of normal. Since behaviour was reinstated by L-DOPA or by apomorphine the authors prefer a dopaminergic mechanism, in line with the decreased dopamine disappearance after AMPT seen in hypoxic rats.

Carbon dioxide. Leonard & Rigter (1973) reported that exposure of rats to CO_2 immediately after a passive avoidance step-through task would yield amnesia when tested 24 h later. Biochemical measurements on other rats showed increased catecholamine concentrations in the hippocampus, suggestive of a failure of normal release. However, much larger changes in serotonin were seen, thus rendering the amine of importance indeterminate.

Isolation. Goldberg, Insalaco, Hefner & Salama (1973) found isolation (single caging as opposed to group, five per cage, housing) of mice slowed learning of two-way active avoidance response, especially in the CF-1 strain. No effect on levels of noradrenaline was seen as a result of isolation but the turnover of the amine was reduced in the CF-1 strain (which showed the biggest effect of isolation on learning behaviour) but not in other strains. Most of these effects were exceedingly small, however, a point made by the authors themselves (*ibid.*, p. 1056). Similar results were reported by Valzelli (1973) for noradrenaline turnover with no effect on the steady-state level of the amine (Valzelli, 1967). However, if the isolation was prolonged (34 days as opposed to 28), lower concentrations of noradrenaline were seen (Essman & Smith, 1967). Isolation of young rats (i.e. weaning at 18 days of age), however, has been reported to increase noradrenaline concentrations in brain (Geller, Yuwiler & Zolman, 1965). This is reviewed further in de Freudis (1979).

Uraemic endotoxaemia. Essman & Heldman (1972) reported that mice subjected to nephrectomy showed highly elevated blood urea-nitrogen levels 6 to 7 h post-operatively. At the same time point, learning of a passive avoidance step-through task was impaired and noradrenaline levels in the forebrain were decreased while those in the cerebellum were elevated. The causal relationship of these changes to behaviour is unknown.

Conclusion. A number of manipulations alter noradrenaline levels and affect learning or retention of passive or two-way active avoidance tasks. Probably the most interesting of underlying neurobiological mechanisms is that of the vasopressin interaction with terminals of the dorsal bundle noradrenergic system. Further developments in this fascinating field are eagerly awaited.

Summary

(1) Use of noradrenaline agonist drugs has yielded contradictory evidence with regard to noradrenaline and learning. Peripheral administration may alter learning and indeed this may be the primary site of action of amphetamine. Central effects of agonists relate mainly to the emission of an already learned behaviour rather than to the learning process *per se*.
(2) Use of noradrenaline antagonist drugs has been only sparse. Some authors find no effect of either α- or β-blockers; others report an inhibitory effect of β-blockers on retention without affecting performance of an already learned task; while yet others find α-blockers affect performance of an already learned task, with more complex and less direct effect on learning.
(3) Noradrenaline synthesis inhibitors impair 24 h retention of a task

whether administered before, during or after training on the task. Short-term memory appears intact suggesting a role in memory consolidation processes. However, the broad spectrum of activity of some inhibitors makes it unwise to ascribe their action to an effect of noradrenergic systems, since other, more selective inhibitors fail to cause amnesia despite equally severe noradrenaline depletion.

(4) Neither electrolytic lesion of the LC nor 6-OHDA lesion of the ascending noradrenergic dorsal bundle affects acquisition learning of a wide range of positively or negatively reinforced tasks. This is most convincing evidence against a role for dorsal bundle noradrenaline in learning or memory. A very few tasks are impaired, relating more to their demands on selective attention than on learning *per se*. Peripheral noradrenergic systems and an interaction with corticosteroids are clearly relevant to learning of fear-motivated tasks. No role for the cerebellar projection in motor learning was found. Insufficient evidence exists for a decision concerning the ventral bundle system.

(5) Evidence from self-administration studies suggests that noradrenergic systems are not involved in central reward of pharmacological agents (with the exception of ethanol).

(6) An interaction with peptide systems, especially vasopressin, is clearly indicated but the behavioural significance of this (learning v. selective attention) remains open.

Dopamine and learning behaviour

Agonists

A similar ambiguity about the predictions to be expected with dopamine receptor agonists and learning is to be found as seen with noradrenaline agonists (see p. 192). Non-contingent dopamine receptor stimulation might increase or decrease learning rate.

L-DOPA has been reported to be effective in normal cats in altering learning of a delayed response task (see p. 179) (Kitsikis, Roberge & Frenette, 1972*b*) and to improve the performance of spontaneously slow learners (Roberge, Roy & Boisvert, 1978). Only minimal effects, however, were seen in the reversal of the delayed response habit (Roberge, Boisvert & Everett, 1980).

Amphetamine has been reported to improve performance on a number of tasks. For example, better suppression of punished responding has been seen many times (Geller & Seifter, 1960) and this appears to be dopaminergic in nature since the dopamine blocker haloperidol but not the noradrenaline blockers, propranolol, phenoxybenzamine or phentolamine, reversed this effect (Lazareno, 1979). A related psychomotor stimulant drug, methylphenidate, has been found to produce faster learning of a two-way active avoidance task when compared to saline injections (Yeudall & Walley, 1977).

The above drugs release noradrenaline as well as dopamine and so cannot dissociate one amine from the other. To do this we must turn to the effects of direct dopamine receptor agonists, such as apomorphine, on learning.

Only a few studies are available. Fernandez-Tome, Sanchez-Blasquez & del Rio (1979) found that apomorphine administered either before or immediately after training on a passive avoidance step-through task yielded marked amnesia when tested for retention 24 h later. This effect could be prevented by pre-treatment with a dopamine receptor blocker such as clozapine. This would suggest that heightened activity in dopaminergic systems at the time of learning can interfere with retention. In contradiction to this is the report of Gozzani & Izquierdo (1976) that post-trial injection of apomorphine following training on a two-way active avoidance task failed to impair retention 7 days later, despite doses of apomorphine being used (0.5 or 4.0 mg kg^{-1}) which were effective in causing amnesia in the passive avoidance task of Fernadez-Tome *et al.* (1979).

Electrical stimulation of the source nucleus of the dopaminergic systems has been used to investigate its role in learning and memory. Although this lacks the pharmacological specificity of direct dopamine receptor agonists some interesting data have emerged. Routtenberg & Holzman (1973) found that 60 Hz electric stimulation of the substantia nigra during training on a step-down passive avoidance task produced retention deficits when tested 24 h later. Similar stimulation in nondopaminergic areas around the substantia nigra failed to cause amnesia. Again, consistent with the apomorphine data, it would seem that heightened dopaminergic activity during learning can impair subsequent retention. To demonstrate that the electrical stimulation was having its disruptive effects via activation of dopaminergic systems, Fibiger & Phillips (1976) lesioned the ascending dopaminergic fibre bundle on one side of the brain with injection of 6-OHDA. They found that this abolished the usual effect of unilateral electrical stimulation on subsequent retention if the stimulation were on the side of the brain with the lesion. If the stimulation was applied to the side of the brain contralateral to the 6-OHDA injection, amnesia on the passive avoidance task was still seen compared to unstimulated control rats. This confirms that the disruptive effects of electrical substantia nigra stimulation on memory are indeed mediated through dopaminergic systems. Indeed, electrical stimulation of the dopaminergic system in the caudate nucleus also causes amnesia of passive avoidance tasks (Gold & King, 1972; Wyers, Deadwyler, Hirsuna & Montgomery, 1973). This latter effect may not be dopaminergic, however, since 6-OHDA destruction of the dopamine input to this structure does not affect the ability of electrical stimulation in it to cause amnesia (Fibiger & Phillips, 1976).

In summary, two diverse lines of evidence (agonists and electrical stimulation) combine to suggest that heightened activity in the dopaminergic systems during learning may lead to a retention failure. This conclusion must, however, be distinguished from the related one that activity in dopaminergic systems is normally involved in learning in the intact rat. Since Fibiger & Phillips (1976) found that even rats with bilateral 6-OHDA lesion of the ascending dopaminergic systems could successfully learn and retain the passive avoidance task, this would imply that the dopaminergic system is not normally active during learning and is not necessary for memory consolidation to occur. Only if extra activity is induced in the dopaminergic system does an effect on learning occur – such as to impair long-term retention.

Antagonists

Learning and memory. Some studies have reported an impairment of long-term retention as a result of post-trial injections of dopamine antagonists such as haloperidol. Thus, Gozzani & Izquierdo (1976) found that 0.5 mg kg^{-1} haloperidol injected immediately after training of rats on a two-way active avoidance task impaired retention of the task when tested 7 days later. On the other hand, Fernandez-Tome *et al.* (1979) examined 24 h retention of step-through passive avoidance learning and found 0.5 mg kg^{-1} haloperidol to be without effect, although a higher dose (2.0 mg kg^{-1}) did disrupt retention. Similarly, high doses of chlorpromazine or clozapine also impaired retention. Thus, some evidence for a role of brain dopaminergic systems in post-trial memory consolidation may be gleaned from these experiments.

Other authors have administered neuroleptic drugs during the training period. A drawback of this approach is that the well-known motor effects of the drugs in reducing motor activity and inhibiting voluntary movement (see p. 126) may disrupt the task rather than any role of dopamine in learning or memory. Ranje & Ungerstedt (1977*a*) found that 0.5 mg kg^{-1} spiroperidol prevented rats from learning a brightness discrimination in an underwater T-maze through which they had to swim in order to reach the surface. At this dose of drug the time to swim the maze was no slower than controls, suggesting that a motor impairment could not be invoked to explain the poorer learning. However, a spatial position habit in the same apparatus was not impaired by doses of spiroperidol up to 10 times greater (up to 5 mg kg^{-1}) suggesting that the deficity was not that of a global learning failure, but somehow specific to the brightness discrimination.

Fibiger, Zis & Phillips (1975*b*) found that haloperidol would completely block the learning of a one-way active avoidance task as measured by the number of avoidance responses that the rat made. However, if the animal was then tested drug-free, near perfect avoidance performance was seen, suggesting that learning had taken place but that the motoric effects of the drug had prevented the rat from expressing it by performing the requisite avoidance response. Similar findings were reported by Beninger, Mason, Phillips & Fibiger (1980), who found that rats trained under 0.5 or 1.0 mg kg^{-1} pimozide failed to show learning of a one-way active avoidance response. Nonetheless when tested drug-free on food-reinforced lever-pressing, they showed that they had learned the significance of the tone used as the warning stimulus in the avoidance task, since marked suppression of ongoing lever-pressing occurred when it was presented. Thus, despite not showing any avoidance responses during training the drugged animals were able to learn the significance of the tone suggesting that dopamine antagonists did not prevent learning or memory but rather caused an avoidance impairment by preventing the emission of the required motor response.

Extinction patterns. A theory that has stimulated considerable research interest is the 'anhedonia' hypothesis of Wise (1982). This suggests that dopamine antagonists block central reward processes by blocking dopamine receptors. A blockade of central reward would explain the impaired learning seen in animals under neuroleptics. It would also make a further prediction – that animals performing a task for food or other rewards should gradually cease to emit the response when treated with neuroleptic drugs, despite the continued presentation of the food reward. This is because the central reward processes are postulated to be blocked as a result of central dopamine receptor antagonism and so food reinforcement, although still being presented, is no longer positively reinforcing. Neuroleptics 'take the yummyness out of food' as Wise, Spindler, deWit & Gerber (1978*a*) put it.

Early evidence from Wise *et al.* (1978*a*) seemed to support this. They found that rats lever-pressing on a CRF schedule for food reward would gradually cease responding over days if treated with 0.5 or 1.0 mg kg^{-1} pimozide. This effect was gradual and increased with repeated exposure to the situation. It is thus unlike that expected from a motor impairment in which a sudden decrement in response rate immediately after the first drug injection would be expected, rather than the progressive decrease seen by Wise *et al.* (1978*a*). A similar effect was seen in rats running an alley-way to reach food reward at the end. Administration of either 0.5 or 1.0 mg kg^{-1} pimozide led to a gradual slowing of the running speed which got bigger and bigger with subsequent trials in the apparatus. In both situations, omitting the food reward in normal, undrugged rats (i.e. extinction) also led to a progressive cessation of responding. Moreover, the curve of response cessation with time under pimozide was identical to that seen when food was omitted in undrugged rats. The identity of result suggested to Wise *et al.* (1978*a*) an identity of mechanism. That is, both dopamine antagonist drugs and omitting the food acted to stop the central dopaminergic reward system from conducting nervous activity. An additional line of evidence for the identical nature of dopamine blockade with reward omission was that rats showing a response decrement as a result of reward omission continued to show that decrement, and even increased it, when tested under pimozide with food pellets being delivered once more. This suggested the interchangeability of the manipulations.

Gerber, Sing & Wise (1981) reported the same pattern of results of CRF lever-pressing for water reward in thirsty rats.

However, this apparent similarity between the effects of reward omission and the effects of neuroleptic drugs turned out to be illusory and not to hold when examined in more general situations. Thus, Wise, Spindler & Legault (1978*b*) used saccharin reward, rather than food pellets. Now, omission of saccharin reward led to a gradual and progressive decline in response rate over days whereas administration of 0.5 or 1.0 mg kg^{-1} pimozide with the continuation of saccharin reward led to an immediate and sudden drop in response rate. Further, response rate then actually increased on day 2 of drug administration (see Wise *et al.*, 1978*b*, Fig. 4, p. 81).

Mason, Beninger, Fibiger & Phillips (1980) confirmed that, with food reward on a CRF schedule, pimozide did indeed lead to a gradual decline in lever-pressing over days. However, they found the decline with 1.0 mg kg^{-1} to be faster than that with extinction. Thus, while both manipulations do indeed cause a decline in response rates, they do so to a different degree and hence there is no reason to believe that they are acting through the same mechanism (viz. reduction in central dopaminergic transmission). Further, Mason *et al.* (1980) found that although animals trained for 3 days under extinction showed a further decline on the fourth day when tested under drug with food pellets, the converse transfer – from 3 days of drug with food to drug-free extinction on the fourth day – led to a marked *increase* in responding, not the continued decrease predicted by the anhedonia hypothesis. Thus, the similarity between extinction and neuroleptic treatment break down when other situations are examined.

More evidence along these lines was provided by Phillips & Fibiger (1979), who examined rats lever-pressing on a VI schedule. Haloperidol (0.1 mg kg^{-1}) caused a 60 % decrease in lever-pressing, which was similar in extent and progression over time to that caused by extinction. However, if these two manipulations were combined, that is rats were tested in extinction while receiving haloperidol, a further decrease was seen down to about 10 % of normal. This is directly contrary to the predictions of the anhedonia hypothesis. If extinction and neuroleptics are doing the same thing – reducing transmission in central dopaminergic systems – they should fail to add together when applied simultaneously. Haloperidol could not further block reward if no food pellets were being given. Nonetheless, the two did add in practice, strongly suggesting that they were *not* acting by similar mechanisms. Additionally, on a VI 4 min schedule the animal's response rate under haloperidol started to decline in the first 5 min. That was before the animal even received a food pellet on the VI schedule. According to the anhedonia hypothesis, the decline in response rate is due to the animal learning that food pellets are no longer 'yummy' as a result of drug treatment and hence extinguishing the behaviour resulting from lack of reward. Response decrements could only occur after the rat had sampled the no-longer-rewarding consequences of food delivery. However, in practice the response rate starts to decline earlier in time, before the first food pellet is even presented. This again suggests that haloperidol was doing something more than just blocking the central reward systems activated by food presentation. Similar results were obtained on a VI 2.5 min by Gray & Wise (1980).

Tombaugh, Tombaugh & Anisman (1979) again found that pimozide caused a decrease in food reinforced lever-pressing over days and that this was superficially similar to extinction. Again, however, rats trained for 3 days of drug with food failed to show a further decrease when tested on a fourth day of drug-free extinction. In fact, a marked *increase* in response rate was seen, in contradiction to the predictions of the anhedonia hypothesis. They also found that pimozide impaired the actual consumption of food pellets and decreased the total number eaten during a fixed duration session. As might be expected as a consequence of this, pimozide also severely impaired the acquisition learning of a CRF lever-press task (Tombaugh *et al.*, 1979; Wise & Schwartz, 1981).

Ettenberg, Koob & Bloom (1981) found that the nature of the response also altered the effects that neuroleptic drugs have on responding. Thus, lever-pressing was markedly attenuated by as little as 0.1 mg kg^{-1} α-flupenthixol whereas even 0.8 mg kg^{-1} of the drug failed to reduce a nose-poking response by more than 50 %. The anhedonia hypothesis does not predict this dependence on the nature of the response.

Greenshaw, Sanger & Blackman (1981) examined a FI schedule for food reward and found a decline in response rate with increasing doses of pimozide. A decline in responding could also be caused by omission of the food pellet, and within-session records of responding appeared similar for the two manipulations. However, when electrical brain stimulation was used as the reward, the effect of reward omission was much more severe than even the largest dose of pimozide tested (2 mg kg^{-1}). Further, under 2 mg kg^{-1} of drug, a sudden cessation of FI lever-pressing for brain stimulation was seen, compared to the more gradual and progressive decline in extinction when reward was omitted. The extinction curve was concave whereas all the pimozide curves seemed to be straight or convex. That is, responding during extinction occurred most in the early part of the session and ceased later on. Under pimozide, responding occurred evenly throughout the session albeit at a reduced rate. Thus, on a FI schedule the response patterns generated by reward omission and by neuroleptic drug are *not* identical. Further, Greenshaw *et al.* (1981) report a measure of lever-holding time, which rose markedly in extinction but was little affected by even the largest dose of pimozide, thus suggesting a further difference in the mechanism of action of the two manipulations.

Faustman & Fowler (1981) also find that neuroleptic (haloperidol) and extinction have different effects on lever-holding time on a FR schedule for food reward. The response rate in extinction of FR10 showed a gradual decline over 6 days of testing. On the other hand, the response rate under 0.5 mg kg^{-1} haloperidol dropped immediately and remained low over the 6 days. The response duration (lever-holding time) rose markedly under the drug but was unaffected by extinction on its own. This further indicates that extinction and dopamine blockade by neuroleptic drug are two different processes.

Finally, Tombaugh (1981) found that, although increasing doses of pimozide caused a motor impairment in responding by pigeons on a successive discrimination task, it failed to affect the accuracy of the visual discrimination itself, suggesting that dopamine activity is not necessary for the maintenance of associations between stimuli established on the basis of previous learning. Of interest was the result (Tombaugh, 1981, Fig. 1, p. 138) that the response decrement on the variable-interval schedule was not immediate from the start of the session, as would be indicative of a motor impairment, but developed progressively over time. That discrimination performance remained better than 97 % throughout the session, however, would seem also to exclude a block of reward as the explanation.

Mason *et al.* (1980) examined a DRL schedule of responding for food reward. Despite using a dose of pimozide effective in causing a marked decline in response rate on CRF, no decline in DRL responding, or the accuracy of the temporal discrimination, was seen on any of the 3 days of testing. Extinction in the same DRL paradigm indeed caused the predicted changes in response rate. Thus, a clear dissociation of the effects of neuroleptic drugs and those of extinction has been shown with a DRL schedule.

In summary, neuroleptic drugs in a few limited situations (mainly CRF for food or brain stimulation reward) produce a progressive decline in responding indistinguishable from that seen in extinction. In other situations, however, the two manipulations produce different results (DRL or FI for example). Further, the effects of training on one do not always transfer to testing under the other. Thus, the transfer from drug with food to drug-free extinction fails in all experiments reported to date. Additionally, the manipulations of drug and extinction are additive, the two together producing a more severe effect than either on its own. These last two points are directly contrary to the anhedonia hypothesis. Response rates on VI schedules also start to decline even before the animal has experienced the fact that its food pellet is no longer rewarding. Other response measures, such as lever-holding time, also dissociate the effects of neuroleptic from those of reward omission. Discrimination performance is not affected by neuroleptics in pigeons although acquisition learning of a swim-maze brightness (but not position) discrimination has been reported to suffer. Thus, neuroleptic drugs do not seem to mimic behavioural extinction, and there is no satisfactory evidence that central dopaminergic receptors as such mediate reward to be gleaned from the paradigms described above. I shall now turn to the effects of 6-OHDA lesions.

6-OHDA lesions

Cooper *et al.* (1973*a*) found that depletion of both noradrenaline and dopamine by intracisternal injection of 200 μg 6-OHDA preceded by intraperitoneal pargyline completely blocked acquisition learning of both one-way and two-way active avoidance. However, this seemed to be related to a motor deficit (see Chapter 4), rather than a global learning impairment, since these same animals were able successfully to learn a step-through passive avoidance task. The amine of importance appeared to be dopamine since depletion of noradrenaline by three intracisternal injections of 25 μg 6-OHDA *facilitated* rather than impaired, two-way active avoidance learning. Depletion of dopamine by two injections of 6-OHDA preceded by desipramine (giving a 85% depletion of dopamine but only a 24% loss of noradrenaline) also prevented acquisition learning of two-way active avoidance. These effects seemed to be due to difficulties in performing the response rather than in learning *per se* since dopamine depletion in rats already trained on two-way active avoidance produced a similar failure of the behaviour. In this case, learning has already been completed so a deficit in this process alone could not explain the impaired behaviour.

Neill, Boggan & Grossman (1974) localised this deficit to the ventral anterior striatum by application of crystalline 6-OHDA to this region via indwelling cannulae. Disruption of two-way active avoidance correlated with the degree of dopamine depletion achieved in the forebrain as a result of this technique. This confirms early work by Mitcham & Thomas (1972) using electrolytic lesions in the region of the ventral caudate. These disrupted acquisition learning of both one-way and two-way active avoidance, and interestingly enough also impaired learning of a step-through passive avoidance task. Electrolytic lesion to the source nucleus of the dopaminergic system, the substantia nigra, also impaired acquisition of all these tasks; perhaps more severely than the caudate lesion in the case of one-way active avoidance. This result was also seen by Fibiger, Phillips & Zis (1974), who injected 8 μg 6-OHDA into each substantia nigra and obtained a 96% depletion of striatal tyrosine hydroxylase activity. This completely blocked acquisition learning of one-way active avoidance and also of an approach response to food in the same apparatus. This might be explicable as a learning deficit, since similar lesions inflicted on animals already trained on one-way active avoidance responding produced but slight disruption – contrary to prediction if a motor deficiency were all that was preventing acquisition learning.

In line with a learning deficit, Ranje & Ungerstedt (1977*b*) found that injection of 8 μg 6-OHDA bilaterally into the ascending dopaminergic fibre bundle caused an accumulation of amine fluorescence posterior to the injection site in the dopaminergic fibres and virtually total disappearance of fluorescence in the striatum, accumbens and olfactory tubercle. These lesions completely prevented acquisition learning of either a brightness or a spatial position discrimination task in a submerged swim-maze. A swim-maze had to be used in order to motivate these severely motor-impaired rats to move at all! These effects were claimed to represent an impairment in learning *per se* since similar dopamine depletion of animals already trained on the task produced only a slight worsening of the brightness discrimination. If the failure of acquisition learning had been due to an inability to perform the required response, a similar deficit should be seen in rats lesioned after training was complete. Some worsening of performance did indeed occur (considerably slower swimming and an increase in errors from 0.1 to 0.3 per trial; Ranje & Ungerstedt, 1977*b*, Fig. 5, p. 105) but this level of performance was still considerably better than that of untrained rats.

Shaywitz, Yager & Klopper (1976) found that rats depleted of dopamine by intracisternal injection of 100 μg 6-OHDA preceded by desipramine when neonatal (on day 5 after birth) suffered from an impairment of two-way active avoidance learning when adult. This did not seem to be attributable to a motor deficit since the 6-OHDA animals were *more* active than controls between days 15 and 22 after birth and showed no sign of hypokinesia when adult.

Zis, Fibiger & Phillips (1974) were able to reverse these deficits after 6-OHDA lesion of the ascending dopaminergic systems by injection of L-DOPA on a daily basis. Dopamine depletion completely prevented acquisition learning of a one-way active avoidance task even after 8 days of training. When animals were injected with L-DOPA 30 min prior to each training session, even the 6-OHDA lesioned rats learned the one-way active avoidance task, and at virtually the same rate as unlesioned animals. When dopamine-depleted animals, which had been trained for 8 days on the avoidance task without showing any sign of learning, were administered L-DOPA they acquired the task over the next 4 days of training. This again suggests that dopamine depletion may cause a learning impairment, distinct from a motor deficit, since had the L-DOPA merely reversed an inability to perform the necessary response these rats would have started performing the avoidance task *immediately* upon being placed on the drug and not shown a *gradual* learning curve over the next 4 days.

However, arguing against an explanation of the impaired avoidance behaviours as being due to learning processes is the finding of Price & Fibiger (1975) that if electric footshock were used to motivate these severely akinetic rats to move then subsequent learning of a discriminated (brightness) escape from footshock in a Y-maze proceeded as rapidly in rats with only 1.7 % of the normal dopamine levels in the caudate as it did in controls. With depletions as severe as this, a deficit would surely be expected to be detected and thus the results stand in direct contradiction to those of Ranje & Ungerstedt (1977*a*) described above. This highlights the grave drawbacks of this dopamine-depleted animal. It is virtually immobile in normal circumstances and has to be tube-fed for two or three weeks to keep it alive after the operation since it is aphagic and hypodipsic. The problems of dissociating a true learning deficit from a motor impairment are so strenuous that research has abandoned the bilaterally dopamine depleted rat in favour of more alive preparations.

One promising approach lies in the fact that the most severe motor impairments are caused by depletion of caudate dopamine, originating mainly from the substantia nigra (see p. 5). A second dopaminergic system, the so-called mesolimbic system, originates in the more medial ventral tegmental area (A10) and innervates the accumbens, olfactory tubercle and frontal cortex. 6-OHDA lesions of the A10 area spare caudate dopamine and so do not induce movement problems. Simon, Scatton & Le Moal (1980) found that such a 6-OHDA lesion which depleted frontal cortical dopamine to 30% and accumbens dopamine to 15% caused no impairment on retention of a brightness discrimination task in a T-maze but severely impaired retention of a delayed spatial alternation task in which the rat had to alternate between the left- and the right-hand side goal boxes to receive food pellets. Only with great difficulty could these animals be retrained to perform the spatial alternation successfully. Similar impairment is found after electrolytic A10 lesion (Simon *et al.*, 1979). The unimpaired brightness discrimination argues against a simple motor impairment, as does unaltered FI performance (Simon *et al.*, 1979). So a role of nonstriatal dopaminergic systems, those running from the A10 area to frontal and limbic regions, in learning and memory may be possible. Research has only begun to explore this problem.

In summary, the 6-OHDA lesion technique has failed to yield convincing evidence of a role for dopamine in learning and memory mainly because of the severe motor deficits caused by 6-OHDA injection into the substantia nigra or the ascending dopaminergic bundle. Such deficits render behavioural analysis impossible. However, 6-OHDA lesions of nonstriatal systems, which do not cause movement disorders, may be more useful and are starting to be utilised.

Self-administration

It is in the field of self-administration of dopaminergic drugs that the strongest evidence for a role of dopamine in reward and learning is to be found. Animals will learn to emit a response in order to receive an intravenous infusion of various drugs affecting dopaminergic systems. Since this is the only consequence of the response emitted, it must be that such infusions of dopaminergic drugs are reinforcing as operationally defined. Learning of a response for dopaminergic drug infusion has been demonstrated for the intravenous route of administration only so far. Unlike the story for noradrenergic drug infusions (see section on Noradrenaline and self-administration, p. 226) no successful reports of learning rewarded by direct infusion of dopaminergic drugs into discrete brain regions have been presented. This is an area of intense research interest and such reports are likely to be forthcoming in the near future.

Drugs which will reinforce learning in return for intravenous administration include amphetamine (Pickens & Harris, 1968), and the lever-pressing rate changes as the drug concentration is altered in such a way as to seek to maintain a constant level of amphetamine in the brain (Deneau, Yanagita & Seevers, 1969). That is, lower concentrations of drug in the infusion will cause the rat to lever-press more in order to receive more infusions and thus maintain a constant drug input. Conversely, increasing the concentration of drug in the infusion will cause a decrease in lever-pressing rate as the rat can receive the same amount of drug in fewer infusions. Amphetamine seems to act through dopaminergic systems in this case since dopamine blockers such as pimozide will increase rate, as would be observed had the effective concentration of drug in the infusion been lowered (Yokel & Wise, 1975). This is a much clearer result than that for neuroleptics and ICSS (see Chapter 3) or for other forms of learned behaviour (see p. 233), since an *increase* in responding is observed and this cannot be attributed to motor deficits also caused by the drug. It is these which were found to interfere so irretrievably with the interpretation of the effects of pimozide on ICSS and other learned behaviours. A noradrenergic basis for amphetamine reward seems to be excluded, since Yokel & Wise (1975) found that both the α-antagonist phentolamine and the β-antagonist propranolol decreased response rates in a nonspecific fashion, rather than causing a response rate increase.

Other direct dopamine receptor stimulants such as apomorphine are also self-administered in rats by the intravenous route (Baxter, Gluckman, Stein & Scerni, 1974). That this action is by direct receptor activation is supported by the finding of Baxter, Gluckman & Scerni (1976) that apomorphine self-administration continues after the depletion of presynaptic amine stores with AMPT whereas that of amphetamine showed an increase in the number of lever-presses. Apomorphine self-administration is blocked by dopamine receptor antagonists such as (+)-butaclamol (Yokel & Wise, 1978). Other dopamine agonists such as piribedil are also self-administered (Yokel & Wise, 1978), as are cocaine (Wilson & Schuster, 1973), metamphetamine (Balster & Schuster, 1973), various phenylethylamines (Griffiths, Winger, Brady & Snell, 1976), and mazindol (Wilson & Schuster, 1976). All these drugs seem to interact with dopaminergic systems to a greater or lesser degree of specificity.

Finally, Roberts *et al.* (1977) showed that bilateral 6-OHDA lesion of the nucleus accumbens and olfactory tubercle would reduce cocaine self-administration to about 20 % of pre-lesion rates for 15 or more days post-operatively. Since these 6-OHDA lesions had only slight effects on food responding (for 2–3 days post-operatively) this cannot be explained as a motor deficit and would seem to support the idea of dopaminergic reward in the learning and performance of self-administration. Apomorphine self-administration, as a direct dopamine agonist, was unaffected by the 6-OHDA lesions.

In conclusion, considerable evidence suggests that self-administration of some drugs may involve a dopaminergic component of reward. However, the brain areas of importance have not been exclusively localised (but see Roberts *et al.*, 1977) and it is not clear that the same reward systems involved in drug self-administration need necessarily be involved in ICSS or natural reward and learning.

Interactions with other systems

A few reports are available of an interaction of dopamine with other neurotransmitter systems specifically in the field of learning and memory.

Opiates. Schwartz & Marchok (1974) found that morphine-dependent rats would learn to choose a goal box in a T-maze (which was initially nonpreferred) as a result of administering a morphine injection 20 min later. Administration of AMPT during such training blocked the acquisition of the morphine-rewarded position habit. This seemed to be dopaminergic, since inhibition of noradrenaline synthesis with DDC failed to prevent acquisition of the morphine-rewarded learning, whereas the dopamine receptor blocker haloperidol was effective.

Harris & Snell (1980) found that behaviour maintained on a FI schedule for food reward was decreased by dopaminergic drugs such as pimozide, chlorpromazine, bromocryptine and cocaine. The response suppressant effect of the drugs could be potentiated by the opiate antagonist naltroxone in doses which by themselves had no effect on responding. McMillan (1971) found that the response suppressant effects of chlorpromazine on responding in the pigeon were potentiated by another opiate antagonist, naloxone.

Conversely, the rate depressant effects of apomorphine or haloperidol were differentially affected by chronic morphine administration for 3 days (Glick & Cox, 1977). Thus, apomorphine now induced greater behavioural disruption whereas haloperidol became less effective.

However, given the well-demonstrated interaction between dopaminergic systems and opiates in the field of motor activity (see p. 155) it is uncertain how much of the above evidence reflects a true interaction specific to learning and reward and how much of it is due solely to motor effects.

Cholinergic systems. Spontaneous alternation behaviour in the rat (see Appendix) involves habituation to novel stimuli, a form of learning. Such behaviour is impaired by dopaminergic drugs such as amphetamine (Adkins, Packwood & Marshall, 1969) and also by cholinergic antagonists such as scopolamine (Meyers & Domino, 1964).

Kokkinidis & Anisman (1976) showed that elevation of acetylcholine levels in the brain by administration of the acetylcholine esterase inhibitor physostygmine would reverse the disruptive effects on alternation behaviour normally shown by *d*-amphetamine, suggesting an interaction between acetylcholine and dopaminergic systems in the learning of novel stimuli. However, the finding of Anisman & Kokkinidis (1975) that the disruptive effects of *d*-amphetamine on spontaneous alternation are prevented by noradrenaline synthesis inhibition with a DBH inhibitor suggests the possibility that it is a noradrenaline/acetylcholine interaction, rather than a dopaminergic one. On this point the reader is referred to the section on noradrenaline/acetylcholine interaction (Chapter 4).

Hallucinogens. Zeller *et al.* (1976) found that high doses of the hallucinogen mescaline would improve shuttle box learning in the goldfish (*Carassius auratus*), and similar effects could be obtained by dopaminergic agonists such as apomorphine or L-DOPA. Pharmacologically inactive analogues of mescaline (3,4,5-trimethoxybenzylamine or 2,3,4-trimethoxy-β-phenylethylamine) were ineffective in altering learning rate. More prolonged chronic administration of mescaline led to an impairment of learning because of difficulties in initiating voluntary movement. These are suggestive of an interaction of hallucinogens with dopaminergic systems in learning and memory but, given the interaction in the field of locomotor activity (see p. 158), it is uncertain if the effects are specific to learning or just a reflection of changes in motor performance.

Summary

(1) Electrical stimulation of the ascending dopaminergic systems can cause amnesia if administered during learning. However, it would appear that the dopaminergic system may not be tonically active and normal learning can occur in its absence. Pharmacological activation of dopamine receptors by apomorphine may cause amnesia although there is no general agreement about this.

(2) High doses of dopamine receptor blockers can cause amnesia of

passive avoidance and will markedly impair other learning paradigms, most probably via the motor initiation deficit which they also cause. The similarities between neuroleptic drugs and extinction of rewarded behaviour turn out to be superficial and no evidence for a role of dopamine in learning and reward can be gleaned from these studies. Conversely, this does not necessarily mean that dopamine is *not* involved in reward.

(3) 6-OHDA lesions of the ascending dopaminergic systems uniformly impair learning but this cannot be dissociated from the motor impairment that they also cause. Terminal area lesions of the accumbens and other nonstriatal systems hold more promise.

(4) The strongest evidence for a role of dopamine in learning and reward comes from self-administration of dopamine agonists. However, this need not necessarily mean that dopamine is involved in reward and learning using ICSS or natural reinforcers.

(5) Interactions in learning behaviour with other systems may be seen, but these may merely reflect the well-known interactions with dopaminergic systems in motor activity and need not be specific to learning and memory processes.

Conclusions

Most of the evidence is uniform in excluding forebrain noradrenaline from any critical role in learning. Possible secondary involvement via other cognitive functions such as attention, fear or nonreward will be considered in Chapter 6.

Despite much work, the only firm involvement of dopamine in learning or reward is in the field of self-administration and this may still be a pharmacological peculiarity unrelated to naturally invoked learning mechanisms. It would seem that the time is right for the consideration of noncatecholaminergic neurotransmitter systems in learning and memory.

References

Adkins, J., Packwood, J. W. & Marshall, G. I. Jr (1969). Spontaneous alternation and *d*-amphetamine. *Psychonomic Science*, **17**, 167–8.

Ahlenius, S. (1973). Inhibition of catecholamine synthesis and conditioned avoidance acquisition. *Pharmacology, Biochemistry & Behavior*, **1**, 347–50.

Ahlenius, S. (1974). Effects of L-DOPA on conditioned avoidance responding after behavioural suppression by alpha-methyltyrosine or reserpine in mice. *Neuropharmacology*, **13**, 729–39.

Ahlenius, S. & Engel, J. (1972). Effects of a dopamine (DA)-β-hydroxylase inhibitor on timing behaviour. *Psychopharmacologia*, **24**, 243–6.

Ahlenius, S. & Engel, J. (1973). Antagonism by *dl*-threo-DOPS of the suppression of a conditioned avoidance response induced by a dopamine-beta-hydroxylase inhibitor. *Journal of Neural Transmission*, **34**, 267–77.

Albert, L. H., Emmett-Oglesby, M. & Seiden, L. S. (1977). Effects of schedules of reinforcement of brain catecholamine metabolism in the rat. *Pharmacology, Biochemistry & Behavior*, **6**, 481–6.

Amaral, D. G. & Foss, J. A. (1975). Locus coeruleus lesions and learning. *Science*, **188**, 377–8.

Amit, Z., Brown, Z. W., Levitan, D. E. & Ogren, S. O. (1977). Noradrenergic mediation of the positive reinforcing properties of ethanol. I. Suppression of ethanol consumption in laboratory rats following dopamine-beta-hydroxylase inhibition. *Archives internationales de pharmacodynamie et de thérapie*, **230**, 65–75.

Amit, Z., Levitan, D. E. & Lindros, K. O. (1976). Suppression of ethanol intake following administration of dopamine-beta-hydroxylase inhibitors in rats. *Archives internationales de pharmacodynamie et de thérapie*, **223**, 114–19.

Anisman, H. & Kokkinidis, L. (1975). Effects of scopolamine, *d*-amphetamine and other drugs affecting catecholamines on spontaneous alternation and locomotor activity in mice. *Psychopharmacologia*, **45**, 55–63.

Anlezark, G. M., Crow, T. J. & Greenway, A. P. (1973). Impaired learning and decreased cortical norepinephrine after bilateral locus coeruleus lesions. *Science*, **181**, 682–4.

Audigier, Y., Virion, A. & Schwartz, J. C. (1976). Stimulation of cerebral histamine H_2 receptors by clonidine. *Nature*, **262**, 307–8.

Balster, R. L. & Schuster, C. R. (1973). A comparison of *d*-amphetamine, *l*-amphetamine and methamphetamine self-administration in rhesus monkey. *Pharmacology, Biochemistry & Behavior*, **1**, 67–71.

Bauer, R. H. (1973). Brain norepinephrine and 5-hydroxytryptamine as a function of time after avoidance and footshock. *Pharmacology, Biochemistry & Behavior*, **1**, 615–18.

Baxter, B. L., Gluckman, M. I. & Scerni, R. A. (1976). Apomorphine self-injection is not affected by alpha-methylparatyrosine treatment: support for dopaminergic reward. *Pharmacology, Biochemistry & Behavior*, **4**, 611–12.

Baxter, B. L., Gluckman, M. I., Stein, L. & Scerni, R. A. (1974). Self-injection of apomorphine in the rat: positive reinforcement by a dopamine receptor stimulant. *Pharmacology, Biochemistry & Behavior*, **2**, 387–91.

Beaton, J. M. & Crow, T. J. (1969). The effect of noradrenaline synthesis inhibition on motor activity and lever pressing for food and water in the rat. *Life Sciences*, **8**, 1129–34.

Beer, B. & Lenard, L. G. (1975). Differential effects of intraventricular administration of 6-hydroxydopamine on behaviour in rats in approach and avoidance procedures: reversal of avoidance decrements by diazepam. *Pharmacology, Biochemistry & Behavior*, **3**, 879–86.

Beninger, R. J., Mason, S. T., Phillips, A. G. & Fibiger, H. C. (1980). The use of conditioned suppression to evaluate the nature of neuroleptic-induced avoidance deficits. *Journal of Pharmacology and Experimental Therapeutics*, **213**, 623–7.

Boarder, M. R., Feldon, J., Gray, J. A. & Fillenz, M. (1979). Effect of runway training on rat brain tyrosine hydroxylase: differential effect of continuous and partial reinforcement. *Neuroscience Letters*, **15**, 211–15.

Bohus, B., Kovacs, G. L. & de Wied, D. (1978). Oxytocin, vasopressin and memory: opposite effects on consolidation and retrieval processes. *Brain Research*, **157**, 414–17.

Bolme, P., Corrodi, H., Hokfeldt, T., Lidbrink, P. & Goldstein, M. (1974). Possible involvement of central adrenaline neurons in vasomotor and respiratory control. Studies with clonidine and its interactions with piperoxan and yohimibine. *European Journal of Pharmacology*, **28**, 89–94.

Botwinick, C. Y., Quartermain, D., Freedman, L. S. & Hallock, M. F. (1977). Some characteristics of amnesia induced by FLA 63 an inhibitor of dopamine beta-hydroxylase. *Pharmacology, Biochemistry & Behavior*, **6**, 487–91.

Brown, R. M., Kehr, W. & Carlsson, A. (1975). Functional and biochemical aspects of catecholamine metabolism in brain under hypoxia. *Brain Research*, **85**, 491–509.

Brown, Z. W. & Amit, Z. (1977). The effects of selective catecholamine depletions by 6-hydroxydopamine on ethanol preference in rats. *Neuroscience Letters*, **5**, 333–6.

Brown, Z. W., Amit, Z., Levitan, D. E., Ogren, S. O. & Sutherland, E. A. (1977). Noradrenergic mediation of the positive reinforcing properties of ethanol. II. Extinction of ethanol-drinking behaviour in laboratory rats by inhibition of dopamine beta-hydroxylase. Implications for treatment procedures in human alcoholics. *Archives internationales de pharmacodynamie et de thérapie*, **230**, 76–82.

Bucci, L. (1974). The biphasic effect of small doses of tranylcypromine on the spontaneous motor activity and learned conditioned behaviour in rats. *Pharmacology*, **12**, 354–61.

Bucci, L. & Bovet, D. (1974). The effect of iproniazid and tranylcypromine studied with a dark-avoidance conditioned schedule. *Psychopharmacologia*, **35**, 179–88.

Cairncross, K. D., King, M. G. & Schofield, S. P. (1975). Effect of amitriptyline on avoidance learning in rats following olfactory bulb ablation. *Pharmacology, Biochemistry & Behavior*, **3**, 1063–7.

Carley, J. W. & Haynes, J. R. (1969). Epinephrine and nor-epinephrine effects on approach-avoidance behavior. *Psychological Reports*, **24**, 100–2.

Cedarbaum, J. M. & Aghajanian, G. K. (1976). Noradrenergic neurons of the locus coeruleus: inhibition by epinephrine and activation by the alpha-antagonist piperoxan. *Brain Research*, **112**, 413–19.

Cessens, G., Roffman, M., Kurac, A., Orsulak, P. J. & Schildkraut, J. J. (1980). Alterations in brain norepinephrine metabolism induced by environmental stimuli previously paired with inescapable shock. *Science*, **209**, 1138–40.

Clark, D. W. J., Laverty, R. & Phelan, E. L. (1972). Long-lasting peripheral and central effects of 6-hydroxydopamine in rats. *British Journal of Pharmacology*, **44**, 233–43.

Cohen, R. P. & Hamburg, M. D. (1975). Evidence for adrenergic neurons in a memory access pathway. *Pharmacology, Biochemistry & Behavior*, **3**, 519–23.

Cooper, B. R., Breese, G. R., Grant, L. D. & Howard, J. L. (1973*a*). Effects of 6-hydroxydopamine treatment on active avoidance responding: evidence for the involvement of brain dopamine. *Journal of Pharmacology and Experimental Therapeutics*, **185**, 538–70.

Cooper, B. R., Breese, G. R., Howard, J. L. & Grant, L. D. (1972). Effect of central catecholamine alterations by 6-hydroxydopamine on shuttle box avoidance acquisition. *Physiology and Behavior*, **9**, 727–31.

Cooper, B. R., Grant, L. D. & Breese, G. R. (1973*b*). Comparison of the behavioral depressant affects of biogenic amine depleting agents following various 6-hydroxydopamine treatments. *Psychopharmacologia*, **31**, 95–109.

Cooper, B. R., Howard, J. L., Grant, L. D., Smith, R. D. & Breese, G. R. (1974). Alteration of avoidance and ingestive behavior after destruction of central catecholaminergic pathways with 6-hydroxydopamine. *Pharmacology, Biochemistry & Behavior*, **2**, 639–49.

Coyle, J. T., Wender, P. & Lipsky, A. (1973). Avoidance conditioning in different strains of rats: neurochemical correlates. *Psychopharmacologia*, **31**, 25–34.

Crow, T. J. (1972). Catecholamine containing neurones and electrical self-stimulation. 1. A review of some data. *Psychological Medicine*, **2**, 414–17.

Crow, T. J. (1973). Catecholamine containing neurones and electrical self-stimulation. 2. A theoretical interpretation and some psychiatric implications. *Psychological Medicine*, **3**, 1–5.

Crow, T. J., Longden, A., Smith, A. & Wendlandt, S. (1977). Pontine tegmental lesions, monoamine neurons and varieties of learning. *Behavioral Biology*, **29**, 184–96.

Crow, T. J. & Wendlandt, S. (1976). Impaired acquisition of a passive avoidance response after lesions induced in the locus coeruleus by 6-OH-dopamine. *Nature*, **259**, 42–4.

Cytawa, J., Jurkowlaniec, E. & Biatowas, J. (1980). Positive reinforcement produced by noradrenergic stimulation of the hypothalamus in rats. *Physiology and Behavior*, **25**, 615–19.

Dahlof, C., Engberg, G. & Svensson, T. H. (1981). Effects of beta-adrenoceptor antagonists on the firing rate of noradrenergic neurones in the locus coeruleus of the rat. *Naunyn-Schmiedeberg's Archives of Pharmacology*, **317**, 26–30.

Danscher, G. & Fjerdinstad, F. J. (1975). Diethyldithiocarbamate (antabuse): decrease of brain heavy metal staining pattern and improved consolidation of shuttle box avoidance in goldfish. *Brain Research*, **83**, 143–55.

Davis, W. M. & Smith, S. G. (1977). Catecholaminergic mechanisms of reinforcement: direct assessment by drug-self-administration. *Life Sciences*, **20**, 483–92.

Davis, W. M., Smith, S. G. & Werner, T. E. (1978). Noradrenergic role in the self-administration of ethanol. *Pharmacology, Biochemistry & Behavior*, **9**, 368–74.

de Freudis, F. V. (1979). Environment and central neurotransmitters in relation to learning, memory and behavior. *General Pharmacology*, **10**, 281–6.

de Wit, H. & Wise, R. A. (1977). Blockade of cocaine reinforcement in rats with the dopamine receptor blocker pimozide but not with the noradrenergic blockers phentolamine or phenoxybenzamine. *Canadian Journal of Psychology*, **31**, 195–203.

Deneau, G., Yanagita, T. & Seevers, M. H. (1969). Self-administration of psychoactive substances by the monkey. *Psychopharmacologia*, **16**, 30–48.

DiCara, L. V. & Stone, E. A. (1970). Effect of instrumental heart-rate training on rat cardiac and brain catecholamines. *Psychosomatic Medicine*, **32**, 359–68.

Di Guisto, E. L. (1972). Adrenaline or peripheral noradrenaline depletion and passive avoidance in the rat. *Physiology and Behavior*, **8**, 1059–62.

Di Guisto, E. L. & King, M. G. (1972). Chemical sympathectomy and avoidance learning in the rat. *Journal of Comparative and Physiological Psychology*, **81**, 491–500.

Dismukes, R. K. & Rake, A. V. (1972). Involvement of biogenic amines in memory formation. *Psychopharmacologia*, **23**, 17–25.

Essman, E. B. & Heldman, E. (1972). Impairment of avoidance acquisition and altered regional brain amine levels in mice with uremic endotoxemia. *Physiology and Behavior*, **8**, 143–6.

Essman, W. B. & Smith, G. E. (1967). Behavior and neurochemical differences between differentially housed mice. *American Zoology*, **7**, 793.

Ettenberg, A., Koob, G. F. & Bloom, F. E. (1981). Response artifact in the measurement of neuroleptic-induced anhedonia. *Science*, **213**, 357–9.

Faustman, W. O. & Fowler, S. C. (1981). Use of operant response duration to distinguish the effects of haloperidol from nonreward. *Pharmacology, Biochemistry & Behavior*, **15,** 327–9.

Fernandez-Tome, M. P., Sanchez-Blasquez, P. & del Rio, J. (1979). Impairment by apomorphine of one-trial passive avoidance learning in mice: the opposing roles of the dopamine and noradrenaline systems. *Psychopharmacology*, **61,** 43–7.

Fibiger, H. C. & Mason, S. T. (1978). The effects of dorsal bundle injections of 6-hydroxydopamine on avoidance responding in rats. *British Journal of Pharmacology*, **64,** 601–5.

Fibiger, H. C. & Phillips, A. G. (1976). Retrograde amnesia after electrical stimulation of the substantia nigra: mediation by the dopaminergic nigroneostriatal bundle. *Brain Research*, **116,** 23–33.

Fibiger, H. C., Phillips, A. G. & Zis, A. P. (1974). Deficits in instrumental responding after 6-hydroxydopamine lesions of the nigro-neostriatal dopaminergic projection. *Pharmacology, Biochemistry & Behavior*, **2,** 87–96.

Fibiger, H. C., Roberts, D. C. S. & Price, M. T. C. (1975*a*). On the role of telencephalic noradrenaline in learning and memory. In *Chemical tools in catecholamine research*, ed. G. Jonsson, T. Malmfors & C. Sachs, vol. 1, pp. 349–56. North-Holland Publishing Co.: Amsterdam.

Fibiger, H. C., Zis, A. P. & Phillips A. G. (1975*b*). Haloperidol-induced disruption of conditioned avoidance responding: attenuation by prior training or by anticholinergic drugs. *European Journal of Pharmacology*, **30,** 309–14.

Flexner, J. B. & Flexner, L. B. (1976). Effect of two inhibitors of dopamine-beta-hydroxylase on maturation of memory in mice. *Pharmacology, Biochemistry & Behavior*, **5,** 117–21.

Fulginiti, S., Molina, V. A. & Orsingher, O. A. (1976). Inhibition of catecholamine biosynthesis and memory processes. *Psychopharmacology*, **51,** 65–9.

Fuxe, K. & Hanson, C. F. (1967). Central catecholamine neurons and conditioned avoidance behaviour. *Psychopharmacologia*, **11,** 439–47.

Geller, I. & Seifter, J. (1960). The effects of meprobamate, barbiturates, *d*-amphetamine and promazine on experimentally induced conflict in the rat. *Psychopharmacology*, **1,** 482–92.

Geller, E., Yuwiler, A. & Zolman, J. Z. (1965). Effects of environmental complexity on constituents of brain and liver. *Journal of Neurochemistry*, **12,** 949–55.

Gerber, G. J., Sing, J. & Wise, R. A. (1981). Pimozide attenuates lever pressing for water reinforcement in rats. *Pharmacology, Biochemistry & Behavior*, **14,** 201–5.

Gibbs, M. E. (1976). Modulation of cycloheximide-resistant memory by sympathomimetic agents. *Pharmacology, Biochemistry & Behavior*, **4,** 703–7.

Gibbs, M. E. & Ng, K. T. (1977). Counteractive effects of norepinephrine and amphetamine on ouabain-induced amnesia. *Pharmacology, Biochemistry & Behavior*, **6,** 533–7.

Gibbs, M. E. & Ng, K. T. (1979). Similar effects of a monoamine oxidase inhibitor and a sympathomimetic amine on memory formation. *Pharmacology, Biochemistry & Behavior*, **11,** 335–9.

Gilbert, D. C. (1975). How the cerebellum could memorise movements. *Nature*, **254,** 688–9.

Gilman, S. C. & Fischer, G. J. (1976). The effect of diethyldithiocarbamate on passive avoidance learning by chicks (*Gallus domesticus*). *Psychopharmacology*, **48,** 287–9.

Glick, S. D. & Cox, R. D. (1977). Changes in sensitivity to operant effects of dopaminergic and cholinergic agents following morphine withdrawal in rats. *European Journal of Pharmacology*, **42,** 303–6.

Gold, P. E. & King, R. A. (1972). Caudate stimulation and retrograde amnesia: amnesia threshold and gradient. *Behavioral Biology*, **7,** 709–15.

Gold, P. E. & Murphy, J. M. (1980). Brain noradrenergic responses to training and to amnestic frontal cortex stimulation. *Pharmacology, Biochemistry & Behavior*, **13,** 257–63.

Gold, P. E. & van Buskirk, R. (1978*a*). Post-training brain norepinephrine concentrations: correlation with retention performance of avoidance training and with peripheral epinephrine modulation of memory processing. *Behavioral Biology*, **23,** 509–20.

Gold, P. E. & van Buskirk, R. (1978*b*). Effects of alpha- and beta-adrenergic receptor antagonists on post-trial epinephrine modulation of memory: relationship to post-training brain norepinephrine concentrations. *Behavioral Biology*, **24,** 168–84.

Goldberg, M. E., Insalaco, J. R., Hefner, M. A. & Salama, A. I. (1973). Effect of prolonged isolation on learning, biogenic amine turnover and aggressive behaviour in three strains of mice. *Neuropharmacology*, **12,** 1049–58.

Goldberg, S. G. & Gonzales, F. A. (1976). Effects of propranolol on behaviour maintained under fixed ratio schedules of cocaine injection or food presentation in squirrel monkeys. *Journal of Pharmacology and Experimental Therapeutics*, **198,** 626–34.

Gozzani, J. L. & Izquierdo, I. (1976). Possible peripheral adrenergic and central dopaminergic influences in memory consolidation. *Psychopharmacology*, **49,** 109–11.

Gray, T. & Wise, R. A. (1980). Effects of pimozide on lever pressing behavior maintained on an intermittent reinforcement schedule. *Pharmacology, Biochemistry & Behavior*, **12,** 931–5.

Greenshaw, A. J., Sanger, D. J. & Blackman, D. E. (1981). The effects of pimozide and reward omission on fixed-interval behavior of rats maintained by food and electrical brain stimulation. *Pharmacology, Biochemistry & Behavior*, **15,** 227–33.

Griffiths, R. R., Winger, G., Brady, J. V. & Snell, J. D. (1976). Comparison of behavior maintained by infusions of eight phenylethylamines in baboons. *Psychopharmacology*, **50,** 251–8.

Hall, M. E. (1976). The effects of norepinephrine biosynthesis inhibition on the consolidation of two discrimination escape responses. *Behavioral Biology*, **16,** 145–53.

Hall, M. E. (1977). Enhancement of learning by cycloheximide and DDC: a function of response strength. *Behavioral Biology*, **21,** 41–51.

Hall, M. E. & Mayer, M. A. (1975). Effects of alpha-methyl-*para*-tyrosine on the recall of a passive avoidance response. *Pharmacology, Biochemistry & Behavior*, **3,** 579–82.

Hamburg, M. D. & Cohen, R. P. (1973). Memory access pathway: role of adrenergic versus cholinergic neurons. *Pharmacology, Biochemistry & Behavior*, **1,** 295–300.

Hamburg, M. D. & Kerr, A. (1976). DDC-induced retrograde amnesia prevented by injections of *dl*-DOPS. *Pharmacology, Biochemistry & Behavior*, **5,** 499–501.

Hanson, L. C. (1965). The disruption of conditioned avoidance response following selective depletion of brain catecholamines. *Psychopharmacologia*, **8,** 100–10.

Hanson, L. C. (1967). Biochemical and behavioural effects of tyrosine hydroxylase inhibition. *Psychopharmacologia*, **11,** 8–17.

Harris, R. A. & Snell, D. (1980). Interactions between naltrexone and non-opiate drugs evaluated by schedule-controlled behaviour. *Neuropharmacology*, **19,** 1087–93.

Haycock, J. W., van Buskirk, R. & McGaugh, J. L. (1976). Facilitation of retention performance in mice by post-training diethyldithiocarbamate. *Pharmacology, Biochemistry & Behavior*, **5,** 525–8.

Haycock, J. W., van Buskirk, R. & McGaugh, J. L. (1977). Effects of catecholaminergic drugs upon memory storage processes in mice. *Behavioral Biology*, **20,** 281–310.

Haycock, J. W., van Buskirk, R., Ryan, J. R. & McGaugh, J. L. (1977). Enhancement of retention with centrally administered catecholamines. *Experimental Neurology*, **54,** 199–208.

Heffner, T. G. & Seiden, L. S. (1980). Synthesis of catecholamines from [^{3}H]tyrosine in brain during the performance of operant behaviour. *Brain Research*, **183,** 403–19.

Holz, W. C., Pendleton, R. G., Fry, W. T. & Gill, C. A. (1977). Epinephrine and recovery from punishment. *Journal of Pharmacology and Experimental Therapeutics*, **202,** 379–87.

Howard, J. L. & Breese, G. R. (1974). Physiological and behavioral effects of centrally-administered 6-hydroxydopamine in cats. *Pharmacology, Biochemistry & Behavior*, **2,** 651–61.

Howard, J. L., Grant, L. D. & Breese, G. R. (1974). Effects of intracisternal 6-hydroxydopamine treatment on acquisition and performance of rats in a double T-maze. *Journal of Comparative and Physiological Psychology*, **86,** 995–1007.

Hraschek, A. & Endroczi, E. (1979). Effects of systemic and intracerebral administration of adrenergic receptor blocking drugs on conditioned avoidance behavior and maze learning in rats. *Psychoneuroendocrinology*, **3,** 271–7.

Hraschek, A., Pavlik, A. & Endroczi, E. (1977). Brain catecholamine metabolism and avoidance conditioning in rats. *Acta physiologica Academiae scientiarum hungaricae*, **49,** 119–23.

Hurwitz, D. A., Robinson, S. M. & Barofsky, I. (1971*a*). The influence of training and avoidance performance on disulfiram-induced changes in brain catecholamines. *Neuropharmacology*, **10,** 447–52.

Hurwitz, D. A., Robinson, S. M. & Barofsky, I. (1971*b*). Behavior decrements and brain catecholamine changes in rats exposed to hypobaric hypoxia. *Psychopharmacologia*, **19,** 26–33.

Iversen, L. L. (1975). Dopamine receptors in the brain. *Science*, **188,** 1084–9.

Izquierdo, I., Beamish, D. G. & Anisman, H. (1979). Effect of an inhibitor of dopamine-beta-hydroxylase on the acquisition and retention of four different avoidance tasks in mice. *Psychopharmacology*, **63,** 173–8.

Jurkowlaniec, E. & Biatowas, J. (1981). Rewarding effect of noradrenergic stimulation of the amygdala in food deprived rats. *Physiology and Behavior*, **27,** 27–31.

Kety, S. S. (1970). The biogenic amines in the central nervous system: their possible roles in arousal, emotion and learning. In *The neurosciences*, ed F. O. Schmitt, pp. 324–36. Rockefeller University Press: New York.

Kety, S. S. (1972). The possible roles of the adrenergic systems of the cortex in learning. *Research Publications of the Association for Research in Nervous and Mental Diseases*, **50,** 376–89.

Kiianmaa, K., Fuxe, K., Jonsson, G. & Ahtee, L. (1975). Evidence for involvement of central NA neurones in alcohol intake. Increased alcohol consumption after degeneration of the NA pathway to the cerebral cortex. *Neuroscience Letters*, **1,** 41–5.

Kitsikis, A., Roberge, A. G. & Frenette, G. (1972*a*). Effect of L-dopa on delayed response and visual discrimination in cats and its relation to brain chemistry. *Experimental Brain Research*, **15,** 305–17.

Kitsikis, A., Roberge, A. G. & Frenette, G. (1972*b*). Differential effects of L-DOPA on behaviour. *Experimental Brain Research*, **15,** 315–21.

Kokkinidis, L. & Anisman, H. (1976). Interaction between cholinergic and catecholaminergic agents in a spontaneous alternation task. *Psychopharmacology*, **48**, 261–70.

Komissarov, I. V. & Talalaenko, A. N. (1974). Effect of catecholamines, serotonin, and some amino acids, injected into the amygdala on the conditioned avoidance reflex and motor activity of rats. *Bulletin of Experimental Biology and Medicine*, **77**, 143–6.

Koob, G. F., Kelley, A. E. & Mason, S. T. (1978). Locus coeruleus lesions: learning and extinction. *Physiology and Behavior*, **20**, 709–16.

Kostowski, W. & Plaznik, A. (1978). Effect of lesions of the ventral noradrenergic bundle on the two-way avoidance behavior in rats. *Acta physiologica polonica*, **29**, 509–14.

Kovacs, G. L., Bohus, B. & Versteeg, D. H. (1979*a*). Facilitation of memory consolidation by vasopressin: mediation by terminals of the dorsal noradrenergic bundle? *Brain Research*, **172**, 73–85.

Kovacs, G. L., Bohus, B. & Versteeg, D. H. (1979*b*). The effects of vasopressin on memory processes: the role of noradrenergic neurotransmission. *Neuroscience*, **4**, 1529–37.

Kovacs, G. L., Bohus, B., Versteeg, D. H., de Kloet, E. R. & de Wied, D. (1979*c*). Effect of oxytocin and vasopressin on memory consolidation: sites of action and catecholaminergic correlates after local microinjection into limbic–midbrain structures. *Brain Research*, **175**, 303–14.

Kovacs, G. L., Vecsei, L., Szabo, G. & Telegdy, G. (1977). The involvement of catecholaminergic mechanisms in the behavioral action of vasopressin. *Neuroscience Letters*, **5**, 337–44.

Kovacs, G. L., Versteeg, D. H. G., de Kloet, E. R. & Bohus, B. (1981). Passive avoidance performance correlates with catecholamine turnover in discrete limbic brain regions. *Life Sciences*, **28**, 1109–16.

Krantz, K. D. & Seiden, L. S. (1968). Effects of diethyldithiocarbamate on the conditioned avoidance response of the rat. *Journal of Pharmacy and Pharmacology*, **20**, 166–7.

Kruk, Z. L. & Millar, J. (1979). The effect of bilateral ventral noradrenaline bundle lesions on lever pressing for food in rats. *Psychopharmacology*, **64**, 41–3.

Laverty, R. & Taylor, K. M. (1968). Propranolol uptake into the central nervous system and the effect on rat behavior and amine metabolism. *Journal of Pharmacy and Pharmacology*, **20**, 605–9.

Lazareno, S. (1979). *d*-Amphetamine and punished responding: the role of catecholamines and anorexia. *Psychopharmacology*, **66**, 133–42.

Leconte, P. & Hennevin, E. (1981). Post-learning paradoxical sleep, reticular activation and noradrenergic activity. *Physiology and Behavior*, **26**, 587–94.

Lenard, L. G. & Beer, B. (1975*a*). 6-Hydroxydopamine and avoidance: possible role of response suppression. *Pharmacology, Biochemistry & Behavior*, **3**, 873–8.

Lenard, L. G. & Beer, B. (1975*b*). Relationship of brain levels of norepinephrine and dopamine to avoidance behaviour in rats after intraventricular administration of 6-hydroxydopamine. *Pharmacology, Biochemistry & Behavior*, **3**, 895–99.

Lenard, L. G. & Beer, B. (1975*c*). Modification of avoidance behavior in 6-hydroxydopamine-treated rats by stimulation of central noradrenergic and dopaminergic receptors. *Pharmacology, Biochemistry & Behavior*, **3**, 887–93.

Leonard, B. E. & Rigter, H. (1973). Changes in brain monoamine metabolism associated with CO_2-induced amnesia in rats. *British Journal of Pharmacology*, **48**, 351P–2P.

Levine, T. E., Erinoff, L., Dregits, D. P. & Seiden, L. S. (1980*a*). Effects of neonatal and adult 6-hydroxydopamine treatment on random-interval behavior. *Pharmacology, Biochemistry & Behavior*, **12**, 281–5.

Levine, T. E., McGuire, P. S., Heffner, T. G. & Seiden, L. S. (1980*b*). DRL performance in 6-hydroxydopamine-treated rats. *Pharmacology, Biochemistry & Behavior*, **12,** 287–91.

Lewy, A. J. & Seiden, L. S. (1972). Operant behavior changes norepinephrine metabolism in rat brain. *Science*, **175,** 454–6.

Lord, B. J., King, M. G. & Pfister, H. P. (1976). Chemical sympathectomy and two-way escape and avoidance learning. *Journal of Comparative and Physiological Psychology*, **90,** 303–16.

McMillan, D. E. (1971). Interactions between naloxone and chlorpromazine on behavior under schedule control. *Psychopharmacologia*, **19,** 128–33.

Margules, D. L. (1968). Noradrenergic basis of inhibition between reward and punishment in amygdala. *Journal of Comparative and Physiological Psychology*, **66,** 329–34.

Margules, D. L. (1971*a*). Localization of anti-punishment actions of norepinephrine and atropine in amygdala and entopeduncular nucleus of rats. *Brain Research*, **35,** 177–84.

Margules, D. L. (1971*b*). Alpha and beta adrenergic receptors in amygdala: reciprocal inhibitors and facilitators of punished operant behavior. *European Journal of Pharmacology*, **16,** 21–6.

Martinez, J. L., Jensen, R. A., Messing, R. B., Vasquez, B. J., Soumireu-Mourat, B., Geddes, D., Liang, K. C. & McGaugh, J. L. (1980*a*). Central and peripheral actions of amphetamine on memory storage. *Brain Research*, **182,** 157–66.

Martinez, J. L., Vasquez, B. J., Rigter, H., Messing, R. B., Jensen, R. A., Laing, K. C. & McGaugh, J. L. (1980*b*). Attenuation of amphetamine-induced enhancement of learning by adrenal demedullation. *Brain Research*, **195,** 433–43.

Mason, S. T. (1976). Noradrenaline pathways and learning behaviour in the rat. Ph.D. thesis, Cambridge University.

Mason, S. T. (1979). Noradrenaline: reward or extinction? *Neuroscience and Biobehavioural Reviews*, **3,** 1–10.

Mason, S. T., Beninger, R. J., Fibiger, H. C. & Phillips, A. G. (1980). Pimozide-induced suppression of responding: evidence against a block of food reward. *Pharmacology, Biochemistry & Behavior*, **12,** 917–23.

Mason, S. T. & Corcoran, M. E. (1978). Depletion of brain noradrenaline, but not dopamine, by intracerebral 6-hydroxydopamine potentiates convulsions induced by electroshock. *Journal of Pharmacy and Pharmacology*, **31,** 209–11.

Mason, S. T., Corcoran, M. E. & Fibiger, H. C. (1979*a*). Noradrenaline and ethanol intake in the rat. *Neuroscience Letters*, **12,** 137–42.

Mason, S. T. & Fibiger, H. C. (1978*a*). Noradrenaline and partial reinforcement in rats. *Journal of Comparative and Physiological Psychology*, **92,** 1110–18.

Mason, S. T. & Fibiger, H. C. (1978*b*). 6-OHDA lesion of the dorsal noradrenergic bundle alters extinction of passive avoidance. *Brain Research*, **152,** 209–14.

Mason, S. T. & Fibiger, H. C. (1979*a*). Noradrenaline and avoidance learning in rats. *Brain Research*, **161,** 321–34.

Mason, S. T. & Fibiger, H. C. (1979*b*). Noradrenaline and selective attention. *Life Sciences*, **25,** 1949–56.

Mason, S. T. & Fibiger, H. C. (1979*c*). The dorsal noradrenergic bundle and varieties of passive avoidance. *Psychopharmacology*, **66,** 179–82.

Mason, S. T. & Fibiger, H. C. (1979*d*). Noradrenaline, fear and extinction. *Brain Research*, **165,** 47–56.

Mason, S. T. & Fibiger, H. C. (1979*e*). Noradrenaline and extinction of conditioned taste aversion in the rat. *Behavioral and Neural Biology*, **25,** 206–16.

Mason, S. T. & Iversen, S. D. (1974). Learning impairment in rats after 6-hydroxydopamine-induced depletion of brain catecholamines. *Nature*, **248**, 697–8.
Mason, S. T. & Iversen, S. D. (1975). Learning in the absence of forebrain noradrenaline. *Nature*, **258**, 422–4.
Mason, S. T. & Iversen, S. D. (1977*a*). Effects of selective forebrain noradrenaline loss on behavioral inhibition in the rat. *Journal of Comparative and Physiological Psychology*, **91**, 165–73.
Mason, S. T. & Iversen, S. D. (1977*b*). Behavioural basis of the dorsal bundle extinction effect. *Pharmacology, Biochemistry & Behavior*, **7**, 373–9.
Mason, S. T. & Iversen, S. D. (1977*c*). An investigation of the role of cortical and cerebellar noradrenaline in associative motor learning in the rat. *Brain Research*, **134**, 513–27.
Mason, S. T. & Iversen, S. D. (1978*a*). Central and peripheral noradrenaline and resistance to extinction. *Physiology and Behavior*, **20**, 681–6.
Mason, S. T. & Iversen, S. D. (1978*b*). The dorsal noradrenergic bundle, extinction and non-reward. *Physiology and Behavior*, **21**, 1043–5.
Mason, S. T. & Iversen, S. D. (1978*c*). Reward, attention and the dorsal noradrenergic bundle. *Brain Research*, **150**, 135–48.
Mason, S. T. & Lin, D. (1980). Noradrenaline and selective attention in the rat. *Journal of Comparative and Physiological Psychology*, **94**, 819–32.
Mason, S. T. & Robbins, T. W. (1979). Noradrenaline and conditioned reinforcement. *Behavioral and Neural Biology*, **25**, 523–34.
Mason, S. T., Roberts, D. C. S. & Fibiger, H. C. (1979*b*). Interaction of brain noradrenaline and the pituitary–adrenal axis in learning and memory. *Pharmacology, Biochemistry & Behavior*, **10**, 11–16.
Meligeni, J. A., Ledergerber, S. A. & McGaugh, J. L. (1978). Norepinephrine attenuation of amnesia produced by diethyldithiocarbamate. *Brain Research*, **149**, 155–64.
Merlo, A. B. & Izquierdo, I. (1967). The effect of catecholamines on learning in rats. *Medicina et pharmacologia experimentalis*, **16**, 343–9.
Meyers, B. & Domino, E. F. (1964). The effect of cholinergic blocking drugs on spontaneous alternation in rats. *Archives internationales de pharmacodynamie et de thérapie*, **150**, 525–9.
Miller, F. P., Cox, R. H. & Maickel, R. P. (1970). Effects of altered brain norepinephrine levels on continuous avoidance responding and action of amphetamine. *Neuropharmacology*, **9**, 511–21.
Mitcham, J. C. & Thomas, R. K. Jr (1972). Effects of substantia nigra and caudate nucleus lesions on avoidance learning in rats. *Journal of Comparative and Physiological Psychology*, **81**, 101–7.
Moore, K. E. & Rech, R. H. (1967). Reversal of alpha-methyltyrosine-induced behavioural depression with dihydroxyphenylalanine and amphetamine. *Journal of Pharmacy and Pharmacology*, **19**, 405–7.
Morris, M. D., Tremmel, F. & Gebhart, G. F. (1979). Forebrain noradrenaline depletion blocks the release by chlordiazepoxide of behavioral extinction in the rat. *Neuroscience Letters*, **12**, 343–8.
Myers, R. D. & Melchior, C. L. (1975). Alcohol drinking in the rat after destruction of serotonergic and catecholaminergic neurons in the brain. *Research Communications in Chemistry, Pathology and Pharmacology*, **10**, 363–78.
Neill, D. B., Boggan, W. O. & Grossman, S. P. (1974). Impairment of avoidance

performance by intrastriatal administration of 6-hydroxydopamine. *Pharmacology, Biochemistry & Behavior*, **2,** 97–103.

Oei, T. P. (1978). Central catecholamines and peripheral noradrenaline depletion: effects on one-way trace-conditioning. *Pharmacology, Biochemistry & Behavior*, **8,** 25–9.

Oei, T. P. & King, M. G. (1978). Central catecholamine and peripheral noradrenaline depletion by 6-hydroxydopamine and active avoidance learning in rats. *Journal of Comparative and Physiological Psychology*, **92,** 94–108.

Oei, T. P. & Ng, C. P. (1978). 6-Hydroxydopamine induced catecholamine depletion and passive avoidance learning in rats. *Pharmacology, Biochemistry & Behavior*, **8,** 553–6.

Ogren, S. O., Archer, T. & Ross, S. B. (1980). Evidence for a role of the locus coeruleus noradrenaline system in learning. *Neuroscience Letters*, **20,** 351–6.

Ogren, S. O. & Fuxe, K. (1974). Learning, brain noradrenaline and the pituitary–adrenal axis. *Medical Biology*, **52,** 399–405.

Ogren, S. O. & Fuxe, K. (1977). On the role of brain noradrenaline and the pituitary–adrenal axis in avoidance learning. I. Studies with corticosterone. *Neuroscience Letters*, **5,** 291–6.

Olpe, H. R. & Baltzer, V. (1981). Vasopressin activates noradrenergic neurons in the rat locus coeruleus: a microiontophoretic investigation. *European Journal of Pharmacology*, **73,** 377–8.

Osbourne, R. H. & Kerkut, G. A. (1972). Inhibition of noradrenaline biosynthesis and its effects on learning in rats. *Comparative and General Pharmacology*, **3,** 359–62.

Owens, S., Boarder, M. R., Gray, J. A. & Fillenz, M. (1982). Acquisition and extinction of continuously and partially reinforced running in rats with lesion of the dorsal noradrenergic bundle. *Behavioural Brain Research*, **5,** 11–42.

Palfai, T., Brown, O. M. & Walsh, T. J. (1978). Catecholamine levels in the whole brain and the probability of memory formation are not related. *Pharmacology, Biochemistry & Behavior*, **8,** 717–21.

Pappas, B. A. & Sobrian, S. K. (1972). Neonatal sympathectomy by 6-hydroxydopamine in the rat: no effects on behavior but changes in endogenous brain norepinephrine. *Life Sciences*, **11,** 653–9.

Peterson, D. W. & Laverty, R. (1976). Operant behavioral and neurochemical effects after neonatal 6-hydroxydopamine treatment. *Psychopharmacology*, **50,** 55–60.

Phillips, A. G. & Fibiger, H. C. (1979). Decreased resistance to extinction after haloperidol: implications for the role of dopamine in reinforcement. *Pharmacology, Biochemistry & Behavior*, **10,** 751–60.

Pickens, R. & Harris, W. C. (1968). Self-administration of *d*-amphetamine by rats. *Psychopharmacologia*, **12,** 158–63.

Plech, A., Herman, Z. S. & Wierzbicki, A. (1975). Role of the noradrenergic and the dopaminergic receptors in the mechanism of the avoidance reflex in rats. *Acta medica polona*, **16,** 225–30.

Prado de Carvalho, L. & Zornetzer, S. F. (1981). The involvement of the locus coeruleus in memory. *Behavioral and Neural Biology*, **31,** 173–86.

Price, M. T. C. & Fibiger, H. C. (1975). Discriminated escape learning and response to electric shock after 6-hydroxydopamine lesions of the nigroneostriatal dopaminergic projections. *Pharmacology, Biochemistry & Behavior*, **3,** 285–90.

Price, M. T. C., Murray, G. N. & Fibiger, H. C. (1977). Schedule dependent changes in operant responding after lesions of the dorsal tegmental noradrenergic projection. *Pharmacology, Biochemistry & Behavior*, **6,** 11–15.

Rainbow, T. C., Adler, J. E. & Flexner, L. B. (1976). Comparison in mice of the amnestic effects of cycloheximide and 6-hydroxydopamine in a one-trial passive avoidance task. *Pharmacology, Biochemistry & Behavior*, **4**, 347–9.

Randt, C. T., Quartermain, D., Goldstein, M. & Anagoste, B. (1971). Norepinephrine biosynthesis inhibition: effects on memory in mice. *Science*, **172**, 498–9.

Ranje, C. & Ungerstedt, U. (1977*a*). Discriminative and motor performance in rats after interference with dopamine neurotransmission with spiroperidol. *European Journal of Pharmacology*, **43**, 39–46.

Ranje, C. & Ungerstedt, U. (1977*b*). Lack of acquisition in dopamine denervated animals tested in an underwater Y-maze. *Brain Research*, **134**, 95–111.

Richardson, J. S., Stacey, P. D., Cerauskis, P. W. & Musty, R. E. (1971). Propranolol interfered with inhibitory behavior in rats. *Journal of Pharmacy and Pharmacology*, **23**, 459–60.

Ridley, R. M., Haystead, T. A . J., Baker, H. F. & Crow, T. J. (1981). A new approach to the role of noradrenaline in learning: problem-solving in the marmoset after alpha-noradrenergic receptor blockade with aceperone. *Pharmacology, Biochemistry & Behavior*, **14**, 849–55.

Risner, M. E. & Jones, B. E. (1976). Role of noradrenergic and dopaminergic processes in amphetamine self-administration. *Pharmacology, Biochemistry & Behavior*, **5**, 477–82.

Roberge, A. G., Boisvert, C. & Everett, J. (1980). Monoamine roles in retention and reversal of delayed response in cats. *Pharmacology, Biochemistry & Behavior*, **12**, 229–34.

Roberge, A. G., Roy, J.-P. & Boisvert, C. (1978). Effect of metergoline on delayed response in cats and its relation to the metabolism of dopamine and serotonin in neostriatal and mesolimbic neurons. *Experimental Brain Research*, **32**, 19–31.

Roberts, D. C. S. & Bloom, F. E. (1981). Adrenal steroid-induced changes in beta-adrenergic receptor binding in rat hippocampus. *European Journal of Pharmacology*, **74**, 37–42.

Roberts, D. C. S., Corcoran, M. E. & Fibiger, H. C. (1977). On the role of ascending catecholaminergic systems in intravenous self-administration of cocaine. *Pharmacology, Biochemistry & Behavior*, **6**, 615–20.

Roberts, D. C. S. & Fibiger, H. C. (1976). Conditioned taste aversion induced by diethyldithiocarbamate (DDC). *Neuroscience Letters*, **2**, 339–42.

Roberts, D. C. S. & Fibiger, H. C. (1977). Evidence for interactions between central noradrenergic neurons and adrenal hormones in learning and memory. *Pharmacology, Biochemistry & Behavior*, **7**, 191–4.

Roberts, D. C. S., Price, M. T. C. & Fibiger, H. C. (1976). The dorsal tegmental noradrenergic projection: an analysis of its role in maze learning. *Journal of Comparative and Physiological Psychology*, **90**, 363–72.

Rose, S. P. R. (1981). What should a biochemistry of learning and memory be about? *Neuroscience*, **6**, 811–21.

Routtenberg, A. & Holzman, N. (1973). Memory disruption by electrical stimulation of substantia nigra, pars compacta. *Science*, **181**, 83–6.

Ruiz, M. & Monti, J. M. (1975). Reversal of 6-hydroxydopamine-induced suppression of a CAR by drugs facilitating central catecholaminergic mechanisms. *Pharmacology*, **13**, 281–6.

Sastry, B. S. R. & Phillis, J. W. (1977). Evidence that clonidine can activate histamine H_2-receptors in rat cerebral cortex. *Neuropharmacology*, **16**, 223–5.

Schoenfeld, R. I. & Seiden, L. S. (1967). Alpha-methyltyrosine: effects on fixed ratio schedules of reinforcement. *Journal of Pharmacy and Pharmacology*, **19**, 771–2.

Schoenfeld, R. I. & Seiden, L. S. (1969). Effect of alpha-methyltyrosine on operant behavior and brain catecholamine levels. *Journal of Pharmacology and Experimental Therapeutics*, **167**, 319–27.

Schoenfeld, R. I. & Uretsky, N. J. (1972). Operant behavior and catecholamine-containing neurons: prolonged increase in lever-pressing after 6-hydroxydopamine. *European Journal of Pharmacology*, **20**, 357–62.

Schwartz, A. S. & Marchok, P. O. (1974). Depression of morphine-seeking behaviour by dopamine inhibition. *Nature*, **248**, 257–8.

Segal, M. & Edelson, A. (1978). Effects of priming stimulation of catecholamine containing nuclei in rat brain on runway performance. *Brain Research Bulletin*, **3**, 203–6.

Sessions, G. R., Kant, G. J. & Koob, G. F. (1976). Locus coeruleus lesions and learning in the rat. *Physiology and Behavior*, **17**, 853–9.

Shaywitz, B. A., Yager, R. D. & Klopper, J. H. (1976). Selective brain dopamine depletion in developing rats: an experimental model of minimal brain dysfunction. *Science*, **191**, 305–8.

Simon, H., Scatton, B. & Le Moal, M. (1979). Definitive disruption of spatial delayed alternation in rats after lesions in the ventral mesencephalic tegmentum. *Neuroscience Letters*, **15**, 319–24.

Simon, H., Scatton, B. & Le Moal, M. (1980). Dopaminergic A10 neurones are involved in cognitive functions. *Nature*, **286**, 150–1.

Singh, K. P., Bhandari, D. S. & Mahawar, M. M. (1971). Effects of propranolol (a beta adrenergic blocking agent) on some central nervous system parameters. *Indian Journal of Medical Research*, **59**, 786–94.

Sladek, J. R. & Walker, P. (1977). Serotonin-containing neuronal perikarya in the primate locus coeruleus and subcoeruleus. *Brain Research*, **134**, 359–66.

Solanto, M. V. & Hamburg, M. D. (1979). DDC-induced amnesia and norepinephrine: a correlated behavioral–biochemical analysis. *Psychopharmacology*, **66**, 167–70.

Stahl, S. M., Narotsky, R. A., Boshes, B. & Zeller, E. A. (1974). The effects of perturbation of cerebral amine metabolism on operationally defined learning and memory processes in goldfish. *Biological Psychiatry*, **9**, 295–323.

Stahl, S. M., Zeller, E. A. & Boshes, B. (1971). On the effect of modulation of cerebral amine metabolism on the learning and memory of goldfish (*Carassius auratus*). *Transactions of the American Neurological Association*, **96**, 310–12.

Stein, L., Belluzzi, J. D. & Wise, C. D. (1975). Memory enhancement by central administration of norepinephrine. *Brain Research*, **84**, 329–35.

Stephenson, R. M. & Andrew, R. J. (1981). Amnesia due to beta-antagonists in a passive avoidance task in the chick. *Pharmacology, Biochemistry & Behavior*, **15**, 597–604.

Taboada, M. E., Souto, M., Hawkins, H. & Monti, J. M. (1979). The actions of dopaminergic and noradrenergic antagonists on conditioned avoidance responses in intact and 6-hydroxydopamine-treated rats. *Psychopharmacology*, **62**, 83–8.

Thut, P. D. (1977). Effect of L-DOPA analogues in Sidman avoidance performance in mice. *Life Sciences*, **21**, 423–31.

Tilson, H. A., Rech, R. H. & Sparber, S. B. (1975). Release of ^{14}C-norepinephrine into the lateral cerebroventricle of rats by exposure to a conditioned aversive stimulus. *Pharmacology, Biochemistry & Behavior*, **3**, 385–92.

Tombaugh, T. N. (1981). Effects of pimozide on nondiscriminated and discriminated performance in the pigeon. *Psychopharmacology*, **73**, 137–41.

Tombaugh, T. N., Tombaugh, J. & Anisman, H. (1979). Effects of dopamine receptor blockade on alimentary behaviors: home cage food consumption, magazine training, operant acquisition and performance. *Psychopharmacology*, **66**, 219–25.

Tyrer, P. (1972). Propranolol in alcohol addiction. *Lancet*, **ii**, 707.

Vachon, L. & Roberge, A. G. (1981). Involvement of serotonin and catecholamine metabolism in cats trained to perform a delayed response task. *Neuroscience*, **6**, 189–94.

Valzelli, L. (1967). Drugs and aggressiveness. *Advances in Pharmacology*, **5**, 79–108.

Valzelli, L. (1973). The 'isolation syndrome' in mice. *Psychopharmacologia*, **31**, 305–20.

Walsh, T. J. & Palfai, T. (1981). Reversal of guanethidine- and diethyldithiocarbamate-induced amnesia by peripherally-administered catecholamines. *Pharmacology, Biochemistry & Behavior*, **14**, 713–18.

Weiss, J. M., Stone, E. A. & Harrell, N. (1970). Coping behavior and brain norepinephrine levels in rats. *Journal of Comparative and Physiological Psychology*, **72**, 153–60.

Weissman, A. & Koe, B. K. (1965). Behavioral effects of L-alpha-methyltyrosine, an inhibitor of tyrosine hydroxylase. *Life Sciences*, **4**, 107–48.

Wilson, M. C. & Schuster, C. R. (1973). The effects of stimulants and depressants on cocaine self-administration behavior in rhesus monkeys. *Psychopharmacologia*, **31**, 291–304.

Wilson, M. C. & Schuster, C. R. (1976). Mazindol self-administration in the rhesus monkey. *Pharmacology, Biochemistry & Behavior*, **4**, 207–10.

Wise, R. A. (1982). Neuroleptics and operant behaviour: the anhedonia hypothesis. *Behavioral and Brain Sciences*, **5**, 39–88.

Wise, R. A. & Schwartz, H. V. (1981). Pimozide attenuates acquisition of lever-pressing for food in rats. *Pharmacology, Biochemistry & Behavior*, **15**, 655–6.

Wise, R. A., Spindler, J., deWit, H. & Gerber, G. J. (1978*a*). Neuroleptic-induced 'anhedonia' in rats: pimozide blocks reward quality of food. *Science*, **201**, 262–4.

Wise, R. A., Spindler, J. & Legault, L. (1978*b*). Major attenuation of food reward with performance-sparing doses of pimozide in the rat. *Canadian Journal of Psychology*, **32**, 77–85.

Wyers, E. J., Deadwyler, S. A., Hirsuna, N. & Montgomery, D. (1973). Passive avoidance retention and caudate stimulation. *Physiology and Behavior*, **11**, 809–19.

Yeudall, L. T. & Walley, R. E. (1977). Methylphenidate, amygdalectomy and active avoidance performance in the rat. *Journal of Comparative and Physiological Psychology*, **91**, 1207–19.

Yokel, R. A. & Wise, R. A. (1975). Increased lever pressing for amphetamine after pimozide in rats: implication for a dopamine theory of reward. *Science*, **187**, 547–9.

Yokel, R. A. & Wise, R. A. (1978). Amphetamine-type reinforcement by dopaminergic agonists in the rat. *Psychopharmacology*, **58**, 289–96.

Yonkov, D. & Roussinov, K. (1979). Influence of diethyldithiocarbamate (DDC) on the memory-facilitating effect of some central stimulants. *Acta physiologica et pharmacologica bulgarica*, **5**, 41–6.

Zeller, E. A., Couper, G. S., Huprikar, S. V., Mellow, A. M. & Moody, R. R. (1976). Mescaline: its effects on learning rate and dopamine metabolism in goldfish (*Carassius auratus*). *Experientia*, **32**, 1453–4.

Zis, A. P., Fibiger, H. C. & Phillips, A. G. (1974). Reversal by L-dopa of impaired learning due to destruction of dopaminergic nigro-neostriatal projection. *Science*, **185**, 960–2.
Zornetzer, S. F., Abraham, W. C. & Appleton, R. (1978). Locus coeruleus and labile memory. *Pharmacology, Biochemistry & Behavior*, **9**, 227–34.
Zornetzer, S. F. & Gold, M. S. (1976). The locus coeruleus: its possible role in memory consolidation. *Physiology and Behavior*, **16**, 331–6.
Zornetzer, S. F., Gold, M. S. & Hendrickson, J. (1974). Alpha-methyl-*p*-tyrosine and memory: state-dependency and memory failure. *Behavioral Biology*, **12**, 135–41.

6

Catecholamines and cognitive behaviour

Introduction

The format of this chapter must needs be somewhat different from that of preceding ones. This is because of the patchy and episodic nature of the research available to date in the area of catecholamines and cognitive behaviour. Thus, with the exception of selective attention, it is not possible to examine the relative roles of noradrenaline and dopamine in a specified behaviour. This is simply because only one of the two amines has actually been studied in the appropriate context. There are just no data at all on the other amine. For example, research on fear and anxiety has concentrated on the possible role of noradrenaline to the total exclusion of any consideration of dopamine. The approach adopted in this chapter then will have to lack the symmetry possible in previous ones and simply cover what higher level, cognitive behaviours have actually been examined for each amine.

Noradrenaline and cognitive behaviour

Extinction

The paradigm. Extinction refers to the behavioural paradigm in which animals, having been trained on a particular task for a given reinforcement, are then faced with an identical situation except that reinforcement is no longer presented. A typical situation might involve training a rat to press a lever for continuous reinforcement of food pellets. Once the animal has acquired the task and is exhibiting high rates of responding, the feeder is disconnected so that food is no longer presented; although other aspects of the situation remain the same. Animals in extinction will soon cease to emit the no-longer reinforced response. This may be regarded merely as a form of learning – learning *not* to respond – but as will be seen it is a far more complicated process. Although central learning processes are invoked during extinction, so too are many other psychological mechanisms, and extinction is of interest because it gives us access to central functions *other than* those involved merely in learning.

Positive reinforcement. The first demonstration of a role for brain noradrenaline in extinction behaviour was reported by Mason & Iversen (1975). They trained rats depleted of brain noradrenaline by either adult injections of 6-hydroxydopamine (6-OHDA) into the dorsal bundle or by neonatal intraperitoneal injection of 6-OHDA (see p. 213) on an L-shaped runway task for food reward. Acquisition learning of these animals was unaltered, being as efficient as saline-injected controls (see Fig. 5.6, p. 214). However, when these animals were tested in extinction, that is without food being present in the goal box at the end of the alley-way, a difference in behaviour emerged (see Fig. 6.1). The noradrenaline depleted rats continued running to the now-empty goal box more quickly than did the controls. Eventually, after a number of trials, the rats ceased to run, and

Fig. 6.1. Extinction of an alley-way response previously trained for food reward in rats treated with 6-hydroxydopamine either when neonatal (NPT) or by adult dorsal bundle injections (DB). The trial number on which the extinction criterion was reached is plotted against overall rank order. Note that both noradrenaline depleted groups take longer than their controls to extinguish and so cluster towards the upper right-hand corner of the graph while the controls cluster towards the bottom left-hand side of the graph. (Redrawn from Mason & Iversen, 1975.)

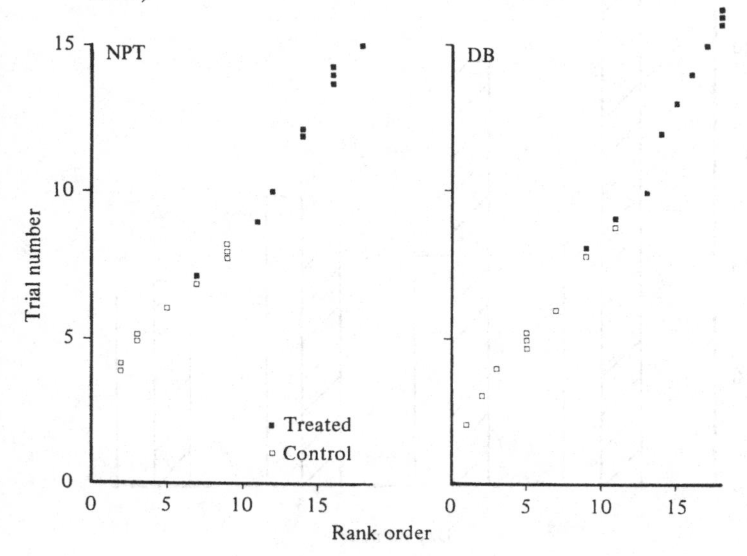

this extinction criterion took more trials to reach with the noradrenaline depleted rats than with controls. This phenomenon of faster running to a now-empty goal box is an example of resistance to extinction and was the first demonstration of what later came to be called the dorsal bundle extinction effect (DBEE).

Rather than seek to elucidate *how* such resistance to extinction comes about behaviourally, I shall now list the many other examples in which such resistance to extinction has also been observed and then return to the psychological mechanisms below (see p. 269).

Another situation in which resistance to extinction has been seen after forebrain noradrenaline depletion is in continuous reinforcement (CRF) lever-pressing for food reward. Thus, Mason & Iversen (1977*a*) found that dorsal bundle injections of 6-OHDA led to unaltered acquisition learning of a CRF lever-press task but again caused increased resistance to extinction when the feeder was disconnected (see Fig. 6.2). Both an

Fig. 6.2. Extinction of a CRF lever-press response previously trained for food reward in adult dorsal bundle 6-hydroxydopamine rats (Treated) and their controls for four successive days of extinction testing. (Redrawn from Mason & Iversen, 1977*a*.)

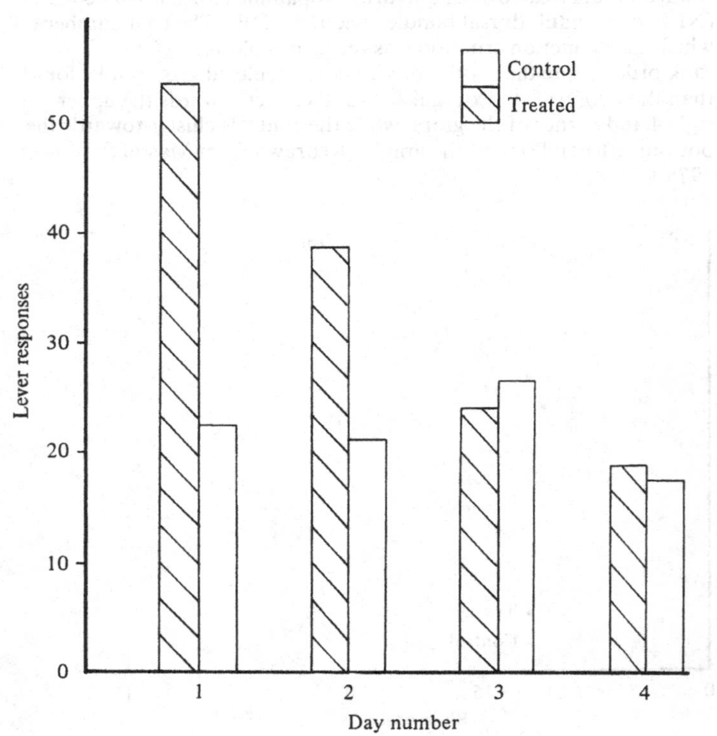

increased rate of lever-pressing per minute was seen and a greater persistance in that, if animals were allowed to continue until they had not pressed for 2 consecutive minutes, the noradrenaline depleted rats took longer to reach this extinction criterion than did the saline-injected controls. Such resistance to extinction was seen on the first 2 days of extinction but by the third day even the noradrenaline depleted rats had decreased their responding to low levels. This raises an important point – it is not that noradrenaline depleted rats cannot extinguish, it is merely that they take longer. Even noradrenaline depleted rats will eventually abandon the unrewarded response.

The resistance to extinction found in a runway task for food reward by Mason & Iversen (1975) was confirmed by Owen, Boarder, Gray & Fillenz (1982) using similar dorsal bundle 6-OHDA lesions and a straight runway.

The resistance to extinction found in a CRF lever-press task by Mason & Iversen (1977*a*) has also been seen by Thornton, Goudie & Bithell (1975) using the neonatal intraperitoneal technique of 6-OHDA administration to deplete brain noradrenaline selectively.

Other operant schedules may also show resistance to extinction after forebrain noradrenaline depletion. Thus, Mason & Iversen (1978*a*) and Mason (1979) found that a fixed-interval schedule (FI, see Appendix) for food reward generated resistance to extinction (see Appendix) in noradrenaline depleted rats compared to saline-injected controls. This is shown in Fig. 6.3. Acquisition learning of this schedule was not affected (see p. 216).

Other workers (Tremmel, Morris & Gebhart, 1977) have seen resistance to extinction after electrolytic lesion of the dorsal bundle in a go/no-go alternation task. Here, reward is available on every second presentation of a retractable lever. On the nonrewarded trials, pressing the lever fails to yield a food pellet. Rats come to pattern their responses such that rapid responding occurs on the rewarded trials, but the latency to lever-press on an unrewarded trial is much longer. When none of the trials is rewarded, noradrenaline depleted rats show more rapid responding than controls on the last few days of a 6 day extinction period (see Fig. 6.4).

Finally, resistance to extinction was seen on a complex motor manipulative task by Mason & Iversen (1977*b*). Rats were trained to run through a tunnel connecting start and goal boxes in order to reach food reward present in the goal box. After overtraining on the running response the tunnel was blocked by a plastic ball with a handle attached. The ball could not be pushed through the tunnel and had, instead, to be pulled out by the handle in order to clear the tunnel and allow the rat to reach food reward. Acquisition learning of this task did not differ as a result of severe depletion of forebrain noradrenaline, either by adult dorsal bundle or by neonatal peripheral injections of 6-OHDA (see p. 217). However, when reward was no longer present in the goal box the noradrenaline depleted rats continued to run rapidly through the tunnel to the now-empty food cup and took more unrewarded trials to cease pulling the ball out of the tunnel than did saline-injected controls (see Fig. 6.5). Thus, resistance to extinction is seen irrespective of the motor nature of the response required. Both simple running and complex motor-manipulative responses show increased persistence after noradrenaline depletion of the forebrain. Lever-pressing on a number of continuously reinforced or consistently cued schedules such as CRF, FI or go/no-go alternation also demonstrates the DBEE.

Negative reinforcement. The DBEE has also been seen after negatively reinforced original learning. Ashford & Jones (1976) reported that

Fig. 6.3. Extinction of a fixed-interval lever-press response previously trained for food reward in adult dorsal bundle 6-hydroxydopamine (6-OHDA) rats and their controls for four successive days of extinction testing. (Redrawn from Mason & Iversen, 1978*a*.)

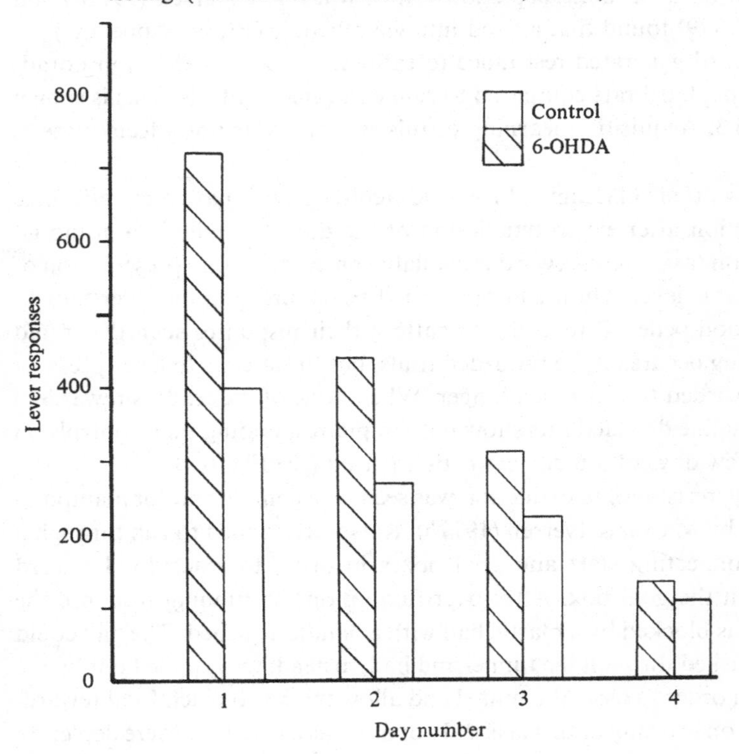

noradrenaline depletion by injection of 6-OHDA into the amygdala caused resistance to extinction of one-way active avoidance behaviour (see Appendix). This is shown in Fig. 6.6, and was confirmed by Fibiger & Mason (1978) for 6-OHDA injection into the dorsal noradrenergic bundle.

Increased responding in extinction of a two-way active avoidance task (see Appendix) was found by Mason & Fibiger (1979*a*) after dorsal bundle 6-OHDA injections, and confirmed by Plaznik & Kostowski (1980) with electrolytic lesion of the locus coeruleus (LC) (see Fig. 6.7).

Passive avoidance also shows resistance to extinction after forebrain noradrenaline loss (Mason & Fibiger, 1978*a*), again indicating that the nature of the response is not of crucial importance in determining the DBEE. Resistance to extinction has been reported after taste aversion learning (see Appendix) (Mason & Fibiger, 1979*b*). These instances of the DBEE are listed in Table 6.1.

Partial reinforcement. Situations in which the DBEE does *not* occur seem to include partially reinforced schedules (PRF, see Appendix) in which not

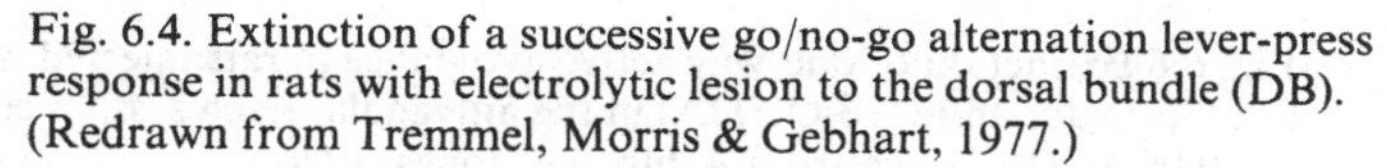
Fig. 6.4. Extinction of a successive go/no-go alternation lever-press response in rats with electrolytic lesion to the dorsal bundle (DB). (Redrawn from Tremmel, Morris & Gebhart, 1977.)

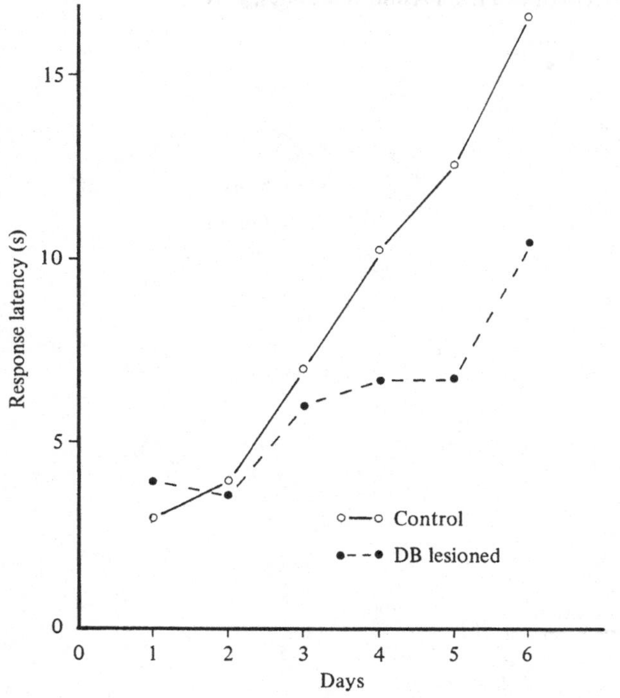

Fig. 6.5. Extinction of a motor manipulative task previously trained for food reward in rats treated with 6-hydroxydopamine (6-OHDA), either when neonatal (NPT) or by adult dorsal bundle administration (DB). Time to enter goal box is shown for five extinction trials per day for three successive days. (Redrawn from Mason & Iversen, 1977*b*.)

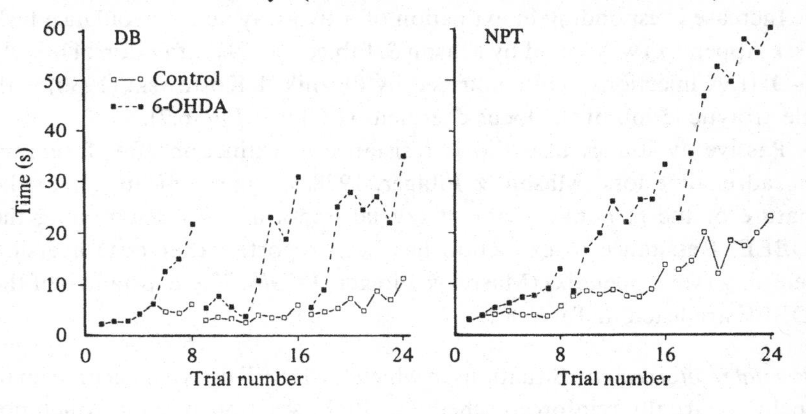

Fig. 6.6 Extinction of one-way active avoidance in rats injected with 6-hydroxydopamine (6-OHDA) in the amygdala. Each session consisted of 10 trials. (Redrawn from Ashford & Jones, 1976.)

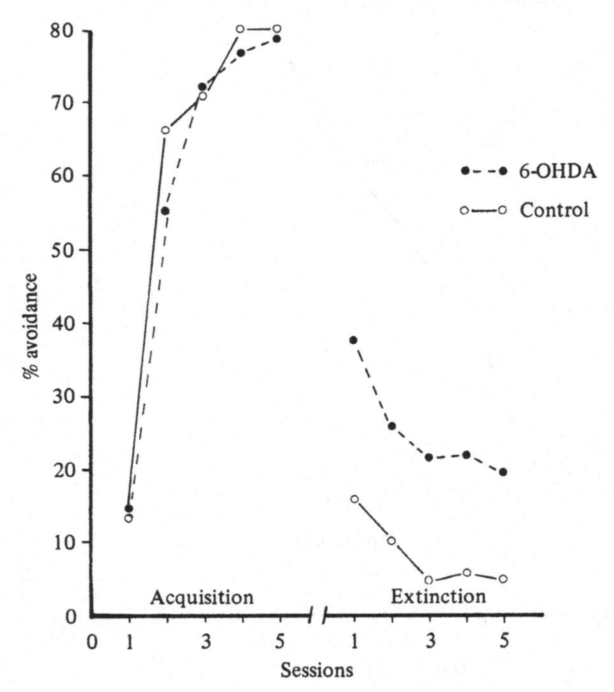

every response leads to reinforcement. Thus, extinction of a fixed-ratio schedule (FR, see Appendix) is not affected by noradrenaline depletion (Peterson & Laverty, 1976) nor is extinction of a variable-interval schedule (VI, see Appendix) (Price, Murray & Fibiger, 1977). Variable ratio (VR, see Appendix) training leads to similar extinction in both control and dorsal bundle 6-OHDA lesioned rats (Mason & Fibiger, 1978*b*) and partial reinforcement in a straight alley-way abolishes the resistance to extinction seen in dorsal bundle 6-OHDA lesioned rats seen after CRF training in the same situation (Owens *et al.*, 1982). Extinction of Sidman avoidance behaviour (see Appendix), which, due to its complexity of response-reinforcement contingencies, might bear some resemblance to a PRF schedule as far as the rat is concerned, is also not altered by noradrenaline depletion (Mason & Fibiger, 1979*c*).

Fig. 6.7. Extinction of two-way active avoidance in rats with electrolytic lesion to the locus coeruleus (LC) and control groups (C). After training to an acquisition criterion, two-week retention was not different between the two groups but the LC group took more trials to reach an extinction criterion ($P < 0{\cdot}05$). (Redrawn from Plaznik & Kostowski, 1980.)

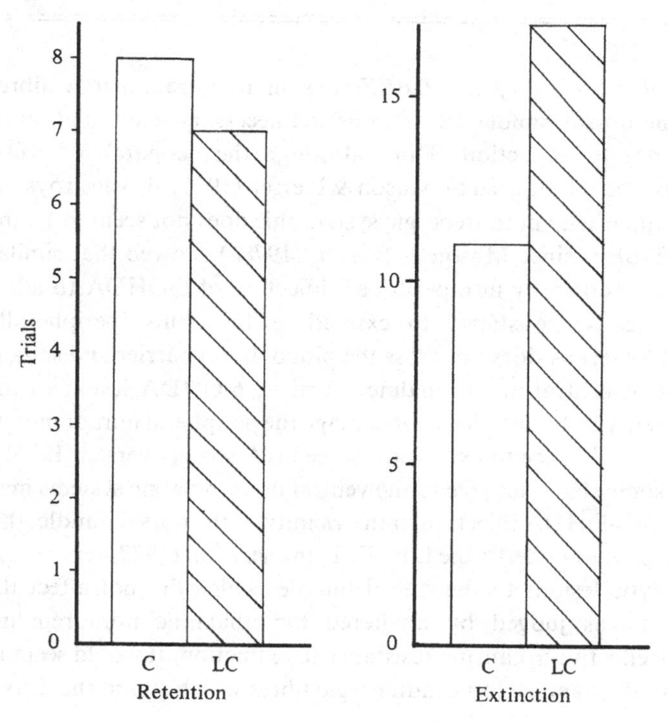

Table 6.1. *Instances of the dorsal bundle extinction effect*

Paradigm	Reference
Positive reinforcement	
Runway for food reward (CRF)	Mason & Iversen (1975); Owens *et al.* (1982)
Lever-pressing for food reward (CRF)	Mason & Iversen (1977*a*); Thornton *et al.*, (1975)
Lever-pressing for food reward (fixed interval)	Mason & Iversen (1978*a*); Mason (1979)
Lever-pressing for food reward (go/no-go alternation)	Tremmel *et al.* (1977)
Motor manipulative response for food reward (CRF)	Mason & Iversen (1977*b*)
Negative reinforcement	
One-way active avoidance	Ashford & Jones (1976); Fibiger & Mason (1978)
Two-way active avoidance	Mason & Fibiger (1979*a*); Plaznik & Kostowski (1980)
Passive avoidance	Mason & Fibiger (1978*a*)
Taste aversion	Mason & Fibiger (1979*b*)
Conditioned suppression to a tone on a CRF lever-press baseline	Mason & Fibiger (1979*c*)

Neurochemical substrate of the DBEE. Lesion to noradrenergic fibres running in the dorsal bundle seems to be the necessary manipulation to cause resistance to extinction. Thus, although the neonatal 6-OHDA administration technique used by Mason & Iversen (1975) also destroys the peripheral sympathetic noradrenergic system this does not seem to be the basis of the DBEE, since Mason & Iversen (1978*b*) showed that similar destruction on its own by intraperitoneal injection of 6-OHDA to adult rats did not cause resistance to extinction. In adults, peripherally administered 6-OHDA does not cross the blood–brain barrier and so fails to cause loss of central noradrenaline. Further, 6-OHDA lesion to the dorsal noradrenergic bundle does not damage the peripheral noradrenergic system and yet resistance to extinction is seen (Mason & Iversen, 1975).

Although some slight damage to the ventral noradrenergic systems may also occur with 6-OHDA injection in the vicinity of the dorsal bundle, this too does not seem to underlie the DBEE. Tremmel *et al.* (1977) report that small electrolytic lesions to the dorsal bundle, which did not affect the ventral systems as judged by unaltered hypothalamic noradrenaline content, were effective in causing resistance to extinction. It would seem to be specifically the damage to noradrenergic fibres which run in the dorsal bundle since Mason & Fibiger (1979*d*) found that kainic acid lesion of the cell bodies around the dorsal bundle which spared the noradrenergic fibres themselves to a great extent did not cause resistance to extinction, nor did injection of the serotonergic neurotoxin 5,7-dihydroxytryptamine. Additionally, noradrenaline loss can be prevented by pre-treatment with desipramine (DMI) 30 min prior to 6-OHDA injection. This does not alter any small nonspecific effects of injecting 6-OHDA in the vicinity of the dorsal bundle on nonnoradrenergic elements. However, DMI pre-treatment *was* effective in preventing the usual resistance to extinction, hand-in-hand with its protection of noradrenergic neurones (Mason & Fibiger, 1979*d*). A word of caution as to the origin of the noradrenergic fibres may be deserved, however, since Koob, Kelley & Mason (1978) found that electrolytic lesion to the LC did *not* cause resistance to extinction on a CRF lever-press task. This may indicate that the crucial noradrenergic fibres running in the dorsal bundle do not originate from the LC, or it might indicate that the portion of the LC spared by the small lesions of Koob *et al.* (1978), viz. the anterior pole, may be the source of the noradrenergic fibres responsible for the DBEE. Further research is required to answer this question.

The forebrain area of the dorsal bundle involved in the DBEE is totally unknown at this time. Since limbic electrolytic lesions often lead to resistance to extinction (Douglas, 1967; Dickinson, 1974), it is tempting to speculate that it might be the noradrenergic innervation to these areas – septum, hippocampus or possibly the amygdala – which is important for the DBEE, but no hard evidence is yet available.

Other noradrenaline manipulations and extinction. Merlo & Izquierdo (1967) reported that intraperitoneal injection of noradrenaline would speed up extinction of avoidance behaviour. It is uncertain how much noradrenaline, if any, would get into the brain after peripheral injection so more interesting is the finding of Torda (1976) that injection of noradrenaline directly into the hypothalamus would speed up extinction of one-way active avoidance in rats. Finally, Ellison, Handel, Rodgers & Weiss (1975) found that the antidepressant drug protriptyline slowed down extinction of avoidance behaviour. Since this class of drug acts in part to reduce the firing of the LC cells (Nyback, Walters, Aghajanian & Roth, 1975), this may be further demonstration of resistance to extinction as a result of reduced activity in the noradrenergic system.

Mechanisms of the DBEE

Various psychological mechanisms have been suggested to underlie the

DBEE. I shall briefly examine some of them here and also the evidence pertaining to them.

Hyperactivity. Perhaps the simplest mechanism might be that of a general increase in motor activity. This would explain the faster running to the now-empty goal box during extinction and also the more persistent lever-pressing in CRF extinction. However, this explanation may quickly be discarded. Hyperactivity would suggest faster running during the acquisition phase of a runway response as well as during its extinction. Such is not seen (Mason & Iversen, 1975; Owens *et al.*, 1982). Additionally, when noradrenaline depleted rats are examined for locomotor activity in photocell cages (see p. 96), Mason (1977) found no activity greater than that of controls during any part of a 1 h test period (see Fig. 6.8*a, b*). Similar activity testing in a simple, unstructured environment such as an open field also found no difference in locomotor activity between noradrenaline depleted and control animals (Fig. 6.8*c, d*). Crow *et al.* (1978) also found no difference in locomotor activity between controls and rats depleted of forebrain noradrenaline by 6-OHDA injections near LC. Hyperactivity thus does not appear to be an explanation of the DBEE.

Internal inhibition. A deficit in the brain mechanism involved in the inhibition of behaviour might also lead to increased resistance to extinction. If noradrenaline depleted rats were impaired in inhibiting the emission of a response then this would lead to increased responding in situations where controls were actively inhibiting behavioural output, such as extinction situations. However, this fails to explain the generally unimpaired acquisition of passive avoidance (see Appendix) by noradrenaline depleted rats (Roberts & Fibiger, 1977; Mason & Fibiger, 1979*e*), since this task also requires the ability to inhibit the naturally prepotent step-down or step-through response. Further, Mason & Iversen (1977*c*) examined a differential reinforcement of low rates of responding to an operant task (DRL, see Appendix), which requires the rat to inhibit the emission of a prepotent response, lever-pressing, for the DRL interval. No deficit in either the learning or the performance of a DRL schedule was seen in animals depleted of forebrain noradrenaline, either by adult dorsal bundle 6-OHDA injections or by peripheral neonatal 6-OHDA administration. This is shown in Fig. 6.9. A final task requiring behavioural inhibition was described by Mason & Iversen (1977*c*). This was a conditioned taste aversion (see Appendix) in which inhibition of the prepotent response of saccharin consumption is required from the rat. Again, despite severe noradrenaline depletion, both control and lesioned rats learned the taste aversion as normal. These examples of unimpaired inhibition of responding seem to rule out an explanation of the DBEE as a failure of internal inhibition.

Perseveration. A related, but conceptually distinct, mechanism to internal inhibition is that of perseveration. Here, although the capacity for the inhibition of behaviour *per se* is intact the animal is slower to switch from

Fig. 6.8. Locomotor activity in photocell cages for rats treated with 6-hydroxydopamine (6-OHDA) when neonatal and either satiated (*a*) or 24 h food deprived (*b*). Locomotor activity of the animals in an open field is shown for crossings of squares in the centre of the apparatus (*c*) and for those adjacent to the wall (*d*). (Redrawn from Mason, 1977.)

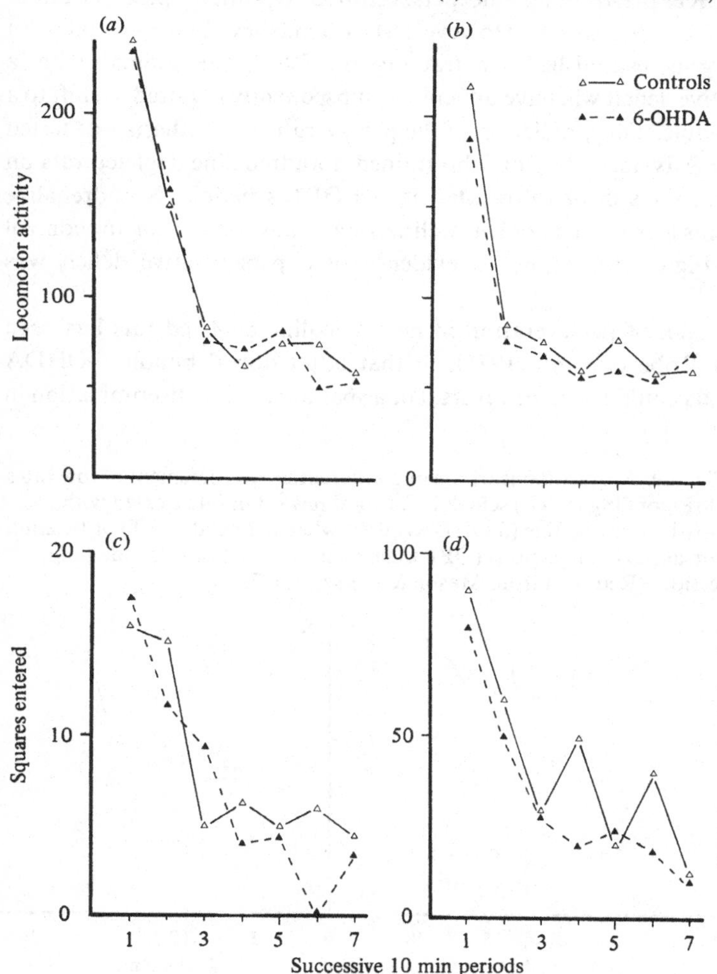

one response strategy to another. Thus, if trained on one response, the lesioned animal is envisaged as being deficient in abandoning this response and embarking on another. If no response has been established as prepotent by previous training, the animal will experience no difficulty in mastering a response, even if that response requires the inhibition of emission of behaviour. This might be described as a deficit in changing behaviour rather than a deficit in withholding behaviour. Such a perseverative deficit does not predict an impairment in passive avoidance, or in taste aversion. It need not necessarily predict a deficit in DRL schedules. If the animal is placed immediately onto the DRL schedule without pre-training on CRF, then no tendency to high response rates will be prepotent. Thus, without CRF pre-training, the perseverative hypothesis predicts unimpaired DRL acquisition. However, if a tendency for high rates of responding is established by training on CRF, the animal with a perseverative deficit will have difficulty if subsequently required to shift to a DRL schedule. This prediction of the perseverative hypothesis was tested by Mason & Iversen (1977*c*), who trained noradrenaline depleted rats on CRF for 15 days prior to transfer onto a DRL schedule. Noradrenaline depleted rats learned the DRL as well as, or actually *better than*, the control rats (see Fig. 6.10). Thus, no evidence of a perseverative deficit was obtained.

Similar lack of perseveration in noradrenaline depleted rats has been shown by Roberts *et al.* (1976), in that adult dorsal bundle 6-OHDA lesioned rats could learn the reversal of a spatial position discrimination in

Fig. 6.9. Acquisition learning of a differential reinforcement of low rates of responding (DRL) schedule for food reward in rats treated with 6-hydroxydopamine (6-OHDA) either when neonatal (NPT) or by adult dorsal bundle injections (DB). Each trial consisted of a 25 min daily session. (Redrawn from Mason & Iversen, 1977*c*.)

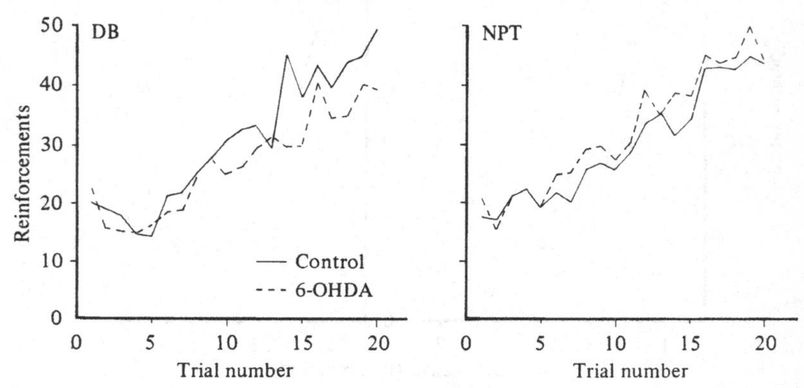

a T-maze as rapidly as controls. Since this requires an animal trained, say, to turn left, to start turning right, this would be an ideal situation to reveal a perseverative deficit if such existed.

Motivation. An increase in hunger might be expected to generate faster running and greater lever-pressing. Perhaps the lesioned rats have higher levels of motivation to perform the task as a result of the lesion. This has been tested in a number of ways. First, greater hunger might be expected to lead to greater food intake. Just such a phenomenon has been reported after lesion of the ventral noradrenergic bundle (see p. 358). However, when dorsal bundle lesioned rats, or those with lesions in the LC, are examined, no increase in food intake is seen (Sessions, Kant & Koob, 1976; Mason, Roberts & Fibiger, 1979). A specific test of the motivational hypothesis was undertaken by Mason (1979), who looked at the effects of pre-loading rats with free food on lever-pressing during a subsequent FI session. In normal rats such pre-loading with free food decreases the response rate in the FI schedule. Exactly the same decrease in responding was also seen in noradrenaline depleted animals. Had the lesion caused an increase in motivation, pre-loading would have been expected to have been less effective in reducing food rewarded lever-pressing (see Fig. 6.11). This direct examination of the demotivating effects of free food suggests that the DBEE cannot result from an increase in motivational levels.

Fig. 6.10. Acquisition learning of a DRL schedule (see Fig. 6.9) for food reward after extensive training on continuous reinforcement (CRF) (15 daily sessions of 25 min each). (Redrawn from Mason & Iversen, 1977*c*.)

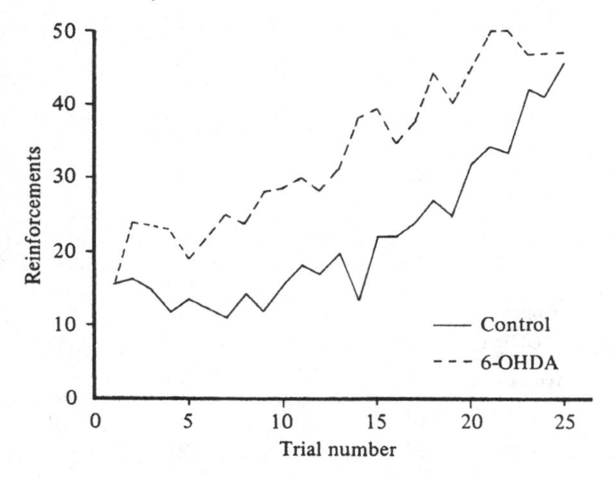

Secondary reinforcement. Initially neutral stimuli which are repeatedly associated with primary reward (food, electric shock, etc.) eventually acquire reinforcing properties themselves. This is called secondary reinforcement and the DBEE might be the result of an enhancement in the ability of stimuli to acquire secondary reinforcing power or an enhancement in the ability of such secondary reinforcers, once established, to maintain responding during extinction. To test this hypothesis, Mason & Robbins (1979) examined the effect that dorsal bundle 6-OHDA lesions had on the Hill paradigm. This consisted of pairing the click of an automatic feeder and the flash of the magazine light with the delivery of food on a VR schedule. After initial training in which such pairing occurred, the animal was given a choice between two levers, one of which produced the flash of light and the click of the feeder and the other which had no consequence. No food or primary reinforcer was presented during this stage. Normal rats display a preference for the lever which causes the presentation of the secondary reinforcing light flash and click. This then is a direct measure of the secondary reinforcing power of these initially neutral stimuli. No alteration was seen in the preference as a result of severe depletion of forebrain noradrenaline (see Fig. 6.12). Thus, it would not

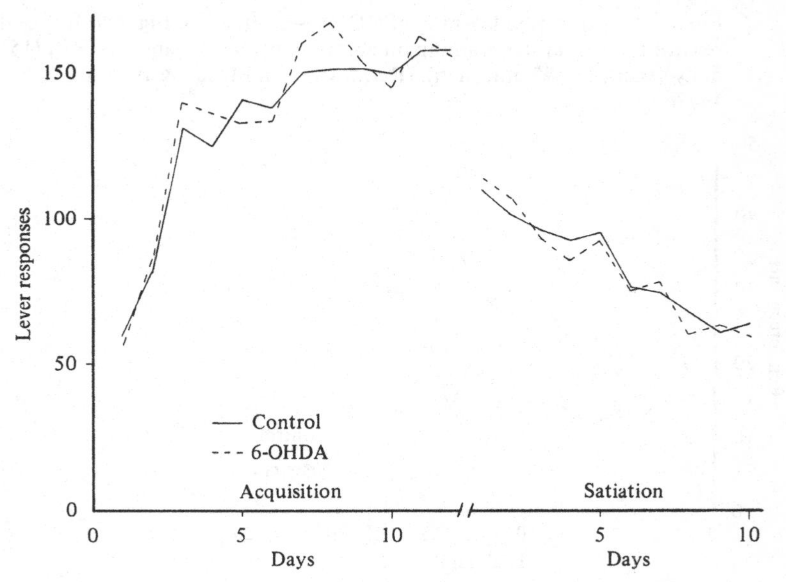

Fig. 6.11. Acquisition and satiation testing on a fixed-interval schedule of rats treated with adult dorsal bundle 6-hydroxydopamine (6-OHDA) injections. A 30 min period of access to free food was allowed immediately prior to each satiation test. (Redrawn from Mason, 1979.)

appear that the DBEE can be explained by an enhancement of secondary reinforcement.

Frustrative nonreward. A role for noradrenaline in nonreward processes is examined in greater detail later in this chapter (see p. 290). However, a brief consideration as to whether such a role could explain the DBEE will be presented here. Amsel (1958, 1962) suggested that the omission of an expected reward has a number of psychological consequences. One effect was to energise ongoing behaviour. This was called the response invigorating effect of nonreward. However, with repeated occurrence of nonreward, a second effect, the response suppressant effect, started to predominate and ultimately caused the extinction of the behaviour. This response suppres-

Fig. 6.12. Measurement of the degree of secondary reinforcing power acquired by initially neutral stimuli after pairing with reward. Responses on the lever which gave the conditioned stimuli (CR) and responses on the lever which had no effect (NCR) are shown for control (C) and 6-hydroxydopamine adult dorsal bundle treated (T) rats where the CR lever produced only the click of the feeder, only the flash of the magazine light, or both. (Redrawn from Mason & Robbins, 1979.)

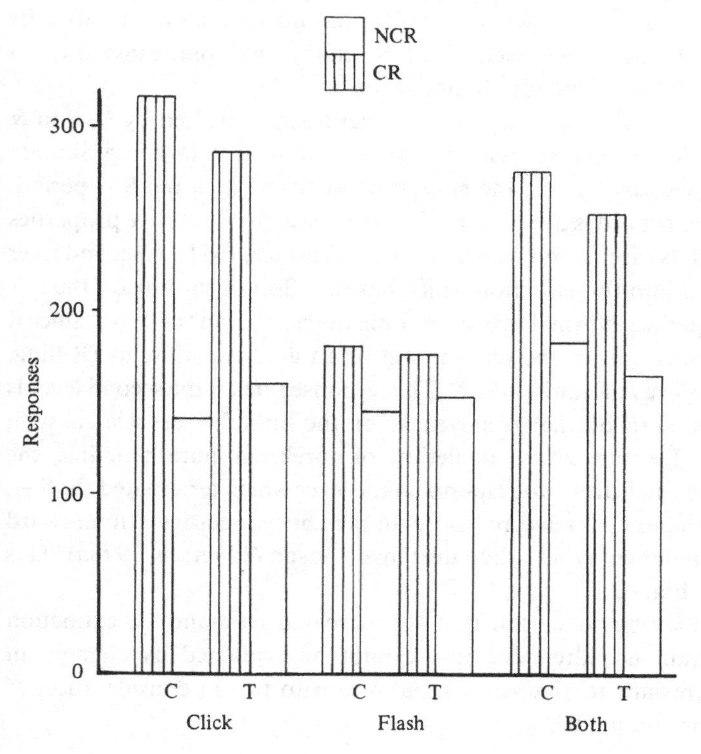

sion was believed to come about as a consequence of the aversive nature of nonreward. These two properties of nonreward, the invigorating and the aversive nature, might be altered by noradrenaline depletion. Either an increase in the response invigorating or a decrease in the aversiveness of nonreward might produce the observed resistance to extinction. Mason & Iversen (1978*a*) tested these possibilities by direct measurement of the properties of nonreward. The response invigorating effects were assayed in a double FI schedule in which the response rate in the second interval varies depending on whether reward was presented at the end of the first interval or not. When the usual reward is omitted at the end of the first interval, a response invigoration is seen, with much higher levels of responding occurring in the second interval than are seen when reward is presented. Identical response invigoration was seen in dorsal bundle 6-OHDA lesioned rats and controls, despite severe loss of forebrain noradrenaline (see Fig. 6.13). This suggests that the response invigorating effects of nonreward are not altered by noradrenaline loss. A similar conclusion was reached by Owens *et al.* (1982), who used a runway analogue of the double FI described above, in which reward was presented intermittently in the first goal box of a double runway and the running speed in the second alley-way of the runway was faster after reward omission than after its presentation (Amsel & Roussell, 1952). No change in response invigoration was seen after forebrain noradrenaline loss.

The other aversive property of nonreward was measured by Mason & Iversen (1978*a*) using a successive visual discrimination task in a Skinner box (see Appendix). Here, the stimulus associated with the S− period, during which reward is not presented, comes to acquire aversive properties as a result of its association with nonreward (Terrace, 1971). A second lever is introduced into this situation which has the effect of turning off the S− for a short period. Normal rats soon come to respond on this lever, since it has as its consequence the termination of an aversive stimulus (Rilling, Askew, Ahlskog & Kramer, 1969). The response rate on the second lever is a direct measure of the aversiveness of the stimulus associated with nonreward. Despite severe depletion of forebrain noradrenaline, the lesioned rats still learned to respond on the lever which terminated the S−, suggesting that the aversive, response suppressant properties of nonreward were also unaltered by noradrenaline loss (Mason & Iversen, 1978*a*). This is shown in Fig. 6.14.

Thus, it may be concluded that the increased resistance to extinction after forebrain noradrenaline loss cannot be explained by changes in central nonreward mechanisms. I will return to this in consideration of Gray's theory on p. 290.

Selective attention

Introduction. Central noradrenergic systems have been implicated in the cognitive processes underlying attentional behaviour (Mason, 1981). It has been suggested that neural activity in the dorsal noradrenergic bundle serves to screen out irrelevant sensory stimuli impinging on the organism (Mason, 1980). Animals with the dorsal bundle destroyed after 6-OHDA injection will consequently sample more stimuli in the environment than controls. In the present context of mechanisms of the DBEE this may be expected to increase resistance to extinction, since Mackintosh (1965) has suggested that the rate of extinction is determined, in part, by the number of cues which the animal has associated with reward during acquisition. If noradrenaline depleted rats sample more cues during acquisition than do controls, more cues will become associated with reward during this process

Fig. 6.13. Measurement of the response invigorating effects of nonreward in adult dorsal bundle 6-hydroxydopamine (6-OHDA) treated rats. Responses per fixed interval are shown when that interval followed one in which the expected reward was presented (Reward) and when it was omitted (Nonreward). (Redrawn from Mason & Iversen, 1978*a*.)

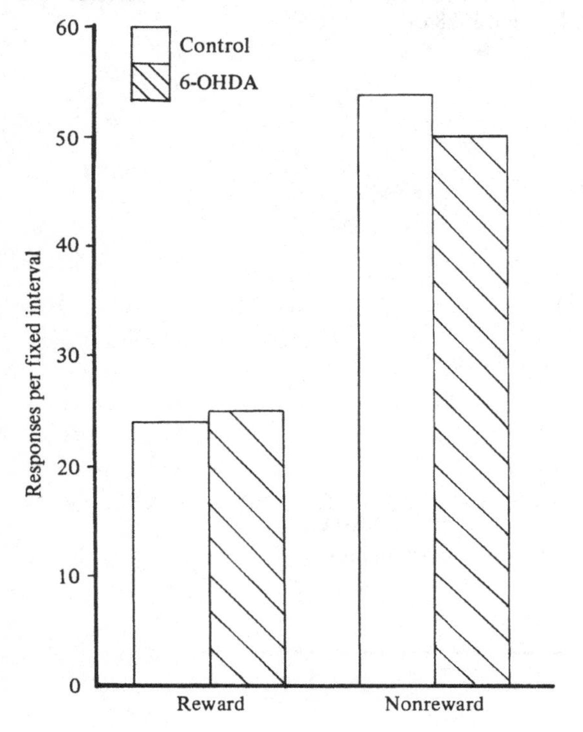

and so extinction will take longer. It is assumed by Mackintosh (1965) that each association of a stimulus with reward has to be broken before responding will cease in extinction.

The attentional mechanism of the DBEE can explain resistance to extinction after both positive and negative reinforcers. It does not matter what reinforcer lays down the stimulus–reward link, merely that noradrenaline depleted rats are sampling more stimuli as a result of the lesion and so will attach more stimuli to reward. The attentional mechanism of the DBEE can explain the absence of the DBEE after PRF schedules. According to Mackintosh (1974), partial reinforcement of a response causes normal rats to sample a large number of cues in acquisition. This is because animals are trying to find a single cue which predicts reward consistently. On a schedule like a FI, the passage of time is such a cue and, once the animal attends to this dimension and learns that it does indeed consistently predict reinforcement, it ceases to search for new cues. Thus,

Fig. 6.14. Measurement of the aversiveness of nonreward in adult dorsal bundle 6-hydroxydopamine (6-OHDA) treated rats. Responses per S – period on the lever which terminated that S – stimulus for 10 s are shown for the first 5 days after the introduction of this lever. (Redrawn from Mason & Iversen, 1978*a*.)

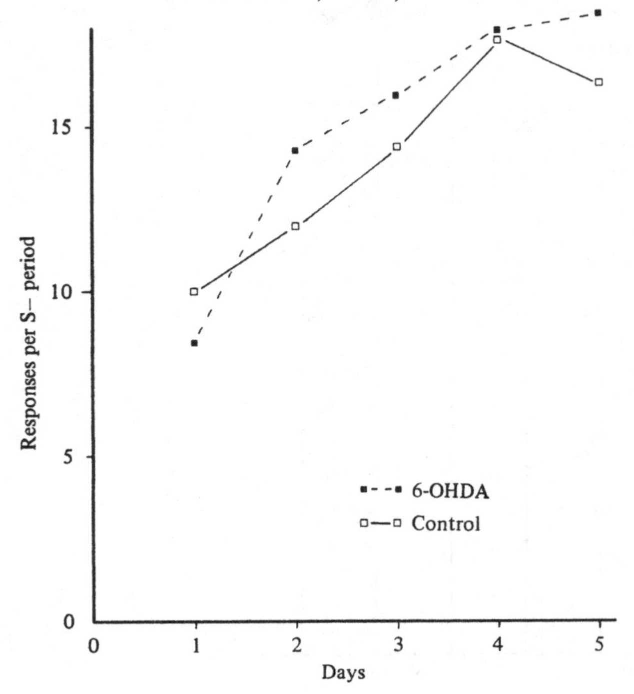

during acquisition learning the animal on FI (or CRF) will have attended to relatively few cues present in the experimental situation, since it will rapidly have found the appropriate one to predict reward. On a PRF schedule, such as VR4, where reward follows sometimes the third lever-press, sometimes the fifth and so on, there is no cue available to predict reward all the time. The animal will thus continue to sample one cue after another in a vain attempt to find a consistent predictor of reward. Thus, at the end of acquisition, a large number of cues will have been attached to reward. This explains the well-known fact that PRF schedules take much longer to extinguish than CRF ones. With reference to the effect of noradrenaline depletion, Mason (1980, 1981) suggests that this is doing exactly what the behavioural manipulation of partial reinforcement also does, namely to increase cue sampling. It might be expected that these two manipulations would fail to add. That is, if animals are already sampling many cues as a result of noradrenaline depletion, then PRF could not cause them to sample even more. Therefore, it was predicted on the attentional theory that a dorsal bundle lesion would not cause resistance to extinction after PRF acquisition training (Mason & Fibiger, 1978*b*). As can be seen from the section above (p. 265) such is the observed result. Thus, the attentional theory of the DBEE can explain the absence of the DBEE following PRF. A similar argument can explain the absence of the DBEE after previous experience of extinction, such as on a successive discrimination schedule or on multiple CRF acquisition–extinction sessions (Mason, 1978).

As a result of the finding that noradrenaline depletion increased resistance to extinction, and the possible explanation of this effect in terms of stimulus sampling, a number of lines of evidence have emerged linking noradrenaline with selective attention. I shall now examine them in detail.

Discrimination behaviour. If animals with noradrenaline depletion are unable to filter out irrelevant stimuli they should be impaired in learning to attend to a specific stimulus dimension. Such a paradigm is represented by successive or simultaneous stimulus discrimination tasks. Mason & Iversen (1977*a*) found that dorsal bundle 6-OHDA treated rats were markedly impaired in the acquisition of a successive visual discrimination task (see Fig. 6.15). Although able to learn to lever-press under a CRF schedule as well as did controls, considerable impairment was seen when the schedule was changed to one in which lever-pressing was rewarded only when the house-light was illuminated, and during other periods of time when the houselight was off lever-pressing did not bring a food delivery. Here the animal is required to attend to only one stimulus in order to structure behaviour appropriately, namely the houselight. To do this, normal animals have to filter out all the other (to this task 'irrelevant') stimuli also present in the environment. The noradrenaline depleted rats are unable, or slower, to do this and thus performance is impaired on this task. This is not a general learning impairment, since acquisition of many other tasks, for example the CRF which preceded the successive discrimination described by Mason & Iversen (1977*a*), is not impaired.

A further prediction of the attentional theory is that reversal of a stimulus discrimination should also be impaired for similar reasons, and so it is (Mason & Iversen, 1978*b*). It should be noted, however, that by reason of its innate salience (Mason, 1980) the spatial dimension does not have to be learned to be attended to and so spatial discriminations and their reversals will not be affected on the attentional theory. And so it is found by Roberts *et al.* (1976) for acquisition and reversal of a spatial T-maze task (see Appendix).

Other discrimination tasks which are impaired include a successive visual–auditory task (Mason & Fibiger, 1979*f*) and a simultaneous brightness discrimination in a T-maze (Mason & Lin, 1980).

Distractability. Reaching right to the heart of the attentional theory of noradrenaline function is the finding by Roberts *et al.* (1976) that

Fig. 6.15. Acquisition of a successive light–dark operant discrimination task for food in adult dorsal bundle 6-hydroxydopamine (6-OHDA) treated rats. Values are the ratio of lever-presses in the light period (L) which were rewarded on a continuous reinforcement (CRF) schedule) to those in a dark period (D) (which were never rewarded) for 12 days of training. (Redrawn from Mason & Iversen, 1977*a*.)

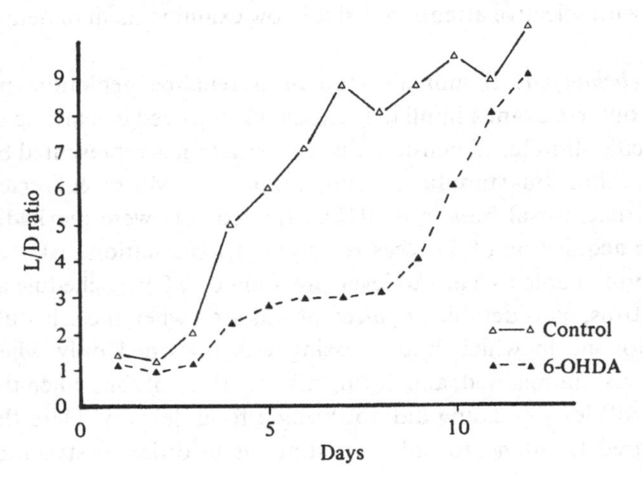

noradrenaline depleted rats are more distractable by irrelevant stimuli presented during the performance of a previously learned response. Roberts *et al.* (1976) trained rats to run a L-shaped runway for food reward and then presented flashing lights above the stem of the T-maze and placed sandpaper on its floor. These visual and tactile stimuli had no relevance to the task, since food was still presented as usual. Normal rats upon encountering the novel stimuli for the first time took a while to explore them and only then did they run on to the goal box. Over the next few trials, controls soon learned that these stimuli were irrelevant, ignored them and ran right by. Noradrenaline depleted rats took much longer to learn to ignore these irrelevant stimuli and so their performance of the food-rewarded task was disrupted more than that of controls (see Fig. 6.16). This increased distractability and slowness to come to ignore irrelevant stimuli is most directly that required on the attentional model of noradrenaline function (Mason, 1980, 1981).

Similar enhanced distractability to novel stimuli has been seen by Koob *et al.* (1978) after electrolytic lesions of the LC using the paradigm of

Fig. 6.16. Distraction by novel stimuli as reflected by the disruption of running in an alley-way for food reward for rats treated with adult dorsal bundle 6-hydroxydopamine (6-OHDA) in the first five trials after the introduction of flashing light above the alley-way and sandpaper on its floor. (Redrawn from Roberts *et al.*, 1976.)

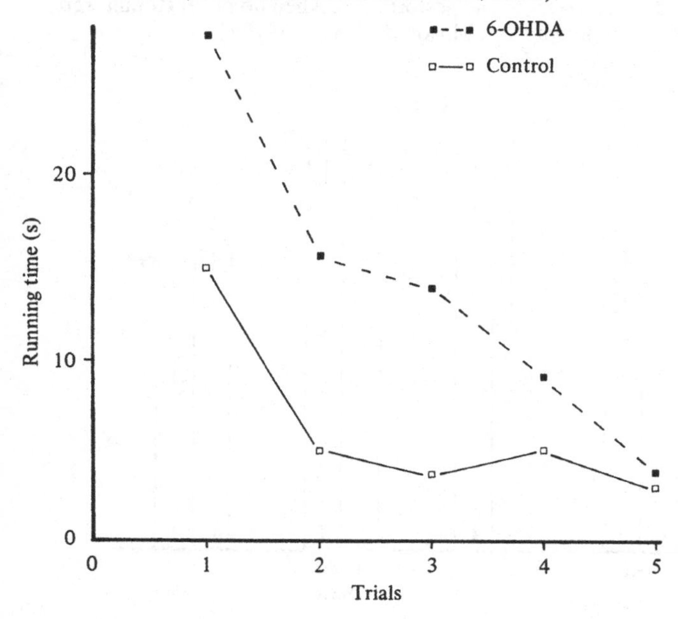

Roberts *et al.* (1976). Also, Mason & Fibiger (1978*c*) found that CRF lever-pressing for food reward was more disrupted by flashing lights placed above the head of the animal in noradrenaline depleted rats than in controls (see Fig. 6.17). Oke & Adams (1978) found that novel visual marking added to the black or white floors of a visual T-maze discrimination disrupted the performance of noradrenaline depleted rats more than controls. These findings all speak very directly for a role of forebrain noradrenaline in stimulus sampling.

Nonreversal shift. If noradrenaline depleted rats are slow to come to ignore stimulus dimensions irrelevant to the task in hand it might be expected that they would show changes in a nonreversal shift paradigm (Mackintosh & Holgate, 1968). This involves training the animal on a stimulus discrimination with two dimensions simultaneously present from the start of the task. Dimension *A* is relevant and the sole consistent predictor of reward. Dimension *B* does not predict reward and the two values of it merely occur randomly from trial to trial. Normal animals learn that dimension *B* is irrelevant and soon cease to attend to it. Upon nonreversal shift dimension

Fig. 6.17. Distraction by novel stimuli as reflected by the disruption of lever-pressing on a continuous reinforcement (CRF) schedule for rats treated with adult dorsal bundle 6-hydroxydopamine (T) and their controls (C). The bigger the suppression ratio the more disruption was caused. CRF refers to the performance when no novel stimuli were present. (Redrawn from Mason & Fibiger, 1978*c*.)

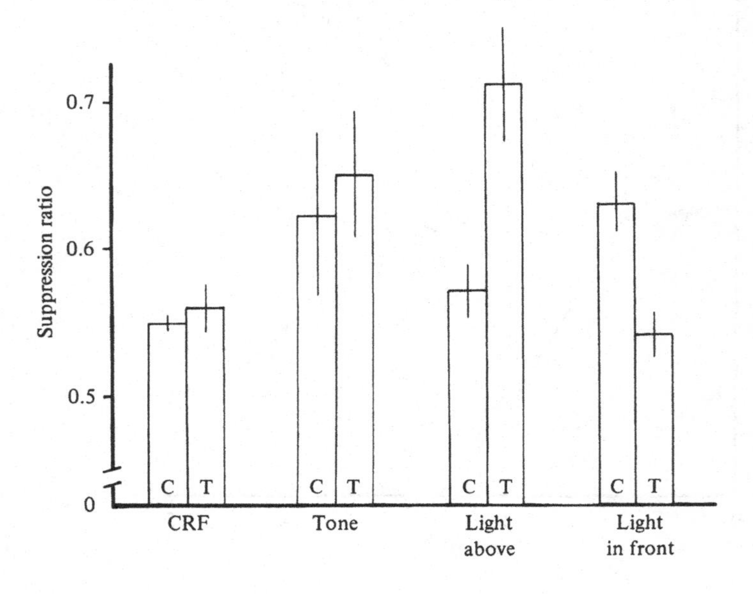

A becomes irrelevant and nonpredictive of reward while dimension *B* suddenly comes to cue reinforcement. Normal rats must 'unlearn' that dimension *B* is irrelevant and start attending to it in order to solve the task. Noradrenaline depleted rats may be at an advantage in this situation, since they were slow to ignore dimension *B* during acquisition as a result of their lesion and so will be quicker than controls to start using it as the discriminative cue after the nonreversal shift. Such a result is indeed found for a shift from a spatial position irrelevant/brightness relevant T-maze task to a spatial position relevant/brightness irrelevant task (Mason & Fibiger, 1979*f*; Mason & Lin, 1980). This is particularly interesting since it is one of few situations in which rats perform better as a result of a brain lesion (see Fig. 6.18).

Fig. 6.18. Nonreversal shift in rats treated with adult 6-hydroxydopamine (6-OHDA) dorsal bundle injections. Following pre-training on a spatial position irrelevant/brightness relevant discrimination in a Y-maze the animals were shifted to a task in which now the spatial cue was the sole predictor of reinforcement and the brightness cue was unpredictive although still present. (Redrawn from Mason & Fibiger, 1979*f*.)

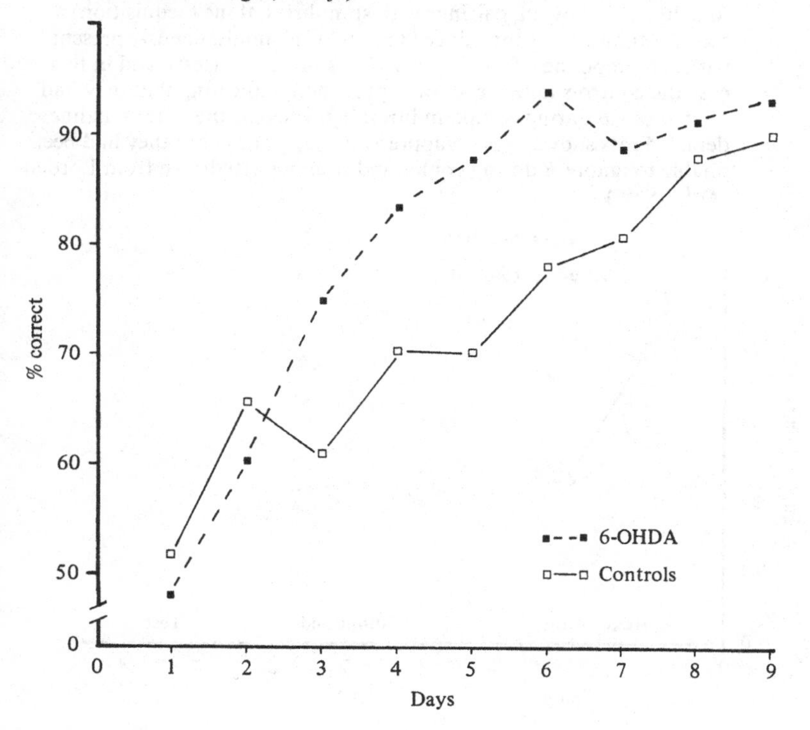

Blocking. If the second dimension in a stimulus discrimination task is introduced only late in training, i.e. training starts with dimension *A* only and *B* is not present until later, the animal may ignore dimension *B* entirely when it is added as it conveys no additional information about the task in hand not already present in dimension *A*. When tested upon dimension *B* in the absence of dimension *A*, normal animals are unable to perform the task since they have not attended to *B* during training. This is called the blocking effect (Kamin, 1969). Noradrenaline depleted rats, if unable to ignore dimension *B* as irrelevant upon its introduction during training, will associate it with reward to some extent and so be better able to perform the task when dimension *A* is withdrawn. This attenuation of the blocking effect as a result of noradrenaline depletion has indeed been seen (Fig. 6.19) by Lorden, Richert, Dawson & Pelleymounter (1980). It is thus strong evidence for a role of noradrenaline in selective attention, as suggested by Mason (1980, 1981).

Fig. 6.19. Attenuation of blocking by dorsal bundle 6-OHDA lesions. The disruption of a continuous reinforcement lever-press response caused by the presentation of stimuli previously paired with electric footshock. Following pairings with stimulus *A* alone (acquisition) a second stimulus was introduced (*B*) and was simultaneously present with *A* (compound). Finally, only *B* was presented (test), and in this case the controls failed to show suppression, indicating that they had filtered out *B* during compound training, whereas the noradrenaline depleted rats showed good suppression, suggesting that they had been unable to ignore *B* during compound training. (Redrawn from Lorden *et al.*, 1980.)

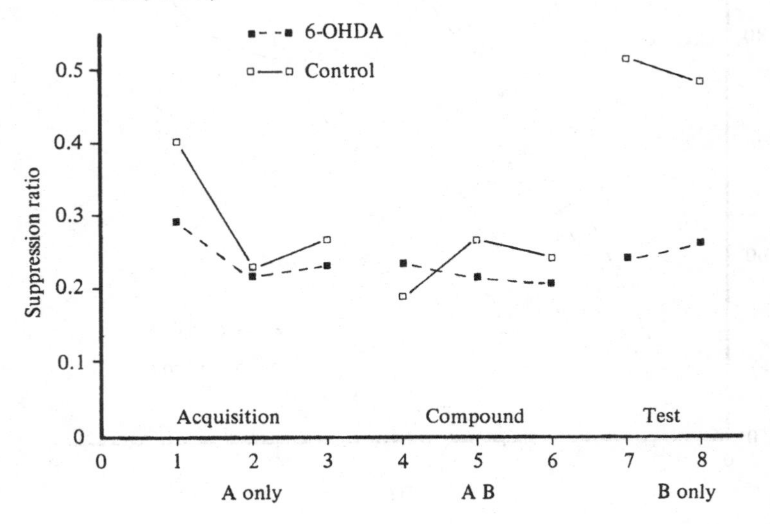

Latent inhibition. A final paradigm tapping very directly into the attentional functioning of the animal is that of latent inhibition. Here animals are exposed to a stimulus in the absence of reward for a number of trials. Since it signifies nothing to do with reward, normal rats gradually learn to ignore this stimulus. When it is subsequently used as the S+ in a discrimination task, such animals have to 'unlearn' that it is irrelevant in order to solve the task. They are slow to do this compared to naive animals and this is called the latent inhibition effect (Mellgren & Ost, 1971). Noradrenaline depleted rats will be slower to learn that the stimulus is irrelevant upon its initial presentation in the absence of reward and so will find it easier to use it as the discriminative cue in a subsequent discrimination task. They will show a weaker latent inhibition effect. Such an experimental finding of attenuation of the latent inhibition effect has been reported (see Fig. 6.20) (Mason & Fibiger, 1979*f*; Mason & Lin, 1980). It is thus very strong evidence supporting a role of noradrenaline in attention.

Summary. It would seem that there is much evidence which serves to implicate noradrenaline in attentional functioning. This theory is also the best able to explain the occasional effect which noradrenaline depletion by 6-OHDA had on acquisition learning of tasks (see Chapter 5, p. 217). It is mainly those tasks which require the animal to attend to one, and only one, stimulus dimension which are impaired by noradrenaline loss. Other learning tasks in which selective attention is not taxed do not show acquisition impairments. For the purpose of this book the full ramifications of the attentional theory of noradrenaline function have not been spelled out and interested readers are referred to Mason (1980, 1981) for more detailed coverage of the questions of innate salience, learning to ignore and motivational intensity.

Fear and anxiety

On a biochemical level there is evidence that anti-anxiety drugs such as the benzodiazepines may interact with noradrenergic systems. Thus, Taylor & Laverty (1969, 1973) found that various benzodiazepines acted to reduce the turnover of noradrenaline which had been elevated in response to stressful electric footshock. Similar results were obtained by Corrodi, Fuxe, Lidbrink & Olson (1971) using the disappearance of fluorescence after synthesis inhibition to assay amine turnover rate. Diazepam has also been reported to reduce noradrenaline synthesis (Biswas & Carlsson, 1978). Further, the β-receptor blocking agent propranolol has been claimed to have anti-anxiety effects clinically (Gottshalk, Stone & Glesser, 1974).

These indirect indications were given more body by the findings of Redmond *et al.* (1976*a*) that monkeys (*Macaca arctoides*) with bilateral electrolytic lesions of the LC failed to show fear reactions when threatened by the experimenter or when presented with a rubber snake which is normally fear-inducing to a monkey. The converse of this finding was also seen (Redmond, Huang, Snyder & Maas, 1976*b*), that electrical stimulation of the LC through indwelling chronic electrodes could elicit behavioural signs of fear. These consisted of grimacing and scratching, which, in the

Fig. 6.20. Attenuation of latent inhibition by dorsal bundle 6-hydroxydopamine (6-OHDA) lesions. The effect of pre-exposure to initially irrelevant stimuli in the absence of food served to retard the subsequent learning of a stimulus discrimination using these stimuli (compare unexposed and pre-exposed groups of controls). 6-OHDA depletion of forebrain noradrenaline not only slowed learning of the discrimination *per se* but also prevented stimulus pre-exposure from further slowing learning (compare unexposed and pre-exposed 6-OHDA groups). (Redrawn from Mason & Fibiger, 1979*f*.)

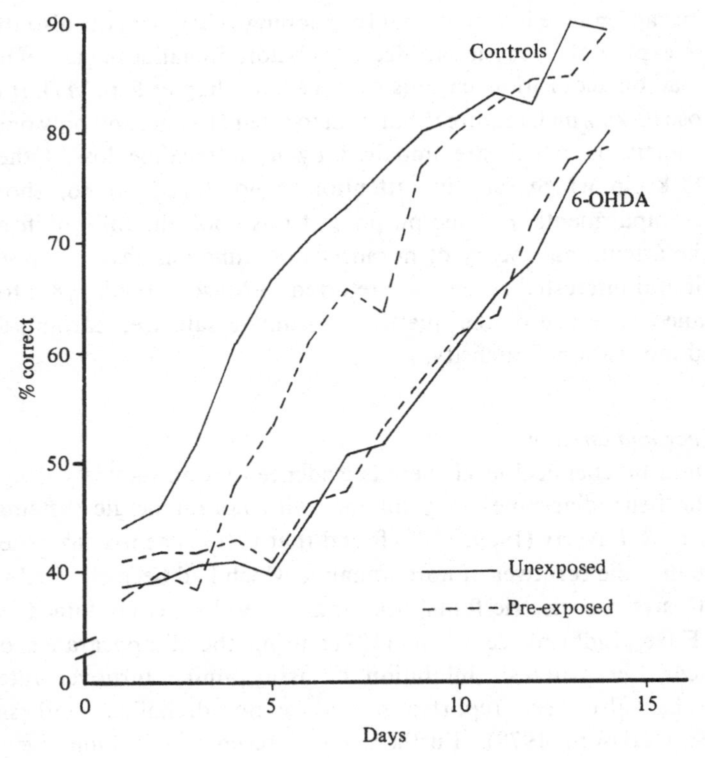

wild, are seen in threatening or fearful situations. In human patients undergoing neurosurgery, electrical stimulation in the vicinity of the LC (although this may be closer to the central grey; F. G. Graeff, personal communication) will trigger feelings of fear and death. Small lesions in this area are reported to have a calming effect (Nashold, Wilson & Slaughter, 1974).

Finally, Davis, Redmond & Baraban (1979) found that noradrenaline agonist and antagonist drugs affect the potentiated startle paradigm. In this situation the startle response of the rat to loud tones is found to be larger in magnitude if tested in the presence of stimuli (illumination of the houselights) previously paired with electric footshock than if tested in their absence. If the animals were tested under either clonidine or propranolol, a reduction in the amount that fear-associated stimuli served to potentiate the startle response was seen. Clonidine, among other actions, will decrease the firing of LC cells by an α_2-receptor agonist action. Propranolol will decrease postsynaptic transmission in the noradrenergic system by β-blockade. Drugs which, among other effects, increase firing of the LC cells (such as piperoxan and yohimbine, which act via blockade of a tonically active α_2-receptor) were found to increase the amount by which fear-associated stimuli potentiated the startle response (see Fig. 6.21). The α_1-blocker, WB 4101, was ineffective. Such evidence can be explained by a role for noradrenaline in fear. However, since the fear was induced by stimuli associated with previous footshock a noradrenaline-induced alteration in the sampling of those stimuli as implied by the attentional theory (see p. 277) might also have a similar effect. Because of the complexity of the behavioural situation, no firm conclusions as to the psychological mechanism responsible for the change in startle response can be made. The short-comings of electrolytic lesion to the LC have been detailed elsewhere (see p. 205). Thus, it cannot be claimed from electrolytic lesion to the LC that the effect is necessarily mediated through noradrenergic systems. Electrical stimulation in the vicinity of this nucleus has a similar drawback. It will activate not only noradrenergic systems but also all other nonnoradrenergic fibres of passage that run through this area. Without a close and detailed mapping study (such as that carried out by Corbett & Wise (1979) for ICSS) it is not even certain that the best site for eliciting the behavioural effect is actually the LC and not some closely adjacent (and nonnoradrenergic) cell cluster. Thus, neither the lesion nor the stimulation studies prove a role for noradrenaline in fear.

This assumes importance since more selective lesion techniques using 6-OHDA fail to find the predicted reduction in fear. Mason & Fibiger (1979*c*) found that despite better than 95 % reduction in the forebrain noradrenergic projections of the LC no deficit in the learning or performance of the fear-motivated Sidman avoidance schedule was to be seen. Neither was a conditioned suppression paradigm changed by severe noradrenaline loss. Here animals receive pairings of tone and shock in one apparatus and then are presented with the tone on its own during the performance of another behaviour, in this case CRF lever-pressing for food reward. The degree of disruption of ongoing behaviour caused by the tone is a direct measure of the amount to which it has been associated with the fearful electric footshock. Any reduction in fear as a consequence of noradrenaline loss would be expected to attenuate the suppression caused by the tone. No changes were seen despite severe noradrenaline loss (Mason & Fibiger, 1979*c*).

Acquisition learning of other tasks motivated by and sensitive to fear is also unimpaired by noradrenaline loss due to 6-OHDA dorsal bundle lesion. Thus, acquisition learning and performance on one-way active avoidance is unaltered by severe noradrenaline loss caused by 6-OHDA treatment (Fibiger & Mason, 1978). Similarly acquisition and performance

Fig. 6.21. Potentiated startle and its manipulation by noradrenergic drugs. Piperoxan and yohimbine increased, propranolol decreased and WB 4101 failed to affect the magnitude of the startle response measured in the presence of stimuli previously paired with electric footshock. (Redrawn from Davis *et al.*, 1979.)

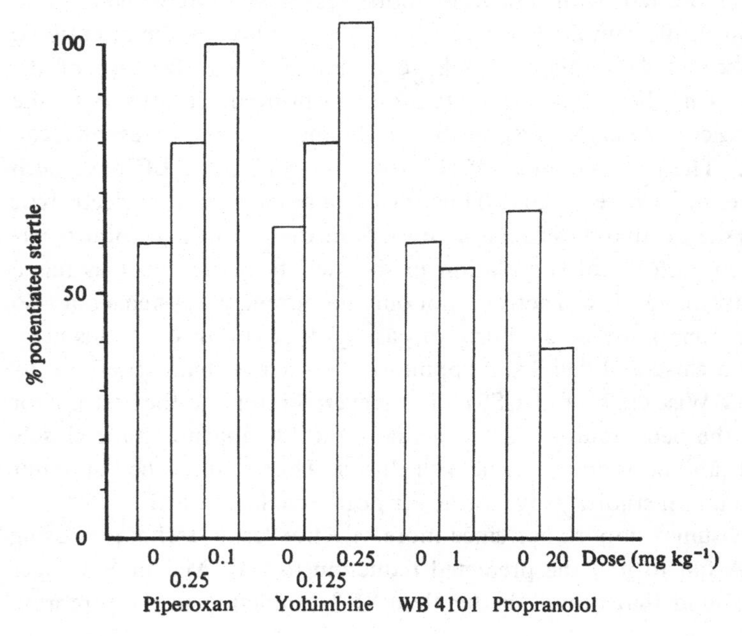

of two-way active avoidance is not slowed by noradrenaline depletions in excess of 95 % (Mason & Fibiger, 1979*a*). Acquisition learning and performance on a variety of passive avoidance tasks is not generally altered by loss of the noradrenergic innervation to the forebrain (Fibiger, Roberts & Price, 1975; Roberts & Fibiger, 1977; Mason *et al.*, 1979; Mason & Fibiger, 1979*e*). Moreover, a social interaction measure of fear is not altered by 6-OHDA injections in the LC and dorsal bundle (Crow *et al.*, 1978; File, Deakin, Longden & Crow, 1979).

What might be the solution to these apparently contradictory sets of data? First, previous suggestions of a species difference (Mason & Fibiger, 1979*g*), in which it was noted that the experiments implicating noradrenaline in fear were carried out on monkeys whereas all the negative findings result from rat work, may be excluded with the report by Davis *et al.* (1979) of a change in rodent fear behaviour as measured by potentiated startle after noradrenaline agonists and antagonists. Secondly, it may be that stimulus sampling, rather than primary fear mechanisms, were being examined in those experiments in which noradrenaline manipulations did alter behaviour. It has been mentioned (p. 277) that a very probable role for forebrain noradrenaline in attention has been documented. The fear induced in the experiments of Redmond *et al.* (1976*a*) used secondary conditioned stimuli such as the sight of a rubber snake or implied threat from a human experimenter rather than direct pain or electric footshock. Such 'cognitive' types of threat may be susceptible to attentional alterations in their sampling or perception of significance or salience, which primary negative reinforcers such as the footshock used in the experiments of Mason & Fibiger (1979*a*, 1979*f*) were not. This is further supported by the paradigm of Davis *et al.* (1979), which again found positive effects of noradrenaline manipulations in situations using stimuli (houselight) which are aversive only because of their previous association with shock. Alterations in the sampling, interpretation of the significance or salience of these stimuli would thus *indirectly* affect their fear-inducing power.

Thirdly, the resolution may lie in the specificity of the various manipulations used in noradrenergic systems. Electrolytic lesion and electrical stimulation of the LC lack the neurochemical specificity that 6-OHDA lesion possesses (as outlined above and on p. 205). Drugs such as clonidine, piperoxan and yohimbine, which certainly alter the firing of LC cells (Cedarbaum & Aghajanian, 1976) may also have many other effects in the central nervous system and their action on fear may indeed be due to one of these other, nonnoradrenergic effects. Clonidine, for example, is also an H_2 histamine agonist and may interact with adrenaline receptors. 6-OHDA lesions seem, on the basis of considerable investigation (see Table

5.1, p. 212) to be much more specific for noradrenergic systems. Only further research will resolve these complex issues.

It is probably premature to exclude a role of the LC in fear behaviour. However, on the basis of 6-OHDA lesion work it would not appear to be involved in primary fear but more related to 'cognitively-induced' fear, depending on the sampling, interpretation and salience of stimuli in the environment.

Frustration and nonreward

Introduction. There is evidence on a behavioural level of certain marked similarities between punishment and the omission of an expected reward (Adelman & Maatsch, 1956; Brown & Wagner, 1964; Azrin, Hutchinson & Hake, 1966; Uhl, 1967; Wagner, 1969; Wong, 1971). Such omission has been suggested by Amsel (1958, 1962) to cause a behavioural state of frustration which possesses aversive properties in its own right (Rilling *et al.*, 1969; Terrace, 1971). Given the postulations regarding a role for noradrenaline in fear and anxiety discussed above, it is not surprising that similar suggestions for a role for noradrenaline in frustrative nonreward have also been made. These are outlined most formally by Gray (1976, 1977, 1978, 1981), although it is important to note that some evolution and change in the precise form of the theory has occurred over the years. The evidence relating to a role for noradrenaline in nonreward is mainly indirect, and few experimental tests have been undertaken. The reader is hence asked to bear with me if the direction of the experiments to be described below appears to be circuitous; in cumulation it may suggest a role for noradrenaline in frustrative nonreward.

Septal driving. Much of the evidence for a role of noradrenergic systems in nonreward has come from the electrophysiological paradigm of septal driving. This involves electrically stimulating in the septum while recording in the hippocampus. Various frequencies of electrical stimulation are imposed on the septum with increasing amplitude. The stimulation voltage in the septum at which the electrical activity recorded in the hippocampus starts to be driven by the septum, i.e. to show the same frequency as that used to stimulate in the septum, is noted. Some frequencies of septal stimulation will cause entrainment of the hippocampal rhythm at low voltages. It is generally found that septal driving of the hippocampal rhythm occurs most easily at 7.7 Hz, but at frequencies different, in either direction, from 7.7 Hz, greater stimulating voltages are needed.

This paradigm is of interest since Gray (1970) has suggested that septo-hippocampal activity at 7.7 Hz is involved in the coding of nonreward. Neurochemical manipulations which are found to alter the ease of septal driving at 7.7 Hz would then also be expected to affect the coding of nonreward. I shall briefly review the evidence which led Gray (1970) to implicate this particular frequency of septo-hippocampal activity in nonreward.

Gray & Ball (1970) recorded from chronic electrodes implanted in the hippocampal formation of the rat during various behavioural situations. They examined the hippocampal frequency obtained as the rat ran along an alley-way towards water reward in the goal box and upon omission of that expected reward. During exploration of the novel runway upon first introduction to it they observed a hippocampal frequency of 7.5 to 8 Hz. When the animal had been trained to run rapidly to the end of the runway to reach water reward they found a frequency of 8 to 10 Hz during the performances of this learned response. During consumption of the water reward in the goal box the hippocampal electrical activity showed a frequency of 6 to 7.5 Hz. When the animal had run up to the goal box and found the expected water reward not present there, a rhythm of 7 to 8 Hz was seen. Gray & Ball (1970) suggest that this frequency of 7 to 8 Hz (average 7.7 Hz) is of physiological significance in the coding of nonreward. A drug manipulation which specifically alters the ease of septal driving of the hippocampus at 7.7 Hz without affecting septal driving at surrounding frequencies is the anxiolytic drug amobarbital. It selectively raises the threshold of (makes more difficult) septal driving at 7.7 Hz. This might suggest on Gray's model that coding for nonreward becomes more difficult under amobarbital. It is thus of great interest that the drug also slows down extinction of learned behaviours such as alley-way running (Blough, 1956; Miller, 1961; Barry, Wagner & Miller, 1962). Such increased resistance to extinction would indeed be expected if the physiological processes coding for nonreward were being disrupted by the drug.

Further support for the role of 7.7 Hz hippocampal activity in nonreward come from the finding of Gray (1972) that artificial driving of the hippocampus at 7.7 Hz speeded up extinction of a water-rewarded running response. This may be explained by 7.7 Hz driving mimicking the physiological consequences of nonreward.

Other evidence comes from the behavioural paradigm known as the partial reinforcement extinction effect (PREE). Rats trained on PRF schedules take longer to extinguish than animals trained on CRF. I have described the attentional explanation of this phenomenon on p. 278. An explanation in terms of frustrative nonreward is also possible. It runs like this. During acquisition learning, partially reinforced rats experience frustration as a result of the expected reward sometimes not occurring – as it is a *partially* reinforced schedule. Since this occurs during acquisition training the state of frustration will become associated with reward. This is called counter-conditioning. This contrasts with animals trained on CRF, which never experience frustration during acquisition. During extinction, frustration occurs in both groups as a result of total omission of the expected reward. In the CRF animals frustration has an unconditioned, innate effect of eventually causing responding to cease, i.e. the aversive or response suppressant effects of nonreward (Amsel, 1958, 1962) described earlier (p. 275). However, in PRF animals the state of frustration has been associated during acquisition with reward and so when it occurs during extinction it actually serves to maintain responding for a while. Because of counter-conditioning in acquisition of PRF, the frustrative state has become associated with the approach response. Thus, PRF animals take longer to extinguish than CRF animals.

Amobarbital, by blocking nonreward coding via the 7.7 Hz hippocampal rhythm, would, on this model, have different effects depending on *when* during training or extinction it was given. If amobarbital were given to CRF-trained rats during extinction it should increase resistance to extinction by blocking the nonreward coding mechanism which normally acts to decrease responding resulting from the aversive, response suppressant effects of nonreward. This has been seen to occur (Blough, 1956; Miller, 1961; Barry *et al.*, 1962). If amobarbital were given during acquisition training on PRF, it should prevent the counter-conditioning of nonreward to the approach tendency which, when tested in extinction, is what causes PRF rats to be more resistant to extinction than CRF rats. Thus, PRF and CRF rats should now extinguish as rapidly as each other and such has been reported. That is, amobarbital during PRF acquisition blocks the PREE (Gray, 1969; Ison & Pennes, 1969).

If nonreward is coded for by the 7.7 Hz hippocampal rhythm, one should be able to mimic it even on a CRF schedule where food always occurs by electrically stimulating at 7.7 Hz on a randomly selected half of the trials. This Gray (1970) found led to greater resistance to extinction when compared to unstimulated CRF rats – exactly as if they had been trained on PRF despite always receiving the food pellets. The converse is also true: if the 7.7 Hz rhythm is blocked by electrical stimulation of the septum at 200 Hz during PRF training, nonreward cannot be coded and so the PREE fails to occur. That is, despite receiving food pellets on only half the acquisition trials, the rats extinguish as quickly as CRF-trained animals (Gray, Araujo-Silva & Quintao, 1972*a*). Blockade of nonreward coding by surgical lesion to the medial septal area also has this effect on preventing the usual increase in resistance to extinction caused by PRF training (Gray,

Quintao & Araujo-Silva, 1972*b*). Finally, recordings of hippocampal activity during the reinforced and nonreinforced trials on PRF acquisition show that, during nonreinforced trials, the activity is closer to 7.7 Hz than during reinforced ones and that the closer it is to 7.7 Hz the greater the subsequent resistance to extinction (Gray, 1972).

All of the above seems to establish that 7.7 Hz activity in the hippocampal formation may code for nonreward. So where does noradrenaline fit in? The link between noradrenaline and nonreward is made by the finding of Gray, McNaughton, James & Kelly (1975) that bilateral 6-OHDA lesion to the dorsal noradrenergic bundle raised the threshold for septal driving of the hippocampus at 7.7 Hz without much apparent effect on driving at other frequencies (see Fig. 6.22*a*). This would imply a direct role for noradrenaline in the coding of nonreward, given the connection apparently established above of 7.7 Hz hippocampal activity and nonreward.

A role for noradrenaline in the 7.7 Hz frequency of septal driving was confirmed by McNaughton *et al.* (1977), who found that synthesis inhibition of noradrenaline with FLA 63 also increased the threshold of 7.7 Hz driving, with no effect on it at other frequencies (see Fig. 6.22*b*). Finally, however, a complication was added by McNaughton, Kelly & Gray (1980), who found that unilateral lesion to the dorsal bundle with 6-OHDA or temporary prevention of conduction using the local anaesthetic procaine would raise the threshold at 7.7 Hz for septal driving, but on the side of the brain *contralateral* to the lesion! No effect of blockade of impulse traffic by either lesion or anaesthetic was seen with the ipsilateral dorsal bundle (see Fig. 6.22*c*). The explanation of this is unclear but suggests that not everything is yet known about the precise circuitries being tapped by septal driving of the hippocampal rhythm. Nonetheless, a strong case is made for a role of noradrenaline in facilitating septal driving at 7.7 Hz. If, and the evidence reviewed above suggests that it may be the case, 7.7 Hz is involved in the coding for nonreward, then noradrenaline may have a role in nonreward processes.

Behavioural tests. Relatively few direct tests of a role for brain noradrenaline in nonreward have been carried out. Both Mason & Iversen (1978*a*) and Owens *et al.* (1982) agree that the response invigorating effects of nonreward are not affected by 6-OHDA lesion to the forebrain noradrenergic innervation (described on p. 276). This is of little importance to the latest manifestation of Gray's theory, since it is now postulated that only the associability of nonreward with environmental stimuli is altered as a result of noradrenaline loss. The unconditioned effects of nonreward, such as is the response invigorating effect, occur prior to that point in the mechanism (Gray, 1978, 1981). The change in the point at which noradrenergic systems are postulated to act in nonreward mechanisms is shown diagrammatically in Fig. 6.23. It has also been described on p. 276 how one test of the aversiveness of nonreward and its associability with environmental stimuli (the S− of a successive discrimination schedule) found no effect as a result of severe loss of forebrain noradrenaline (Mason & Iversen, 1978*a*). This would seem to argue against a role for noradrenergic systems in nonreward mechanisms *per se*.

A further prediction of the nonreward hypothesis is that where the PREE is due to counter-conditioning of frustration to an approach tendency then noradrenaline loss, by preventing the association of primary frustration with secondary environmental stimuli, should reduce resistance to extinction after PRF acquisition training. Owens *et al.* (1982) report that, after 50 training trials in a straight runway for food reward, PRF rats were more resistant to extinction than CRF rats. Noradrenaline depleted rats were more resistant to extinction than controls after CRF (the DBEE) but less resistant to extinction than controls after PRF. This would be analogous to the effects of amobarbital on PRF training and would be

Fig. 6.22. Threshold for driving of hippocampal theta rhythm by electrical stimulation in the septum. (*a*) In rats depleted of forebrain noradrenaline by adult dorsal bundle 6-hydroxydopamine (6-OHDA) injection. Note the minimum at 130 ms interpulse interval (7.7 Hz) in control animals is abolished by noradrenaline depletion. (*b*) In rats depleted of noradrenaline by synthesis inhibition with FLA 63. (*c*) In unilateral 6-OHDA lesioned rats. Note the loss of the 7.7 Hz minimum in *contralaterally* lesioned (Contra) animals. Ipsi = ipsilaterally lesioned. (*a*) and (*c*) redrawn from McNaughton *et al.*, 1980 and (*b*) from McNaughton *et al.*, 1977.

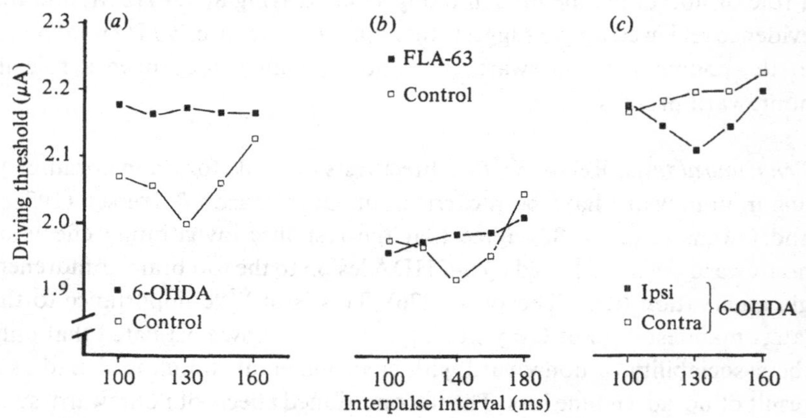

consistent with the notion that both minor tranquillisers and noradrenaline loss block the processing of nonreward. However, this result has not always been obtained. Owens *et al.* (1982) also report a second experiment in which 100 training trials were used and now noradrenaline depleted rats trained on PRF were just as resistant to extinction as controls trained on PRF. Similarly, Mason & Fibiger (1978*b*) examined extinction rates after acquisition training on the PRF schedule of VR4. Although this may not be the classic instance for the demonstration of the PREE it is nonetheless PRF in terms of its operational definition and here too noradrenaline depleted rats trained on PRF extinguished no faster than control rats.

Thus, two out of three experiments currently available in the literature suggest that noradrenaline loss causes resistance to extinction after CRF but does not affect, *by an increase or by a decrease*, extinction rates after PRF training. It would seem that only slight evidence can be adduced to support a role for noradrenergic systems in nonreward from experiments on partial reinforcement and extinction.

Conclusion. It is probably premature to conclude that noradrenergic systems play no role in nonreward. However, as in the case of fear and anxiety, it may be an indirect involvement more related to their function in the control of selective attention (see p. 277).

Summary

(1) A role for noradrenaline in extinction behaviour has been demonstrated in a number of positively and also negatively

Fig. 6.23. Summary of Gray's various theories of the role of noradrenaline (NA) in frustrative nonreward. (Redrawn from Mason, 1981.)

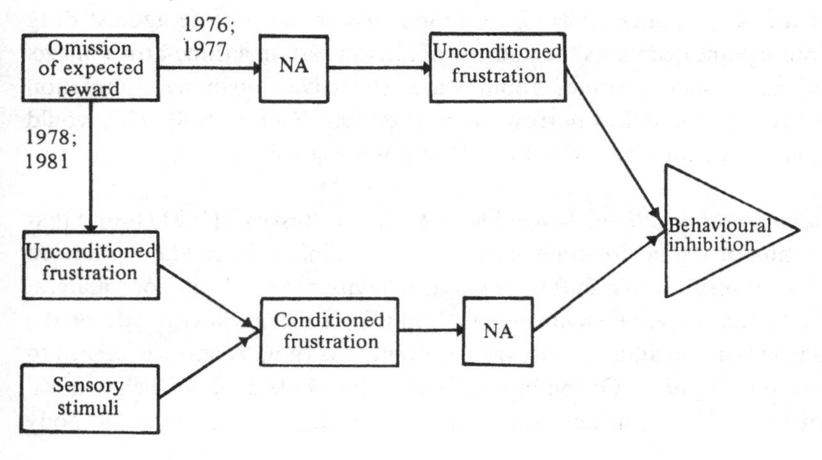

reinforced tasks. Loss of forebrain noradrenaline serves to increase resistance to extinction (the DBEE) after continuously, but not after partially reinforced acquisition learning.

(2) Noradrenergic systems do not seem to be involved in the DBEE by way of hyperactivity, perseveration, internal inhibition, motivation or secondary reinforcement.

(3) A role for noradrenaline in selective attention can explain the parameters of the DBEE and is supported by the findings on distractability, blocking, latent inhibition, nonreversal shifts and stimulus discrimination tasks.

(4) Noradrenergic systems do not seem to be primarily involved in fear and anxiety, except to the extent that changes in the sampling, interpretation or salience of stimuli which are secondary conditioned fear-inducers are concerned.

(5) Noradrenergic systems may be involved in nonreward. This argument rests mainly on an indirect chain of evidence. If the 7.7 Hz hippocampal rhythm codes for nonreward then, since noradrenaline manipulations alter septal driving at the 7.7 Hz frequency, this implies that noradrenergic systems are involved in nonreward. Direct behavioural tests of this postulate remain ambiguous.

Dopamine and cognitive behaviour

Attention

The data on the part played by dopamine in attention are much more scattered. Nonetheless, a role does seem to be emerging from recent studies.

Katz & Schmaltz (1981) have found that the dopamine agonist drug apomorphine decreases the number of alternations in a three-arm Y-maze. This they suggest comes about since 'excessive dopamine stimulation results in pathological narrowing of attention' (*ibid.* p. 588). This would appear to be only one of a number of possible explanations.

Intercerebral injections. Joyce, Davis & van Hartesveldt (1981) found that injection of 100 μg dopamine into various regions of the caudate–putamen in the rat resulted in a shift of ongoing behaviour towards the contralateral side of the animal's environment. Stimuli on the ipsilateral side of the environment, including those on the animal's own body surface, ceased to elicit any response. Grooming, including the whole complex behavioural sequence of face washing, shifted to involve only the contralateral body half after dopamine injection. The dorsal caudate was found to be more sensitive to dopamine injections than the ventral caudate or extrastriatal regions. Complementing this, Hoyman (1980) found that injections of the dopamine blocking agent haloperidol (51 μg) caused contralateral sensory neglect as shown by failure to respond on a tactile discrimination task when the tactile stimuli were presented to the body surface opposite the brain side of haloperidol administration. No effect on correct responses to ipsilaterally presented tactile stimuli occurred, indicating the laterality of this effect.

These results suggest that dopamine activity may have the effect of enhancing the processing of sensory stimuli by the animal, especially those relating to its own body surface. It would appear to be a crossed system, since manipulation of the left caudate affects the responsiveness of the right-hand side of the animal's body. Microinjection of a dopamine blocker caused failure to process incoming stimuli (sensory neglect) while microinjection of dopamine itself enhanced behaviours directed to the contralateral side of the body, to the exclusion of ipsilaterally directed behaviours. Further confirmation of this pattern is obtained from the effects of 6-OHDA lesion to ascending dopaminergic systems.

6-OHDA lesions. Sechzer, Ervin & Smith (1973) found that 6-OHDA injections into the lateral hypothalamus in the ascending course of the dopaminergic fibres abolished the visual placing response. This response, requiring the integration of visual input with motor coordination, is seen when the rat is lowered towards a surface head-first. When close to the horizontal surface, but before the vibrissae actually touch, the animal will extend its forepaws in preparation for contact. Lowering of the hindpaws also occurs. Animals treated with 6-OHDA failed to display this response although motor responses of hopping and righting appeared to be unaffected.

Marshall, Richardson & Teitlebaum (1974) also found deficits in orientating towards visual, olfactory or tactile stimuli presented to the contralateral side of the animal. Orientation to ipsilateral stimuli of all modalities was normal. Following bilateral injections of 6-OHDA into the nigrostriatal bundle, these authors also confirmed the failure of visual placing reactions. Some recovery of orientation capacity occurred by four months after the lesion but this was very limited and slow in appearance.

Ljungberg & Ungerstedt (1976) describe the phases of development of this sensory neglect after 6-OHDA lesion to the ascending dopaminergic system. One to two days after the lesion, while the dopaminergic fibres are actually in the process of degenerating, variable reactions were seen with some hyperresponsiveness to sensory stimuli. By days 2 to 5 after lesion, behaviour had stabilised and sensory neglect was present to contralateral visual, tactile, auditory, olfactory and painful stimuli (see Fig. 6.24). Ipsilateral stimulus presentation elicited essentially normal responses. Some recovery was reported by day 9 after the lesion, with olfaction and pain responsiveness recovering before visual and auditory responses (see Fig. 6.25). Touch, either light or heavy, failed to recover for the whole duration of the six month experiment (not shown).

Marshall (1979) described the stages of recovery of tactile responsiveness in detail. Thus, after injection of 8 μg 6-OHDA following desmethylimipramine pre-treatment (to prevent noradrenaline loss), 8 out of 26 rats showed rapid recovery of contralateral responsiveness to tactile stimulation. A further 7 failed to show any recovery over the whole 28 day post-operative test period (see Fig. 6.26). Some rats did not show sensory neglect and a marked correlation between the degree of dopamine loss in the striatum and the degree of sensory neglect was found (Spearman $\rho = 0.66$). Dopamine levels in accumbens, olfactory tubercle and cortex did not correlate with the degree of sensory impairment. Recovery of orientation to tactile stimuli occurred first in the front parts of the animal's body with progressive spread to more posterior portions with the passage of time.

Marshall, Levitan & Stricker (1976) found that recovery could be

Fig. 6.24. Sensory neglect caused by unilateral 6-hydroxydopamine (6-OHDA) lesion to the ascending dopaminergic systems in rats on the side ipsilateral (L) and that contralateral (R) to the 6-OHDA injection. (Redrawn from Ljungberg & Ungerstedt, 1976.)

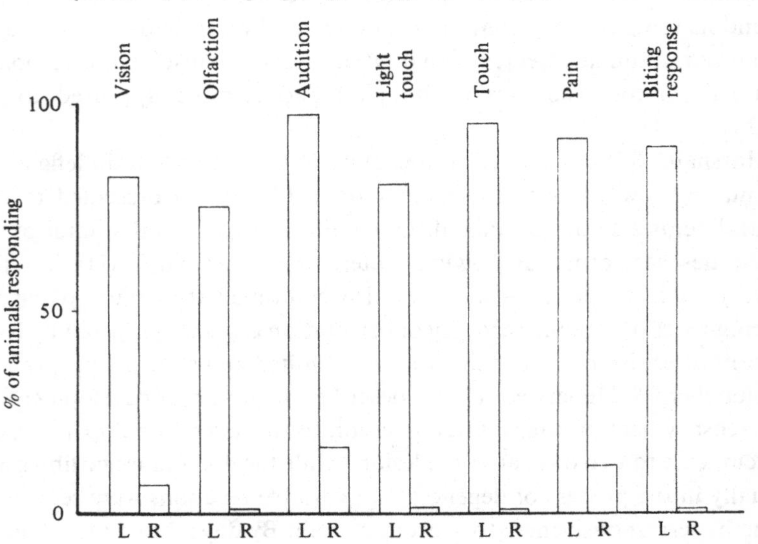

Fig. 6.25. Time course of recovery from sensory neglect in animals described in Fig. 6.24. (Redrawn from Ljungberg & Ungerstedt, 1976.)

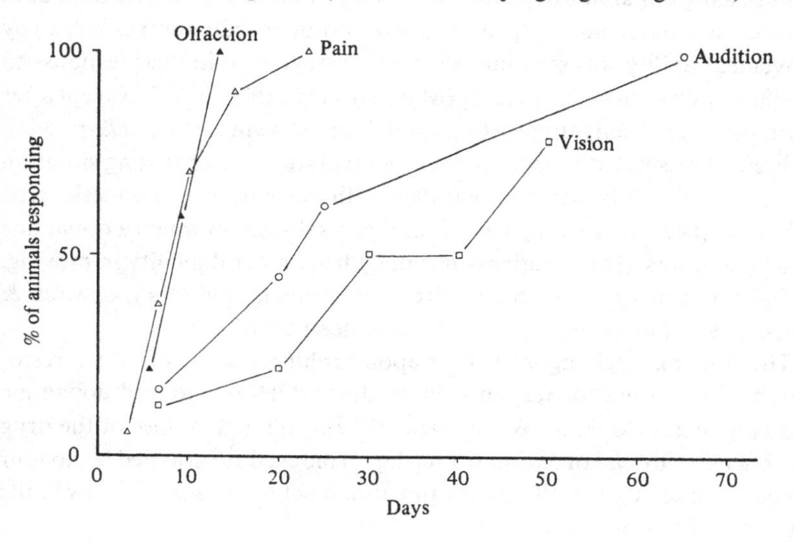

Fig. 6.26. Sensory neglect caused by unilateral 6-hydroxydopamine lesion to ascending dopaminergic systems correlated with degree of dopamine (DA, μg per g tissue weight) depletion in the striatum for groups of rats with differing severity of lesion. (Redrawn from Marshall, 1979.)

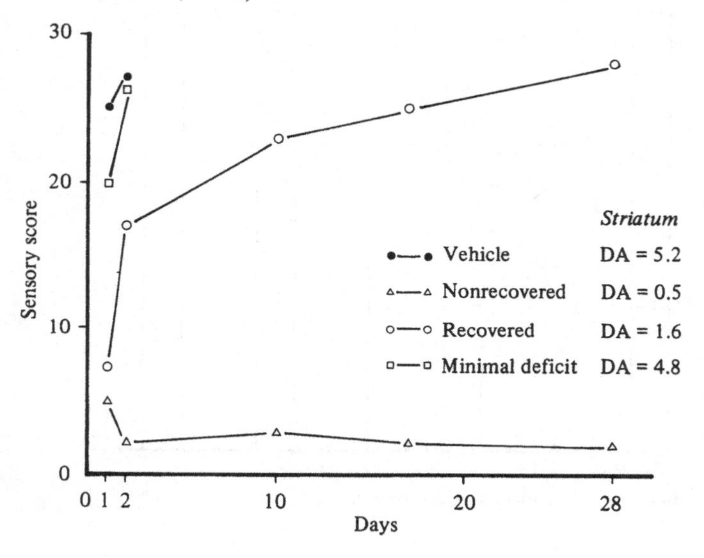

stimulated by stress-induced activation of the animal. Thus, immediately after placing dopamine depleted rats in a water bath, a cooled ice bath or in a colony of cats (!) much improved sensorimotor coordination as judged by movement ability was seen but no great improvement in their response to tactile stimuli or in visual placing (Marshall *et al.*, 1976, p. 542), except after the most severe activating procedure of forced swimming (*ibid.*, p. 544). This effect dissipated very soon after removal from the activating situation (see Fig. 6.27). This may have parallels with the 'paradoxical kinesia' seen in human patients suffering from Parkinson's disease in which a dopamine loss also occurs. These patients normally have great difficulty in moving, but at times of stress, such as a fire in their living quarters (Schwaub & Zeiper, 1965), may demonstrate unusual fleetness of foot.

The dopaminergic agonist drug apomorphine may also induce restoration of sensorimotor responsiveness after 6-OHDA-induced dopamine loss (Fig. 6.28) (Marshall & Gotthelf, 1979). Too high a dose of the drug (6.2 mg kg^{-1} instead of 0.05 or 0.1 mg kg^{-1}) induced stereotyped behaviour instead, indicating that dopamine function must lie between narrow limits for optimal sensory processing.

The area of importance for sensory neglect after dopamine loss appears to be the neostriatum. Thus, as described above, dopamine levels in this

Fig. 6.27. Effects of the activating treatment of forced swimming on the movement deficit in animals with bilateral 6-hydroxydopamine lesion of the ascending dopaminergic system. (Redrawn from Marshall *et al.*, 1976.)

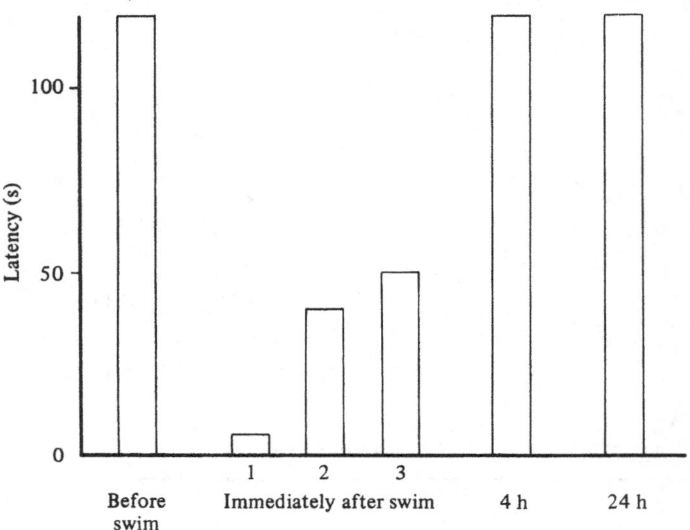

area, but not in the adjacent tubercle, accumbens or cortex, correlated with the degree of sensory neglect (Marshall, 1979). Further, Marshall, Berrios & Sawyer (1980) found that local injection of 6-OHDA into the striatum, but not into the accumbens, tubercle, lateral septum of frontal cortex, would also cause sensory neglect. Electrolytic lesions of these areas were also ineffective in causing sensory neglect, while lesions in the striatum caused sensory neglect organised in a topographical fashion over the animal's body surface on the contralateral side. Lesion in the anterior striatum induced sensory neglect in the front parts of the rat with normal response to tactile stimulation of the hind-quarters. Lesion in the more posterior parts of the striatum induced loss of responses in the hind-parts and middle of the animal.

After 6-OHDA injections into the ventral tegmental area which depleted dopamine throughout the forebrain, Marshall *et al.* (1980) found that microinjection of 6 μg of the dopamine agonist apomorphine into the striatum (see Fig. 6.29), but not into the cortex, accumbens or lateral septum, would restore sensorimotor integration for about 2 h.

Confirmation of the role of the striatum in sensory neglect after nigrostriatal lesion and its recovery came from Kozowlski & Marshall (1981), who showed that only the striatum and the central nucleus of the amygdala, but not the accumbens or tubercle, demonstrated recovery of

Fig. 6.28. Improvement in sensory neglect caused by intraperitoneal injection of apomorphine. Sal = saline. (Redrawn from Marshall & Gotthelf, 1979.)

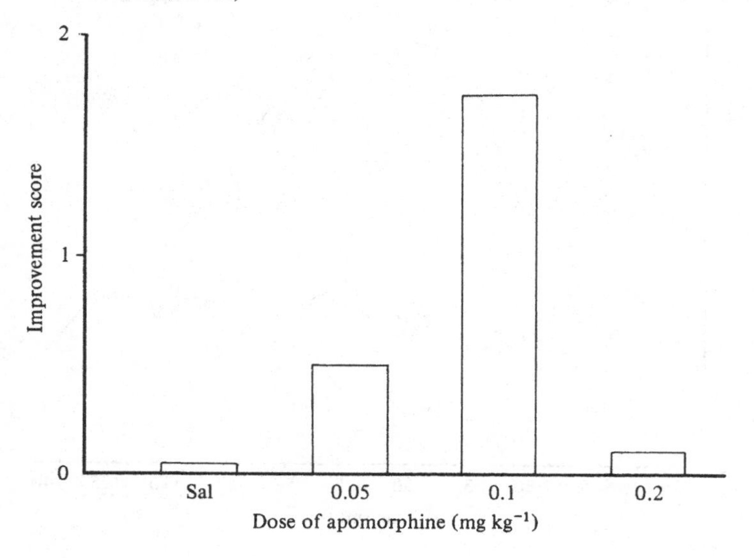

metabolic activity (by the 2-deoxyglucose method) at the same time as behavioural recovery from sensory neglect.

Further evidence for a role of dopaminergic systems in sensorimotor integration comes from the report of Uguru-Okorie & Arbuthnott (1981) that unilateral 6-OHDA lesion to the ascending dopaminergic fibres in the lateral hypothalamus would impair lever-pressing for food reward *only* when performed contralateral to the preferred paw. Rats tend to have a preference for using one paw or the other to press a lever and, if depleted of dopamine on the side contralateral to this, have to relearn to use the other paw to perform the response. If lesioned ipsilateral to the preferred paw, lever-pressing continued at the usual rate and without requiring a change in paw usage. As Uguru-Okorie & Arbuthnott point out (1981, p. 465) this may complicate the interpretation of some of the ICSS experiments which made use of the ipsilateral versus contralateral 6-OHDA lesion technique (see Chapter 3, p. 83).

Conclusion. Considerable evidence has emerged in recent years that dopaminergic systems, especially in the striatum and arranged in a

Fig. 6.29. Improvement in sensory neglect caused by intrastriatal injection of apomorphine (APO) in various doses. C/I ratio expresses contralateral to ipsilateral orientation and a return to normality is shown as it approaches unity. 'Pre' refers to pre-injection performance. Sal-saline. (Redrawn from Marshall *et al.*, 1980.)

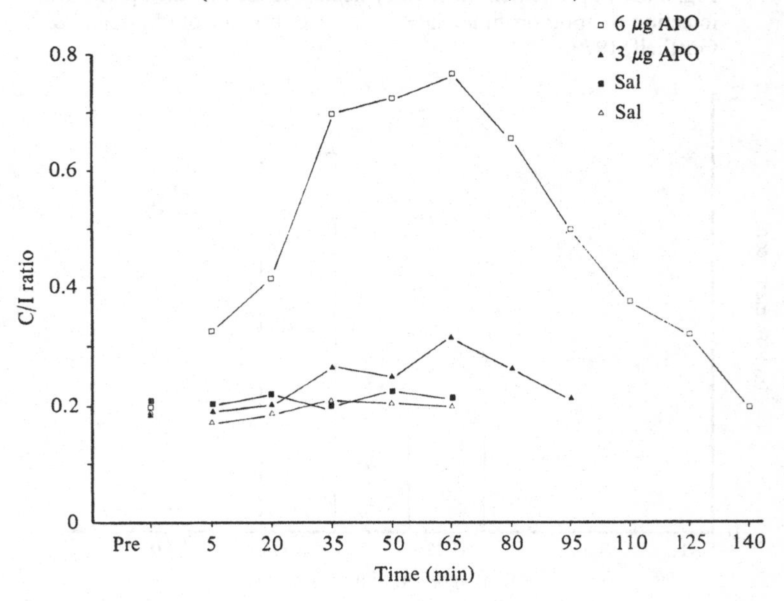

somatotopic fashion, may be of importance for the processing of sensory information, not only from the body surface but also in the visual, auditory and olfactory modalities as well. Thus dopamine appears to have a role opposite to that of noradrenaline in attentional behaviour. Dopamine loss results in sensory inattention while noradrenaline loss results in the sampling of too many stimuli (see p. 277).

Incentive motivation

Some theorists have speculated (Iversen & Koob, 1977; Iversen, 1980; Iversen & Fray, 1980) that the nonstriatal dopaminergic systems may play a role in the integration and expression of approach behaviour by the rat. This may conveniently be termed incentive motivation.

6-OHDA lesions. Selective destruction of the dopaminergic innervation to the nucleus accumbens can be caused by injection of 8 μg 6-OHDA bilaterally and yields depletions better than 80 % (Koob, Riley, Smith & Robbins, 1978). Such lesions have already been described to result in locomotor hypoactivity and a block of the locomotor stimulant, but not the stereotypical, response to amphetamine (see p. 140). More complex alterations in behaviour may also result. Koob *et al.* (1978) report that lesioned rats explore less when introduced to an activity cage. If food is present in this situation, normal rats will first explore and only then eat. Accumbens dopamine depleted rats will eat immediately and not explore. Thus, hyperphagia is seen in this situation but not in free-feeding in the home cage (which is not a novel environment) (Koob *et al.*, 1978). A possible inflexibility of behaviour was also seen in that food-associated water drinking was reduced by the lesion, since these animals failed to switch consumatory behaviours between eating and drinking as did normal rats.

A similar failure to switch between food and water was seen in an adjunctive drinking paradigm (Robbins & Koob, 1980) in which free presentations of food at irregular intervals can cause marked water consumption above and beyond that seen in the absence of food. 6-OHDA lesion to the dopaminergic innervation of the accumbens reduced adjunctive drinking with no direct effect on thirst *per se* as assessed by drinking in the absence of a competing activity such as food consumption. A similar deficit in switching behaviours was seen in a two-choice water tube situation in which 6-OHDA lesioned rats drank for longer from their preferred tube whereas normals partook of more frequent and shorter bouts of drinking more equally from the two available tubes.

Deficits in the exploration of the environment have been reported after destruction of the mesolimbic dopaminergic systems by 6-OHDA injection in the lateral hypothalamus (Jeste & Smith, 1980) or by electrolytic lesion of the A10 cell group (Gaffori, Le Moal & Stinus, 1980). In the latter case, locomotor activity was increased although exploration, in terms of the number of contact with objects in an open field, was reduced.

A similar failure to organise behaviour in the most efficient manner has been reported by Iversen (1980) for accumbens dopamine depleted rats tested on the retrieval of food pellets from multiple food cups. Normal animals soon learn the pattern of baited food cups and visit them sequentially and without repetition to emptied ones. Accumbens dopamine depleted rats continued to visit unbaited food cups and indulged in irrelevant behaviours such as rearing (Iversen, 1980). Stinus, Gaffori, Simon & Le Moal (1978) report that hoarding of food disappears after electrolytic lesion of A10 and food consumption becomes less well coordinated, with greater spillage than normal.

Conclusion. All of the above lines of evidence suggest, albeit in a somewhat indirect fashion, that nonstriatal dopaminergic systems may play a role in the organisation of behaviour in terms of the incentive motivational significance of stimuli in the rat's environment. As such, extrastriatal dopaminergic systems may complement striatal dopaminergic systems by giving cognitive significance to stimuli whose access to the sensorium is enhanced on a lower level of processing by the striatal dopaminergic systems (see section on sensory neglect, p. 297). More evidence for the role that dopaminergic systems have in determining the cognitive significance of stimuli is presented in the next section in relation to secondary conditioning.

Secondary conditioning

It has been proposed that dopaminergic systems play a role in the establishment and expression of secondary reinforcement (Hill, 1970). This is the phenomenon whereby initially neutral stimuli, by their repeated association with primary reinforcers such as food or water, come to acquire reinforcing properties in their own right (Wike, 1966). The evidence for a role of dopamine is circumstantial, using psychomotor stimulant drugs such as amphetamine and pipradrol, which among other actions serve to increase the release of dopamine in the brain. More direct evidence is provided by the specific dopamine receptor blocker pimozide.

Robbins (1975) reported that animals trained to press equally on two levers on a VR schedule would associate the click of the food mechanism and the flash of the magazine light which accompanied food delivery with primary reinforcement. This was shown during extinction testing by giving the animal a choice between one lever which produced the conditioned reinforcers (light, flash and click) on a VR schedule and a second lever which did not have any effect. No food was ever presented during this preference testing. Normal rats showed a small preference to respond on the lever giving the secondary conditioned stimuli compared to that having no effect. This preference was markedly enhanced by doses of the dopamine-releasing drug pipradrol (see Fig. 6.30), especially at doses of 5 and 10 mg kg^{-1}. This was not just a general release of locomotor activity, since responding on the lever which did not yield the secondary conditioned stimuli was not increased by any dose of the drug. This suggests that enhanced activity in dopaminergic systems may make more effective the expression of the already established secondary reinforcing properties of stimuli. Robbins (1975) reports experiments which exclude the possibility that pipradrol merely increases responding for any novel stimulus.

Robbins (1976) showed that pipradrol would enhance the secondary conditioned properties of initially neutral stimuli as measured by a second

Fig. 6.30. Effects of pipradrol in enhancing responding on a lever which gives conditioned stimulus presentation (CR) compared to that on a lever which has no effect (NCR) for animals under drug and when under saline (control). (Redrawn from Robbins, 1975.)

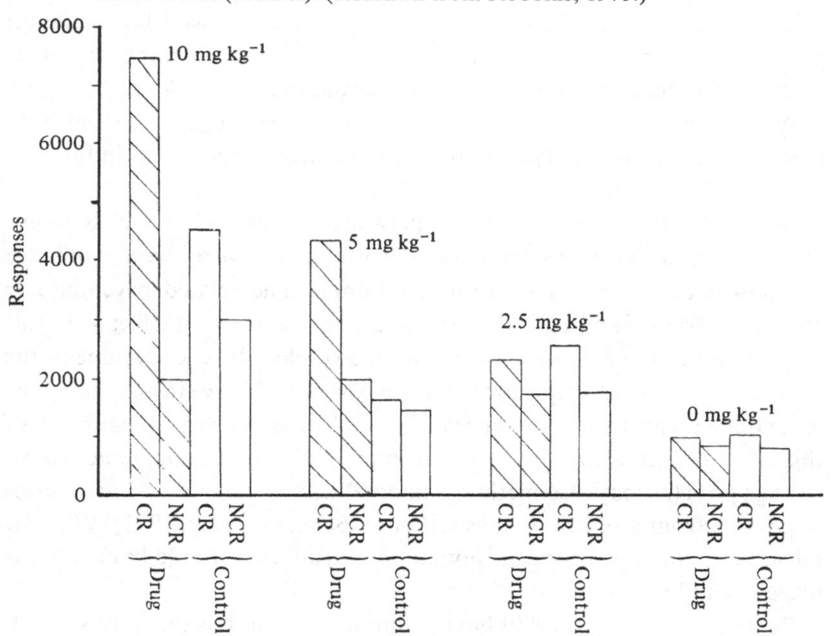

technique, that of reinforcing new learning. Animals experienced pairing of a light flash during the presentation of a dipper containing water. After 120 such pairings, the light flash was used to reinforce the learning *de novo* of a two-lever spatial discrimination task. No primary reinforcer (water) was given during this stage. For half the rats pressing on the left lever produced the light flash on a VR2 schedule. The rats which were tested under 10 mg kg^{-1} pipradrol learned to press this, and not the right lever. The other half of the rats received the secondary conditioned stimulus (light flash) only after pressing the right lever, and, under pipradrol, response totals on this, and not on the left lever, rose over days of testing. Animals under saline failed to learn this novel spatial discrimination task. Demonstration of the power to motivate new learning is hence strong evidence for an effect of pipradrol on secondary conditioning and, to the extent that pipradrol acted through dopaminergic systems, of a role for dopamine in this behaviour.

Robbins & Koob (1978) demonstrated similar new learning under 10 mg kg^{-1} pipradrol for secondary conditioned stimuli which had been associated with brain stimulation reward. Animals were trained to panel push to receive lateral hypothalamic ICSS when either white noise was present or the magazine light behind the panel was illuminated. After pairing these initially neutral stimuli with ICSS the rat was required to learn a novel spatial discrimination task between the left- and the right-hand side lever in order to receive the secondary conditioned stimuli. In this new task, animals under either 5 or 10 mg kg^{-1} pipradrol showed high rates of responding to the lever giving the conditioned stimuli but not to the other lever which had no programmed consequence. It would appear that pipradrol enhances the secondary conditioned properties of stimuli no matter what reinforcer (water, food or ICSS) has been used initially to establish those properties.

Using the two-lever preference paradigm described above, Robbins (1978) showed that methylphenidate as well as pipradrol would enhance the power of secondary conditioned stimuli. The related psychomotor stimulant drugs *d*-amphetamine and nomifensine were not effective in this respect (see Fig. 6.31). Since amphetamine would release dopamine to the same degree as would pipradrol, and yet was not effective in enhancing the secondary conditioned properties of stimuli, the neurochemical basis of this effect is uncertain. It may be that pipradrol releases dopamine from a storage pool (as may also methylphenidate) whereas amphetamine releases dopamine from a newly synthesised pool (Scheel-Kruger, 1971). Why the release of dopamine in one fashion but not in another should be effective is presently unknown.

Beninger & Phillips (1980) have examined the establishment of secondary conditioning by the use of a paradigm in which food pellets are paired with a tone, either under a drug or under saline. The animal is then tested for its preference in the absence of food for a lever which produces the tone versus a second lever which has no stimulus consequence. As a result of pairing of tone and food, a marked preference for the lever which subsequently delivers the tone is seen. This does not occur if food pellets and tones have been presented in an uncorrelated fashion during training. The presence of the dopamine receptor blocker pimozide during the tone–food pairing prevents preference for the tone lever occurring when tested either drug free or under a second dose of pimozide. This the authors interpret as evidence that activity in dopaminergic systems is necessary for the establishment of secondary conditioning. This is then the other side of the coin to the results of Robbins (1978), who found that increased dopamine activity as a result of pipradrol would enhance the expression of

Fig. 6.31. Effects of various doses of pipradrol, methylphenidate, *d*-amphetamine and nomifensine on responding on a lever which gives conditioned reinforcement (CRF) stimulus presentation. (Redrawn from Robbins, 1978.)

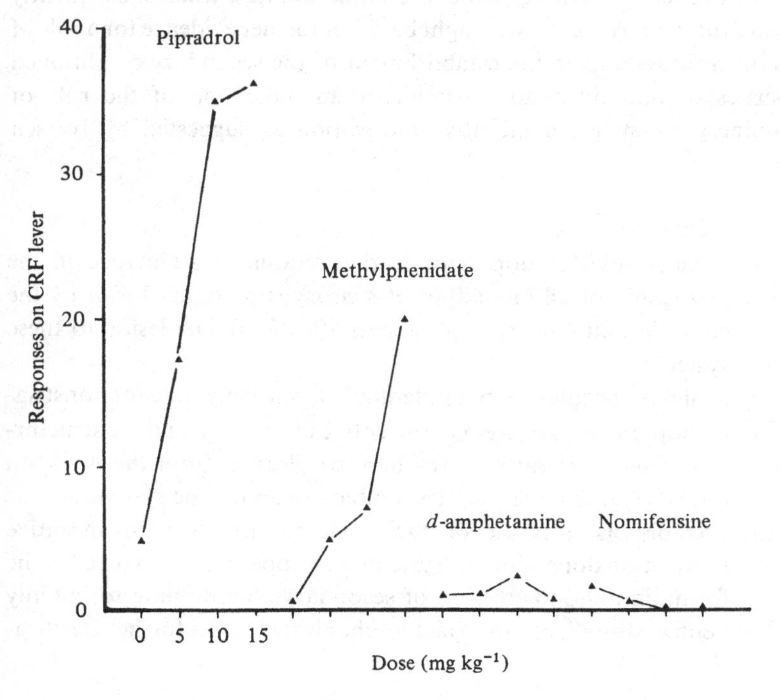

the secondary conditioned properties of already established secondary reinforcers.

Beninger, Hansen & Phillips (1980) found that administration of pipradrol during the final test phase in the absence of food would also enhance the preference for the lever giving secondary conditioned stimuli. This is very similar to the demonstration of Robbins & Koob (1978) that animals would learn a new task for secondary conditioned stimuli and that this was enhanced by pipradrol. This examines the expression, rather than the acquisition, of secondary reinforcing power by initially neutral stimuli.

A further demonstration of this effect was shown by the partial blockade by pimozide of the beneficial effect that tone–food pairings had on the subsequent learning of an operant discrimination task using the tone as the S+ (Beninger & Phillips, 1981). Normal rats which receive 80 pairings of tone with food learn a subsequent VR lever-press task using the tone as the S+ more quickly than rats which received only pre-exposure to the pellets without the tone. This may be due to the incentive motivational properties which the tone acquires as a result of its pairing with food (Bolles, 1972). Presentation of tone–food pairings while the rats were under 1 mg kg^{-1} of the dopamine blocker pimozide greatly attenuated (but did not entirely prevent) this faster learning of the discrimination task when subsequently trained drug-free. As such, this might be either further evidence for a role of dopaminergic systems in the establishment of the secondary conditioned properties of initially neutral stimuli or an indication of the role of dopaminergic systems in incentive motivation as suggested by Iversen (1980).

Summary

(1) A clear role for dopamine in the striatum has emerged in the processing of all modalities of sensory stimuli, as shown by the contralateral sensory neglect seen after 6-OHDA lesion to these systems.
(2) A highly speculative, but potentially fascinating, role for nonstriatal dopaminergic systems in incentive motivation and the structuring of behavioural patterns may be gleaned from the work on 6-OHDA lesions of the mesolimbic dopaminergic systems.
(3) Possibly as a result of their role in attention or incentive motivation dopaminergic systems also appear to be involved in the formation and expression of secondary conditioning to initially neutral stimuli, but the exact mechanism requires further clarification.

References

Adelman, H. M. & Maatsch, J. L. (1956). Learning and extinction based upon frustration, food reward and exploratory tendency. *Journal of Experimental Psychology*, **52**, 311–15.

Amsel, A. (1958). The role of frustrative nonreward in noncontinuous reward situations. *Psychological Bulletin*, **55**, 102–19.

Amsel, A. (1962). Frustrative nonreward in partial reinforcement and discrimination learning: some recent history and theoretical extension. *Psychological Review*, **69**, 306–28.

Amsel, A. & Roussell, J. (1952). Motivational properties of frustration: I. Effect on a running response of addition of frustration to the motivational complex. *Journal of Experimental Psychology*, **43**, 363–8.

Ashford, J. & Jones, B. J. (1976). The effects of intra-amygdaloid injections of 6-hydroxydopamine on avoidance responding in rats. *British Journal of Pharmacology*, **56**, 255–61.

Azrin, N. H., Hutchinson, R. R. & Hake, D.F. (1966). Extinction-induced aggression. *Journal of the Experimental Analysis of Behavior*, **9**, 191–204.

Barry, H., Wagner, A. R. & Miller, N. E. (1962). Effects of alcohol and amobarbital on performance inhibited by experimental extinction. *Journal of Comparative and Physiological Psychology*, **55**, 464–8.

Beninger, R. J., Hansen, D. E. & Phillips, A. G. (1980). The effects of pipradrol on the acquisition of responding with conditioned reinforcement: a role for sensory preconditioning. *Psychopharmacology*, **69**, 235–42.

Beninger, R. J. & Phillips, A. G. (1980). The effect of pimozide on the establishment of conditioned reinforcement. *Psychopharmacology*, **68**, 147–53.

Beninger, R. J. & Phillips, A. G. (1981). The effects of pimozide during pairing on the transfer of classical conditioning to an operant discrimination. *Pharmacology, Biochemistry & Behavior*, **14**, 101–5.

Biswas, B. & Carlsson, A. (1978). On the mode of action of diazepam on brain catecholamine metabolism. *Naunyn-Schmiedeberg's Archives of Pharmacology*, **303**, 73–8.

Blough, D. S. (1956). Technique for studying the effects of drugs on discrimination in the pigeon. *Annals of the New York Academy of Sciences*, **65**, 334–44.

Bolles, R. C. (1972). Reinforcement, expectancy and learning. *Psychological Review*, 79, 394–409.

Brown, R. T. & Wagner, A. R. (1964). Resistance to punishment and extinction following training with shock or reinforcement. *Journal of Experimental Psychology*, **68**, 503–7.

Cedarbaum, J. M. & Aghajanian, G. K. (1976). Noradrenergic neurons of the locus coeruleus: inhibition by epinephrine and activation by piperoxan. *Brain Research*, **112**, 413–20.

Corbett, D. & Wise, R. A. (1979). Intracranial self-stimulation in relation to the ascending noradrenergic fibres of the pontine tegmentum and caudal midbrain: a moveable electrode study. *Brain Research*, **177**, 423–36.

Corrodi, H., Fuxe, K., Lidbrink, P. & Olson, L. (1971). Minor tranquillizers, stress and central catecholamine neurones. *Brain Research*, **29**, 1–16.

Crow, T. J., Deakin, J. F. W., File, S. E., Longden, A. & Wendlandt, S. (1978). The locus coeruleus noradrenergic system – evidence against a role in attention, habituation, anxiety and motor activity. *Brain Research*, **155**, 249–61.

Davis, M., Redmond, D. E. & Baraban, J. M. (1979). Noradrenergic agonists and antagonists: effects on conditioned fear measured by the potentiated startle paradigm. *Psychopharmacology*, **65,** 111–18.

Dickinson, A. (1974). Response suppression and facilitation following septal lesions in rats: a review and a model. *Physiological Psychology*, **2,** 444–56.

Douglas, R. J. (1967). The hippocampus and behavior. *Psychological Bulletin*, **67,** 416–42.

Ellison, G., Handel, J., Rodgers, R. & Weiss, J. (1975). Tricyclic antidepressants: effects on extinction and fear learning. *Pharmacology, Biochemistry & Behavior*, **3,** 7–11.

Fibiger, H. C. & Mason, S. T. (1978). The effect of dorsal bundle injections of 6-OHDA on avoidance responding in rats. *British Journal of Pharmacology*, **64,** 601–6.

Fibiger, H. C., Roberts, D. C. S. & Price, M. T. C. (1975). On the role of telencephalic noradrenaline in learning and memory. In *Chemical tools in catecholamine research*, ed. G. Jonsson, T. Malmfors & C. Sachs, pp. 349–56. North-Holland Publishing Co.: Amsterdam.

File, S. E., Deakin, J. F. W., Longden, A. & Crow, T. J. (1979). Investigation of the role of the locus coeruleus in anxiety and agonistic behavior. *Brain Research*, **169,** 411–20.

Gaffori, O., Le Moal, M. & Stinus, L. (1980). Locomotor hyperactivity and hypoexploration after lesion of the dopaminergic-A10 area in the ventral mesencephalic tegmentum (VMT) of rats. *Behavioural Brain Research*, **1,** 313–29.

Gottschalk, L. A., Stone, W. N. & Glesser, G. C. (1974). Peripheral versus central mechanisms accounting for antianxiety effects of propranolol. *Psychosomatic Medicine*, **36,** 47–56.

Gray, J. A. (1969). Sodium amobarbital and effects of frustrative non reward. *Journal of Comparative and Physiological Psychology*, **69,** 55–64.

Gray, J. A. (1970). Sodium amobarbital, the hippocampal theta rhythm and the partial reinforcement extinction effect. *Psychological Review*, **77,** 465–80.

Gray, J. A. (1972). Effects of septal driving of the hippocampal theta rhythm on resistance to extinction. *Physiology and Behavior*, **8,** 481–90.

Gray, J. A. (1976). The behavioural inhibition system: a possible substrate for anxiety. In *Theoretical and experimental basis of the behaviour therapies*, ed. R.S. Feldman & P.L. Broadhurst, pp. 3–41. John Wiley & Sons: New York.

Gray, J. A. (1977). Drug effects on fear and frustration: possible limbic site of action of minor tranquillizers. In *Handbook of psychopharmacology*, ed. L.L. Iversen, S.D. Iversen & S.H. Snyder, vol. 8, pp. 433–50. Plenum Press: London.

Gray, J. A. (1978). The neuropsychology of anxiety. *British Journal of Pharmacology*, **69,** 417–34.

Gray J. A. (1981). *The neuropsychology of anxiety: An enquiry into the functions of the septo-hippocampal system.* Oxford University Press: Oxford.

Gray, J. A., Araujo-Silva, M. T. & Quintao, L. (1972*a*). Resistance to extinction after partial reinforcement training with blocking of the hippocampal theta rhythm by septal stimulation. *Physiology and Behavior*, **8,** 497–502.

Gray, J. A. & Ball, G. G. (1970). Frequency-specific relation between hippocampal theta rhythm, behavior and amobarbital action. *Science*, **168,** 1246–8.

Gray, J. A., McNaughton, N., James, D. T. D. & Kelly, P. H. (1975) Effect of minor tranquillizers on hippocampal theta rhythm mimicked by depletion of forebrain noradrenaline. *Nature*, **258,** 424–6.

Gray, J. A., Quintao, L. & Araujo-Silva, M.T. (1972*b*). The partial reinforcement extinction effect in rats with medial septal lesions. *Physiology and Behavior*, **8,** 491–6.

Hill, R. T. (1970). Facilitation of conditioned reinforcement as a mechanism of psychomotor stimulation. In *Amphetamines and related compounds*, ed. E. Costa & S. Garattini, pp. 781–95. Raven Press: New York.

Hoyman, L. (1980). Tactile discrimination performance deficits following unilateral microinjections of catecholaminergic blockers in the rat. *Physiology and Behavior*, **23,** 1057–63.

Ison, J. R. & Pennes, E. S. (1969). Interaction of amobarbital sodium and reinforcement schedule in determining resistance to extinction of an instrumental running response. *Journal of Comparative and Physiological Psychology*, **68,** 215–19.

Iversen, S. D. (1980). Brain chemistry and behaviour. *Psychological Medicine*, **10,** 522–39.

Iversen, S. D. & Fray, P. J. (1980) Brain catecholamines in relation to affect. In *The neural basis of behaviour*, ed. A. Beckman, pp. 113–25. Spectrum Publications: New York.

Iversen, S. D. & Koob, G. F. (1977). Behavioural implications of dopaminergic neurones in the mesolimbic system. In *Advances in biochemical psychopharmacology*, vol. 16, eds. E. Costa & G. L. Gessa, pp. 209–14. Raven Press: New York.

Jeste, D. V. & Smith, G. P. (1980) Unilateral mesolimbicocortical dopamine denervation decreases locomotion in the open field and after amphetamine. *Pharmacology, Biochemistry & Behavior*, **12,** 453–7.

Joyce, J. N., Davis, R. M. & van Hartesveldt, C. (1981). Behavioral effects of unilateral dopamine injection into dorsal or ventral striatum. *European Journal of Pharmacology*, **72,** 1–10.

Kamin, L. J. (1969). Predictability, surprise, attention and conditioning. In *Punishment and aversive behavior*, ed. B. A. Campbell & R. N. Church, pp. 279–96. Appleton-Century-Crofts: New York.

Katz, R. J. & Schmaltz, K. (1981). Dopaminergic involvement in attention: a novel animal model. *Progress in Psychopharmacology*, **4,** 585–90.

Koob, G. F., Kelley, A. F. & Mason, S. T. (1978). Locus coeruleus lesions: learning and extinction. *Physiology and Behavior*, **20,** 709–16.

Koob, G.F., Riley, S.J., Smith, S.C. & Robbins, T.W. (1978). Effects of 6-hydroxydopamine lesions of the nucleus accumbens septi and olfactory tubercle on feeding, locomotor activity and amphetamine anorexia in the rat. *Journal of Comparative and Physiological Psychology*, **92,** 917–27.

Kozowlski, M. R. & Marshall, J. F. (1981). Plasticity of neostriatal metabolic activity and behavioral recovery from nigrostriatal injury. *Experimental Neurology*, **74,** 318–23.

Ljungberg, T. & Ungerstedt, U. (1976). Sensory inattention produced by 6-hydroxydopamine-induced degeneration of ascending dopamine neurons in the brain. *Experimental Neurology*, **53,** 585–600.

Lorden, J. F., Richert, E. J., Dawson, R. & Pelleymounter, M. A. (1980). Forebrain norepinephrine and the selective processing of information. *Brain Research*, **190,** 569–73.

McNaughton, N., James, D. T. D., Stewart, J., Gray, J. A., Valero, I. & Drewnowski, A. (1977). Septal driving of hippocampal theta rhythm as a function of frequency in the male rat: effects of drugs. *Neuroscience*, **2,** 1019–27.

McNaughton, N., Kelly, P. H. & Gray, J. A. (1980). Unilateral blockade of the dorsal ascending noradrenergic bundle and septal elicitation of the hippocampal theta rhythm. *Neuroscience Letters*, **18,** 67–72.

Mackintosh, N. J. (1965). Selective attention in animal discrimination learning. *Psychological Bulletin*, **64,** 124–50.

Mackintosh, N. J. (1974). *The psychology of animal learning*. Academic Press: New York.

Mackintosh, N. J. & Holgate, V. (1968). Effects of inconsistent reinforcement on reversal and non-reversal shifts. *Journal of Experimental Psychology*, **76**, 154–9.

Marshall, J. F. (1979). Somatosensory inattention after dopamine-depleting intracerebral 6-OHDA injections: spontaneous recovery and pharmacological control. *Brain Research*, **177**, 311–24.

Marshall, J. F., Berrios, N. & Sawyer, S. (1980). Neostriatal dopamine and sensory inattention. *Journal of Comparative and Physiological Psychology*, **94**, 833–46.

Marshall, J. F. & Gotthelf, T. (1979). Sensory inattention in rats with 6-hydroxydopamine-induced degeneration of ascending dopaminergic neurons: apomorphine-induced reversal of deficits. *Experimental Neurology*, **65**, 398–411.

Marshall, J. F., Levitan, D. & Stricker, E. M. (1976). Activation-induced restoration of sensorimotor functions in rats with dopamine-depleting brain lesions. *Journal of Comparative and Physiological Psychology*, **90**, 536–46.

Marshall, J. F., Richardson, J. S. & Teitlebaum, P. (1974) Nigrostriatal bundle damage and the lateral hypothalamic syndrome. *Journal of Comparative and Physiological Psychology*, **87**, 808–30.

Mason, S. T. (1977). Noradrenaline pathways and learning behaviour in the rat. PhD Thesis, Cambridge University.

Mason, S. T. (1978). Parameters of the dorsal bundle extinction effect: previous extinction experience. *Pharmacology, Biochemistry & Behavior*, **8**, 655–9.

Mason, S. T. (1979). The dorsal bundle extinction effect: motivation or attention? *Physiology and Behavior*, **23**, 43–51.

Mason, S. T. (1980). Noradrenaline and selective attention: a review of the model and the evidence. *Life Sciences*, **27**, 617–31.

Mason, S. T. (1981). Noradrenaline in the brain: progress in theories of behavioural function. *Progress in Neurobiology*, **16**, 263–303.

Mason, S. T. & Fibiger, H. C. (1978*a*). 6–OHDA lesion to the dorsal noradrenergic bundle alters extinction of passive avoidance. *Brain Research*, **152**, 209–14.

Mason, S. T. & Fibiger, H. C. (1978*b*). Noradrenaline and partial reinforcement in rats. *Journal of Comparative and Physiological Psychology*, **92**, 1110–18.

Mason, S. T. & Fibiger, H. C. (1978*c*). Evidence for a role of brain noradrenaline in attention and stimulus sampling. *Brain Research*, **159**, 421–6.

Mason, S. T. & Fibiger, H. C. (1979*a*). Noradrenaline and avoidance learning in the rat. *Brain Research*, **161**, 321–34.

Mason, S. T. & Fibiger, H. C. (1979*b*). Noradrenaline and extinction of conditioned taste aversion in the rat. *Behavioral and Neural Biology*, **25**, 205–16.

Mason, S. T. & Fibiger, H. C. (1979*c*). Noradrenaline, fear and extinction. *Brain Research*, **165**, 47–56.

Mason, S. T. & Fibiger, H. C. (1979*d*). Neurochemical basis of the dorsal bundle extinction effect. *Pharmacology, Biochemistry & Behavior*, **10**, 373–80.

Mason, S. T. & Fibiger, H. C. (1979*e*). The dorsal noradrenergic bundle and varieties of passive avoidance. *Psychopharmacology*, **66**, 179–82.

Mason, S. T. & Fibiger, H. C. (1979*f*). Noradrenaline and selective attention. *Life Sciences*, **25**, 1949–56.

Mason, S. T. & Fibiger, H. C. (1979*g*). Anxiety: the locus coeruleus disconnection. *Life Sciences*, **25**, 2121–7.

Mason, S. T. & Iversen, S. D. (1975). Learning in the absence of forebrain noradrenaline. *Nature*, **258**, 422–4.

Mason, S. T. & Iversen, S. D. (1977*a*). Effects of selective forebrain noradrenaline loss on behavioural inhibition in the rat. *Journal of Comparative and Physiological Psychology*, **91**, 165–73.

Mason, S. T. & Iversen, S. D. (1977*b*) An investigation of the role of cortical and cerebellar noradrenaline in associative motor learning. *Brain Research*, **134**, 513–27.

Mason, S. T. & Iversen, S. D. (1977*c*). Behavioural basis of the dorsal bundle extinction effect. *Pharmacology, Biochemistry & Behavior*, **7**, 373–9.

Mason, S. T. & Iversen, S. D. (1978*a*). The dorsal noradrenergic bundle, extinction and non reward. *Physiology and Behavior*, **21**, 1043–5.

Mason, S. T. & Iversen, S. D. (1978*b*). Reward, attention and the dorsal noradrenergic bundle. *Brain Research*, **150**, 135–48.

Mason, S. T. & Lin, D. (1980). Noradrenaline and selective attention in the rat. *Journal of Comparative and Physiological Psychology*, **94**, 819–32.

Mason, S. T. & Robbins, T. W. (1979). Noradrenaline and conditioned reinforcement. *Behavioral and Neural Biology*, **25**, 523–34.

Mason, S. T., Roberts, D. C. S. & Fibiger, H. C. (1979). Interaction of brain noradrenaline and the pituitary–adrenal axis in learning and extinction. *Pharmacology, Biochemistry & Behavior*, **10**, 11–16.

Mellgren, R. L. & Ost, J. W. P. (1971). Discriminative stimulus preexposure and learning of an operant discrimination in the rat. *Journal of Comparative and Physiological Psychology*, **77**, 179–87.

Merlo, A. B. & Izquierdo, I. (1967). The effect of catecholamines on learning in rats. *Medicina et pharmacologia experimentalis*, **16**, 343–9.

Miller, N. E. (1961). Some recent studies of conflict behavior and drugs. *American Psychologist*, **16**, 12–24.

Nashold, B. S., Wilson, W. P. & Slaughter, G. (1974). The midbrain and pain. In *Advances in neurology*, vol. 4, pp. 191–6. Raven Press: New York.

Nyback, H., Walters, J. R., Aghajanian, G. K. & Roth, R. H. (1975). Tricyclic antidepressants: effects on the firing rate of brain noradrenergic neurons. *European Journal of Pharmacology*, **32**, 302–12.

Oke, A. F. & Adams, R. N. (1978). Selective attention dysfunctions in adult rats neonatally treated with 6-hydroxydopamine. *Pharmacology, Biochemistry & Behavior*, **9**, 429–32.

Owens, S., Boarder, M. R., Gray, J. A. & Fillenz, M. (1982). Acquisition and extinction of continuously and partially reinforced running in rats with lesions of the dorsal noradrenergic bundle. *Behavioural Brain Research*, **5**, 11–42.

Peterson, D. W. & Laverty, R. (1976). Operant behavioural and neurochemical effects after neonatal 6-hydroxydopamine treatment. *Psychopharmacology*, **50**, 55–60.

Plaznik, A. & Kostowski, W. (1980). Locus coeruleus lesions and avoidance behavior in rats. *Acta neurobiologica experimentalis*, **40**, 217–25.

Price, M. T. C., Murray, G. N. & Fibiger, H. C. (1977). Schedule dependent changes in operant responding after lesions of the dorsal tegmental noradrenergic projection. *Pharmacology, Biochemistry & Behavior*, **6**, 11–15.

Redmond, D. E., Huang, Y. H., Snyder, D. R., Baulu, J. & Maas, J. W. (1976*a*). Behavioral changes following lesions of the locus coeruleus in *M. arctoides*. *Society for Neuroscience Abstracts*, **2**, 114.

Redmond, D. E., Huang, Y. H., Snyder, D. R. & Maas, J. W. (1976*b*). The behavioral effects of stimulation of the locus coeruleus in the stumptail monkey (*Macaca arctoides*). *Brain Research*, **116**, 502–10.

Rilling, M., Askew, H. R., Ahlskog, J. E. & Kramer, T. J. (1969). Aversive properties of the negative stimulus in a successive discrimination. *Journal of Experimental Analysis of Behavior*, **12,** 917–32.

Robbins, T. W. (1975). The potentiation of conditioned reinforcement by psychomotor stimulant drugs: a test of Hill's hypothesis. *Psychopharmacology*, **45,** 103–14.

Robbins, T. W. (1976). Relationship between reward-enhancing and stereotypical effects of psychomotor stimulant drugs. *Nature*, **264,** 57–9.

Robbins, T. W. (1978). The acquisition of responding with conditioned reinforcement: effects of pipradol, methylphenidate, *d*-amphetamine and nomifensine. *Psychopharmacology*, **58,** 79–87.

Robbins, T. W. & Koob, G. F. (1978). Pipradrol enhances reinforcing properties of stimuli paired with brain stimulation. *Pharmacology, Biochemistry & Behavior*, **8,** 219–22.

Robbins, T. W. & Koob, G. F. (1980). Selective disruption of displacement behaviour by lesions of the mesolimbic dopamine system. *Nature*, **285,** 409–12.

Roberts, D. C. S. & Fibiger, H. C. (1977). Evidence for interactions between central noradrenergic neurons and adrenal hormones in learning and memory. *Pharmacology, Biochemistry & Behavior*, **7,** 191–4.

Roberts, D. C. S., Price, M. T. C. & Fibiger, H. C. (1976). The dorsal tegmental noradrenergic projection: analysis of its role in maze learning. *Journal of Comparative and Physiological Psychology*, **90,** 363–72.

Scheel-Kruger, J. (1971). Comparative studies of various amphetamine analogues demonstrating different interactions with the metabolism of catecholamines in the brain. *European Journal of Pharmacology*, **14,** 47–59.

Schwaub, R. S. & Zeiper, I. (1965). Effects of mood, motivation, stress and alertness on the performance in Parkinson's disease. *Psychiatria et neurologia*, **150,** 345–57.

Sechzer, J. A., Ervin, G. N. & Smith, G. P. (1973). Loss of visual placing in rats after lateral hypothalamic microinjections of 6-hydroxydopamine. *Experimental Neurology*, **41,** 723–37.

Sessions, G. R., Kant, G. J. & Koob, G. F. (1976) Locus coeruleus lesions and learning in the rat. *Physiology and Behavior*, **17,** 853–9.

Stinus, L., Gaffori, O., Simon, H. & Le Moal, M. (1978). Disappearance of hoarding and disorganization of eating behavior after ventral mesencephalic tegmentum lesions in rats. *Journal of Comparative and Physiological Psychology*, **92,** 289–96.

Taylor, K. M. & Laverty, R. (1969). The effect of chlordiazepoxide, diazepam and nitrazepam on catecholamine metabolism in regions of the rat brain. *European Journal of Pharmacology*, **8,** 296–301.

Taylor, K. M. & Laverty, R. (1973). The interaction of chlordiazepoxide, diazepam and nitrazepam with catecholamine and histamine in regions of the rat brain. In *The benzodiazepenes*, ed. S. Garratini, E. Mussini & L. O. Randall, pp. 191–202. Raven Press: New York.

Terrace, H. S. (1971). Escape from S –. *Learning and Motivation*, **2,** 148–63.

Thornton, E. W., Goudie, A. J. & Bithell, V. (1975). The effects of neonatal 6-hydroxydopamine induced sympathectomy on response inhibition in extinction. *Life Sciences*, **17,** 363–8.

Torda, C. (1976). Effects of catecholamines on behavior. *Journal of Neuroscience Research*, **2,** 193–202.

Tremmel, F., Morris, M. D. & Gebhart, G. F. (1977). The effect of forebrain norepinephrine depletion on two measures of response suppression. *Brain Research*, **126,** 185–8.

Uguru-Okorie, D. C. & Arbuthnott, G. W. (1981). Altered paw preference after unilateral 6-hydroxydopamine injections into lateral hypothalamus. *Neuropsychologia*, **19,** 463–8.

Uhl, C. N. (1967). Persistence in punishment and extinction testing as a function of percentages of punishment and reward in training. *Psychonomic Science*, **8,** 193–4.

Wagner, A. R. (1969). Frustrative non reward: a variety of punishment. In *Punishment and aversive behavior*, ed. B. A. Campbell & R. M. Church, pp. 157–81. Appleton-Century-Crofts: New York.

Wike, E. L. (1966) *Secondary reinforcement*. Harper & Row: New York.

Wong, P. T. P. (1971). Coerced approach to shock, punishment of competing responses and resistance to extinction in the rat. *Journal of Comparative and Physiological Psychology*, **76,** 275–81.

7

Catecholamines and vegetative behaviour

Introduction

Various homeostatic behaviours such as eating and drinking, sleep, and temperature and blood pressure regulation have been suggested to involve catecholaminergic systems. I shall not deal with temperature regulation or the control of blood pressure or cerebral blood flow here, as they belong more appropriately in a treatise on the physiological, rather than the behavioural functions of catecholamines. I shall examine here, in addition to consummatory behaviour and sleep, two other categories of behaviour often associated with vegetative processes, namely sexual behaviour and aggression.

This chapter will be divided into sections concerned with each individual vegetative behaviour and within each section the role of noradrenaline and then of dopamine will be considered.

Catecholamines and sleep

Noradrenaline

Release of noradrenaline. The early evidence implicating noradrenaline in sleep mechanisms came in part from the measurement of the metabolism of noradrenaline during normal sleep, or during the absence of sleep (especially deprivation of rapid eye movement (REM) sleep) which can be caused in experimental animals.

Hartmann & Schildkraut (1973) found that the levels of a metabolite of noradrenaline, 3-methoxy-4-hydroxyphenylglycol (MHPG), in the urine of patients showed a strong negative correlation with the amount of REM sleep that the subjects demonstrated each night. These patients were hospitalised for depression and once a week electroencephalic recordings were taken throughout the night in order to determine the proportion of time spent in REM sleep compared to slow-wave sleep (SWS). Twenty-four hour urine samples were also collected and assayed for MHPG. Those patients who spent a high percentage of their sleep time in REM sleep had the lowest levels of MHPG in their urine (Pearson correlation coefficient $= -0.70$, $P < 0.01$). No correlation was seen between the amount of SWS and MHPG levels in the urine. This would suggest a noradrenergic role in REM sleep (but not SWS) with noradrenergic systems being least active during REM sleep. It will be seen that more recent research has supported this conclusion, but not without considerable detours and red herrings being encountered on the way.

A technique much used to increase the amount of time spent in REM sleep is to deprive rats of REM sleep for a period of days. This is achieved by placing them on small platforms surrounded by a pool of water. These 'islands' are so small that when muscle tension is lost in the limbs during REM sleep they fail to support the animal and so it tumbles into the water and awakens. SWS can still occur (albeit somewhat reduced) as muscle tension is not completely abolished during this stage of sleep. Virtually total deprivation of REM sleep can be achieved by this technique. Pujol, Mouret, Jouvet & Glowinski (1968) deprived rats of REM sleep for 91 h in this way and then injected [^{3}H]noradrenaline into the ventricles. The animals were allowed to sleep in a normal cage for the next 5 h. During this period following REM deprivation, a marked rebound in the amount of time spent in REM sleep is seen. This corresponded to an increased turnover in noradrenaline as shown by greater decrease in the brain content of radioactivity in animals undergoing REM rebound as compared to undeprived controls receiving [^{3}H]noradrenaline injection. This seemed to give exactly the opposite answer to the first experiment described (Hartmann & Schildkraut, 1973). Now, it appeared that noradrenaline release increased during REM sleep.

The resolution of these opposing results was suggested by Mark, Heiner, Mandel & Godin (1969). They pointed out that the procedure used to deprive the animal of REM sleep was extremely stressful. Indeed, within 4 days of REM deprivation, a marked increase in the weight of the adrenal glands was seen, a sure sign of stress. They also showed that animals kept, not on small islands, but on 'flower pots' (somewhat larger) showed much less REM deprivation (46 % of time in sleep was in REM compared to 0 % in the 'island' group) but still showed the same degree of reduction of [^{3}H]noradrenaline content in the brain when administered intraventricularly after 5 or 6 days of deprivation. Thus, despite much less REM deprivation, noradrenaline metabolism showed the same change, suggesting that it was actually reflecting some other factor, such as the stress of the situation. It is known that many different forms of stress will increase noradrenaline metabolism quite readily (electric footshock, Korf, Aghajanian & Roth, 1973; cold stress, Zigmond, Schon & Iversen, 1974; tail-pinch, Antelman, Szechtman, Chin & Fisher, 1975*b*).

This conclusion was supported by Stern, Miller, Cox & Maickel (1971), who used a stress control group which was immersed in cold water for 1 h per day for each of the 5 days that other rats were kept on REM deprivation 'islands'. These stress control rats were, however, allowed to sleep normally and so did not suffer from REM deprivation, Nonetheless, this group showed a similar increase in noradrenaline turnover as did the REM deprived group, as judged by the increase in the noradrenaline levels in whole brain following pargyline administration. Further, Schildkraut & Hartmann (1972) found that the increase in noradrenaline metabolism was largest during the time that the rats were on the 'islands' (and so not experiencing REM sleep at all) rather than in the period immediately after removal from the 'islands' (when the REM rebound occurs). Thus, the increase in noradrenaline metabolism did not correlate with the period when REM sleep was happening.

In conclusion, it would appear that the increase in noradrenaline activity caused by the 'island' procedure results from the stress engendered by the situation, rather than the rebound in REM sleep which subsequently occurs. Thus, no evidence actually indicates that noradrenaline activity *increases* during REM. This is further supported by the results of chronic single unit recording from locus coeruleus (LC) cells during stages of the sleep–wake cycle.

Single unit recording. The first study to look at the electrical activity of LC cells during the sleep–wake cycle was reported by Chu & Bloom (1973). They found that LC cells in cats chronically implanted with epoxy-coated steel microelectrodes showed about 4 impulses s^{-1} during SWS, rising to 6–12 impulses s^{-1} during wakefulness. During REM sleep these cells tended to fire during the rapid eye movements themselves or during the ponto-geniculate-occipital (PGO) waves seen in the electroencephalogram (EEG). Thus, during REM sleep these cells had an average activity of 10 impulses s^{-1}. This would appear to suggest, yet again, that noradrenergic systems become more active during REM sleep. However, later work had reversed this conclusion.

In the cat, not all cells in the 'locus coeruleus' are noradrenergic. Nonaminergic cells are found intermixed with noradrenergic ones throughout this nucleus and especially towards its medial edge. Chu & Bloom (1974) report a more detailed mapping of cells in the cat LC and find that those which increase their firing during REM sleep lie on the medial part of the LC. Cells more lateral and dorsal may be found which decrease their activity during REM sleep (Chu & Bloom, 1974, Fig. 8, p. 538). The authors used the passage of a small current after completion of recording to produce a small lesion which, upon post-mortem histology, would reveal the position of the cells from which they had been recording. Some noradrenergic and some nonnoradrenergic cells would often be found in this vicinity, especially for medial placements. Since the current used to mark the tip of the recording electrode destroyed the exact cells from which electrical events were being picked up it was impossible to say whether they were noradrenergic or used some other transmitter. Since many cells which increased their firing during REM sleep were found to be located distinctly outside the boundaries of the LC (see Fig. 7.1) and which, therefore, were clearly not noradrenaline containing, it would seem most likely that those medial cells within the LC which also increased their firing during REM sleep were also nonnoradrenergic. They would represent a lateral spread of

Fig. 7.1. Single units recorded in the pons of the cat brain. Cells shown by solid squares increased their firing from slow-wave sleep to rapid eye movement sleep, open squares did not. V4, fourth ventricle; Mes V, mesencephalic trigeminal nucleus; BC, brachium conjunctivum; LC, locus coeruleus; TV, ventral tegmental nucleus; MLB, medial longitudinal fasciculus; NR, raphe nucleus. (Redrawn from Chu & Bloom, 1974.)

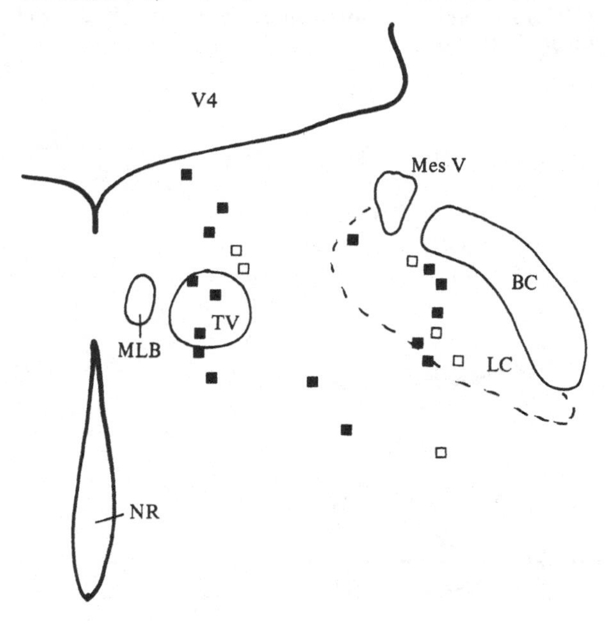

the nonnoradrenergic cells of the dorsal and ventral tegmental nuclei of Gudden into the supposedly noradrenergic LC.

This is supported by the results of Hobson, McCarley & Wyzinski (1975), who found that cells in the posterolateral part of the cat LC mainly (13 out of 21) decreased their firing during REM sleep (see Fig. 7.2), while those of the nonnoradrenergic gigantocellular tegmental field close to the LC increased their firing. The clearest evidence comes for chronic single unit recordings from the *rat* LC. Here, there are no nonnoradrenergic cells within the confines of the nucleus and the use of injection of dye stuff, rather than microlesioning, permits better definition of which cells were being recorded. Jones, Foote, Segal & Bloom (1978) found that almost all cells in the rat LC decrease their firing during sleep and are most active during wakefulness. Aston-Jones & Bloom (1981) found that changes in LC activity may precede shifts in the stages of sleep by some 500–1000 ms, and so may be of causative significance. A similar picture emerges from the squirrel monkey (*Saimiri speciosus*) (Foote, Bloom & Schwartz, 1978), which again has only noradrenergic cells in its LC (but compare the stumptailed macaque monkey (*Macaca arctoides*) which has some serotonergic cells here as well; Sladek & Walker, 1977).

In summary, release/turnover and single unit recording studies have now

Fig. 7.2. Firing rates of cells in the locus coeruleus (LC) and gigantocellular tegmental field (FTG) during waking (W), slow-wave sleep (SWS) and rapid eye movement sleep (REM). (Redrawn from Hobson *et al.*, 1975.)

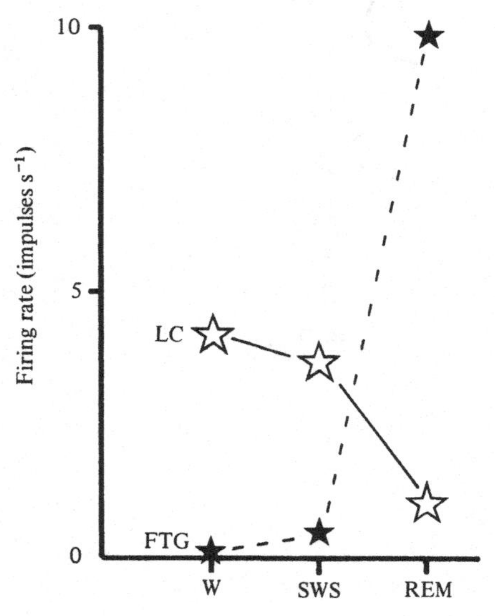

reached a consensus that noradrenergic cells in the LC *decrease* their firing during REM sleep. Findings suggesting an opposite effect must now be acknowledged to be due either to the stress of REM deprivation procedures *or* to recording from nonnoradrenergic cells in species with 'mixed' LCs.

Noradrenergic agonists. Drugs which stimulate noradrenergic receptors, either directly or indirectly, have been reported to alter the states of waking and sleep. The first agonist which I will consider will be noradrenaline itself.

Chambers & Roberts (1968) found that intraventricular injection of noradrenaline would cause sedation and motor akinesia. Such effects were not found after peripheral administration, arguing for a role of central noradrenergic systems in arousal. EEG synchronisation, to give activity similar to that seen in SWS, was also reported (Hansen & Whishaw, 1973).

Clarenbach & Cramer (1972) used peripheral administration of noradrenaline (3.75 μmol) in the newborn chick, where the blood–brain barrier has not yet developed (Spooner & Winters, 1966), and found REM 30 to 60 min after injection. Peripheral administration of the nonphysiological noradrenaline precursor dihydroxyphenylserine (DOPS) in the rat also results in an increase in REM sleep (Havlicek, 1967).

Torda (1968) found that injection of noradrenaline into either the hypothalamic preoptic area or the mesencephalic reticular activating system would make a transition from SWS or REM sleep into wakefulness more rapid. This may be because of a localised effect on one of two opposing noradrenergic systems or a possible activation by noradrenaline of dopamine receptors.

Jones (1972) found that injection of the noradrenaline uptake inhibitor desipramine (5 mg kg^{-1}) in cats caused behavioural and EEG signs of SWS very soon after administration together with decreased waking. This manipulation might be expected to increase the availability of noradrenaline at the synapse by preventing reuptake inactivation. The monoamine oxidase inhibitor (MAOI) drug pargyline also led to a decrease in waking with increased SWS. Inhibition of MAO might also be expected to increase the availability of noradrenaline in the synapse by preventing its enzymatic breakdown. However, two other MAOIs tranylcypromine and phenipra-zine led to opposite effects with an increase in arousal and waking. All of the above drugs caused a suppression of REM sleep.

The α_2-agonist clonidine causes a profound sedation when injected intraperitoneally (Laverty & Taylor, 1969).

An increase in REM sleep was seen by Enslen, Milon & Wurzner (1980) in the offspring of rats treated with caffeine during gestation. An increase in dopamine but not noradrenaline content of the LC was also seen. Caffeine is an inhibitor of phosphodiesterase and may act to increase intracellular concentrations of the second messenger involved in catecholaminergic neurotransmission.

In summary, injection of noradrenaline or noradrenergic agonists appear to decrease waking by causing a state of sedation and SWS. It is more uncertain what effects these manipulations have on REM sleep, since some experiments find an increase in REM sleep after noradrenaline injection in chicks (Clarenbach & Cramer, 1972) or DOPS injections in rats (Havlicek, 1967); whereas others find manipulations such as uptake inhibition or MAO inhibition, which might be expected to increase noradrenaline availability, actually decrease REM sleep (Jones, 1972). Fuxe *et al.* (1974) report that the α_2-antagonist piperoxan increased waking time by more than 30 % at a dose of 5 mg kg^{-1}. No change in REM sleep was seen. Since piperoxan increases the firing of LC by blocking a tonically active α_2 inhibitory input (Cedarbaum & Aghajanian, 1976), this may well be producing its effect indirectly by actually *increasing* the release of noradrenaline in the forebrain.

Synthesis inhibition and LC lesion. Synthesis inhibition of both dopamine and noradrenaline by α-methyl-*para*-tyrosine (AMPT) has been found to decrease both waking and REM sleep time, with the animal showing prolonged SWS periods (Torda, 1968). Three other studies have, however, found AMPT to *increase* the duration of REM sleep (Hartmann, Bridwell & Schildkraut, 1971*a*; King & Jewett, 1971; Stern & Morgane, 1973). Either way, the use of AMPT does not allow us to dissociate the contribution of noradrenaline from that of dopamine. Inhibition of noradrenaline biosynthesis alone by disulfiram also induces a state of sedation (Roll, 1970). Disulfiram has also been reported to reduce the number of rapid eye movements during otherwise normal REM sleep periods (Harner & Dorman, 1970). The problems of neurochemical specificity of disulfiram and the related diethyldithiocarbamate (DDC) have been discussed elsewhere (see Chapter 5, p. 205). Attempts have also been made to cause a selective loss of brain noradrenaline by electrolytic lesion to the LC.

Jouvet (1969) reported that LC lesion in the cat very severely disrupted REM sleep without any change in SWS. This appeared to correlate with the degree of LC damage and hence with the amount of brain noradrenaline lost (see Table 7.1). Although lesions placed laterally, medially, rostrally or caudally to the LC did not alter REM sleep, those placed ventrally to the LC in the vicinity of the nucleus reticularis pontis caudalis were also effective in suppressing REM sleep (Jouvet, 1969). Given the presence of many nonnoradrenergic cells in the feline LC, together with the effectiveness of more ventral lesions clearly outside the LC proper, caution must be expressed right from the start about the noradrenergic nature of this effect.

Jones, Bobillier, Pin & Jouvet (1973) found that electrolytic lesion to the anterior pole of LC and/or the caudal part of the ascending dorsal noradrenergic bundle would reduce the duration of EEG waking from about 50 % of total recording time to less than 27 %. However, in apparent contradiction to the results of Jouvet (1969), an increase in REM sleep was seen. The lesions resulted in only a 55 % decrease in the noradrenaline content of the forebrain and so may not have been extensive enough to alter REM sleep.

Jalowiec *et al.* (1973) found that electrolytic lesion in the dorsal tegmentum which depleted forebrain noradrenaline to between 30 and 50 % of normal had no effect on waking, SWS or REM sleep. Again, the poor depletion obtained by these lesions makes interpretation uncertain.

Perhaps the most convincing of the negative studies which fail to find an effect of LC lesion on REM sleep was reported by Jones, Harper & Halaris (1977). Their electrolytic lesion resulted in a depletion of forebrain noradrenaline by better than 85 %. Nonetheless, EEG waking was normal by 12 h post-operative and respresented the same amount of recording time as before the lesion was inflicted. Furthermore, REM sleep was seen within 48 h after lesion and in total time represented the same amount as recorded prior to lesion. However, muscle atonia usually seen during REM sleep *was* permanently abolished and PGO spiking activity was reduced by over 50 %. Similar loss of muscle atonia after large LC lesions has also been seen by

Table 7.1. *Effects of locus coeruleus (LC) lesions on sleep in cats*

Group	% of recording time		% LC destroyed	% brain NA
	in SWS	in REM		
Controls	43±2	9±0	0	96.5±8
Caudal to LC	46±4	8.5±2	11±6	78±4
Total LC	40±3	0.2±0.1	70±4.5	43±6
Lateral to LC	42±6	4.3±1	27±5	85±11

Values are means ± SD. NA = noradrenaline.
From Jouvet, 1969.

Braun & Pivik (1981) in the rabbit. Thus, although the LC may not be necessary for the occurrence of REM sleep it may be involved in certain features of that state such as PGO spiking activity and muscle atonia.

The greatest problem with most of the above studies utilising electrolytic lesions is that they have been carried out in the cat, whose LC is not a pure noradrenergic nucleus as it is in, say, the rat. Even the region of the LC proper has recently been shown to contain serotonergic cell bodies in the cat (Wiklund, Leger & Persson, 1981) which may account for up to 1 in 10 cell bodies in that nucleus (Leger, Wiklund, Descarries & Persson, 1979). As such, electrolytic lesions are most unreliable as evidence for a role of noradrenaline in sleep. More selective noradrenaline loss may be induced by the neurotoxin 6-hydroxydopamine (6-OHDA), to which I now turn.

6-OHDA lesions. A few 6-OHDA lesion studies have been carried out in the cat. Thus Zolovick *et al.* (1973) injected 6-OHDA directly into the vicinity of the dorsal noradrenergic bundle in the dorsolateral pontine tegmentum. This caused an increase in SWS at the expense of waking and REM sleep time. However, post-mortem amine assay revealed large losses of serotonin as well as depletion of noradrenaline. Thus, the amine of importance cannot be asçertained. Laguzzi, Petitjean, Pujol & Jouvet (1972) found a similar reduction in REM sleep after intraventricular injections of large doses of 6-OHDA in the cat (2.5 or 5 mg). A permanent reduction of PGO spiking to 50 % of its pre-operative frequency was also seen. However, amine assays also revealed a loss of serotonin (Petitjean *et al.*, 1972) and this seems to be a consistent action of the large doses of 6-OHDA used in the cat. It is not generally seen with 6-OHDA in the rat. As such, it renders 6-OHDA studies in the cat of marginal value in demonstrating a role specifically for noradrenaline in REM sleep.

6-Hydroxydopamine lesions in the rat, which do not affect serotonin, also change sleep patterns. Thus, Hartmann, Chung, Draskoczy & Schildkraut (1971*b*) report that administration of two doses of 200 μg and 300 μg respectively, of 6-OHDA (which depleted brain noradrenaline to 27 % of control and would also be expected to affect brain dopamine) lead to a permanent *increase* in REM sleep. Both the number of REM episodes and the duration of each was increased by this treatment.

However, such results were not confirmed by Matsuyama, Coindet & Mouret (1973), who found a transient decrease in REM sleep after intracisternal injection of 500 μg of 6-OHDA into the rat. Despite a permanent decrease in brain amines the decrease in REM sleep recovered to normal within a mere 8 days.

In newborn chicks (Clarenbach & Cramer, 1972) injection of 750 μmol of

6-OHDA led to a drop in the percentage of time spent in REM from 9.3 % in controls to 7 % in 6-OHDA-treated chicks ($P < 0.05$). This also resulted in an apparent supersentivity in that noradrenaline injected peripherally caused a larger increase in REM time (19.9 %) in lesioned chicks than it did in intact controls (15 %).

Hansen & Whishaw (1973) report a decrease in EEG waking and an increase in SWS following intraventricular injection of 400 μg 6-OHDA. However, the animals were behaviourally awake all the while that the cortex was showing SWS EEG patterns.

Up to now the 6-OHDA techniques used have either resulted in depletion of serotonin as well as noradrenaline (in the cat) or have inevitably depleted dopamine as well (in the rat with intraventricular or intracisternal administration). A number of studies have avoided these problems and achieved loss of only noradrenaline.

Panksepp *et al.* (1973) injected 8 μg 6-OHDA into the trajectory of fibres in the ventral noradrenergic bundle and found a small increase in both REM sleep time and SWS. Considerable loss of noradrenaline was found in cortical regions, so the behavioural change cannot be ascribed solely to ventral noradrenaline damage, but may also be due to lesion of the dorsal noradrenergic bundle. Lidbrink & Fuxe (1973) injected 6-OHDA into the dorsal noradrenergic bundle and found a reduction in EEG waking for 8 days after the lesion, with a corresponding increase in SWS, especially stage I SWS. No significant change in REM sleep duration was seen despite better than an 80 % depletion of forebrain noradrenaline. One paradox of this study (Lidbrink, 1974) was that no change was seen in behavioural waking despite the animals' EEG showing slow-wave activity. Further, by one month after the lesion the EEG waking time had returned to normal despite the permanent loss of noradrenaline. It would thus appear that normal waking (both behavioural and EEG) can be maintained in the continued absence of forebrain noradrenaline.

In summary, the results from 6-OHDA lesion have been conflicting and generally negative. Some studies have found an increase in REM sleep while others have reported a decrease. This may reflect damage to serotonergic or dopaminergic systems since pure noradrenaline lesion of the dorsal bundle has found no change in REM sleep. Transient decreases in waking seem to be fairly common in these studies and, since they are also found with dorsal bundle lesion, may be caused by noradrenaline loss. However, behavioural waking, as distinct from the desynchronised EEG pattern, is not affected and even EEG waking returns to normal within one month. Thus, only a slight role of noradrenaline in waking is indicated and no support for a role of noradrenaline in REM sleep has been forthcoming from 6-OHDA lesion studies. I shall now turn to a possible involvement of dopamine in sleep.

Dopamine

Release/levels. Baekelund, Schenker, Schenker & Lasky (1969) found that the amount of dopamine in the overnight and 1 h post-sleep urine of 20 normal male subjects showed a positive correlation with the amount of time spent in EEG waking and a marked negative correlation with the time in SWS. No relation to REM sleep was observed. Since similar correlations were seen with urinary levels of noradrenaline and adrenaline, the significance of these results is uncertain. This is so especially since it is still unclear how much, if any, of the amines measured in urine actually comes from central, as opposed to peripheral, noradrenergic nerve activity.

More direct measurement of central dopamine function was achieved in cats by Kovacevic & Radulovacki (1976). They sacrificed animals immediately after a period of either wakefulness, SWS or REM sleep and found that the dopamine concentration in the hippocampus increased by 180 % during SWS. Both the levels of dopamine and those of its metabolite, homovanillic acid (HVA) fell markedly in the striatum and thalamus during SWS. Of interest was the fact that no great changes in noradrenaline levels were seen in these brain structures during SWS. REM sleep was not, however, examined in these experiments. A possible confound with these changes in dopamine levels and the attribution of them to sleep processes *per se* is, of course, that locomotor activity and movement drops to nil during SWS compared to wakefulness. Given the well-documented involvement of dopamine in motor behaviour (see Chaper 4), it is premature to regard these data as evidence for a role of dopamine in sleep.

A possible way around this problem was reported by Ghosh, Hrdina & Ling (1976) by the use of the REM sleep deprivation technique. REM sleep deprivation in rats for 4 days by the island technique elevated striatal dopamine levels by 73 % and REM sleep deprivation for 10 days increased them to 133 %. No stress control groups were examined so, considering the limitations of the REM sleep deprivation technique mentioned above (see p. 317), it cannot be concluded that dopamine is involved in REM sleep as opposed to the stress which accompanies such REM sleep deprivation.

In summary, the studies of dopamine levels during sleep stages, or after deprivation, has yielded suggestive but by no means conclusive evidence of an involvement of dopamine in sleep. More detailed investigations with better behavioural controls are urgently needed. Of the evidence already available it is that from the effects of dopaminergic agonists which is the more persuasive.

Dopamine agonists. Cianchetti, Masala, Mangoni & Gessa (1980) found that intravenous infusion of the dopamine receptor stimulant apomorphine into human volunteers led to a reduction in both deep SWS and REM sleep. The lighter stages of SWS came to predominate during the infusion. This effect of apomorphine could be blocked by prior administration of the dopamine receptor antagonist drugs haloperidol or sulpiride. A similar effect of a large dose of apomorphine in rats was reported by Kafi & Gaillard (1976). They found that 4.3 mg kg^{-1} apomorphine increased waking time at the expense of SWS and REM sleep. However, a lower dose of apomorphine (0.86 mg kg^{-1}) had a tendency to increase REM sleep, although this failed to reach significance. This is of importance in the light of reports from Mereu *et al.* (1979) that very low doses of apomorphine (0.1 mg kg^{-1}) would increase REM sleep in the rat.

In line with this is the report of Hall & Keene (1975) that infusion of dopamine into the caudate of the cat would induce behavioural and EEG sleep and lead to spindling activity in the EEG typically seen during SWS. Electrical stimulation of the caudate had the same effect.

In summary, it would appear that a low dose of apomorphine may induce sleep and increase REM time. This may be acting through the caudate but more detailed mapping studies would be required to make it more certain. Higher doses of apomorphine, however, seem to reduce SWS and abolish REM sleep with a consequent enhancement of waking. Thus, a biphasic response to apomorphine is indicated.

Dopamine antagonists. Sagales & Erill (1975) found that the dopamine receptor blocker pimozide produced virtually no effects on sleep patterns in six human volunteers. Thus, no changes in waking, SWS or REM sleep duration were seen, despite quite large doses of the drug being used. Further, no changes in sleep states has been seen in Parkinsonian patients, in whom a deficit of dopamine transmission in the nigrostriatal pathway is well established (Stern, Roffwarg & Duvoisin, 1968; Kales *et al.*, 1971). Further, haloperidol and sulpiride were without effect on sleep patterns of human volunteers in the study of Cianchetti *et al.* (1980) described above, and Hartmann, Zwilling & Koski (1973) found pimozide to be without effect on the sleep stages in the rat.

Against this, however, must be set the report of Kafi & Gaillard (1976) that spiroperidol decreases total sleep and REM time, while chlorpromazine enhanced REM sleep. Although both agents block dopamine receptors it must be noted that spiroperidol may also interact with serotonin receptors in the frontal cortex and chlorpromazine has distinct noradrenergic effects. Thus, the nondopaminergic side effects of these drugs may account for their effect on sleep.

In summary, most pure antagonists of dopamine receptors fail to produce any alteration in any component of sleep in either animals or man. No evidence of an involvement of dopamine in sleep can thus be gleaned from the effects of antagonists.

Summary

(1) Release of noradrenaline during sleep and single unit recording from LC neurones agree that noradrenergic systems are most active during waking, become less active during SWS and virtually turn off during REM sleep. This need not, of course, indicate any causative action in sleep states.

(2) Noradrenergic agonists and noradrenaline itself generally induce a state of sedation and SWS. Their effect on REM sleep remains unclear. The noradrenergic antagonist piperoxan increases waking, perhaps by blocking a tonically inhibitory α_2 input to the LC and thus increasing its firing.

(3) Neither synthesis inhibition nor electrolytic LC lesion has yielded convincing evidence for a role of noradrenaline in REM sleep: the first due to the neurochemical impurity of the effects of disulfiram and diethyldithiocarbamate, the latter due to the presence of nonnoradrenergic cells in the LC of the species (the cat) on which most of the initial noradrenaline and sleep studies have been carried out.

(4) 6-OHDA lesions, if carried out in the rat so as not to affect serotonin, do not greatly alter REM sleep although a transient decrease in EEG waking is seen. This, however, is not accompanied by any effect on behavioural waking and even the EEG recovers within one month.

(5) Some changes in dopamine levels/metabolites have been seen during sleep but confounds such as reduced motor activity render their interpretation ambiguous.

(6) Dopamine agonists show a biphasic response, with low doses increasing SWS and REM sleep while higher doses may enhance wakefulness.

(7) No effect of a number of dopamine blockers has been found on any stage of sleep. Those which do alter sleep may do so via their nondopaminergic side-effects.

(8) In conclusion, dopamine may not be tonically involved in the causation of sleep, although induction of additional activity in this system by exogenous stimulation may modulate sleep patterns. Noradrenaline, despite early work in the cat, does not seem to be much involved in REM sleep but may have a role to play in the maintenance of cortical EEG arousal but not of behavioural waking.

Catecholamines and aggression

Noradrenaline

A number of lines of evidence combine to suggest a fairly prominent role for noradrenaline in aggressive behaviour. It must not, however, be assumed that aggression is a unitary behaviour. 'Aggression' lumps together such diverse paradigms as footshock-induced fighting with conspecifics, isolation-induced fighting with conspecifics, predatory behaviours such as mouse-killing (muricide) or frog-killing (ranicide) by rats, hyperreactivity to handling produced by septal and other brain lesions and defence of territory against an intruder. It is very likely that these very different behavioural situations invoke differing psychological processes and so might be expected to have different neurochemical substrates. With this caveat in mind, I will now examine the role of noradrenaline in these types of vegetative behaviours.

Release/levels. Goldberg & Salama (1969) tested rats to determine those which would spontaneously kill a mouse introduced into the same cage (killers) and those which would not (nonkillers). They then sacrificed the animals and measured the levels of both noradrenaline and serotonin in the forebrain and hindbrain and also examined the turnover of noradrenaline in terms of its disappearance rate after synthesis inhibition with AMPT. No changes in the levels of serotonin were found but a 25 % increase in noradrenaline levels in the forebrain of killer rats was detected. The turnover studies revealed a higher rate of synthesis of noradrenaline in the forebrain as well.

Modigh (1973) failed to find any change in the levels of noradrenaline in mice brain in animals made aggressive by 8 weeks of isolation housing. If, however, the animals were allowed to fight for 30 min, a marked decrease in brain noradrenaline was seen in all brain areas examined. This would suggest that the changes in noradrenaline metabolism may relate more to the stress involved in the actual fighting than to central processes involved in aggression in the absence of the opportunity to express it. It should be noted that, in the Goldberg & Salama (1969) experiment, 'killers' were assayed within 24 h of a previous muricide test and so the effects of that stress might have persisted.

This interpretation is supported by the finding of Henley, Moisset & Welch (1973) that the uptake of noradrenaline was markedly increased in synaptosomes prepared from mice which were either fighting spontaneously with cage-mates or had been rendered aggressive by isolation housing for 5 to 9 months and then allowed to fight for 10 to 15 min per day for 14 days. This was an effect of fighting rather than isolation *per se* since the control group used was of undisturbed isolated mice. Essentially the same results were obtained by Hadfield & Weber (1975): again fighting led to an increase in noradrenaline uptake as judged by the maximum velocity (V_{max}) of uptake which reflects the number of such uptake sites (V_{max} nonfighting controls = 66.7 nmol g^{-1} protein (5 min)$^{-1}$ compared to V_{max} fighting = 173 nmol g^{-1} protein (5 min)$^{-1}$).

This effect may be induced merely by the 'psychological' stress of witnessing fighting and need not involve actually fighting and its consequent increase in motor output. Thus, Welch & Welch (1968) found about a 30 % reduction in the noradrenaline content of the pons and medulla in mice placed in a basket above a cage containing two fighting mice compared to undisturbed controls. No physical contact or actual fighting could occur in the experimental group, merely the psychological stimuli associated with the witnessing.

Tizabi, Massari & Jacobowitz (1980) found that different brain regions may show different changes in noradrenaline metabolism after isolation housing. They report an increase of 43 % in the noradrenaline content of the septum, while that of the olfactory tubercle showed a 37 % decrease. This suggests that the story may be more complex than presented so far, with different noradrenergic brain regions having opposing roles.

Various factors may affect aggressive behaviour. Bernard, Finkelstein & Everett (1975) found conspecific aggression to increase with age in two strains of mice (BALB and ICR). This appeared to be correlated with a decrease in the turnover of brain noradrenaline. However, similar decreases in the turnover of brain dopamine were also seen, so these may simply represent a general effect of age on brain neurotransmitter dynamics and need not underlie the altered aggression.

Electrolytic lesion of the septal area elicits a syndrome called 'septal rage' in which the animal becomes aggressive, difficult to handle and hyperreactive to touch and to attempts to grasp it. This is associated with a small increase in the noradrenaline content of the hindbrain (Salama & Goldberg, 1973) and a marked increase in the turnover of noradrenaline here. Again it is uncertain whether this change in noradrenaline dynamics is the causative factor underlying septal hyperreactivity or only an incidental epiphenomenon. Since whole brain noradrenaline concentrations are also elevated by electroconvulsive shock (ECS), which acts to decrease muricide behaviour (Vogel & Haubrich, 1973), it is clear that all types of aggressive behaviour do not share a common noradrenergic mechanism.

Finally, changes in amine metabolism associated with aggression have been seen in species other than rats and mice. Thus, Kostowski, Tarchalska-Krynska & Markowska (1975) found that ants (*Formica rufa*) displaying intraspecific aggression had lower levels of noradrenaline in the brain. Similar effects were seen following interspecific aggression against a beetle (*Geotrupes* sp.). However, increases in both serotonin and adrenaline were also seen, which brings into question the specificity of the amine involved. Further, the stress involved in the actual attack periods, rather than the central processes involved in aggression, might have been responsible for the changes in noradrenaline levels.

In summary, although noradrenaline has frequently been implicated in aggression on the basis of changed levels or turnover, it has generally been confounded by the stress element present in the test situation, which is known to activate noradrenergic systems (see p. 317). The question of whether the observed changes in noradrenergic systems in any way actually cause the aggressive behaviour has not, of course, been answered by these correlative studies and for this we must turn to noradrenergic drug and 6-OHDA lesion studies.

Noradrenergic drugs. A careful distinction of the precise type of 'aggression' must be made in these studies – whether it is intraspecific fighting after footshock or isolation, or predatory aggression, or the hyperreactivity that results from brain damage such as septal lesions.

Lawrence & Haynes (1970) found that intraperitoneal injection of noradrenaline in mice increased dominance in a social dominance situation while similar injection of adrenaline decreased dominance and elicited a submissive behavioural pattern. Since it is uncertain how much, if any, noradrenaline or adrenaline would get into the brain when injected peripherally, these results do not further our understanding of *central* noradrenergic systems and aggression but they do serve to demonstrate the well-known role of peripheral noradrenergic systems in aggression. This role must be borne in mind when dealing with any manipulation which would affect peripheral, as well as central noradrenergic systems.

Aggressive displays may also be observed in other species of animals. Thus, Siamese fighting fish (*Betta splendens*) if presented with a mirror or with a live conspecific will display aggressive behaviour for a lengthy period of time. Marrone, Pray & Bridges (1968) reported that introduction of noradrenaline into the water of the aquarium holding a single male fish would cause the spontaneous performance of an aggressive display. However, a detailed analysis by Baenninger (1968) in the more realistic situation involving either a live conspecific or a mirror to elicit display showed that noradrenaline suppressed aggressive behaviour. A similar effect was noted for adrenaline as well. Some noradrenaline may enter the brain in fish after peripheral administration so this experiment does not allow us to dissociate peripheral from central aminergic systems, nor effects of adrenaline from those of noradrenaline.

To overcome the confound of the peripheral noradrenergic system, Geyer & Segal (1974) injected noradrenaline intraventricularly into unrestrained rats via chronically implanted cannulae. They found that central noradrenaline infusion (as opposed to the peripheral route described above) reduced shock-induced fighting. In this paradigm, a pair of rats is subjected to inescapable footshock. In many cases this triggers the rats to adopt an antagonistic posture with rearing from the floor and facing each other (Ulrich & Azrin, 1962). Drug manipulations of one, or both, of these animals may alter the frequency with which this behaviour is observed following the delivery of footshock (see Fig. 7.3). Geyer & Segal (1974) found that infusions of saline were without effect on the incidence of shock-induced fighting while injections of dopamine increased the number of attacks. The latter effect was truly dopaminergic, and not the result of uptake of dopamine into noradrenergic neurones with consequent increase in noradrenaline synthesis, since it still occurred after noradrenaline uptake inhibition with imipramine (see Fig. 7.3). Similar findings were reported by Leaf (1970), who found that injections of noradrenaline into the amygdala via chronically implanted cannulae in rats caused a reduction in muricide. Also consistent with this is the report of Anand, Gupta & Bhargava (1977) that inhibition of noradrenaline synthesis with DDC increased shock-induced fighting while increase of the synaptic availability of noradrenaline by the uptake inhibitor imipramine reduced shock-induced fighting. The dopaminergic agonist apomorphine increased fighting, as did dopamine itself (Geyer & Segal 1974, see above). The precursor of both noradrenaline and dopamine, L-DOPA, had the net effect of increasing fighting, but its mixed neurochemical actions make this difficult to interpret.

The α_2-antagonist drug piperoxan, which among other actions increases the firing rate of LC cells (Cedarbaum & Aghajanian, 1976), has been found to inhibit fighting in isolated mice (Buus Larsen, 1978). This effect of piperoxan was blocked by the α_2-agonist clonidine (see Fig. 7.4). Paradoxically, clonidine on its own, which should decrease firing of LC cells on this model, also acted to decrease aggression. However, since clonidine also has a powerful sedating effect, this may not be a specific effect on aggression but rather a decrease in all types of behaviour. The α-blockers phenoxybenzamine and aceperone also reduce aggression in isolated mice (Buus Larsen, 1974). Since these agents will block both α_1- and α_2-receptors they may be working in a fashion similar to that of piperoxan to increase LC firing and hence the availability of noradrenaline to stimulate postsynaptic β-receptors.

Ranicide in rats was found to be reduced by injection of noradrenaline into the hypothalamus, as was muricide behaviour (Bandler, 1970). Compared to hypothalamic sites which were sensitive to cholinergic stimulation and which, in this case, increased predatory aggressive behaviours, only 4 out of 23 sites were also responsive to noradrenaline (*ibid.*, p. 415). Other sites in the unspecified 'ventral midbrain' were also found (Bandler, 1971) at which noradrenaline administration in crystalline form served to facilitate predatory aggression (ranicide and muricide). The

Fig. 7.3. (*a*) Attacks caused by intraventricular administration of noradrenaline (NA), dopamine (DA) or saline (Sal); (*b*) following pre-treatment with 5 mg kg^{-1} imipramine. (Redrawn from Geyer & Segal, 1974.)

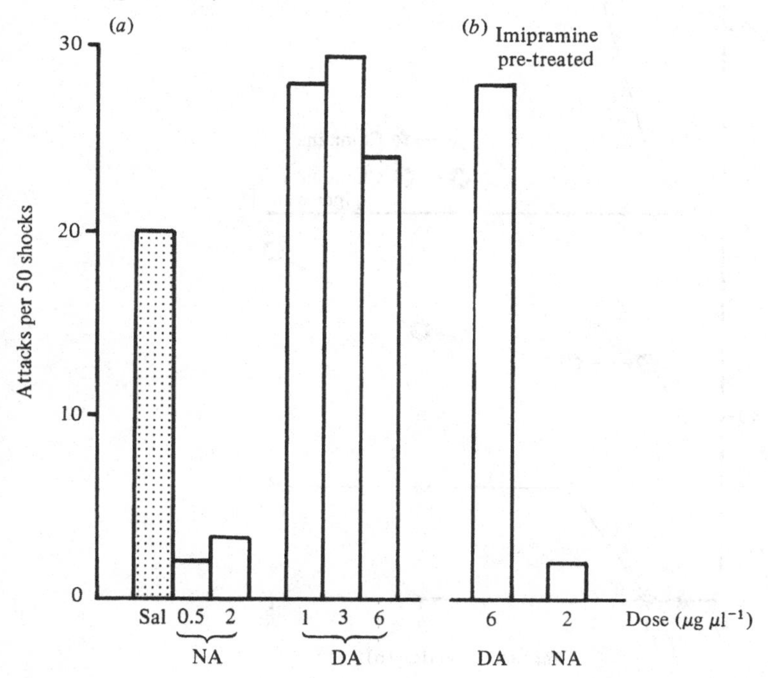

lack of either stereotaxic coordinates or histological reconstruction of the precise site of implantation (Bandler, 1971) renders this finding equivocal in the extreme. The verbal description of the sites as being in the ventral tegmental area and the interpeduncular nucleus raises the possibility of an interaction with the dopaminergic cells of A10. Negative sites were reported in adjacent areas such as the substantia nigra, red nucleus and medial lemniscus.

The receptor type involved in the noradrenergic system in the hypothalamus may be of an α-nature, since Albert & Richmond (1977) reported that injection of either of the α-antagonists tolazoline or phentolamine would cause increased reactivity to handling whereas the β-antagonist propranolol was without effect. Similar increases in both muricide and

Fig. 7.4. Percent of rats showing aggression after injection (subcutaneous) with clonidine or the combination of clonidine and piperoxan. (Redrawn from Buus Larsen, 1978.)

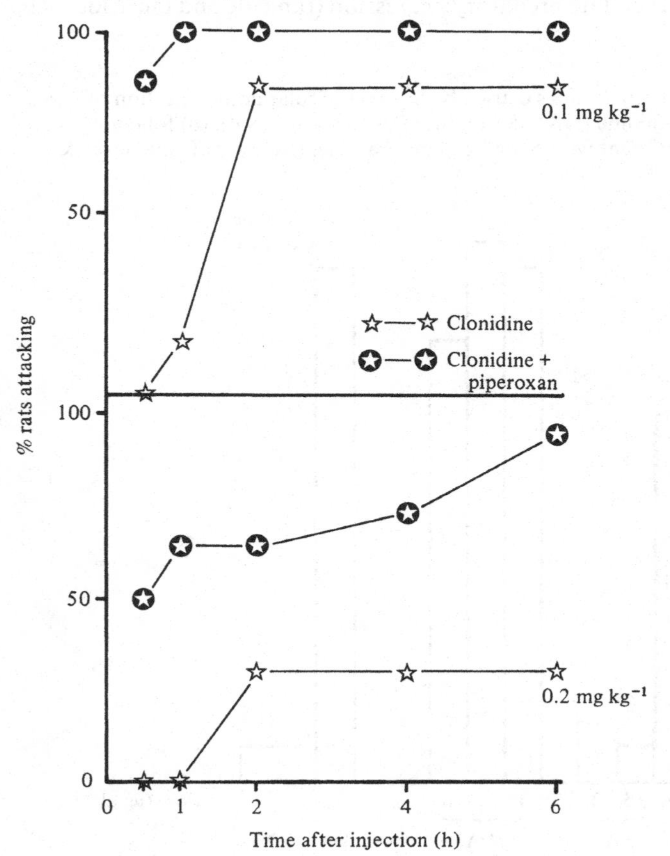

intraspecific aggression in response to tolazoline and phentolamine were also reported (Albert, Wong, Brayley & Fibiger, 1979). The dopamine receptor blocker haloperidol produced only a small effect, arguing for the role of noradrenaline in this effect. (Haloperidol has some weak noradrenergic blocking action as well as its more powerful dopamine receptor blockade.)

Some doubt was cast on the noradrenergic nature of this effect with the finding that 6-OHDA lesion to the dorsal and ventral bundles failed to alter reactivity by itself, and did not alter the ability of tolazoline injected into the medial hypothalamus to increase reactivity. Even injection of 6-OHDA directly into the medial hypothalamus itself failed to increase reactivity. These findings proved negative despite a reduction in hypothalamic noradrenaline to 50 % of normal in the latter case. This is a not very impressive depletion and may well explain the absence of behavioural effect. The further failure of the agonist clonidine to reverse the effects of tolazoline-induced α-blockade on increasing reactivity might be the result of clonidine being an α_2-agonist while the increase in reactivity might be because of α_1-blockade by the nonselective α_1- and α_2-antagonist tolazoline. Further, no dose–response curve for increasing doses of clonidine was run, so the negative conclusion is especially weak (a point noted by Albert *et al.* 1979, p. 8).

In summary, despite the inconsistencies raised in the study by Albert *et al.* (1979) most other work shows surprising unanimity that increasing noradrenaline function reduces and decreasing noradrenaline function enhances aggressive behaviour, whether this be the well-studied intraspecific shock- or isolation-induced fighting or the less well-examined killing behaviours. It should be remembered in the above that manipulation such as the α_2-antagonist piperoxan, which might superficially be thought to reduce noradrenaline function, might actually increase it overall because of its effect of increasing the firing of LC cells as a result of blockade of inhibitory 'autoreceptors'.

With this in mind, the data take on a remarkable degree of agreement that increases in noradrenaline functional activity decrease aggression and vice versa. This conclusion is extended and further supported by the consideration of the effects of 6-OHDA lesions of the noradrenergic systems.

6-OHDA. Early 6-OHDA lesions which often depleted dopamine and noradrenaline together soon showed a marked increase in aggressive behaviour in a number of situations. Thus, Stern, Hartmann, Draskoczy & Schildkraut (1972) found that intraventricular injection of 200 and then

300 μg 6-OHDA markedly increased shock-induced aggression in rats, especially at high shock intensities. A depletion of brain noradrenaline to 35 % of normal was shown by post-mortem assay but depletions of dopamine would also be expected as a result of this regime of 6-OHDA. Greater irritability was also anecdotally observed when handling these lesioned rats, as was reported initially by Evetts, Uretsky, Iversen & Iversen (1970). The increased irritability lasted for more than 4 months (Nakamura & Thoenen, 1972) and was inversely correlated with the levels of noradrenaline remaining in the brain after 6-OHDA lesion by the intraventricular route. Correlations were particularly good in the hypothalamus and less so in pons–medulla and rest of brain.

While Eichelman, Thoa & Ng (1972) confirmed the increase in shock-induced fighting after a single 200 μg dose of 6-OHDA they found no change in muricide. Post-mortem amine assays confirmed a 70 % depletion of brain dopamine as well as a 80 % loss of noradrenaline, thus rendering the amine of importance indeterminate, as indeed it was in all the above reports. This was generally supported by Banerjee (1974) who found that two injections of 200 μg 6-OHDA intraventricularly failed to reduce muricide in rats already showing this behaviour, nor did it induce muricide in rats classed as 'friendly' in their previous interactions with mice. However, in 4 out of 12 rats previously indifferent to mice, muricide did emerge after the lesion. This argues for a slight, albeit rather weak, inhibitory role of catecholamines in muricide. Again, the amine of importance was not determined.

Rather paradoxically, Crawley & Contrera (1976) found that a single intraventricular injection of 6-OHDA significantly reduced isolation-induced fighting in mice. This appeared to correlate with the lowering of brain noradrenaline (Spearman $\rho = 0.73$, $P < 0.025$). Since the animals tested were selected for greater fighting behaviour prior to lesion it may be that the only change which could have been seen would have to be a decrease. Nonetheless, this study is unusual in going against a clear trend of increased aggression following 6-OHDA lesions which depleted both dopamine and noradrenaline.

Forms of noxious stimulation other than electric footshock will also induce fighting in 6-OHDA-treated rats. Thus, Mine *et al.* (1981) report that tail-pinch will elicit more than an hour of nearly continuous fighting.

In an attempt to identify the amine of importance Thoa, Eichelman & Ng (1972) pre-treated rats with the noradrenaline uptake inhibitor desipramine prior to 6-OHDA administration. This prevented both the noradrenaline loss usually seen *and* the increase in shock-induced fighting.

Brain dopamine depletions occurred as before, thus implicating noradrenaline in particular. This is supported by the finding of Thoa, Eichelman, Richardson & Jacobowitz (1972) that 6-hydroxydopa administered intraventricularly depleted hypothalamic and cortical noradrenaline without affecting brain dopamine and led to a progressive increase in shock-induced fighting starting about 4 to 6 days post-operatively (see Fig. 7.5*a*). Increased irritability and resistance to handling was also seen. This effect was long lasting and returned to normal only some 70 days after lesions (Richardson & Jacobowitz, 1973). Even then some slight residual difference in emotionality may be seen (Fig. 7.5*b*).

The recovery from 6-OHDA-induced increases in aggression and reactivity seems to interact with environmental factors since Coscina, Goodman, Godse & Stancer (1975) found that daily handling for 6 days after intracisternal injection of 300 μg of 6-OHDA greatly reduced the increased resistance to capture caused by the neurotoxin. Despite this effect of handling, the depletions of brain noradrenaline remained similar to those in nonhandled 6-OHDA groups. Thus, repeated testing after the lesion may cause the increased irritability and resistance to capture to wane (Coscina, Seggie, Godse & Stancer, 1973, see Fig. 7.6).

More localised 6-OHDA lesions to determine which of the two noradrenergic systems is involved in aggression have also been carried out. Jimerson & Reis (1973) found that injection of 6-OHDA into the lateral hypothalamus reduced the frequency of ranicide in rats. However, this may well be because of the loss of dopamine resulting from these lesions rather than noradrenaline depletion. This is especially so since Oishi & Ueki

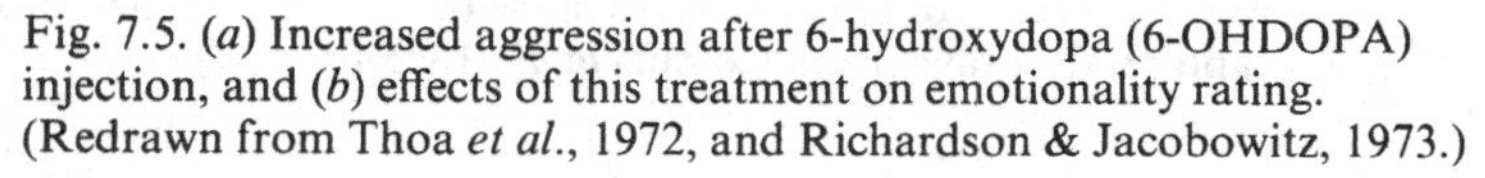
Fig. 7.5. (*a*) Increased aggression after 6-hydroxydopa (6-OHDOPA) injection, and (*b*) effects of this treatment on emotionality rating. (Redrawn from Thoa *et al.*, 1972, and Richardson & Jacobowitz, 1973.)

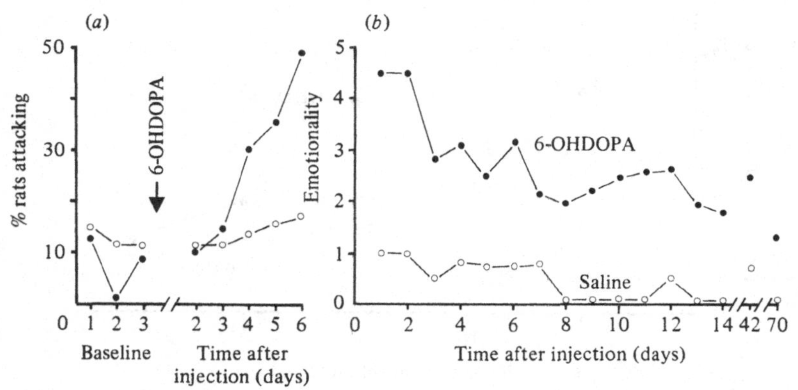

(1978) found that electrolytic lesion of the dorsal bundle, which did not affect brain dopamine but reduced cortical noradrenaline by 50 % with lesser effect on hypothalamic noradrenaline, greatly increased muricide in rats already possessing an ablation of the olfactory bulbs. Dorsal bundle lesion in normal rats did not alter muricide. However, this may simply reflect the absence of the behaviour prior to dorsal bundle lesion.

Crow *et al.* (1978) and File, Deakin, Longden & Crow (1979) injected 6-OHDA into the dorsal bundle and into the vicinity of the LC and then examined the animals in a social interaction situation. They found a marked increase in 'aggressive' behaviours such as 'boxing' and 'rolling'. This would further implicate the dorsal noradrenergic system arising from the LC in intraspecific aggression. They did not look at muricide or other predatory behaviours.

Finally, Kostowski *et al.* (1980) found that electrolytic lesion to the ventral bundle actually decreased shock-induced fighting in rats. Muricide was, however, not altered by these lesions. Since Kostowski *et al.* (1978) found that bilateral electrolytic lesion to the source nucleus of the dorsal bundle system, the LC, was effective in increasing shock-induced fighting in rats, it would appear to be the dorsal bundle system rather than the ventral bundle noradrenergic fibres which accounts for the uniformly reported enhancement of this behaviour after whole brain noradrenaline loss with 6-OHDA. These authors also found apomorphine-induced fighting or 'bizarre social behaviour' (see p. 340) to be enhanced after noradrenaline loss.

Divergent results may be obtained in animals housed in colonies, rather

Fig. 7.6. Effects of 6-hydroxydopamine (6-OHDA) on resistance to capture. (Redrawn from Coscina *et al.*, 1975.)

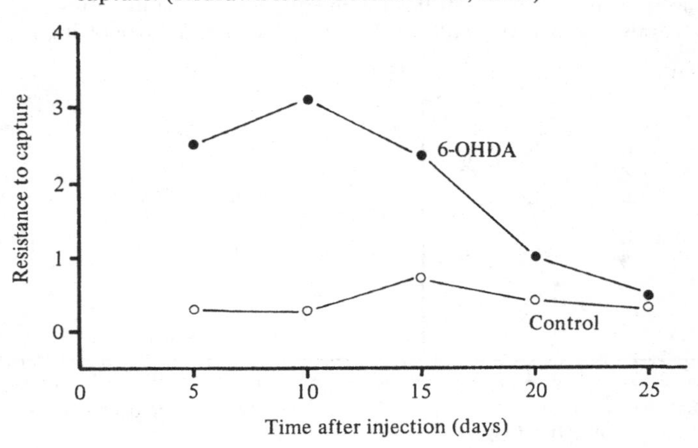

than the usual laboratory cages. Thus, Ellison (1976) reported that 6-OHDA lesion to the noradrenergic system caused aggressive and competitive measures to fall after treatment and the animals remained isolated in their burrows. Their social dominance position dropped also. This contrasts to the hyperirritability seen after similar treatment with 6-OHDA in cage-housed rats. Further, Eison, Stark & Ellison (1977) found that electrolytic lesion to the LC of colony-housed rats caused similar inactivity, with animals remaining isolated in their burrows. Further, these lesioned rats appeared to fight *less* than sham-operated controls in the colony environment.

In summary, virtually all the studies using 6-OHDA lesions report an increase in shock-induced fighting in rats. Depletion of just noradrenaline is effective in causing this change, as is either lesion to the LC or to the dorsal noradrenergic bundle. It would thus appear that the dorsal noradrenergic system has a role as inhibitor to shock-induced aggression in the rat. Electrolytic lesion to the ventral bundle may have an opposite effect although further investigations with more selective 6-OHDA lesions would be useful.

Predatory aggression seems to be distinctly different from shock-induced fighting. Thus, muricide is not altered by ventral bundle lesions which decrease shock-induced fighting. Further, ranicide is decreased, not increased, by injection of 6-OHDA into the hypothalamus. Muricide is not affected by dorsal bundle electrolytic lesion or intraventricular 6-OHDA which does increase shock-induced fighting. A systematic examination of dorsal and ventral bundle 6-OHDA lesions on shock-induced fighting, predatory aggression (both muricide and rancide) and also reactivity to handling and capture would be most profitable.

Dopamine

A smaller body of evidence suggests some role for dopamine in aggression. It comes from many and diverse sources and will be dealt with in a single section.

Modigh (1973) found an increase in HVA, a metabolite of dopamine, in the brain of mice made aggressive by isolation housing. Hutchins, Pearson & Sharman (1975) failed to confirm this but found, instead, an increase in another dopamine metabolite, 3,4-dihydroxyphenylacetic acid (DOPAC), in the striatum of isolated mice. This seemed to reflect the act of fighting *per se*, rather than any psychological processes involved in the induction of aggression, since it was only observed after actual fighting had occurred. Whether it was specific to aggressive behaviour was also questioned by the finding of a similar increase if isolated mice were simply placed into a novel environment without another mouse with which to engage in fighting. As such, it may be related to the well-known role of dopamine in motor behaviours (see Chapter 4).

Analogous to the change in noradrenaline uptake caused by fighting (see p. 330), Hadfield & Rigby (1976) found an increase in the uptake of dopamine by striatal synaptosomes from rats after only 30 s of shock-induced fighting. Again, this fails to distinguish between the motor activity of fighting as opposed to involvement of central processes in aggression *per se*.

Barr, Sharpless & Gibbons (1979) screened rats on intraspecific aggression and muricide for 'high' and 'low' fighters. Those which consistently attacked a conspecific showed higher levels of dopamine in the hypothalamus than those which were nonfighters. No difference was seen in dopamine levels in the amygdala or olfactory bulbs. Killer and nonkiller rats (muricide) showed identical dopamine levels in all brain areas. Similar results were found by Bernard *et al.* (1975) in mice made aggressive by isolation, but not by Da Vanzo, Daugherty, Ruckart & Kang (1966) nor by Valzelli (1974) for similarly isolated mice. The status of changes in dopamine levels in the brains of aggressive v. non-aggressive animals remains controversial.

The causative role of some of these changes in brain dopamine is called into question by Everett (1977), who found that as the male of the ICR strain of mouse aged a marked increase in aggressive behaviour occurred and this was accompanied by an increase in brain dopamine content. However, in females of this strain no increase in aggression was seen with age despite an equally marked increase in brain dopamine.

Baggio & Ferrari (1980) report that the potent dopamine agonist drug, *N*,*n*-propylnorapomorphine (NPA) increased shock-induced aggression in the rat, as did apomorphine itself. Predatory aggression (attack on live turtles, Bandler & Moyer, 1970) was decreased by NPA and by larger doses of apomorphine, while isolation-induced aggression was also reduced by NPA. These effects of NPA on isolation-induced aggression could be blocked by prior administration of haloperidol, testifying to the dopaminergic nature of the effect.

The neuroleptic drugs spiramide and trifluoperazine were also found to be effective in reducing amphetamine-induced aggression in mice (Hasselager, Rolinski & Randrup, 1972). Blockade of noradrenergic systems with the α-blockers phenoxybenzamine and aceperone was not effective. Apomorphine, a direct dopamine receptor stimulant, can induce a syndrome known as 'bizarre social behaviour', which consists of pairs of rats rearing against each other with aggressive and even fatal attacks

(McKenzie & Karpowicz, 1970). This seems superficially similar to the behaviour seen in shock-induced aggression.

There would thus seem to be some evidence for a role of dopamine in aggression. Again, aggression is not found to be a unitary behaviour, since activation of dopaminergic systems may increase shock-induced aggression while decreasing predatory and isolation-induced aggression. However, much additional research is required to elucidate the precise nature of this dopaminergic involvement.

Summary

(1) Increased synthesis and uptake of noradrenaline in the brain seem to relate mainly to the stress (both physical and 'psychological') of the fighting experience and no clear demonstration of a role in central processes of aggression independent of its expression have been reported.

(2) A consistent picture has emerged with drugs which affect noradrenergic systems. Manipulations which increase noradrenaline activity (central injection of noradrenaline itself, or the α_2-blocker piperoxan) seem to decrease shock-induced aggression and muricide. Drugs which decrease activity in noradrenergic systems (such as the postsynaptic α-blockers tolazoline and phentolamine) seem to increase intraspecific aggression, muricide and irritability to handling. However, β-blockers are without effect.

(3) 6-OHDA lesions are virtually uniform in demonstrating an increase in intraspecific aggression and resistance to capture after noradrenaline depletions. Changes in muricide, however, are less frequently seen. More localised lesion to specific noradrenergic systems suggests that it is the dorsal noradrenergic system originating in the LC which has the inhibitory effect on aggression, while the ventral bundle noradrenergic system may even have an opposing effect.

(4) Perhaps the best evidence for a role of dopamine in aggression comes from the effects that apomorphine and other dopamine agonists have an increasing shock-induced aggression. However, predatory aggression and isolation-induced fighting may be reduced indicating that 'aggression' is by no means a unitary concept.

Catecholamines and sexual behaviour

Noradrenaline

Introduction. It is now well established that there is a noradrenergic link in the control of secretion of pituitary hormones such as follicle stimulating hormone and luteinising hormone which are involved in sexual physiology. Intraventricular 6-OHDA has been found to alter serum levels of both follicle stimulating hormone and luteinising hormone (Kitchen, Ruf & Younglai, 1974) and to block the release of prolactin in response to stress (Fenske & Wuttke, 1976). These effects lie within the purlieu of physiology rather than behaviour, and as such will not be further explored. It is important, however, to remember that many of the effects of catecholaminergic manipulations on the behavioural side of sex may result ultimately from the changes in circulating titres of these hormones.

Female lordosis behaviour. Everitt, Fuxe, Hokfeldt & Jonsson (1975) found that the α_2-blocking agents piperoxan and yohimbine would significantly increase lordosis behaviour in ovariectomised female rats maintained on daily oestradiol injections. They increased the lordosis quotient (ratio of those mounts by the male which elicited a lordosis response to the total number of mounts). The duration or intensity of lordosis behaviour were not, however, altered (see Fig. 7.7). The noradrenaline synthesis inhibitor FLA 63 caused a small decrease in the lordosis quotient and intensity of lordosis behaviour (but only of females treated with progesterone). Since reduction of noradrenaline activity by synthesis inhibition reduced lor-

Fig. 7.7. Effects of α_2-blocking drugs on lordosis behaviour (LD) in the female rat. Saline (Sal), yohimbine (Yoh) and piperoxan (Pip). (Redrawn from Everitt *et al.*, 1975.)

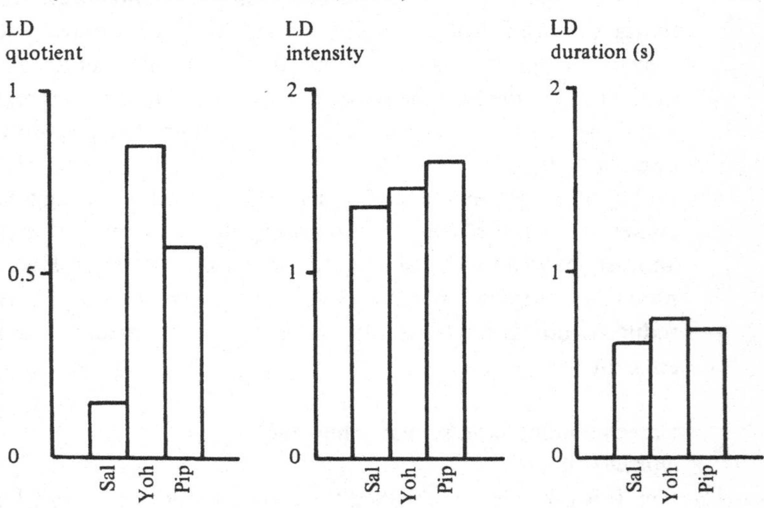

dosis, it is suggested by the Everitt *et al.* that the increases seen after both piperoxan and yohimbine may be because of their action in increasing the firing of LC cells through blockade of inhibitory α_2 'autoreceptors'. This is supported by the finding that the 'postsynaptic' α-blocker phenoxybenzamine was without marked effect on lordosis.

Ward, Crowley, Zemlan & Margules (1975) report that the β-blocker LB-46 will increase lordosis behaviour in female rats if injected directly into brain areas such as the hypothalamus but not if injected into sites in the caudate or the thalamus. Similar injection of the α-blocker phentolamine failed to affect lordosis at any site. This does not seem to support the idea from the work of Everitt *et al.* (1975) that noradrenergic systems play a facilitatory role in lordosis. Nonetheless, the effects of synthesis inhibition of noradrenaline with, in this case, U 14,624, was confirmed to reduce lordosis in another species, the female guinea pig, by Nock & Feder (1979).

Crowley, Feder & Morin (1976) found that the postsynaptic α-blocker phenoxybenzamine would similarly abolish lordosis behaviour in female guinea pigs rendered sexually receptive with oestrogen and progesterone. It proved possible to restore lordosis after noradrenaline synthesis inhibition by the α-agonist clonidine. Since the dose of clonidine used was 1 mg kg^{-1} it appears likely that this was having its effect by direct stimulation of postsynaptic α-receptors, rather than by decreasing the firing activity of LC cells through presynaptic α_2-receptors. Clonidine can induce lordosis in normal animals and this effect is blocked by the postsynaptic α-blocker phenoxybenzamine (Crowley, Nock & Feder, 1978*a*). Again, the dose of clonidine used was 1 mg kg^{-1} which probably has a postsynaptic effect. On the other hand, a species difference may be apparent, since Davis & Kohl (1977) found that clonidine suppresses lordosis behaviour in the female rat. This may, however, also be reconciled with the above by postulating that the rat experiments were actually reducing firing of LC cells with clonidine and hence reducing noradrenaline activity. This appears to be supported by the finding (Davis & Kohl, 1977) that the clonidine suppression of lordosis could be blocked by the presynaptic α_2-blocker yohimbine but not by the postsynaptic α-blocker phenoxybenzamine.

Thus, in summary, an attempt to draw together the various effects of noradrenergic drugs on female lordosis behaviour would suggest that noradrenaline has a facilitatory effect on lordosis acting through post-synaptic α-receptors. Presynaptic α_2-agents may modify this behaviour by altering the firing of LC cells. However, the finding that a β-blocker can induce lordosis when injected into the hypothalamus (Ward *et al.*, 1975) implies that the picture is not as simple as this.

6-OHDA lesions have also been used to examine the role of catecholamines in female sexual behaviour. Unfortunately, to date, no specific depletion of either noradrenaline or dopamine on their own have been examined, so some ambiguity exists concerning the amine of importance in these 6-OHDA studies.

Herndon *et al.* (1978) used two intraventricular injections of 200 μg of 6-OHDA to deplete brain dopamine by 55 % and brain noradrenaline by 66 % in female rats. This resulted in a marked increase in lordosis behaviour, both in terms of the frequency and intensity and in terms of the maintained duration. This was confirmed by Caggiula *et al.* (1979), who used pargyline pre-treatment to improve the catecholamine depletions with 6-OHDA. Loss of caudate dopamine in excess of 85 % and in cortical noradrenaline in excess of 95 % were found and a dramatic increase in lordosis emerged. In neither study did an increase in soliciting behaviour occur (seen as hopping and darting in the rat). This raises the possibility that the increase in lordosis reflects simply a decreased motor activity with consequent increase in behaviours requiring immobility (Caggiula *et al.*, 1979).

The increase in lordosis seems to reflect the loss of brain dopamine rather than noradrenaline, since Herndon (1976) found that electrolytic lesion in the vicinity of the dorsal and ventral bundles led to a virtual abolition of lordosis behaviour in the female. Similar effects were found by Wright & Everitt (1977) for large electrolytic or 6-OHDA-induced noradrenaline loss. However, the ventral bundle appears to be the substrate of this effect, since Hansen, Stanfield & Everitt (1980) found that either electrolytic or 6-OHDA lesion to the ventral bundle, which did not impinge on the dorsal system, would still impair lordosis. This would make the lesion studies compatible with the conclusion of the above drug studies that noradrenaline systems are facilitatory to female lordosis behaviour.

Male sexual behaviour. Clark, Caggiula, McConnell & Antelman (1975) found that electrolytic lesions in the vicinity of the dorsal noradrenergic bundle would lead to a marked increase in sexual behaviour in the male rat. Particularly clear was the shortening of the post-ejaculatory inhibitory period and hence an increase in the number of ejaculatory episodes in a fixed time period. This effect, however, does not appear to be the result of loss of brain noradrenaline since Clark (1980) used injection of 6-OHDA into the dorsal noradrenergic bundle to produce depletions of forebrain noradrenaline more profound than those found after electrolytic lesion but observed no change in male sexual behaviour (electrolytic lesions of Clark *et al.* (1975) depleted cortical noradrenaline to 37 % of control while the 6-OHDA lesions of Clark (1980) depleted cortical noradrenaline to 17 %). Additionally, Sessions, Salwitz & Kant (1976) found that bilateral electrolytic lesion to the LC failed to affect male sexual behaviour in the rat. Finally, Hansen, Kohler & Ross (1982) used peripheral injection of the novel neurotoxin DSP4 to deplete both central and peripheral noradrenaline (although peripheral noradrenaline may recover to some extent after 10 days or so). Despite severe depletion of forebrain noradrenergic systems as shown by decrease in [^{3}H]noradrenaline uptake in the hippocampus, no facilitation of male sexual behaviour was seen. They confirmed that electrolytic lesion in the course of the dorsal bundle did indeed increase the postejaculatory pause. Since a similar facilitation was seen after injection of ibotenic acid near the dorsal bundle, which destroyed cell bodies but did not affect the noradrenergic fibres of passage as judged by unaltered [^{3}H]noradrenaline uptake in the hippocampus, they concluded that a nonnoradrenergic system in the vicinity of the dorsal bundle was the substrate of the effect.

Thus, in conclusion there appears to be little evidence that noradrenergic systems play a role in male sexual behaviour, as opposed to the suggestive evidence that they enhance female lordosis responses, I shall now turn to examine the role of the other catecholamine dopamine.

Dopamine

Female sexual behaviour. Bensch, Lescure, Robert & Faure (1975) reported that the intensity of fluorescence of dopaminergic systems in the median eminence of female rabbits was increased 20–30 min after mating, suggesting a possible response of dopaminergic systems. However, this reflects the consequences, rather than the causes of sexual behaviour.

Everitt, Fuxe, Hokfeldt & Johnsson (1975) found that the dopamine receptor antagonist pimozide caused a marked *increase* in all measures of lordosis behaviour in female rats. The inhibitor of synthesis of both noradrenaline and dopamine, AMPT, increased lordosis quotients. This was also seen by Everitt, Fuxe & Hokfeldt (1974) for pimozide. 6-OHDA lesions of both noradrenergic and dopaminergic systems may increase female lordosis behaviour by their action of dopaminergic systems (see above studies by Herndon *et al.*, 1978, and Caggiula *et al.*, 1979).

The converse has been reported for dopaminergic agonist drugs. Thus, Everitt *et al.* (1974) found that piribedil lowered female rats receptivity. However, Hamburger-Bar & Rigter (1975) found the dopamine agonist apomorphine increased female sexual behaviour (see Fig. 7.8). This effect appeared to last for some time (perhaps 24 h) and may reflect an interaction with pituitary hormone release. Apomorphine also affects serum levels of prolactin (Smalstig, Sawyer & Clemens, 1974), alters growth hormone titres (Brown, Krieger, van Woert & Ambani, 1974) and may release follicle stimulating hormone from the pituitary (Choudhury, Sharpe & Brown, 1973).

Male sexual behaviour. Tagliamonte, Fratta, del Fiaco & Gessa (1974) and Paglietti, Pellegrini Quarantotti, Mereu & Gessa (1978) reported that apomorphine in the male rat acted to increase copulatory behaviour. This was seen at 0.5 mg kg^{-1} of the drug since the higher dose (5 mg kg^{-1}) induced stereotyped behaviours which prevented the occurrence of copulation. Malmnas (1973) also found that apomorphine enhanced sexual behaviour in male rats. Both the studies agree that dopamine blockade with haloperidol or pimozide decreases sexual behaviour in the male rat.

Pimozide also blocks male sexual behaviour activated by electric footshock (Antelman, Herndon, Caggiula & Shaw, 1975*a*) and haloperidol also blocks the hypersexuality seen in rats after temporal lobe seizures (Andy & Velamati, 1978). It should be noted that the effects of the neuroleptic drugs (pimozide, haloperidol) in the male rat are exactly opposite to those reported in female sexual behaviour. Other dopamine blockers such as spiperone and clozapine also reduce male sexual responsiveness (Baum & Starr, 1980).

Finally, L-DOPA facilitates male sexual behaviour in cats (Malmnas, 1976) and since this effect could be blocked by prior administration of pimozide it would appear, in this instance, to be acting through dopaminergic systems.

Fig. 7.8. Effect of apomorphine on female lordosis behaviour (LD). (Redrawn from Hamburger-Bar & Rigter, 1975.)

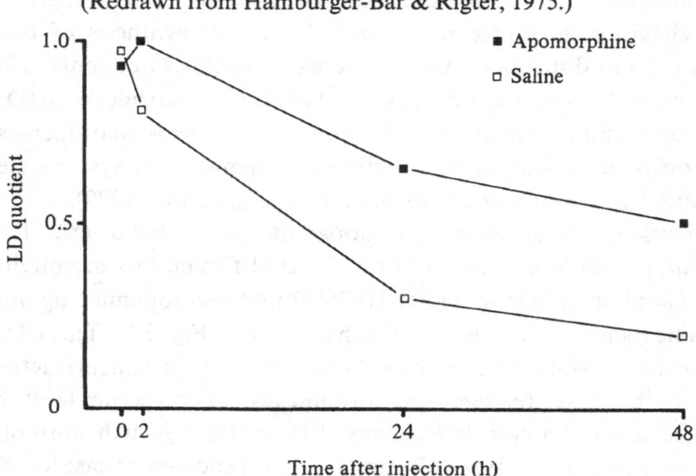

In summary, an opposite pattern of effects is seen in the male rat compared to the female. In the male, dopamine agonists increase and dopamine antagonists decrease sexual behaviour. In the female, dopamine antagonists clearly increase such behaviour while dopamine agonists have more ambiguous effects.

Summary

(1) Noradrenaline synthesis inhibition and postsynaptically acting α-antagonist drugs reduce female lordosis behaviour. Presynaptic α_2-antagonists increase this behaviour, perhaps by increasing the firing of LC cells. High doses of the α-agonist clonidine will increase lordosis, which could be a postsynaptic effect, while lower doses may reduce lordosis, perhaps by reducing the firing of LC cells.

(2) Few 6-OHDA lesions specific to noradrenergic systems are available but those which are suggest that electrolytic or 6-OHDA lesion to the ventral noradrenergic bundle will reduce lordosis in the female.

(3) Although electrolytic lesion in the vicinity of the dorsal noradrenergic bundle has been reported to increase male sexual behaviour, neither more selective 6-OHDA lesions here nor electrolytic lesion of the source nucleus, the LC, has any effect on male sexual behaviour.

(4) Dopamine blockers increase lordosis in the female while reducing male sexual behaviour. Dopamine agonist drugs are more uncertain in their effects on female lordosis but clearly act to increase male copulatory behaviour. A clear sexual dimorphism in the action of dopaminergic system is thus demonstrated.

Catecholamines and eating behaviour

Noradrenaline

Perhaps one of the clearest roles for the catecholamine neurotransmitters lies in the field of ingestive behaviour. I will start by looking at the control of eating.

Release/levels. Martin & Myers (1975) examined the release of noradrenaline in the hypothalamus during eating. This they did by the push–pull technique, in which radioactively labelled noradrenaline was injected into the hypothalamus and 30 min later, when the injected noradrenaline had been distributed throughout the pools of endogenous noradrenaline, the animal was presented with food, having previously been fasted. During the eating bout the hypothalamic area was perfused at the rate of 0.33 $\mu l\ s^{-1}$ with artificial cerebrospinal fluid (CSF) via two concentric cannulae, a 23 gauge outer one and a 28 gauge inner one. Samples were collected for each 30 min of perfusion and then counted to determine how much of the previously introduced [^{14}C]noradrenaline had been washed out in that period. Some samples were subjected to thin-layer chromatography to determine how much of the radioactivity had remained as noradrenaline and how much had been metabolised to noradrenaline derivatives and other compounds. A similar push–pull perfusion was carried out when rats were lever-pressing in order to receive 45 mg food pellets on a FR 6 schedule (see Appendix). During the feeding periods, and to a lesser extent also during the lever-pressing periods, the efflux of [^{14}C]noradrenaline increased significantly compared to the control period preceding or following it (see Fig. 7.9). Similar increases in noradrenaline release during feeding have also been reported for the monkey by McQueen, Armstrong, Singer & Myers (1976). The noradrenaline release could be altered by local hypothalamic perfusion of either glucose or insulin. This suggests a role for noradrenaline in feeding, but the finding that the efflux of dopamine is also increased in this paradigm (Martin & Myers, 1976) brings into question the specificity of the amine involved. Further, it is unclear how much the release resulted from the motor act of feeding and how much from the central state of hunger. Such dissociations require more sophisticated behavioural analysis. A further caveat with the push–pull perfusion technique is the large volume of fluid being forced through a small area of brain and the inevitable and considerable damage which occurs to that area renders the physiological relevance of the finding uncertain.

Other less invasive techniques have generally been adopted to investigate this question. Thus, van der Gugten & Slangen (1975) report that animals on a restricted food access schedule showed higher uptake of noradrenaline in the hypothalamus upon sacrifice than did animals maintained *ad lib.* on food. A correlation was noted between the time of sacrifice and the food intake pattern on previous days at the time. Thus, animals sacrificed at the beginning of the night showed the highest uptakes and this was also the period when food intake was increased.

Krieger (1974) found that food and water restriction resulted in an increase in the noradrenaline content of the hippocampus by about 30 %. Since increases in serotonin content of hypothalamus and hippocampus were also seen, the amine of importance is uncertain. It is also confounded by the fact that both food and water deprivation occurred together.

Kreiger, Crowley, O'Donohue & Jacobowitz (1980) remedied this by investigating food deprivation on its own and found that the serotonin content of the hippocampus was increased, thus indicating that the effect observed in the previous study was indeed caused by food deprivation. No changes were seen in the noradrenaline content of either of two hypothalamic nuclei, the paraventricular nucleus or the median eminence, and the noradrenaline content of the hippocampus was not reported. Some reduction in the dopamine content of the median eminence, however, was apparent.

Other authors have also noted changes in other brain areas. Stachowiak, Bialowas, Jurkowski & Mirosz (1979) found that 48 h food deprivation

Fig. 7.9. [^{14}C]Noradrenaline washout after intraventricular injection measured as disintegrations per minute (d.p.m.) under resting conditions or during feeding and lever-pressing. (Redrawn from Martin & Myers, 1975.)

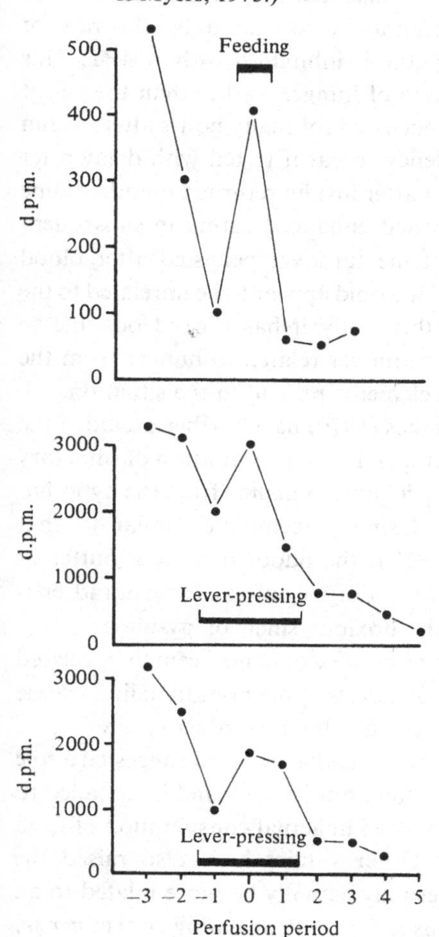

decreased both noradrenaline and dopamine levels in the arcuate and ventromedial hypothalamic nuclei but increased dopamine content of the central amygdaloid nucleus. Depletion of hypothalamic noradrenaline content is commonly observed as Glick, Waters & Milloy (1973) saw after 2 days of food deprivation. Changes in catecholaminergic enzymes have also been reported with food deprivation. Ismahan, Parvez & Parvez (1979) found a slight decline in activity of whole-brain MAO at 60 h of food deprivation in young rats (approx. 50 or 80 g, thus past the neonatal stage). Increases in catechol-*o*-methyltransferase were also seen.

Factors which induce feeding, such as challenge with insulin, may also increase noradrenaline activity in the brain. Thus, Bellin & Ritter (1981) found that injection of 2.5 units kg^{-1} insulin would significantly enhance food intake and that this treatment also led to an increase in the noradrenaline turnover in hypothalamus as measured by the rate of disappearance of the amine after synthesis inhibition with AMPT. This effect seemed to be related to the state of hunger, rather than the act of eating, since the enhanced turnover persisted for many hours after insulin injection (as did the enhanced tendency to eat if tested with delay after insulin). Further, a brief feeding bout after insulin returned noradrenaline turnover rates to normal and abolished enhanced eating in subsequent tests. Since the enhanced noradrenaline turnover persisted after blood glucose levels had returned to normal it would appear to be unrelated to the stress of hypoglycaemia. Thus, in this study it has proved possible to dissociate changes in noradrenaline turnover related to hunger from the motor act of eating and from stress elements present in the situation.

However, Myers, McCaleb & Hughes (1979) have further excluded the motor act by the demonstration that merely the presentation of olfactory stimuli would enhance the release of [^{14}C]noradrenaline from the hypothalamus into a push–pull perfusate. Using a technique similar to that described above, Myers *et al.* found that the odour of peanut butter or pyridine, as well as the physical act of feeding, would increase noradrenaline release (see Fig. 7.10). Since the noxious smell of pyridine was as effective as that of peanut butter, the response would not seem to be related to food associated odours and may suggest that the noradrenaline release results from the activation of a nonspecific olfactory system *per se*.

Thus, in summary, the release/levels studies have indeed suggested a role for hypothalamic noradrenaline in eating, but have by and large failed to dissociate the motor components involved in actual consumption of food from the central state of hunger. Other studies have also raised the possibility that the observed changes may actually be more related to an olfactory system in the hypothalamus rather than an eating system *per se*.

Noradrenergic agonists. The earliest, and still perhaps the most persuasive, evidence for a role of noradrenaline in feeding came from the effects of microinjection of noradrenaline agonist drugs into hypothalamic nuclei. The first demonstration of this by now 'classical' effect of noradrenaline on feeding was that of Grossman (1960, 1962*a*) and was soon replicated by Miller, Gottesman & Emery (1964). A most detailed mapping study of the hypothalamic sites into which noradrenaline injection would elicit eating was reported by Booth (1967). In all these studies, cannulae are implanted in rats with their tips aimed at various hypothalamic nuclei. Following one week of recovery the animals were satiated on wet mash for 10 min prior to injection of either saline or noradrenaline in a volume of around 1 μl or so into the hypothalamus via the cannulae. The response is shown in Fig. 7.11, and over the hours following injection more frequent eating was seen after noradrenaline than after saline administration. The most effective sites for the elicitation of feeding were in the anterior hypothalamic areas (see Fig. 7.11). The lateral hypothalamus was generally ineffective, as were most caudal portions of the hypothalamus. The substantia innominata of the anterior hypothalamus was particularly responsive. Sites outside the hypothalamus, such as the nucleus accumbens and the lateral septum were also effective in eliciting eating as were the rostral thalamus, the subfornical area and the olfactory tubercle at the ventral edge of the preoptic area. Changes in drinking may also accompany those seen in eating (Avery, 1971) and are discussed in more detail in the subsequent section on

Fig. 7.10. [^{14}C]Noradrenaline (NA) efflux (expressed as proportion of initial value ($t = 30$) which is set to 1) in response to feeding, odours of peanut or pyridine. (Redrawn from Myers *et al.*, 1979.)

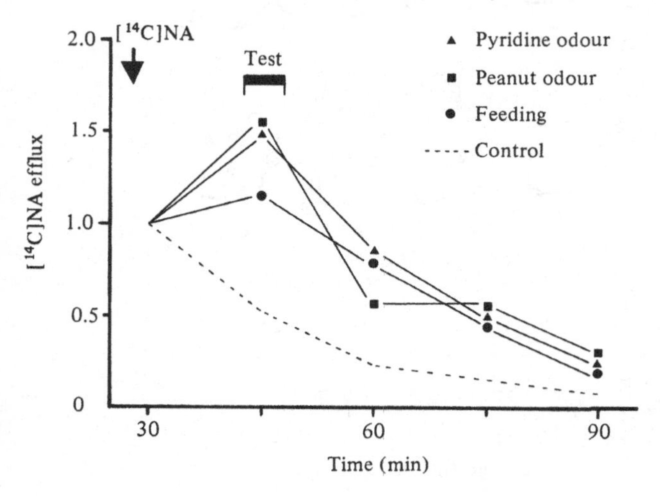

Catecholamines and drinking behaviour (see p. 370). The responses seen are dose-dependent (Leibowitz, 1975*a*), with increased eating occurring at as little as 0.5 μg of noradrenaline injected into the perifornical anterior hypothalamus. The lateral hypothalamus is generally insensitive (Leibowitz, 1975*a*). The pattern of the food intake caused by noradrenaline injection was similar to that seen normally in that the meal sizes were the same and once a bout was initiated it ran almost continuously with only a few interruptions for grooming (Leibowitz, 1975*a*).

Even smaller doses of noradrenaline, which do not of themselves initiate feeding in satiated rats, do serve to increase the meal size when these animals eventually eat spontaneously. Thus, Ritter & Epstein (1972) reported that as little as 2.5 ng noradrenaline injected into the hypothalamus just before a rat initiated spontaneous eating would increase meal size by up to 200 %.

Other agents which stimulate noradrenaline receptors may also initiate feeding. Thus, Broekkamp & van Rossum (1972) reported that feeding could be elicited by intrahypothalamic application of the α_2-agonist

Fig. 7.11. Eating caused by injection of noradrenaline (NA) into sites shown by solid circles in the hypothalamus of the rat. Similar injection at sites shown by hollow circles failed to elicit eating. A = anterior. (Redrawn from Booth, 1967.)

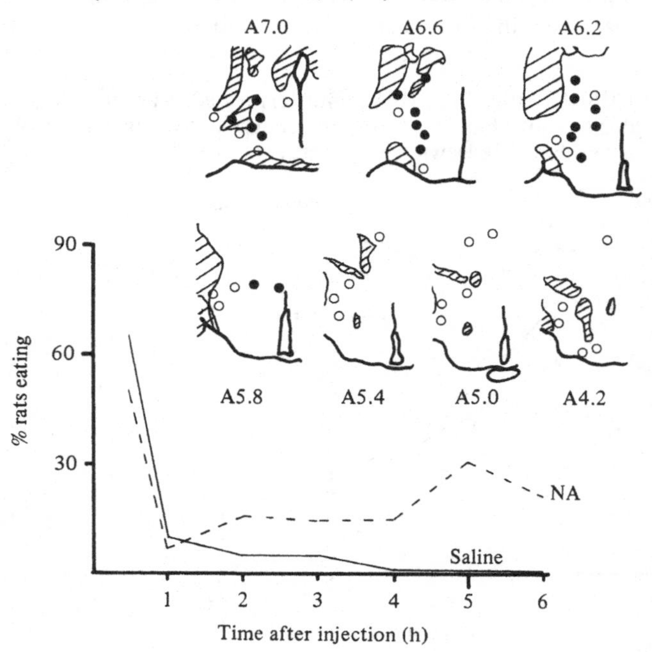

clonidine, in amounts less than a microgram. This was confirmed by Ritter *et al.* (1975). More esoteric agents such as 6-OHDA may also cause eating when injected into the preoptic area of the rat (Evetts, Fitzsimons & Setler, 1972). This is believed to be because of the ability of 6-OHDA to displace noradrenaline from central terminals of the noradrenergic system (see Fig. 7.11). Repeated 6-OHDA injections caused less eating, and post-mortem assay revealed progressive destruction of hypothalamic noradrenergic systems. That the drug was acting via the release of noradrenaline is supported by the finding that pre-treatment with the noradrenaline uptake inhibitor desipramine (DMI) blocked 6-OHDA-induced eating. The released noradrenaline appeared to be acting on an α-receptor since phentolamine also prevented 6-OHDA-induced eating. This is consistent with repeated demonstrations that α-blockers will also prevent eating induced by injection of noradrenaline *per se* into the hypothalamus (Grossman, 1962*b*; Booth, 1968; Slangen & Miller, 1969) and intrahypothalamic phentolamine will reduce natural eating resulting from food deprivation (Broekkamp, Honig, Pauli & van Rossum, 1974).

The noradrenergic pathway which activates these α-receptors has been determined by Leibowitz & Brown (1980*b*), who lesioned various ascend-

Fig. 7.12. (*a*) Eating caused by injection of 6-hydroxydopamine (6-OHDA) into the preoptic area of the rat or by similar injection of Merlis' solution (Saline) containing 1 mg ml^{-1} ascorbic acid; (*b*) effect of repeated injections of 6-OHDA at this site. (Redrawn from Evetts *et al.*, 1972.)

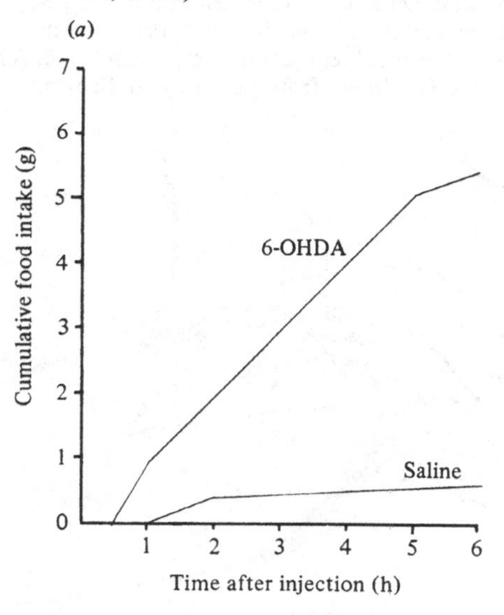

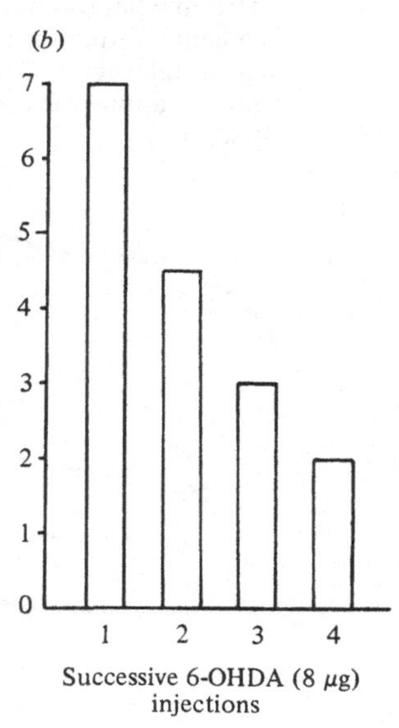

ing noradrenergic fibre systems, both electrolytically and with 6-OHDA, and found that only those lesions which were shown by post-mortem histofluorescence microscopy to have affected the dorsal component of the central tegmental tract (see Fig. 7.13) altered the ability of drugs injected into the paraventricular hypothalamus to elicit eating. Thus, after these lesions, direct agonists such as noradrenaline itself became more effective because of receptor supersensitivity having developed after denervation, while presynaptic drugs such as DMI which induce eating by preventing noradrenaline inactivation by reuptake became ineffective since no noradrenaline was being spontaneously released after lesion.

Eating caused by noradrenaline injection into the hypothalamus may be seen in different species. Thus it has been reported in sheep (Baile, Simpson, Krabill & Martin, 1972; Simpson, Baile & Krabill, 1975) and monkeys (Setler & Smith, 1974; Sharpe & Myers, 1969) as well as the rat. However, in cattle α-agonists and noradrenaline appear to cause satiety and will suppress ongoing feeding (Baile *et al.*, 1972; Simpson *et al.*, 1975). Since noradrenaline injection into the hypothalamus of cats seems to cause sleep as the predominant response, eating may not be observed here either (Myers, 1964).

Other manipulations expected to increase noradrenaline activity may

Fig. 7.13. Representation of the amine pathways (dotted) believed to mediate noradrenaline-induced feeding from paraventricular sites. MFB, medial forebrain bundle; DTB, dorsal tegmental bundle; TR, tegmental radiations; ML, medial lemniscus; dCTT, dorsal central tegmental tract; vCTT, ventral central tegmental tract; A1, A5, A6, A7 and A9, aminergic cell bodies. (Redrawn from Leibowitz & Brown, 1980*b*.)

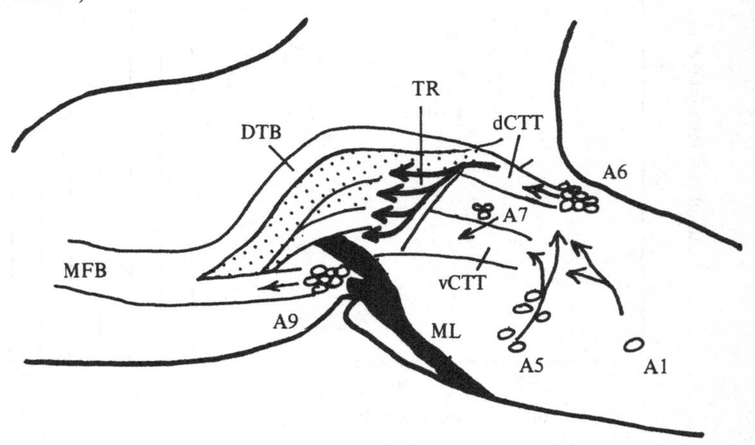

not always elicit feeding either. Thus, Herberg & Stephens (1976) were unable to obtain feeding after injection of the putative second messenger of the noradrenergic system, cAMP, into the accumbens nucleus, despite successful feeding when noradrenaline itself was administered here. Neither did aminophylline, a phosphodiesterase inhibitor which might be expected to increase the intracellular concentration of cAMP by preventing its breakdown, have any effect on eating.

Other factors may influence the ability of injected noradrenaline to elicit feeding. Thus, Margules, Lewis, Dragovich & Margules (1972) found that circadian rhythms were of importance. If noradrenaline were injected into the lateral hypothalamus during the light part of the rats' light–dark cycle it elicited an increase in the consumption of condensed milk, while identical administration during the dark period suppressed consumption. A partial confirmation of this effect was reported by Matthews, Booth & Stolerman (1978), who found that although an increase in eating was elicited by either light-period or dark-period injections of noradrenaline into the paraventricular nucleus of the hypothalamus, a much bigger effect was seen in the light period and the dark period increase was in fact nonsignificant. Of some concern in this study should be the fact that food intake during the dark is spontaneously higher than it is in the light and so a ceiling effect may be preventing the noradrenaline effect from being significant during the dark period. Stern & Zwick (1973) found a significant decrease in food intake when noradrenaline was injected into the ventricle of female rats during the dark period and this contrasted with the significant increase found during light-period administration. They also found that the hormone oestradiol would elicit food intake during the light period but cause a decrease in intake when given in the dark period. Both these effects of oestradiol could be blocked by the α-antagonist phentolamine (as can those of noradrenaline itself).

Intraventricular injection of 2-deoxy-D-glucose may also elicit feeding when injected intraventricularly (Ikeda, Nishikawa & Matsuo, 1980) and this effect fails to occur in spontaneously overweight rats of the Zucker strain.

Injections of noradrenaline into the ventricles may even be effective in causing feeding in rats made anorexic by lesion of the lateral hypothalamus (Berger, Wise & Stein, 1971) and this is consistent with the fact that noradrenaline in the above studies seems to be acting at anterior hypothalamic/preoptic areas rather than in the lateral hypothalamus (with the exception of the study by Margules *et al.* (1972) in which diffusion from the lateral hypothalamic injections sites may nonetheless have occurred).

However, lesion of the posterior ventromedial hypothalamus does block noradrenaline-induced feeding (Herberg & Franklin, 1972).

Noradrenaline-induced feeding differs in a few respects from 'natural' feeding. Injection of noradrenaline does not elicit food-hoarding behaviour, which natural hunger does (Blundell & Herberg, 1970), nor does it motivate lever-pressing for food in a Skinner box with *partially* reinforced schedules (see Appendix) (Coons & Quartermain, 1970) and the feeding is more susceptible to disruption by unpleasant tastes (finickiness; Booth & Quartermain, 1965).

Although direct electrical stimulation near the LC may elicit feeding in some rats and gnawing or licking in others (Micco, 1974), these do not in fact seem to involve noradrenergic systems in the coeruleus itself since 6-OHDA lesion to the ascending noradrenergic fibres failed to abolish these behaviours (van der Kooy, 1979).

In summary, the phenomenon of feeding induction in satiated rats after noradrenaline injection into the anterior hypothalamus is perhaps one of the best-established facts of physiological psychology. The receptors appear to be of an α type since the effect can be mimicked by α_2-agonists such as clonidine and blocked by prior administration of α-antagonists. An interaction with the circadian light–dark cycle is indicated. Species other than the rat also show noradrenaline-induced feeding, although it is possible that cattle may demonstrate an opposite response.

Noradrenergic inhibition of feeding. In addition to the α-receptor-mediated noradrenergic system described above, more recent work has clearly shown the existence of a second noradrenergic system which is inhibitory to feeding.

Margules (1969) reported that administration of noradrenaline into the perifornical hypothalamus of the rat would cause a 50 % reduction in the consumption of condensed milk from 17.7 ml in controls to 7.7 ml after injection of 28 μg noradrenaline (see Fig. 7.14). This was claimed to be an α-receptor-mediated effect since injection of the α-blocker phentolamine (18 μg) increased milk intake (see Fig. 7.14), while injection of the β-blocker propranolol was without effect. However, subsequent workers have almost uniformly characterised these receptors as β in nature (e.g. Leibowitz, 1975*a*). Central administration of the anorexic drug amphetamine, which releases both catecholamines, has also been found to suppress feeding at perifornical sites (Leibowitz, 1975*b*). It is possible that the effect may actually involve adrenaline- and/or dopamine-containing neurones, since injection of adrenaline or dopamine via indwelling cannulae into the perifornical hypothalamus also suppresses feeding (Leibowitz, 1979).

Based on small electrolytic and 6-OHDA lesions of the catecholamine inputs to the perifornical hypothalamus, Leibowitz & Brown (1980*a*) conclude that the noradrenaline/adrenaline input runs in the central tegmental tract, especially in the ventral part of this plexus of fibres, while the dopamine input ascends from A8 and possible A9 dopaminergic cell bodies. Lesions to the dorsal noradrenergic bundle did not affect noradrenaline-induced suppression of feeding (see Fig. 7.15).

The results reported above lead to the possibility that the anorexic action of amphetamine (and for that matter other anorexic drugs) might result from its effect in releasing noradrenaline in the noradrenergic feeding suppression system. This has been tested in a number of ways. Abdallah (1971) found that interruption of both noradrenaline and dopamine (and adrenaline) synthesis by administration of AMPT (16 mg kg^{-1}) reduced the anorexic effect of amphetamine without altering spontaneous food intake. However, higher doses of AMPT were anorexic in their own right (64 mg kg^{-1}). This may well be a dopaminergic effect, since inhibition of noradrenaline and adrenaline synthesis with FLA 63 failed to alter amphetamine anorexia (demonstrated with methylamphetamine (Franklin & Herberg, 1977)). A similar conclusion was reached by Dobrzanski & Doggett (1975), who found that the noradrenaline synthesis inhibitor

Fig. 7.14. Milk consumption caused by (*a*) injection of the α-antagonist phentolamine into the perifornical hypothalamus of the rat and (*b*) the inhibition of intake caused by noradrenaline injection compared to that seen after injection of saline alone (Sal). (Redrawn from Margules, 1969.)

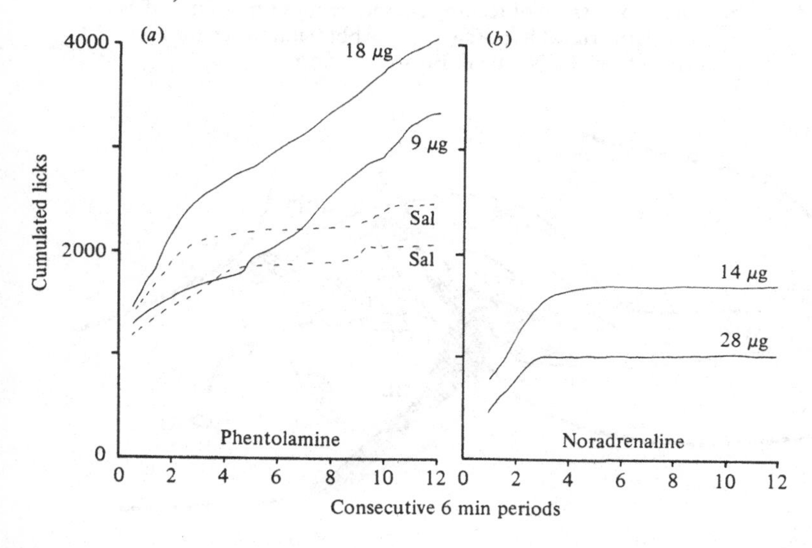

disulfiram or DDC potentiated amphetamine anorexia rather than antagonised it. However, intraventricular injection of noradrenaline also potentiated amphetamine anorexia, which might be interpreted as some evidence that noradrenaline release by amphetamine has some role in its anorexic action. Finally, Carey (1976) reported that electrolytic lesion of the LC attenuated amphetamine anorexia without affecting that caused by the drug fenfluramine. The lesions appear (Carey, 1976, p. 522) to have been large enough to affect ventral noradrenergic systems as well as dorsal ones. More will be said on the effects of noradrenaline lesions on food intake in the next section.

6-OHDA lesions and noradrenergic inhibition of eating. If there is a noradrenergic system in the brain which acts to inhibit eating, as suggested above, then 6-OHDA lesion to it might be expected to lead to hyperphagia and overweight. This section is devoted to the experimental tests of this postulate.

The initial demonstration of the phenomenon was by Ahlskog & Hoebel (1973), who injected 6-OHDA (8 μg in 0.8 μl saline) bilaterally into the fibres of the ventral noradrenergic bundle. This led to a marked increase in food intake (Fig. 7.16*b*) of some 25–40 % and a consequent increase in body weight (Fig. 7.16*a*), often in excess of 50 %. This effect seemed to be noradrenergic since it could be prevented by pre-treatment with DMI which prevents the noradrenergic lesioning effects of 6-OHDA but not any

Fig. 7.15. Representation of amine pathways (dotted) believed to mediate suppression of feeding caused by injection of noradrenaline into the perifornical hypothalamus. Abbreviations as for Fig. 7.13. (Redrawn from Leibowitz & Brown, 1980*a*.)

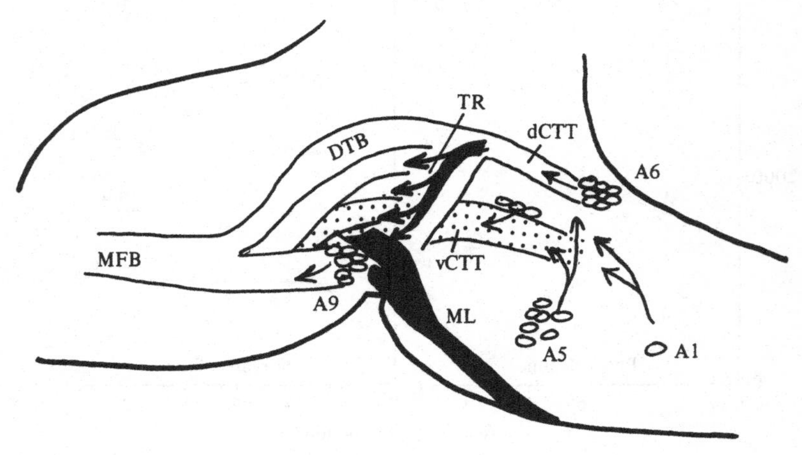

small nonspecific side-effects (but see later, p. 363). Although the initial lesions destroyed not only the ventral noradrenergic bundle but also the dorsal innervation, subsequent 6-OHDA lesion restricted to the dorsal bundle alone failed to induce hyperphagia, indicating a role for the ventral bundle in suppression of food intake. Electrolytic lesion restricted to the ventral bundle which did not affect the noradrenergic innervation via the dorsal bundle to cortex, hippocampus, etc. was by itself adequate to cause hyperphagia. On the other hand, Redmond *et al.* (1977) found that in the stumptailed monkey (*Macaca arctoides*) electrolytic lesion to the LC caused a significant increase in body weight which was not seen with sham procedures. The lesioned monkeys displayed almost constant eating compared to the discrete meals shown by normal and sham monkeys. The authors remark 'They stuffed themselves in some instances to the point of compromising ventilatory capacity and producing marked abdominal distension' (Redmond *et al.*, 1977, p. 1621). Water intake was also markedly elevated. A word of caution as to the neurochemical mechanism is required since it is precisely in this species of monkey that serotonergic cell bodies have been reported in the LC (Sladek & Walker, 1977). Electrolytic lesion to the LC of the rat, which does seem to be a pure noradrenergic nucleus, does *not* yield overeating (Anlezark, Crow & Greenway, 1973; Amaral & Foss, 1975; Sessions, Kant & Koob, 1976) although lesion to the adjacent nonnoradrenergic nucleus tegmentalis

Fig. 7.16. Weight gain (*a*) and hyperphagia (*b*) caused by injection of 6-hydroxydopamine (arrowed) into the ventral bundle. (Redrawn from Ahlskog & Hoebel, 1973.)

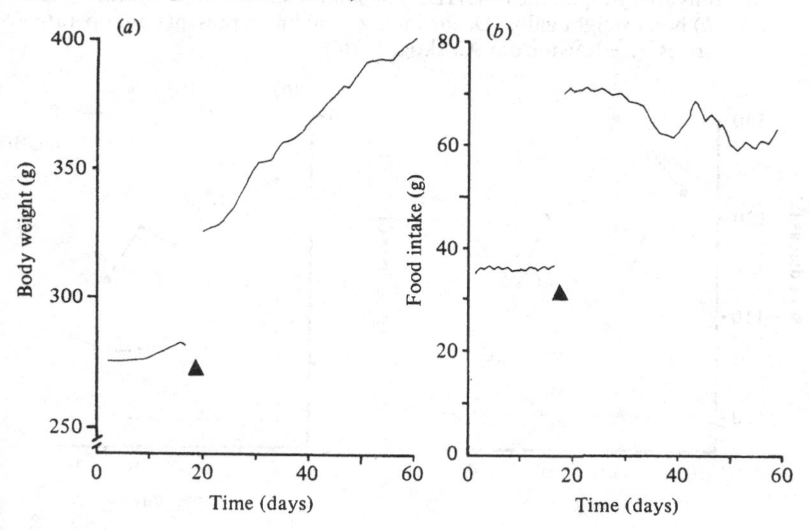

laterodorsalis does yield hyperdipsia (Osumi, Oishi, Fujiwara & Takaori, 1975; Koob, Sessions, Kant & Meyerhoff, 1976; Satoh, Shimizu, Tohyama & Maeda, 1978). Further, neurochemically specific lesion to the dorsal bundle with 6-OHDA confirms the results of Ahlskog & Hoebel (1973) that hyperphagia and weight gain do not occur (Mason & Iversen, 1978; Mason, Roberts & Fibiger, 1979).

Ventral bundle hyperphagia has been seen after 6-OHDA and electrolytic lesion to the ventral bundle (Ahlskog, 1974) and after knife cuts which destroy the noradrenergic innervation to the paraventricular nucleus of the hypothalamus (O'Donohue, Crowley & Jacobowitz, 1978). In the latter case hypophagia was seen for a few days after the lesion which subsequently reversed to become a long lasting hyperphagia/hyperdipsia.

A deficit in the long-term regulation of food intake seems to underlie the ventral bundle effect, since Ahlskog (1976) found that when placed on a calorifically concentrated (high-fat) diet the lesioned rats took twice as long as controls to reduce the amount consumed and hence showed considerable weight gain (Fig. 7.17). Shorter-term regulation of food intake in response to insulin challenge or as a result of adulteration of food with quinine was, however, unaltered by the 6-OHDA lesion. The failure of the 6-OHDA lesion to the ventral bundle to affect the 'finickiness' to food adulteration with quinine differentiates this lesion from electrolytic lesion to the ventromedial nucleus (VMN) of the hypothalamus which also caused hyperphagia, but with markedly enhanced sensitivity to taste stimuli

Fig. 7.17. (*a*) Consumption of high-fat diet after injection of 6-hydroxydopamine (6-OHDA) or saline (Sal) into the ventral bundle; (*b*) body weight gains. Dashed horizontal line represents pre-operative values. (Redrawn from Ahlskog, 1976.)

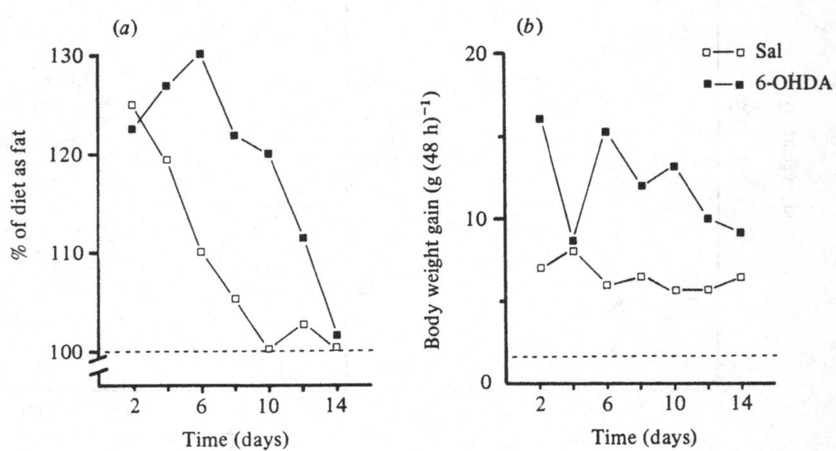

(Ahlskog, Randall & Hoebel, 1975). The overeating after 6-OHDA lesion occurred only at night whereas that seen after VMN lesion occurred equally during the day or the night.

Areas other than the hypothalamus also receive noradrenergic innervation via the ventral bundle and one of these, the amygdala, might also be involved in the control of food intake. Lenard & Hahn (1982) showed that injection of 6-OHDA directly into the amygdala caused an increase in body weight and hyperphagia. This seemed to relate to the destruction of the noradrenergic innervation of the amygdala since hyperphagia did not occur if 6-OHDA was injected after DMI pre-treatment which protected noradrenergic (but not dopaminergic) neurones from destruction.

Problems have arisen with the ventral bundle hyperphagia effect. Thus, some authors may fail to observe the effect. Bellinger, Bernardis & Brooks (1979) used electrolytic lesions in the vicinity of the ventral bundle and found no hyperphagia or increase in body weight when rats were maintained on a normal diet (see also Cox, Kakolewski & Valenstein, 1969; Bellinger & Bernardis, 1976; Bellinger, Bernardis & Goldman, 1976). However, a mild hyperphagia *was* obtained by the above workers when female rats were fed a high-fat diet, and this was echoed by results of Lorden, Oltmans & Margules (1976).

More recently it has been questioned whether ventral bundle hyperphagia is truly noradrenergic. Grossman, Grossman & Halaris (1977) found that knife cuts in the vicinity of the dorsal tegmentum would deplete hypothalamic noradrenaline equally whether they were effective in causing hyperphagia or not (see Fig. 7.18). Thus, some rats showed no change in food intake despite a 45 % decrease in hypothalamic noradrenaline while other animals showed marked hyperphagia with only a 40 % decrease in noradrenaline. Perhaps more compelling, Lorden *et al.* (1976) found overeating and weight gain only after electrolytic lesion in the vicinity of the ventral bundle and *not* after 6-OHDA lesion. A subsequent study (Oltmans, Lorden & Margules, 1977) found that other nonspecific lesion techniques such as injection of copper sulphate or the neurotoxin 5,6-dihydroxytryptamine into this region would yield hyperphagia and weight gain but 6-OHDA at 4 μg or 8 μg per side did not do so despite causing much larger noradrenaline loss in the hypothalamus (Fig. 7.19). Only 12 μg of 6-OHDA (a dose which caused no more noradrenaline loss than did 8 μg) was effective in causing hyperphagia. This dose is higher than is usually used for selective catecholamine lesion and suggests that the ventral bundle hyperphagia may result from noncatecholaminergic damage caused by excessively high doses of 6-OHDA. This fails, however, to explain the report that DMI pre-treatment would prevent the

Fig. 7.18. Depletions of noradrenaline in hypothalamus (Hypoth) or forebrain caused by knife cuts which did (stippled bars) or did not (open bars) cause hyperphagia. (Redrawn from Grossman *et al.*, 1977.)

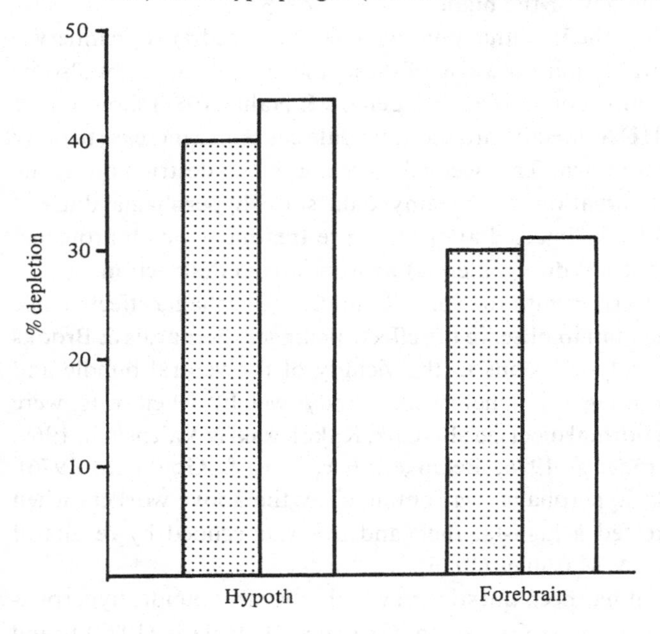

Fig. 7.19. Body weight of rats subjected to lesions in the vicinity of the ventral bundle with 5,6-dihydroxytryptamine (5,6-DHT), copper sulphate ($CuSO_4$), 6-hydroxydopamine (6-OHDA) or electrolytic means (Elec). (Redrawn from Oltmans *et al.*, 1977.)

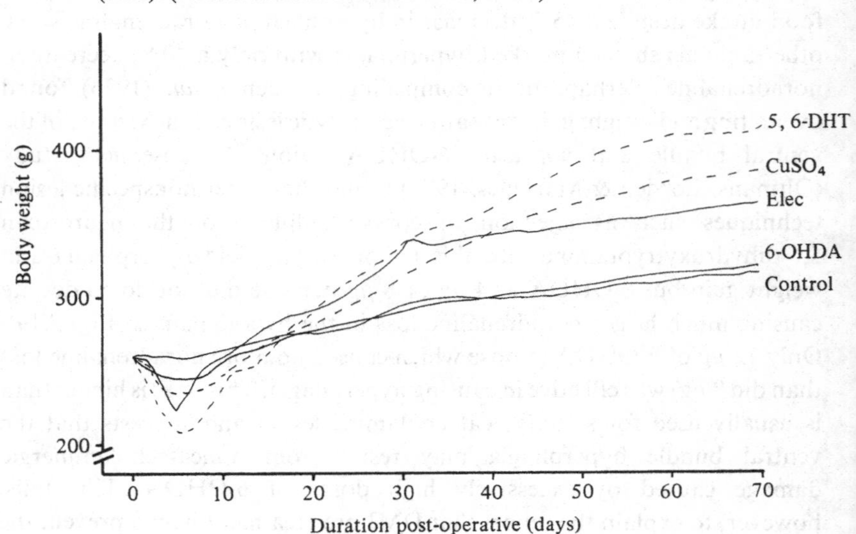

hyperphagia, as well as protecting noradrenergic neurones (Ahlskog & Hoebel, 1973; Ahlskog, 1974). Hernandez & Hoebel (1982) have remarked that DMI by itself may cause *hypo*phagia, and as such simply subtract from the *hyper*phagia caused by nonnoradrenaline lesions to give the appearance of prevention of hyperphagia when tested a short time after the lesion. Subsequent tests reveal that the hyperphagia is, however, still there, despite no loss of noradrenaline. The noradrenergic nature of the ventral bundle hyperphagia is thus uncertain.

Summary

(1) Studies on the release of levels of noradrenaline during fasting or feasting suggested a role, particularly of hypothalamic regions, in ingestive behaviour but failed to separate the olfactory or motor components of the eating response from the central state of hunger.

(2) Clear evidence for a causative role of noradrenaline in eating came with the demonstration that α-receptor agonists and noradrenaline itself would cause eating in satiated animals when injected into hypothalamic areas. Species differences may exist with suppression of eating being seen in cattle, and sleep being predominant in cats following hypothalamic noradrenaline injection.

(3) α-Antagonists such as phentolamine will not only block noradrenaline-induced eating but also attenuate spontaneous eating following food deprivation.

(4) A second noradrenergic system exists which acts to inhibit feeding via β-receptors, as shown by noradrenaline injection into the perifornical hypothalamus. The feeding inhibition system seems to consist of the ventral part of the plexus of noradrenergic fibres in the central tegmental tract. The feeding-initiation system, by contrast, seems to run in the dorsal component of the central tegmental tract and to terminate in the paraventricular hypothalamus.

(5) 6-OHDA lesion to the ventral bundle causes hyperphagia and weight gain but it is still uncertain whether this is truly noradrenergic. Lesion to the dorsal bundle with 6-OHDA or electrolytically to the LC generally fails to affect eating, although hyperdipsia caused by damage to the adjacent, nonnoradrenergic dorsolateral tegmentum may be seen.

Dopamine

Release/levels during eating. Push–pull perfusion experiments carried out by Martin & Myers in 1976 showed that dopamine was released from brain areas during the consumption of food. Microinjection of [^{14}C]dopamine via one part of double push–pull cannulae was performed into various brain areas and then the washout of this radioactivity determined over time. When a steady release had been achieved, the food-deprived rat was allowed access to food and a marked increase in the release of [^{14}C]dopamine observed from sites in the nucleus reuniens and the zona incerta. If the food-deprived rat was allowed to lever-press for food pellets, increased dopamine release was seen from the substantia nigra, the ventromedial and the dorsomedial hypothalamus. This experiment, of course, fails to differentiate whether the release of dopamine was caused by central processes of hunger or by the motor movements made during food consumption. Since many of these sites also showed increased dopamine release during consumption of water, it may be that they are related more to the control of consumatory motor movements than to the central neural substrates of hunger or thirst.

To overcome this limitation, workers have studied the effects of fasting on the levels and metabolism of dopamine in the brain. Fuenmayer (1979) found that 20 h without food increased the concentration of the dopamine metabolite HVA in the striatum of mice. No change in dopamine levels *per se* or in the related metabolite DOPAC was seen. This suggests that the increases in DOPAC seen by Biggio, Porceddu, Fratta & Gessa (1977) and by De Montis, Olianas, Di Lorenzo & Tagliamonte (1978) when rats were fasted and then allowed to eat food were more related to the motor act of food consumption than to the motivational consequences of fasting. A further confound with this latter approach, however, is that fasting may be a stressful experience and stress itself has been observed to increase the metabolism of brain dopamine in a manner unrelated to food intake (Bliss & Ailion, 1971).

The exact brain areas involved in the increased dopamine release seen during food intake have been determined by Heffner, Hartmann & Seiden (1980), who measured the levels of DOPAC in several discrete brain regions after fasted rats had been allowed 4 h of feeding. Increases were seen in the nucleus accumbens, the posterior hypothalamus and the amygdala but not in other dopamine-rich brain areas such as the caudate, septum, frontal cortex or olfactory tubercle. The changes in the amygdala may be related, not to the motor act of food consumption, but to metabolic consequences of ingestion since similar alterations in amygdaloid DOPAC could be caused by passively tube-feeding the animals.

In summary, although release/level studies have shown fasting and feeding to alter dopaminergic systems the mode of action *vis-à-vis* central hunger, consumatory motor acts or post-ingestion metabolic consequences remain to be investigated. More definite evidence has been forthcoming from studies on 6-OHDA lesions and eating (see p. 367).

Agonists. Peripheral injections of apomorphine may cause a suppression of food intake in hungry animals (Barzaghi, Gropetti, Mantegazza & Muller, 1973; Kruk, 1973; Heffner, Zigmond & Stricker, 1977). However, these also induce stereotyped behaviours which may reduce eating by response competition. If an animal is running around the cage it has difficulty eating at the same time. Better evidence for an inhibitory role of dopamine in food intake comes from the effects of central injections of the neurotransmitter. Leibowitz & Rossakis (1979*a*) mapped 24 different brain areas and found that the perifornical region of the hypothalamus was very sensitive to dopamine (and adrenaline) and caused a marked decrease in food intake in 22 h food-deprived rats. Pharmacological characterisation of the receptor type involved indicated (Leibowitz & Rossakis, 1979*b*) that dopamine was most effective, followed by apomorphine and with noradrenaline being considerably less potent. Such inhibition of feeding in response to dopamine could be blocked by pre-treatment with neuroleptic drugs such as haloperidol and pimozide but occurred as normal in the presence of α- or β-noradrenergic blockers such as phentolamine/tolazoline and propranolol/alprenolol/sotalol, respectively.

The results of central injections of dopamine into the perifornical hypothalamus raise the possibility that other anorexic agents may work by activating these self-same dopamine receptors, either directly or indirectly. Thus, Kruk & Zarrindast (1976) showed that mazindol would decrease food intake in 20 h food-deprived rats and that this effect seemed to be dopaminergic since it could be blocked by pre-treatment with pimozide but not by α- or β-blockers.

3′,4′-Dichloro-2-(2-imidazolin-2-yl-thio)-acetophenone (DITA) is a novel anorexic agent which may also act via dopaminergic mechanisms since its anorexic action in mice is blocked by haloperidol but not by serotonergic, α- or β-adrenergic agents (Abdallah, Roby, Boeckler & Riley, 1976).

The mode of action of amphetamine in causing anorexia is uncertain (see section on noradrenergic inhibition of feeding, p. 357). Although dopaminergic blockers such as haloperidol block the anorexic action of amphetamine (Clineschmidt, McGuffin & Werner, 1974; Leibowitz, 1975*c*; Abdallah *et al.*, 1976), it is also affected by both α- and β-adrenergic blockers. Thus, Frey & Schultz (1973) reported that phentolamine, an α-blocker, attenuated amphetamine anorexia. This was not observed by either Leibowitz (1975*c*) or by Abdallah *et al.* (1976). Further, Leibowitz (1975*c*) found that the β-blocker propranolol blocked the anorexic action of amphetamine while Abdallah *et al.* (1976) found it to be without effect and Mantegazza, Naimzada & Riva (1968) reported a potentiation of amphetamine anorexia with propranolol.

In summary, at least a subpopulation of dopaminergic systems, to be found in the perifornical hypothalamus, appear to inhibit feeding. The results of antagonists and 6-OHDA lesions suggests, however, that other dopaminergic systems may have opposing actions.

Antagonists. It has long been known that one of the properties of the neuroleptic drugs is a gross reduction in all forms of motor activity (see Chapter 4) and, as a consequence of this, considerable reduction in free-feeding food intake is also seen. Typical effects on food consumption in a 2 h period after 22 h of deprivation are shown in Table 7.2 (from an experiment by Hynes, Valentino & Lal, 1978). It can be seen that the dopamine blockers, pimozide, haloperidol, benperidol and chlorpromazine steeply reduce eating as the dose is increased. However, neither of the α-blockers phenoxybenzamine or aceperone had such a major effect. This would suggest a role for dopaminergic systems in food intake were it not for the possible confound of reduction in *all* motor behaviours. If the animal is unable to initiate a voluntary movement such as running, grooming, object manipulation and grasping caused by dopamine blockade, it is also going

Table 7.2. *Food consumption after dopamine receptor blockade*

Drug	Dose ($mg\ kg^{-1}$)	Food consumption (% of control)
Saline	—	108±3
Chlorpromazine	2.5	125±4
	10.0	24±10
Haloperidol	1.25	7±2
	5.0	0±1
Benperidol	1.25	4±2
Pimozide	1.25	54±7
	2.5	40±11
Aceperone	0.64	117±5
	2.5	96±4
Phenoxybenzamine	10.0	65±6

From Hynes *et al.* 1978.

to have considerable difficulty eating as much as normal. This was investigated more directly by Rolls *et al.* (1974), who compared the severity of neuroleptics on eating to their disruption of motor activity *per se*. Spiroperidol caused a small decrease in food intake at 0.1 mg kg^{-1} but completely abolished the motor activity required to lever-press for food at as small a dose as 0.06 mg kg^{-1}. This suggests that motor impairments are the primary action of spiroperidol and that all the decrease in eating might arguably have been caused by an inability to perform the necessary movements, unrelated to any central dopaminergic mechanisms of hunger.

Dopamine receptor blockers will also reduce feeding induced by direct electrical stimulation of the brain. Phillips & Nikaido (1975) found that rats electrically stimulated via chronic indwelling electrodes implanted into the lateral hypothalamus would eat upwards of 25 g of food despite being thoroughly satiated immediately prior to the test session. Treatment with haloperidol (0.15 mg kg^{-1}) reduced this stimulation-induced eating by more than 75 %. Again, it is unclear whether this is a direct effect on central hunger mechanisms or a response deficit resulting from the role of dopamine in voluntary movement (see Chapter 4).

Another way to cause satiated rats to eat is to pinch their tails. This bizarre behaviour has been investigated by Antelman & Szechtman (1975), and appears to depend on brain dopamine since it can be blocked by pimozide, haloperidol or spiroperidol but not by the α-blocker phentolamine or the β-blocker sotalol (Antelman *et al.*, 1975*b*). Again, it fails to dissociate the anorexic from the immobility-induced effects of neuroleptic drugs.

In summary, dopamine antagonists cause a clear reduction in food intake but this cannot be ascribed to direct interference with a dopaminergic hunger mechanism, as opposed to the dopaminergic substrate of voluntary movement.

6-OHDA lesion. By far the most direct evidence for a role of dopaminergic systems in food intake has come from 6-OHDA lesion studies. One especial advantage is the possibility they offer through localised injection into specific brain regions of determining *which* dopaminergic system has what effect on eating.

The classic first use of 6-OHDA to investigate behaviour was by Ungerstedt in 1971. He injected 8 μg 6-OHDA bilaterally into the substantia nigra or A10 area or into the dopaminergic fibre bundle as it ascended in the lateral hypothalamus. Body weight was determined daily, as was food and water intake. 6-OHDA injections into the substantia nigra and into the lateral hypothalamus caused severe aphagia and hypodipsia with a concomitant loss of body weight. 6-OHDA injection into the A10 area produced much smaller effects, whereas 6-OHDA injections aimed at the innervation to the nucleus accumbens and the olfactory tubercle produced virtually no prolonged effect on food intake. Ungerstedt (1971) pointed out the similarities with the classic lateral hypothalamic syndrome (Anand & Brobeck, 1951) of aphagia and adipsia caused by electrolytic lesion to this area and speculate that it may be caused by destruction of the dopaminergic fibres which run in the vicinity.

Similar results were quickly obtained by others. Smith, Strohmayer & Reis (1972) found slow recovery from the aphagia started about day 10 after bilateral injection of 6-OHDA into the hypothalamus. Drinking recovered but showed a similar pattern to that seen after electrolytic lateral hypothalamic lesion in which most fluid intake accompanies the consumption of dry food and even 20 h of water deprivation failed to cause much water intake unless dry food was also available. Unfortunately, loss of noradrenaline as well as dopamine occurred after the 6-OHDA injections so the catecholamine of importance could not be determined. That brain dopamine was of more importance was indicated by Zigmond & Stricker (1972), who found that pre-treatment with pargyline increased the degree of dopamine depletion without increasing that of noradrenaline and this was accompanied by the development of total aphagia as opposed to mild hypophagia. Response to metabolic challenges which normally result in increased food intake, such as injection of 2-deoxy-D-glucose, was also deficient after 6-OHDA lesion. However, the pattern of meals in rats recovered from 6-OHDA aphagia is indistinguishable from normals and vagotomy has the same effect of decreasing liquid meal size in both groups (Rowland, 1978).

Oltmans & Harvey (1972) localised the fibres which caused the aphagia by a series of small electrolytic lesions in and around the medial forebrain bundle. Lesions which interrupted the nigrostriatal dopaminergic bundle, as judged by post-mortem assay of dopamine content of the striatum, caused severe aphagia, while those which interrupted noncatecholaminergic fibres running on the adjacent medial forebrain bundle failed to decrease water intake or body weight. However, since ascending noradrenergic and serotonergic fibres also run in the region of the nigrostriatal bundle some loss of those neurotransmitters was also seen. Alheid *et al.* (1977) carried out a series of knife cuts which interrupted the dopaminergic innervation of the striatum to varying degrees. Those which were positioned medial to the striatum caused the greatest dopamine loss and the most severe aphagia and adipsia. However, other cuts more ventral caused similar aphagia but less dopamine loss, while denervation of the anterior striatum caused large dopamine losses with little aphagia. Hypothalamic noradrenaline content correlated equally as well with aphagia as did loss of striatal dopamine. However, Heffner *et al.* (1977) clarified this by the demonstration that only dopamine loss was necessary for aphagia. They protected brain noradrenergic neurones from 6-OHDA lesion by pre-treatment with DMI prior to intraventricular administration of 250 μg 6-OHDA and found marked aphagia despite telencephalic noradrenaline levels between 87 and 96 % of normal. It is thus clear that loss of dopamine is a sufficient condition to cause disruption of food intake.

Severe dopamine loss in the mesolimbic system on its own, caused by injection of 6-OHDA into A10, does not cause aphagia or adipsia (Koob, Stinus & Le Moal, 1981). Neither does injection of 6-OHDA into terminal areas of this system, the nucleus accumbens and the olfactory tubercle (Koob, Riley, Smith & Robbins, 1978; Koob *et al.*, 1981). Thus, the role of the dopaminergic innervation of the striatum is highlighted.

Other techniques for inducing food intake also appear to depend on dopamine mechanisms. Thus, Rowland, Marques & Fisher (1980) found that a single intraventricular injection of 250 μg 6-OHDA caused a total loss of electrically induced eating from electrodes chronically implanted into the lateral hypothalamus. Use of DMI pre-treatment meant that, although brain dopamine was reduced to 0–15 % by this treatment, brain noradrenaline was virtually completely spared. Similar results were seen after injection of 6-OHDA directly into the substantia nigra which caused essentially complete striatal and better than 90 % accumbens depletion of dopamine. Both Rowland *et al.* (1980) and Antelman *et al.* (1975*b*) found that 6-OHDA lesions which depleted brain dopamine abolished, or severely reduced, tail-pinch-induced eating in satiated rats.

In summary, 6-OHDA lesions have yielded clear evidence for a role of catecholamines in food intake and have enabled this effect to be ascribed to dopamine rather than noradrenaline. More localised 6-OHDA lesions, and also knife cuts have suggested that the dopaminergic innervation of the striatum is of more importance for this effect than that to the nucleus accumbens or the olfactory tubercle. 6-OHDA lesions to the source of the striatal dopamine projection arising from the substantia nigra have confirmed this and 6-OHDA lesion to the adjacent A10, which gives rise to the mesolimbic dopaminergic fibres, caused much less disruption of food intake. Some role for dopamine in water intake is also indicated by the disruptions in drinking which accompany, and may outlast, the disruption of food intake seen after 6-OHDA lesions. This will form the substance of the next section on the role of catecholamines in the related ingestive behaviour of drinking.

Catecholamines and drinking behaviour

Noradrenaline

Various approaches have been used to investigate the role of noradrenaline in water intake. Rather than divide them into methodological subsections the knowledge which they have provided for us will be presented as a unity.

Luttinger & Seiden (1981) found that limiting the access of the rat to water for only 10 min per day caused an increase in the metabolism of noradrenaline in the hypothalamus as judged by its faster disappearance after synthesis inhibition with AMPT. This effect was specific for noradrenaline since brain dopamine failed to show any change. Further, other *noradrenergic* regions such as the forebrain and brainstem failed to show any change in noradrenaline turnover. This highlights the role of hypothalamic noradrenaline in water intake. A possible mechanism for this lies in the role of hypothalamic noradrenaline in the control of vasopressin secretion. Vasopressin regulates water balance by decreasing water excretion via the kidneys. Crowley, O'Donohue, George & Jacobowitz (1978*b*) found that lesion to the ventral noradrenergic bundle decreases noradrenaline concentration in the supraoptic and paraventricular nuclei of the hypothalamus and decreases vasopressin levels in the pituitary.

Singer & Montgomery (1968) reported that injection of noradrenaline into the amygdala of rats chronically implanted with cannulae would reduce the amount of drinking after 3 h of water deprivation compared to saline injection. However, injections of noradrenaline into the hypothalamus of rats generally elicit drinking, even in satiated animals (Coury, 1967; Setler, 1973; Leibowitz, 1975*a*, 1980). Similar results have been obtained for intrahypothalamic noradrenaline injections in the gerbil (Block, Vallier & Glickman, 1974).

Peripheral injection of the β-agonist isoproterenol elicits copious drinking in the rat (Lehr, Mallow & Krukowski, 1967). However, this appears to involve a peripheral system (perhaps the renin-angiotensin system: Houpt & Epstein, 1971), since central injections of isoproterenol directly into the hypothalamus failed to elicit drinking (Singer & Armstrong, 1973), although weak drinking could be achieved by injections into the hippocampus (Mountford, 1969) or the septum (Giardina & Fisher, 1971).

6-OHDA lesions which maximise the loss of noradrenaline while sparing forebrain dopamine have been achieved by three intraventricular injections of 25 μg 6-OHDA by Sorenson, Ellison & Masuoka (1972). No change in baseline water intake was seen, but marked reduction in the consumption of both 0.02 % quinine and 3 % sucrose solution was observed in the lesioned rats (see Fig. 7.20). Similar reduction was seen with 4 % but not with 2 % sucrose solutions. Similar results have also been seen by Mason, Roberts & Fibiger (1978) in which dorsal bundle injections of 6-OHDA which depleted forebrain noradrenaline, but not dopamine, caused a reduction in consumption of saccharin, saline and chocolate milk solution compared to control rats. Again, no change occurred in the baseline intake of water *per se*. Similar reduction in saccharin consumption has been seen by Archer, Ogren & Ross (1981) after depletion of brain and peripheral noradrenaline by the novel neurotoxin DSP4. However, since this neurotoxin of uncertain specificity also reduced baseline water intake, Archer *et al.* (1981) attempted to allow for this by calculating a ratio measure of saccharin intake to water intake. This doubtful procedure led them to the paradoxical conclusion that saccharin intake was unaffected by noradrenaline loss.

In summary, injection of noradrenaline into the hypothalamus and possibly limbic areas may elicit drinking. Specific lesion to noradrenergic systems with 6-OHDA does not alter baseline water intake but may affect consumption of flavoured solutions, possibly via an effect on taste sensitivity or neophobia.

Fig. 7.20. Intakes of sucrose, water or quinine by rats injected with various intraventricular doses of 6-hydroxydopamine (6-OHDA). (Redrawn from Sorenson *et al.*, 1972.)

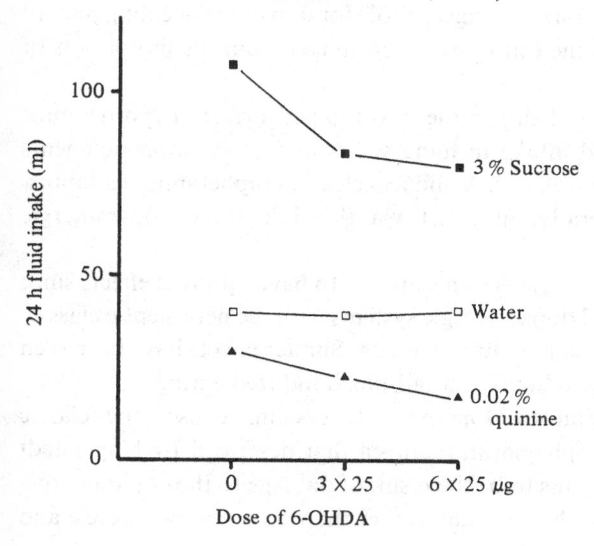

Dopamine

The profound adipsia caused by extensive 6-OHDA lesions of the ascending nigrostriatal dopaminergic systems has been mentioned previously (see p. 367). Brief mention may also be made here of the effects of water deprivation on the dopaminergic systems in the pituitary. Banns *et al.* (1978) found that 1–5 days of water deprivation caused doubling of the levels of dopamine in the intermediate lobe of the pituitary, as well as a rise in the catecholamine content of the median eminence and arcuate nucleus of the hypothalamus. Related results were obtained by Alper, Demarest & Moore (1980), who found 2 days of water deprivation increased the pituitary dopamine concentration but failed to affect that in the median eminence. Food deprivation was without effect on pituitary dopamine. This suggests that water deprivation may activate the tuberohypophyseal dopaminergic system, which has cell bodies in the arcuate nucleus of the hypothalamus and terminals in the posterior pituitary.

Not surprisingly the dopamine antagonist pimozide reduced water intake (Nielson & Lyon, 1973) at doses above 1 mg kg^{-1}. Since these doses start to cause motor impairments it is uncertain whether this effect is directly on central thirst mechanisms or more related to the impairment of voluntary movement caused by neuroleptic drugs. Perhaps more paradoxical is their observation that the dopamine agonist apomorphine also reduced water intake at a dose of 0.8 mg kg^{-1}.

Summary

(1) Release/levels studies suggest a role for dopamine in eating but fail to distinguish the central state of hunger from the motor acts of eating.

(2) Microinjection of dopamine into the perifornical hypothalamus decreases food intake in hungry rats and other anorexic agents (such as mazindol, DITA and less clearly amphetamine) administered peripherally, may act via this inhibitory dopaminergic system.

(3) Other dopaminergic systems appear to have opposite effects since blockade of all dopaminergic systems with the neuroleptic class of drugs dramatically reduces feeding. Similar effects have been seen on electrically induced and tail-pinch induced eating.

(4) 6-OHDA lesion to dopaminergic systems causes the classic syndrome of aphagia and adipsia first described by Ungerstedt (1971). This seems to be the result of damage to the dopaminergic innervation of the striatum, rather than accumbens–tubercle and

it also blocks electrically induced and tail-pinch induced eating behaviour.

(5) Noradrenaline injections into the hypothalamus may elicit drinking, although isoproterenol thirst seen after peripheral administration of the drug does not seem to involve this mechanism. 6-OHDA lesion to dorsal noradrenergic systems does not affect baseline water consumption but decreases intake of flavoured solutions, possibly via effects on taste sensitivity or neophobia.

(6) Dopaminergic systems in the pituitary show changes after water deprivation and may have a role to play in the control of thirst.

References

Abdallah, A. H. (1971). On the role of norepinephrine in the anorexic effect of D-amphetamine in mice. *Archives internationales de pharmacodynamie et de thérapie*, **192,** 72–77.

Abdallah, A. H., Roby, D. M., Boeckler, W. H. & Riley, C. C. (1976). Role of dopamine in the anorexigenic effect of DITA: comparison with *d*-amphetamine. *European Journal of Pharmacology*, **40,** 39–44.

Ahlskog, J. E. (1974). Food intake and amphetamine anorexia after selective forebrain norepinephrine loss. *Brain Research*, **82,** 211–40.

Ahlskog, J. E. (1976). Feeding response to regulatory challenges after 6-hydroxydopamine injection into the brain noradrenergic pathways. *Physiology and Behavior*, **17,** 407–11.

Ahlskog, J. E. & Hoebel, B. G. (1973). Overeating and obesity from damage to a noradrenergic system in the brain. *Science*, **182,** 166–9.

Ahlskog, J. E., Randall, P. K. & Hoebel, B. G. (1975). Hypothalamic hyperphagia: dissociation from hyperphagia following destruction of noradrenergic neurons. *Science*, **190,** 399–401.

Albert, D. J. & Richmond, S. E. (1977). Reactivity and aggression in the rat induced by alpha-adrenergic blocking agents injected into the region ventral to the anterior septum but not into the lateral septum. *Journal of Comparative and Physiological Psychology*, **91,** 886–96.

Albert, D. J., Wong, R. C., Brayley, K. N. & Fibiger, H. C. (1979). Evaluation of adrenergic, cholinergic and dopaminergic involvement in the inhibition of hyperreactivity and interanimal aggression by the medial hypothalamus in the rat. *Pharmacology, Biochemistry & Behavior*, **11,** 1–10.

Alheid, G. F., McDermott, L., Kelly, J., Halaris, A. & Grossman, S. P. (1977). Deficits in food and water intake after knife cuts that deplete striatal DA or hypothalamic NE in rats. *Pharmacology, Biochemistry & Behavior*, **6,** 273–87.

Alper, R. H., Demarest, K. T. & Moore, K. E. (1980). Dehydration selectively increases dopamine synthesis in tuberohypophyseal dopaminergic neurons. *Neuroendocrinology*, **31,** 112–15.

Amaral, D. G. & Foss, J. A. (1975). Locus coeruleus lesions and learning. *Science*, **190,** 399–401.

Anand, B. K. & Brobeck, J. R. (1951). Hypothalamic control of food intake in rats and cats. *Yale Journal of Biology and Medicine*, **24**, 123–140.

Anand, M., Gupta, G. P. & Bhargava, K. P. (1977). Modifications of electroshock fighting by drugs known to interact with dopaminergic and noradrenergic neurones in normal and brain lesioned rats. *Journal of Pharmacy and Pharmacology*, **29**, 437–9.

Andy, O. & Velamati, S. (1978). Temporal lobe seizures and hypersexuality. *Applied Neurophysiology*, **41**, 13–28.

Anlezark, G. M., Crow, T. J. & Greenway, A. P. (1973). Impaired learning and decreased cortical norepinephrine after bilateral locus coeruleus lesions. *Science*, **181**, 682–4.

Antelman, S. M., Herndon, J. G., Caggiula, A. R. & Shaw, D. H. (1975*a*). Dopamine-receptor blockade: prevention of shock-activated sexual behaviour in naive rats. *Psychopharmacology Bulletin*, **11**, 45–6.

Antelman, S. M. & Szechtman, H. (1975). Tail-pinch induces eating in satiated rats which appears to depend on nigrostriatal dopamine. *Science*, **189**, 731–3.

Antelman, S. M., Szechtman, H., Chin, P. & Fisher, A. E. (1975*b*). Tail pinch-induced eating, gnawing and licking behaviour in rats: dependence on the nigrostriatal dopamine system. *Brain Research*, **99**, 319–37.

Archer, T., Ogren, S-O. & Ross, S. B. (1981). *N*-2-chloroethyl-*N*-ethyl-2-bromobenzylamine hydrochloride (DSP4), a new selective noradrenaline neurotoxin, and taste neophobia in the rat. *Physiological Psychology*, **9**, 197–202.

Aston-Jones, G. & Bloom, F. E. (1981). Activity of norepinephrine-containing locus coeruleus neurons in behaving rats anticipates fluctuations in the sleep–waking cycle. *Journal of Neuroscience*, **1**, 876–86.

Avery, D. D. (1971). Intrahypothalamic adrenergic and cholinergic injection: effects on temperature and ingestive behavior in the rats. *Neuropharmacology*, **10**, 753–63.

Baekelund, F., Schenker, V. J., Schenker, A. C. & Lasky, R. (1969). Urinary excretion of epinephrine, norepinephrine, dopamine and tryptamine during sleep and wakefulness. Effects of pentobarbital, pentobarbital plus dextroamphetamine sulfate, and placebo. *Psychopharmacologia*, **14**, 359–70.

Baenninger, R. (1968). Catecholamines and social relations in Siamese fighting fish. *Animal Behavior*, **16**, 442–7.

Baggio, G. & Ferrari, F. (1980). Role of brain dopaminergic mechanisms in rodent aggressive behavior: influence of (±)*N*-*n*-propyl-norapomorphine on three experimental models. *Psychopharmacology*, **70**, 63–8.

Baile, C. A., Simpson, C. W., Krabill, L. F. & Martin, F. H. (1972). Adrenergic agonists and antagonists and feeding in sheep and cattle. *Life Sciences*, **11**, 661–8.

Bandler, R. J. (1970). Cholinergic synapses in the lateral hypothalamus for the control of predatory aggression in the rat. *Brain Research*, **20**, 409–24.

Bandler, R. J. (1971). Chemical stimulation of the rat midbrain and aggressive behaviour. *Nature New Biology*, **229**, 222–3.

Bandler, R. J. & Moyer, K. E. (1970). Animals spontaneously attacked by rats. *Communications in Behavioral Biology*, **5**, 177–82.

Banerjee, U. (1974). Modification of the isolation-induced abnormal behavior in male Wistar rats by destructive manipulation of the central monoaminergic systems. *Behavioral Biology*, **11**, 573–7.

Banns, H., Blatchford, D., Godden, U., Holzbauer, M., Mann, S. P. & Sharman, D. F.

(1978). Pituitary dopamine in rats after decreased water intake. *British Journal of Pharmacology*, **62,** 462P.

Barr, G. A., Sharpless, N. S. & Gibbons, J. L. (1979). Differences in the level of dopamine in the hypothalamus of aggressive and non-aggressive rats. *Brain Research*, **166,** 211–16.

Barzaghi, R., Gropetti, A., Mantegazza, P. & Muller, E. E. (1973). Reduction of food intake by apomorphine: a pimozide-sensitive effect. *Journal of Pharmacy and Pharmacology*, **25,** 909–11.

Baum, M. J. & Starr, M. S. (1980). Inhibition of sexual behaviour by dopamine antagonist or serotonin agonist drugs in castrated male rats given estradiol or dihydrotestosterone. *Pharmacology, Biochemistry & Behavior*, **13,** 57–67.

Bellin, S. I. & Ritter, S. (1981). Insulin-induced elevation of hypothalamic norepinephrine turnover persists after glucorestoration unless feeding occurs. *Brain Research*, **217,** 327–37.

Bellinger, L. & Bernardis, L. (1976). Mesencephalic lesions followed by normophagia but decreased body weight. *Journal of Neuroscience Research*, **2,** 47–56.

Bellinger, L. L., Bernardis, L. L. & Brooks, S. (1979). Mesencephalic lesions in female rats resulting in normophagia and reduced body weight. *Journal of Neuroscience Research*, **4,** 205–14.

Bellinger, L., Bernardis, L. & Goldman, J. (1976). Mesencephalic lesions resulting in normophagia, reduced weight and altered metabolism. *Journal of Neuroscience Research*, **2,** 217–31.

Bensch, C., Lescure, H., Robert, J. & Faure, J. M. A. (1975). Catecholamine histofluorescence in the median eminence of female rabbits activated by mating. *Journal of Neural Transmission*, **36,** 1–18.

Berger, B. D., Wise, C. D. & Stein, L. (1971). Norepinephrine: reversal of anorexia in rats with lateral hypothalamic damage. *Science*, **172,** 281–4.

Bernard, B. K., Finkelstein, E. R. & Everett, G. M. (1975). Alterations in mouse aggressive behavior and brain monoamine dynamics as a function of age. *Physiology and Behavior*, **15,** 731–6.

Biggio, G., Porceddu, M. L., Fratta, W. & Gessa, G. L. (1977). Changes in dopamine metabolism associated with fasting and satiation. In *Advances in biochemical psychopharmacology*, eds. E. Costa & G. L. Gessa. pp. 377-80. Raven Press: New York.

Bliss, E. L. & Ailion, J. (1971). Relationship of stress and activity to brain dopamine and homovanillic acid. *Life Sciences*, **10,** 1161–9.

Block, M. L., Vallier, G. H. & Glickman, S. E. (1974). Elicitation of water ingestion in the Mongolian gerbil (*Meriones unguiculatus*) by intracranial injections of angiotensin II and *l*-norepinephrine. *Pharmacology, Biochemistry & Behavior*, **2,** 235–42.

Blundell, J. E. & Herberg, L. J. (1970). Adrenergic stimulation of the rat diencephalon and its effect on food intake and hoarding activity. *Quarterly Journal of Experimental Psychology*, **22,** 125–32.

Booth, D. (1967). Localization of the adrenergic feeding system in the rat diencephalon. *Science*, **158,** 515–17.

Booth, D. A. (1968). Mechanism of action of norepinephrine in elicitating an eating response on injection into the rat hypothalamus. *Journal of Pharmacology and Experimental Therapeutics*, **160,** 336–48.

Booth, D. A. & Quartermain, D. (1965). Taste sensitivity of eating elicited by chemical stimulation of the rat hypothalamus. *Psychonomic Science*, **3,** 525–6.

Braun, C. M. J. & Pivik, R. T. (1981). Effects of locus coeruleus lesions upon sleeping and waking in the rabbit. *Brain Research*, **230**, 133–51.

Broekkamp, C. L. E., Honig, W. M. M., Pauli, A. I. & van Rossum, J. M. (1974). Pharmacological suppression of eating behavior in relation to diencephalic noradrenergic receptors. *Life Sciences*, **14**, 473–81.

Broekkamp, C. L. E. & van Rossum, J. M. (1972). Clonidine induced intrahypothalamic stimulation of eating in rats. *Psychopharmacologia*, **25**, 162–8.

Brown, W. A., Krieger, D. T., van Woert, M. H. & Ambani, L. M. (1974). Dissociation of growth hormone and cortisol release following apomorphine. *Journal of Clinical Endocrinology and Metabolism*, **38**, 1127–9.

Buus Larsen, J. (1974). Evidence for a noradrenergic and dopaminergic mechanism in hyperactivity produced by 4-alpha-dimethyl-*m*-tyramine (H 77-77) in rats . *Psychopharmacologia*, **37**, 331–40.

Buus Larsen, J. (1978). Piperoxan reduces the effects of clonidine on aggression in mice and on noradrenaline dependent hypermotility in rats. *European Journal of Pharmacology*, **47**, 45–9.

Caggiula, A. R., Herndon, J. G., Scanlon, R., Greenstone, D., Bradshaw, W. & Sharp, D. (1979). Dissociation of active from immobility components of sexual behavior in female rats by central 6-hydroxydopamine: implications for CA involvement in sexual behavior and sensorimotor responsiveness. *Brain Research*, **172**, 505–20.

Carey, R. J. (1976). Effects of selective forebrain depletions of norepinephrine and serotonin on the activity and food intake effects of amphetamine and fenfluramine. *Pharmacology, Biochemistry & Behavior*, **5**, 519–23.

Cedarbaum, J. M. & Aghajanian, G. K. (1976). Noradrenergic neurones of the locus coeruleus: inhibition by epinephrine and activation by piperoxan. *Brain Research*, **112**, 413–20.

Chambers, D. M. & Roberts, D. J. (1968). Some pharmacological effects of noradrenaline and its metabolites injected into the cerebral ventricles of mice. *British Journal of Pharmacology*, **34**, 223P–4P.

Choudhury, S. A. R., Sharpe, R. M. & Brown, P. S. (1973). Pituitary follicle-stimulating hormone activity in rats treated with apomorphine, pimozide and drugs which modify catecholamine levels. *Neuroendocrinology*, **12**, 272–81.

Chu, N. & Bloom, F. E. (1973). Norepinephrine-containing neurons: changes in spontaneous discharge patterns during sleeping and waking. *Science*, **179**, 908–10.

Chu, N. & Bloom, F. E. (1974). Activity patterns of catecholamine-containing pontine neurons in the dorso-lateral tegmentum of unrestrained cats. *Journal of Neurobiology*, **5**, 527–44.

Cianchetti, C., Masala, C., Mangoni, A. & Gessa, G. C. (1980). Suppression of REM and delta sleep by apomorphine in man: a dopamine mimetic effect. *Psychopharmacology*, **67**, 61–5.

Clarenbach, P. & Cramer, H. (1972). Polygraphic sleep pattern in newly hatched chickens: effects of noradrenaline and 6-hydroxydopamine. *Brain Research*, **43**, 695–9.

Clark, T. K. (1980). Male rat sexual behavior compared after 6-OHDA and electrolytic lesions in the dorsal NA bundle region of the midbrain. *Brain Research*, **202**, 429–43.

Clark, T. K., Caggiula, A. R., McConnell, R. A. & Antelman, S. M. (1975). Sexual inhibition is reduced by rostral midbrain lesions in the male rat. *Science*, **190**, 169–171.

Clineschmidt, B. V., McGuffin, J. C. & Werner, A. B. (1974). Role of monoamines in the anorexigenic actions of fenfluramine, amphetamine and *p*-chloroamphetamine. *European Journal of Pharmacology*, **27**, 313–19.

Coons, E. E. & Quartermain, D. (1970). Motivational depression associated with norepinephrine-induced eating from the hypothalamus: resemblance to the ventromedial hyperphagic syndrome. *Physiology and Behavior*, **5**, 687–92.

Coscina, D. V., Goodman, J., Godse, D. D. & Stancer, H. C. (1975). Taming effects of handling on 6-hydroxydopamine induced rage. *Pharmacology, Biochemistry & Behavior*, **3**, 525–8.

Coscina, D. V., Seggie, J., Godse, D. D. & Stancer, H. C. (1973). Induction of rage in rats by central injection of 6-hydroxydopamine. *Pharmacology, Biochemistry & Behavior*, **1**, 1–6.

Coury, J. N. (1967). Neural correlates of food and water intake in the rat. *Science*, **156**, 1763–5.

Cox, V., Kakolewski, J. & Valenstein, E. (1969). Ventromedial hypothalamic lesions and changes in body weight and food consumption in male and female rats. *Journal of Comparative and Physiological Psychology*, **67**, 320–6.

Crawley, J. N. & Contrera, J. F. (1976). Intraventricular 6-hydroxydopamine lowers isolation-induced fighting behavior in male mice. *Pharmacology, Biochemistry & Behavior*, **4**, 381–4.

Crow, T. J., Deakin, J. F. W., File, S., Longden, A. & Wendlandt, S. (1978). The locus coeruleus noradrenergic system – evidence against a role in attention, habituation, anxiety and motor activity. *Brain Research*, **155**, 249–61.

Crowley, W. R., Feder, H. H. & Morin, L. P. (1976). Role of monoamines in sexual behavior of the female guinea-pig. *Pharmacology, Biochemistry & Behavior*, **4**, 67–71.

Crowley, W. R., Nock, B. L. & Feder, H. H. (1978*a*). Facilitation of lordosis behavior by clonidine in female guinea-pig. *Pharmacology, Biochemistry & Behavior*, **8**, 207–9.

Crowley, W. R., O'Donohue, T. L., George, J. M. & Jacobowitz, D. M. (1978*b*). Noradrenergic control of pituitary oxytocin (OXT) and arginine vasopressin (AVP). *Neuroscience Abstracts*, **4**, 1038.

Da Vanzo, J. P. M., Daugherty, M., Ruckart, R. & Kang, L. (1966). Pharmacological and biochemical studies in isolation-induced fighting mice. *Psychopharmacologia*, **9**, 210–19.

Davis, G. A. & Kohl, R. (1977). The influence of alpha-receptors on lordosis in female rat. *Pharmacology, Biochemistry & Behavior*, **6**, 47–53.

De Montis, G., Olianas, M. C., Di Lorenzo, C. & Tagliamonte, A. (1978). Failure of morphine to increase striatal 3,4-dihydroxyphenylacetic acid in fasted rats. *European Journal of Pharmacology*, **47**, 121–3.

Dobrzanski, S. & Doggett, N. S. (1975). Effect of dopamine-beta-hydroxylase inhibitors and centrally administered noradrenaline on (+)-amphetamine anorexia in mice. *British Journal of Pharmacology*, 245P–246P.

Eichelman, B. S., Thoa, N. B. & Ng, K. Y. (1972). Facilitated aggression in the rat following 6-hydroxydopamine administration. *Physiology and Behavior*, **8**, 1–3.

Eison, M. S., Stark, A. D. & Ellison, G. (1977). Opposed effects of locus coeruleus and substantia nigra lesions on social behavior in rat colonies. *Pharmacology, Biochemistry & Behavior*, **7**, 87–90.

Ellison, G. (1976). Monoamine neurotoxins: selective and delayed effects on behavior in colonies of laboratory rats. *Brain Research*, **103**, 81–92.

Enslen, M., Milon, H. & Wurzner, H. P. (1980). Brain catecholamines and sleep states in offspring of caffeine-treated rats. *Experientia*, **36**, 1105–6.

Everett, G. M. (1977). Changes in brain dopamine levels and aggressive behavior with aging in 2 mouse strains. *Experientia*, **33**, 645–6.

Everitt, B. J., Fuxe, K. & Hokfeldt, T. (1974). Inhibitory role of dopamine and 5-hydroxytryptamine in the sexual behaviour of female rats. *European Journal of Pharmacology*, **29**, 187–91.

Everitt, B. J., Fuxe, K., Hokfeldt, T. & Jonsson, G. (1975). Role of monoamines in the control by hormones of sexual receptivity in the female rat. *Journal of Comparative and Physiological Psychology*, **89**, 556–72.

Evetts, K. D., Fitzsimons, J. T. & Setler, P. E. (1972). Eating caused by 6-hydroxydopamine-induced release of noradrenaline in the diencephalon of the rat. *Journal of Physiology*, **223**, 35–47.

Evetts, K. D., Uretsky, N. J., Iversen, L. L. & Iversen, S. D. (1970). Effects of 6-hydroxydopamine on CNS catecholamines, spontaneous motor activity and amphetamine induced hyperactivity in rats. *Nature*, **225**, 961–2.

Fenske, M. & Wuttke, W. (1976). Effects of intraventricular 6-hydroxydopamine injections on serum prolactin and LH levels: absence of stress induced pituitary prolactin release. *Brain Research*, **104**, 63–70.

File, S. F., Deakin, J. F. W., Longden, A. & Crow, T. J. (1979). An investigation of the role of the locus coeruleus in anxiety and agonistic behaviour. *Brain Research*, **169**, 411–20.

Foote, S. L., Bloom, F. E. & Schwartz, A. (1978). Behavioral and electro-encephalographic correlates of locus coeruleus neuronal discharge activity in the unanaesthetised squirrel monkey. *Society for Neuroscience Abstracts*, **4**, 848.

Franklin, K. B. & Herberg, L. J. (1977). Amphetamine induces anorexia even after inhibition of noradrenaline synthesis. *Neuropharmacology*, **16**, 45–6.

Frey, H. & Schultz, R. (1973). On the central mediation of anorexigenic drug effects. *Biochemical Pharmacology*, **22**, 3041–5.

Fuenmayer, L. D. (1979). The effect of fasting on the metabolism of 5-hydroxytryptamine and dopamine in the brain of the mouse. *Journal of Neurochemistry*, **33**, 481–5.

Fuxe, K., Lidbrink, P., Hokfeldt, T., Bolme, P. & Goldstein, M. (1974). Effects of piperoxan on sleep and waking in the rat. Evidence for increased waking by blocking inhibitory adrenaline receptors on the locus coeruleus. *Acta physiologica scandinavica*, **91**, 566–7.

Geyer, M. A. & Segal, D. S. (1974). Shock-induced aggression: opposite effects of intraventricularly infused dopamine and norepinephrine. *Behavioral Biology*, **10**, 99–104.

Ghosh, P. K., Hrdina, P. D. & Ling, G. M. (1976). Effects of REMS deprivation on striatal dopamine and acetylcholine in rats. *Pharmacology, Biochemistry & Behavior*, **4**, 401–5.

Giardina, A. R. & Fisher, A. E. (1971). Effect of atropine on drinking induced by carbachol, angiotensin and isoproterenol. *Physiology and Behavior*, **7**, 653–5.

Glick, S. D., Waters, D. H. & Milloy, S. (1973). Depletion of hypothalamic norepinephrine by food deprivation and interaction with *d*-amphetamine. *Research Communications in Chemical Pathology and Pharmacology*, **6**, 775–8.

Goldberg, M. E. & Salama, A. I. (1969). Norepinephrine turnover and brain monoamine levels in aggressive mouse-killing rats. *Biochemical Pharmacology*, **18**, 532–4.

Grossman, S. P. (1960). Eating or drinking elicited by direct adrenergic stimulation of the hypothalamus. *Science*, **132**, 301–2.

Grossman, S. P. (1962*a*). Direct adrenergic and cholinergic stimulation of hypothalamic mechanisms. *American Journal of Physiology*, **202**, 872–82.

Grossman, S. P. (1962*b*). Effects of adrenergic blocking agents on hypothalamic mechanisms. *American Journal of Physiology*, **202,** 1230–6.

Grossman, S. P., Grossman, L. & Halaris, A. (1977). Effects on hypothalamic and telencephalic NE and 5-HT of tegmental knife cuts that produce hyperphagia or hyperdypsia in the rat. *Pharmacology, Biochemistry & Behavior*, **6,** 101–6.

Hadfield, M. G. & Rigby, W. F. (1976). Dopamine-adaptive changes in striatal synaptosomes after 30 sec of shock-induced fighting. *Biochemical Pharmacology*, **25,** 2752–4.

Hadfield, M. G. & Weber, N. E. (1975). Effect of fighting and diphenylhydantoin on the uptake of ^{3}H-*l*-norepinephrine *in vitro* in synaptosomes isolated from retired male breeding mice. *Biochemical Pharmacology*, **24,** 1538–40.

Hall, R. C. & Keene, P. E. (1975). Dopaminergic and cholinergic interactions in the caudate nucleus in relation to the induction of sleep. *British Journal of Pharmacology*, **54,** 247P–248P.

Hamburger-Bar, R. & Rigter, H. (1975). Apomorphine: facilitation of sexual behavior in female rats. *European Journal of Pharmacology*, **32,** 357–60.

Hansen, M. G. & Whishaw, I. Q. (1973). The effects of 6-hydroxydopamine, dopamine and *d,l*-norepinephrine on food intake and water consumption, self-stimulation, temperature and electroencephalographic activity in the rat. *Psychopharmacologia*, **29,** 33–44.

Hansen, S., Kohler, C. & Ross, S. B. (1982). On the role of the dorsal mesencephalic tegmentum in the control of masculine sexual behavior in the rat: effects of electrolytic lesions, ibotenic acid and DSP4. *Brain Research*, **240,** 311–20.

Hansen, S., Stanfield, E. J. & Everitt, B. J. (1980). The role of the ventral bundle noradrenergic neurones in sensory components of sexual behaviour and coitus-induced pseudopregnancy. *Nature*, **286,** 152–4.

Harner, R. N. & Dorman, R. M. (1970). A role for dopamine in control of sleep and wakefulness. *Transactions of the American Neurological Association*, **95,** 252–4.

Hartmann, E., Bridwell, T. J. & Schildkraut, J. J. (1971*a*). Alpha-methyl-*para*-tyrosine and sleep in the rat. *Psychopharmacologia*, **21,** 157–62.

Hartmann, E., Chung, R., Draskoczy, P. R. & Schildkraut, J. J. (1971*b*). Effects of 6-hydroxydopamine on sleep in the rat. *Nature*, **233,** 425–7.

Hartmann, E. & Schildkraut, J. J. (1973). Desynchronised sleep and MHPG excretion: an inverse correlation. *Brain Research*, **61,** 412–16.

Hartmann, E., Zwilling, G. R. & Koski, S. (1973). *The effects of pimozide on sleep in the rat.* Association for the Psychophysiological Study of Sleep: San Diego, California.

Hasselager, E., Rolinski, Z. & Randrup, A. (1972). Specific antagonism by dopamine inhibitors of items of amphetamine induced aggressive behaviour. *Psychopharmacologia*, **24,** 485–95.

Havlicek, V. (1967). The effects of 3,4-dihydroxyphenylserine on the ECoG of unrestrained rats. *International Journal of Neuropharmacology*, **6,** 83–8.

Heffner, T. G., Hartmann, J. A. & Seiden, L. S. (1980). Feeding increases dopamine metabolism in the rat brain. *Science*, **208,** 1168–70.

Heffner, T. G., Zigmond, M. J. & Stricker, E. M. (1977). Effects of dopaminergic agonists and antagonists on feeding in intact and 6-hydroxydopamine-treated rats. *Journal of Pharmacology and Experimental Therapeutics*, **201,** 386–99.

Henley, E. D., Moisset, B. & Welch, B. L. (1973). Catecholamine uptake in cerebral cortex: adaptive change induced by fighting. *Science*, **180,** 1050–2.

Herberg, L. J. & Franklin, K. B. (1972). Adrenergic feeding: its blockade or reversal by posterior VHM lesions; and a new hypothesis. *Physiology & Behavior*, **8,** 1029–34.
Herberg, L. J. & Stephens, D. N. (1976). Cyclic AMP and central noradrenaline receptors: failure to activate diencephalic adrenergic feeding pathways. *Pharmacology, Biochemistry & Behavior*, **4,** 107–10.
Hernandez, L. & Hoebel, B. G. (1982). Overeating after midbrain 6-hydroxydopamine injection: prevention by central injection of selective catecholamine reuptake blockers. *Brain Research*, **245,** 333–43.
Herndon, J. G. (1976). Effects of midbrain lesions on female sexual behavior in the rat. *Physiology and Behavior*, **17,** 143–8.
Herndon, J. G., Caggiula, A. R., Sharp, D., Ellis, D. & Redgate, E. (1978). Selective enhancement of the lordotic component of female sexual behavior in rats following destruction of central catecholamine-containing systems. *Brain Research*, **141,** 137–51.
Hobson, J. A., McCarley, R. W. & Wyzinski, P. W. (1975). Sleep cycle oscillation: reciprocal discharge by two brainstem neuronal groups. *Science*, **189,** 55–8.
Houpt, K. A. & Epstein, A. N. (1971). The complete dependence of beta-adrenergic drinking on the renal dipsogen. *Physiology and Behavior*, **7,** 897–902.
Hutchins, D. A., Pearson, J. D. & Sharman, D. F. (1975). Striatal metabolism of dopamine in mice made aggressive by isolation. *Journal of Neurochemistry*, **24,** 1151–4.
Hynes, M. D., Valentino, D. & Lal, H. (1978). Role of dopaminergic systems in tolerance to neuroleptic-induced inhibition of food consumption. *Neuropharmacology*, **17,** 507–10.
Ikeda, H., Nishikawa, K. & Matsuo, T. (1980). Feeding responses of Zucker fatty rat to 2-deoxy-D-glucose, norepinephrine, and insulin. *American Journal of Physiology*, **239,** E379–E384.
Ismahan, G., Parvez, H. & Parvez, S. (1979). Influence of progressive starvation upon brain and adrenal monoaminergic activity in developing rats of two different ages. *Biology of the Neonate*, **35,** 224–34.
Jalowiec, J. E., Morgane, P. J., Stern, W. C., Zolovick, A. J. & Panksepp, J. (1973). Effects of midbrain tegmental lesions on sleep and regional brain serotonin and norepinephrine levels in cats. *Experimental Neurology*, **41,** 670–82.
Jimerson, D. & Reis, D. J. (1973). Effects of intrahypothalamic injection of 6-hydroxydopamine on predatory aggression in rat. *Brain Research*, **61,** 141–52.
Jones, B. (1972). The respective involvement of noradrenaline and its deaminated metabolites in waking and paradoxical sleep: a neuropharmacological model. *Brain Research*, **39,** 121–36.
Jones, B. E., Bobillier, P., Pin, C. & Jouvet, M. (1973). The effect of lesions of catecholamine-containing neurons upon monoamine content of the brain and EEG and behavioral waking in the cat. *Brain Research*, **58,** 157–77.
Jones, B. E., Harper, S. T. & Halaris, A. E. (1977). Effects of locus coeruleus lesions upon cerebral monoamine content, sleep-wakefulness states and the response to amphetamine in the cat. *Brain Research*, **124,** 473–96.
Jones, G., Foote, S. L., Segal, M. & Bloom, F. E. (1978). Locus coeruleus neurones in freely behaving rats exhibit pronounced alterations in firing rate during sensory stimulation and stages of sleep-wake cycle. *Society for Neuroscience Abstracts*, **4,** 856.
Jouvet, M. (1969). Biogenic amines and the states of sleep. *Science*, **163,** 32–41.
Kafi, S. & Gaillard, J. M. (1976). Brain dopamine receptors and sleep in the rat: effects of stimulation and blockade. *European Journal of Pharmacology*, **38,** 357–63.
Kales, A., Ansel, R. D., Markham, C. H., Scharf, M. B. & Tan, Y. (1971). Sleep in patients

with Parkinson's disease and normal subjects prior to and following *levo*-dopa administration. *Clinical Pharmacology*, **12,** 397–406.

King, C. D. & Jewett, R. E. (1971). The effects of alpha-methyltyrosine on sleep and brain norepinephrine in cats. *Journal of Pharmacology and Experimental Therapeutics*, **177,** 188–96.

Kitchen, J. H., Ruf, K. B. & Younglai, J. (1974). Effects of intraventricular injection of 6-hydroxydopamine on plasma LH and FSH in male rats. *Journal of Reproduction and Fertility*, **40,** 249–57.

Koob, G. F., Riley, S. J., Smith, S. C. & Robbins, T. W. (1978). Effects of 6-hydroxydopamine lesions of the nucleus accumbens septi and olfactory tubercle on feeding, locomotor activity and amphetamine anorexia in the rat. *Journal of Comparative and Physiological Psychology*, **92,** 917–27.

Koob, G. F., Sessions, G. R., Kant, G. J. & Meyerhoff, J. (1976). Dissociation of hyperdipsia from the destruction of the locus coeruleus in rats. *Brain Research*, **116,** 339–45.

Koob, G. F., Stinus, L. & Le Moal, M. (1981). Hyperactivity and hypoactivity produced by lesions to the mesolimbic dopamine system. *Behavioral Brain Research*, **3,** 341–59.

Korf, J. M., Aghajanian, G. K. & Roth, R. H. (1973). Increased turnover of norepinephrine in the rat cerebral cortex during stress: role of the locus coeruleus. *Neuropharmacology*, **12,** 933–8.

Kostowski, W., Czlonkowski, A., Jerlicz, M., Bidinski, A. & Hauptman, M. (1978). Effect of locus coeruleus lesions on aggressive behavior in rats. *Physiology & Behavior*, **21,** 695–9.

Kostowski, W., Tarchalska-Krynska, B. & Markowska, L. (1975). Aggressive behaviour and brain serotonin and catecholamines in ants (*Formica rufa*). *Pharmacology, Biochemistry & Behavior*, **3,** 717–19.

Kostowski, W., Trzaskowska, E., Jerlicz, M., Bidinski, A. & Hauptman, M. (1980). Effects of lesions of the ventral noradrenergic bundle on aggressive behavior in rats. *Physiology & Behavior*, **24,** 429–33.

Kovacevic, R. & Radulovacki, M. (1976). Monoamine changes in the brain of cats during slow-wave sleep. *Science*, **193,** 1025–7.

Krieger, D. T. (1974). Food and water restriction shifts corticosterone, temperature, activity and brain amine periodicity. *Endocrinology*, **95,** 1195–201.

Krieger, D. T., Crowley, W. R., O'Donohue, T. L. & Jacobowitz, D. M. (1980). Effects of food restriction on the periodicity of corticosteroids in plasma and on monoamine concentrations in discrete brain nuclei. *Brain Research*, **188,** 167–74.

Kruk, Z. L. (1973). Dopamine and 5-hydroxytryptamine inhibit feeding in rats. *Nature New Biology*, **246,** 52–3.

Kruk, Z. L. & Zarrindast, M. R. (1976). Mazindol anorexia is mediated by activation of dopaminergic mechanisms. *British Journal of Pharmacology*, **58,** 367–72.

Laguzzi, R., Petitjean, F., Pujol, J. F. & Jouvet, M. (1972). Effects de l'injection intraventriculaire de 6-hydroxydopamine. II. Sur le cycle veille-sommeils du chat. *Brain Research*, **48,** 295–310.

Laverty, R. & Taylor, K. M. (1969). Behavioural and biochemical effects of 2-(2,6-dichlorophenylamino)-2-amidazoline hydrochloride (ST 155) on the central nervous system. *British Journal of Pharmacology*, **35,** 253–64.

Lawrence, C. W. & Haynes, J. R. (1970). Epinephrine and norepinephrine effects on social dominance behavior. *Psychological Report*, **27,** 195–8.

Leaf, R. C. (1970). Pharmacology, limbic regulation and cerebral function. In *Drugs and cerebral function*, ed. W. L. Smith, pp. 201–14. Thomas: Springfield, Illinois.

Leger, L., Wiklund, L., Descarries, L. & Persson, M. (1979). Description of an indolaminergic cell component in the cat locus coeruleus: A fluorescence histochemical and autoradiographic study. *Brain Research*, **168,** 43–56.

Lehr, D., Mallow, J. & Krukowski, M. (1967). Copious drinking and simultaneous inhibition of urine flow elicited by beta-adrenergic stimulation and contrary effect of alpha-adrenergic stimulation. *Journal of Pharmacology and Experimental Therapeutics*, **158,** 150–63.

Leibowitz, S. F. (1975*a*). Pattern of drinking and feeding produced by hypothalamic norepinephrine injection in the satiated rat. *Physiology and Behavior*, **14,** 731–42.

Leibowitz, S. F. (1975*b*). Catecholaminergic mechanisms of the lateral hypothalamus: their role in the mediation of amphetamine anorexia. *Brain Research*, **98,** 529–45.

Leibowitz, S. F. (1975*c*). Amphetamine: possible site and mode of action for producing anorexia in the rat. *Brain Research*, **84,** 160–9.

Leibowitz, S. F. (1979). Mapping study of brain dopamine- and epinephrine-sensitive sites which cause feeding suppression in the rat. *Brain Research*, **172,** 101–13.

Leibowitz, S. F. (1980). Paraventricular nucleus: a primary site mediating adrenergic stimulation of feeding and drinking. *Pharmacology, Biochemistry & Behavior*, **8,** 163–75.

Leibowitz, S. F. & Brown, L. L. (1980*a*). Histochemical and pharmacological analysis of catecholaminergic projections to the perifornical hypothalamus in relation to feeding inhibition. *Brain Research*, **201,** 315–45.

Leibowitz, S. F. & Brown, L. L. (1980*b*). Histochemical and pharmacological analysis of noradrenergic projections to the paraventricular hypothalamus in relation to feeding stimulation. *Brain Research*, **201,** 289–314.

Leibowitz, S. F. & Rossakis, C. (1979*a*). Mapping study of brain dopamine- and epinephrine-sensitive sites which cause feeding suppression in the rat. *Brain Research*, **172,** 101–13.

Leibowitz, S. F. & Rossakis, C. (1979*b*). Pharmacological characterisation of perifornical hypothalamic dopamine receptors mediating feeding inhibition in the rat. *Brain Research*, **172,** 115–30.

Lenard, L. & Hahn, Z. (1982). Amygdalar noradrenergic and dopaminergic mechanisms in the regulation of hunger and thirst-motivated behavior. *Brain Research*, **233,** 115–32.

Lidbrink, P. (1974). The effect of lesions of ascending noradrenaline pathways on sleep and waking in the rat. *Brain Research*, **74,** 19–40.

Lidbrink, P. & Fuxe, K. (1973). Effects of intracerebral injections of 6-hydroxydopamine on sleep and waking in the rat. *Journal of Pharmacy and Pharmacology*, **25,** 84–7.

Lorden, J., Oltmans, G. A. & Margules, D. L. (1976). Central noradrenergic neurons: differential effects on body weight of electrolytic and 6-hydroxydopamine lesions in rats. *Journal of Comparative and Physiological Psychology*, **90,** 144–55.

Luttinger, D. & Seiden, L. S. (1981). Increased hypothalamic norepinephrine metabolism after water deprivation in the rat. *Brain Research*, **208,** 147–65.

McKenzie, G. & Karpowicz, K. (1970). Aggressive behavior in rats following apomorphine treatment and the effects of age. *Pharmacologist*, **12,** 207.

McQueen, A., Armstrong, S., Singer, G. & Myers, R. D. (1976). Noradrenergic feeding system in monkey hypothalamus is altered by localized perfusion of glucose, insulin, 2-DG and eating. *Pharmacology, Biochemistry & Behavior*, **5,** 491–4.

Malmnas, C. O. (1973). Effects of LSD-25, clonidine and apomorphine on copulatory behaviour in the male rat. *Acta physiologica scandinavica, suppl.*, **395**, 96–103.

Malmnas, C. O. (1976). The significance of dopamine, versus other catecholamines, for L-DOPA induced facilitation of sexual behavior in the castrated male rat. *Pharmacology, Biochemistry & Behavior*, **4**, 521–6.

Mantegazza, P., Naimzada, K. M. & Riva, M. (1968). Effect of propranolol on some activities of amphetamine. *European Journal of Pharmacology*, **4**, 25–32.

Margules, D. L. (1969). Noradrenergic synapses for the suppression of feeding behavior. *Life Sciences*, **8**, 693–704.

Margules, D. L., Lewis, M. J., Dragovich, J. A. & Margules, A. S. (1972). Hypothalamic norepinephrine: circadian rhythms and the control of feeding behavior. *Science*, **178**, 640–3.

Mark, J., Heiner, L., Mandel, P. & Godin, Y. (1969). Norepinephrine turnover in brain and stress reactions in rats during paradoxical sleep deprivation. *Life Sciences*, **8**, 1085–93.

Marrone, R. L., Pray, S. L. & Bridges, C. C. (1968). Norepinephrine elicitation of aggressive display responses in *Betta splendens*. *Psychonomic Science*, **5**, 207–8.

Martin, G. E. & Myers, R. D. (1975). Evoked release of ^{14}C-norepinephrine from rat hypothalamus during feeding. *American Journal of Physiology*, **229**, 1547–55.

Martin, G. E. & Myers, R. D. (1976). Dopamine efflux from the brain stem of the rat during feeding, drinking and lever-pressing for food. *Pharmacology, Biochemistry & Behavior*, **4**, 551–60.

Mason, S. T. & Iversen, S. D. (1978). Reward, attention and dorsal noradrenergic bundle. *Brain Research*, **150**, 135–40.

Mason, S. T., Roberts, D. C. S. & Fibiger, H. C. (1978). Noradrenaline and neophobia. *Physiology and Behavior*, **21**, 353–61.

Mason, S. T., Roberts, D. C. S. & Fibiger, H. C. (1979). Interaction of brain noradrenaline and the pituitary–adrenal axis in learning and extinction. *Pharmacology, Biochemistry & Behavior*, **10**, 11–16.

Matsuyama, S., Coindet, J. & Mouret, J. (1973). 6-Hydroxydopamine intracisternale et sommeil chez le rat. *Brain Research*, **57**, 85–95.

Matthews, J. W., Booth, D. A. & Stolerman, I. P. (1978). Factors influencing feeding elicited by intracranial noradrenaline in rats. *Brain Research*, **141**, 119–28.

Mereu, G. P., Scarnati, E., Paglietti, E., Chessa, P., Di Chiara, G. & Gessa, G. L. (1979). Sleep induced by low doses of apomorphine in rats. *Electroencephalography and Clinical Neurophysiology*, **46**, 241–9.

Micco, D. J. (1974). Complex behaviours elicited by stimulation of the dorsal pontine tegmentum. *Brain Research*, **75**, 172–6.

Miller, N. E., Gottesman, K. S. & Emery, N. (1964). Dose response to carbachol and norepinephrine in rat hypothalamus. *American Journal of Physiology*, **206**, 1384–8.

Mine, K., Nakagawa, T., Fujiwara, M., Ito, Y., Kataoka, Y., Watanabe, S. & Veki, S. (1981). A new experimental model of stress ulcers employing aggressive behavior in 6-OHDA-treated rats. *Physiology and Behavior*, **27**, 715–21.

Modigh, K. (1973). Effects of isolation and fighting in mice on the rate of synthesis of noradrenaline, dopamine and 5-hydroxytryptamine in the brain. *Psychopharmacologia*, **73**, 1–17.

Mountford, D. (1969). Alterations in drinking following isoproterenol stimulation of the hippocampus. *Physiologist*, **12**, 309.

Myers, R. D. (1964). Emotional and autonomic responses following hypothalamic chemical stimulation. *Canadian Journal of Psychology*, **18,** 6–14.

Myers, R. D., McCaleb, M.L. & Hughes, K. A. (1979). Is the noradrenergic 'feeding circuit' in hypothalamus really an olfactory system? *Pharmacology, Biochemistry & Behavior*, **10,** 923–7.

Nakamura, K. & Thoenen, H. (1972). Increased irritability: a permanent behavior change induced in the rat by intraventricular administration of 6-hydroxydopamine. *Psychopharmacologia*, **24,** 359–72.

Nielson, E. B. & Lyon, M. (1973). Drinking behavior and brain dopamine: antagonistic effect of two neuroleptic drugs (pimozide and spiramide) upon amphetamine- or apomorphine-induced hypodipsia. *Psychopharmacologia*, **33,** 299–308.

Nock, B. & Feder, H. H. (1979). Noradrenergic transmission and female sexual behavior of guinea-pigs. *Brain Research*, **166,** 369–80.

O'Donohue, T. L., Crowley, W. R. & Jacobowitz, D. M. (1978). Changes in ingestive behavior following interruption of a noradrenergic projection to the paraventricular nucleus: histochemical analyses. *Pharmacology, Biochemistry & Behavior*, **9,** 99–105.

Oishi, R. & Ueki, S. (1978). Facilitation of muricide by dorsal norepinephrine bundle lesions in olfactory bulbectomised rats. *Pharmacology, Biochemistry & Behavior*, **8,** 133–6.

Oltmans, G. A. & Harvey, J. A. (1972). LH syndrome and brain catecholamine levels after lesions of the nigrostriatal bundle. *Physiology and Behavior*, **8,** 69–78.

Oltmans, G. A., Lorden, J. F. & Margules, D. L. (1977). Food intake and body weight effects of specific and non-specific lesions in the midbrain path of the ascending noradrenergic neurons of the rat. *Brain Research*, **128,** 293–308.

Osumi, Y., Oishi, R., Fujiwara, H. & Takaori, S. (1975). Hyperdipsia induced by bilateral destruction of locus coeruleus in rats. *Brain Research*, **86,** 419–27.

Paglietti, E., Pellegrini Quarantotti, B., Mereu, G. & Gessa, G. L. (1978). Apomorphine and L-DOPA lower ejaculation threshold in the male rat. *Physiology and Behavior*, **20,** 559–62.

Panksepp, J., Jalowiec, J. E., Morgane, P. J., Zolovick, A. J. & Stern, W. C. (1973). Noradrenergic pathways and sleep-waking states in cats. *Experimental Neurology*, **41,** 233–45.

Petitjean, F., Laguzzi, R., Sordet, F., Jouvet, M. & Pujol, J. F. (1972). Effects de l'injection intraventriculaire de 6-hydroxydopamine. I. Sur les monoamines cérébrales du chat. *Brain Research*, **48,** 281–93.

Phillips, A. G. & Nikaido, R. S. (1975). Disruption of brain stimulation-induced feeding by dopamine receptor blockade. *Nature*, **258,** 750–1.

Pujol, J. F., Mouret, J., Jouvet, M. & Glowinski, J. (1968). Increased turnover of cerebral norepinephrine during rebound of paradoxical sleep in the rat. *Science*, **159,** 112–14.

Redmond, D. E., Huang, Y. H., Synder, D. R., Maas, J. W. & Baulu, J. (1977). Hyperphagia and hyperdipsia after locus coeruleus lesions in the stumptailed monkey. *Life Sciences*, **20,** 1619–28.

Richardson, J. S. & Jacobowitz, D. M. (1973). Depletion of brain norepinephrine by intraventricular injection of 6-hydroxydopa: a biochemical, histochemical and behavioral study in rats. *Brain Research*, **58,** 117–33.

Ritter, R. C. & Epstein, A. N. (1972). Control of meal size by central noradrenergic action. *Proceedings of the National Academy of Sciences, USA*, **72,** 3740–3.

Ritter, S., Wise, D. & Stein, L. (1975). Neurochemical regulation of feeding in the rat:

facilitation by alpha-noradrenergic, but not by dopaminergic, receptor stimulants. *Journal of Comparative and Physiological Psychology*, **88,** 778–84.

Roll, S. K. (1970). Intracranial self-stimulation and wakefulness: effect of manipulating ambient brain catecholamines. *Science*, **168,** 1370–2.

Rolls, E. T., Rolls, B. J., Kelly, P.H., Shaw, S. G., Wood, R. J. & Dale, R. (1974). The relative attenuation of self-stimulation, eating and drinking produced by dopamine-receptor blockade. *Psychopharmacologia*, **38,** 219–30.

Rowland, N. (1978). Meal patterns in rats with nigrostriatal dopamine-depleting lesions, subdiaphragmatic vagotomy, and their combination. *Brain Research Bulletin*, **3,** 89–91.

Rowland, N., Marques, D. M. & Fisher, A. E. (1980). Comparison of the effects of brain dopamine-depleting lesions upon oral behaviors elicited by tail pinch and electrical brain stimulation. *Physiology and Behavior*, **24,** 273–281.

Sagales, T. & Erill, S. (1975). Effects of central dopaminergic blockade with pimozide upon the EEG stages of sleep in man. *Psychopharmacologia*, **41,** 53–61.

Salama, A. I. & Goldberg, M. E. (1973). Norepinephrine turnover and brain monoamine levels in septal lesioned aggressive rats. *Life Sciences*, **12,** 521–6.

Satoh, K., Shimizu, N., Tohyama, M. & Maeda, T. (1978). Localization of the micturition reflex centre at dorsal lateral pontine tegmentum of the rat. *Neuroscience Letters*, **8,** 27–33.

Schildkraut, J. J. & Hartmann, E. (1972). Turnover and metabolism of norepinephrine in rat brain after 72 hours on a D-deprivation island. *Psychopharmacologia*, **27,** 17–27.

Sessions, G. R., Kant, G. J. & Koob, G. F. (1976). Locus coeruleus lesions and learning in rats. *Physiology and Behavior*, **17,** 853–9.

Sessions, G. R., Salwitz, J. C. & Kant, G. J. (1976). Lesions of the locus coeruleus and mating behavior in male rats. *Society for Neuroscience Abstracts*, **2,** 725.

Setler, P. E. (1973). The role of catecholamines in thirst. In *The neurophysiology of thirst: new findings and advances in concepts*, ed. A. N. Epstein, H. R. Kissileff & E. Stellar, pp. 279–92. Winston–Wiley: New York.

Setler, P. E. & Smith, G. P. (1974). Increased food intake elicited by adrenergic stimulation of the diencephalon in rhesus monkeys. *Brain Research*, **65,** 459–73.

Sharpe, L. G. & Myers, R. D. (1969). Feeding and drinking following stimulation of the diencephalon of the monkey with amines and other substances. *Experimental Brain Research*, **8,** 295–310.

Simpson, C. W., Baile, C. A. & Krabill, L. F. (1975). Neurochemical coding for feeding in sheep and steers. *Journal of Comparative and Physiological Psychology*, **88,** 176–82.

Singer, G. & Armstrong, S. (1973). Cholinergic and beta-adrenergic compounds in the control of drinking behaviour in the rat. *Journal of Comparative and Physiological Psychology*, **85,** 453–62.

Singer, G. & Montgomery, R. B. (1968). Neurohumoral interaction in the rat amygdala after central chemical stimulation. *Science*, **160,** 1017–18.

Sladek, J. R. & Walker, P. (1977). Serotonin-containing neuronal perikarya in the primate locus coeruleus and subcoeruleus. *Brain Research*, **134,** 359–66.

Slangen, J. L. & Miller, N. E. (1969). Pharmacological tests for the function of hypothalamic norepinephrine in eating behavior. *Physiology and Behavior*, **4,** 543–52.

Smalstig, E. B., Sawyer, B. D. & Clemens, J. A. (1974). Inhibition of rat prolactin release by apomorphine *in vivo* and *in vitro*. *Endocrinology*, **95,** 123–7.

Smith, G. P., Strohmayer, A. J. & Reis, D. J. (1972). Effect of lateral hypothalamic injections of 6-hydroxydopamine on food and water intake in rats. *Nature New Biology*, **235**, 27–9.

Sorenson, C. A., Ellison, G. D. & Masuoka, D. (1972). Changes in fluid intake suggesting depressed appetite in rats with central catecholaminergic lesions. *Nature New Biology*, **237**, 279–81.

Spooner, C. E. & Winters, W. D. (1966). Neuropharmacological profile of the young chick. *International Journal of Neuropharmacology*, **5**, 217–36.

Stachowiak, M., Bialowas, J., Jurkowski, M. & Mirosz, H. (1979). Hunger induced changes in the noradrenaline and dopamine contents in various nuclei of the limbic system in rats. *Polish Journal of Pharmacology and Pharmacy*, **31**, 337–43.

Stern, M., Roffwarg, H. & Duvoisin, R. (1968). The Parkinsonian tremor in sleep. *Journal of Nervous and Mental Diseases*, **147**, 202–10.

Stern, W. C., Hartmann, E. L., Draskoczy, P. R. & Schildkraut, J. J. (1972). Behavioral effects of centrally administered 6-hydroxydopamine. *Psychological Reports*, **30**, 815–20.

Stern, W. C., Miller, F. P., Cox, R. H. & Maickel, R. P. (1971). Brian norepinephrine and serotonin levels following REM sleep deprivation in the rat. *Psychopharmacologia*, **22**, 50–55.

Stern, W. C. & Morgane, P. J. (1973). Effects of alpha-methyl-tyrosine on REM sleep and brain amine levels in the cat. *Biological Psychiatry*, **6**, 301–6.

Stern, J. J. & Zwick, G. (1973). Effects of intraventricular norepinephrine and estradiol benzoate on weight regulatory behavior in female rats. *Behavioral Biology*, **9**, 605–12.

Tagliamonte, A., Fratta, W., del Fiaco, M. & Gessa, G. L. (1974). Possible stimulatory role of brain dopamine in the copulatory behavior of male rats. *Pharmacology, Biochemistry & Behavior*, **2**, 257–60.

Thoa, N. B., Eichelman, B. & Ng, L. K. (1972). Shock-induced aggression: effects of 6-hydroxydopamine and other pharmacological agents. *Brain Research*, **43**, 467–75.

Thoa, N. B., Eichelman, B., Richardson, J. S. & Jacobowitz, D. (1972). 6-Hydroxydopa depletion of brain norepinephrine and the function of aggressive behavior. *Science*, **178**, 75–77.

Tizabi, Y., Massari, V. J. & Jacobowitz, D. M. (1980). Isolation induced aggression and catecholamine variations in discrete brain areas of the mouse. *Brain Research Bulletin*, **5**, 81–6.

Torda, C. (1968). Effect of changes of brain norepinephrine content on sleep cycle in rat. *Brain Research*, **10**, 200–7.

Ulrich, R. E. & Azrin, N. H. (1962). Reflexive fighting in response to aversive stimulation. *Journal of the Experimental Analysis of Behavior*, **5**, 511–20.

Ungerstedt, U. (1971). Adipsia and aphagia after 6-hydroxydopamine induced degeneration of the nigro-striatal dopamine system. *Acta physiologica scandinavica*, suppl., **367**, 95–122.

Valzelli, L. (1974). 5-Hydroxytryptamine in aggressiveness. *Advances in Biochemical Psychopharmacology*, **11**, 255–63.

van der Gugten, J. & Slangen, J. L. (1975). Norepinephrine uptake by hypothalamic tissue from the rat related to feeding. *Pharmacology, Biochemistry & Behavior*, **3**, 855–60.

van der Kooy, D. (1979). An analysis of the behavior elicited by stimulation of the dorsal pons in rats. *Physiology & Behavior*, **23**, 427–32.

Vogel, J. R. & Haubrich, D. R. (1973). Chronic administration of electroconvulsive shock: effects on mouse-killing activity and brain monoamines in rats. *Physiology and Behavior*, **11**, 725–8.

Ward, I. L., Crowley, W. R., Zemlan, F. P. & Margules, D. L. (1975). Monoaminergic mediation of female sexual behavior. *Journal of Comparative and Physiological Psychology*, **88**, 53–61.

Welch, A. S. & Welch, B. L. (1968). Reduction of norepinephrine in the lower brainstem by psychological stimulus. *Proceedings of the National Academy of Sciences, USA*, **60**, 478–81.

Wiklund, L., Leger, L. & Persson, M. (1981). Monoamine cell distribution in the cat brain stem: A fluorescence histochemical study with quantification of indolaminergic and locus coeruleus cell groups. *Journal of Comparative Neurology*, **203**, 613–47.

Wright, C. & Everitt, B. J. (1977). The effects of dorsal noradrenaline bundle lesions on the sexual behaviour of female rats. *Experimental Brain Research*, **28**, R48–R49.

Zigmond, M. J. & Stricker, E. M. (1972). Deficits in feeding behavior after intraventricular injection of 6-hydroxydopamine in rats. *Science*, **177**, 1211–14.

Zigmond, R. E., Schon, F. & Iversen, L. L. (1974). Increased tyrosine hydroxylase activity in locus coeruleus of rat brain after reserpine treatment and cold stress. *Brain Research*, **70**, 547–52.

Zolovick, A. J., Stern, W. C., Jalowiec, J. E., Panksepp, J. & Morgane, P. J. (1973). Sleep-waking patterns and brain biogenic amine levels in cats after administration of 6-hydroxydopamine into the dorsolateral pontine tegmentum. *Pharmacology, Biochemistry & Behavior*, **1**, 557–67.

8

Catecholamines and pathological behaviour

'The Great Mystery cannot be approached by one avenue alone'
Symmachus, *Relationes* **3**

Introduction

This chapter, while in some respects the most important in this book, is in others the least satisfactory. The whole purpose of the study of the role of catecholamines in animal behaviour reviewed in preceding chapters is the light which it can cast on human behaviour and brain function. The most pressing need for such an understanding of the role of brain catecholamines in human behaviour lies in the possibility that malfunction of these systems may be the underlying cause of human mental illnesses and psychopathology. Hence the significance of this chapter.

However, at the same time, the role of catecholamines in pathological behaviour is an area about which very little is firmly established. Whereas in previous chapters a consensus of data and experiments allowed reasonably firm conclusions to be made (albeit with occasional caveats in still controversial areas), in this chapter virtually no conclusions are currently possible. With the notable exception of Parkinsonism, at present we do not know the neurochemical basis of any human neurological or psychopathological illness – for example schizophrenia, depression, senile dementia, epilepsy and so on. All that can be essayed in this chapter is a historical survey of the various theories which have been proposed; and this cannot be exhaustive since virtually every individual neuroscientist has his own distinctive theory of virtually every psychopathological condition. None of the theories proposed has yet been proven (with the single exception of the dopamine theory of Parkinsonism). Many have had currency for some years and then given way to others merely by the passage of time and without convincing *disproof.* A few have failed in the face of overwhelming contrary evidence. Most are still around today, as tantalising, but as impalpable, as they were at the time of their origin. Hence the unsatisfactory nature of this chapter.

Some may say that it is premature even to write this chapter, given the present state of ignorance. However, it is offerred as a historical testament to the state of present development. As a major caveat it must be pointed out that it is precisely in this area that sudden and unpredictable progress is likely and so this chapter in particular is the most likely to go out of date.

Limitations of clinical studies

Why, the perspicacious reader may ask, is this area of catecholamines and human psychopathology so poorly developed compared to the multitudinous data on catecholamines and every conceivable sort of animal behaviour? Much of it lies in the technical and ethical limitations of the sorts of studies which are possible on human subjects, even volunteers of unimpaired intellect, let alone institutionalised mental patients in a position of dependence. Two sorts of evidence for a role of a particular catecholamine in a given mental illness are available. On the one hand, we may seek to determine abnormalities in neurotransmitter function in patients suffering from a given disease. On the other, we may find that a drug is effective in ameliorating or even curing a given psychopathology and then enquire which neurotransmitter systems it may be acting on to cause this improvement.

If we adopt the first approach, we are limited to the indices of neurotransmitter function which we can obtain in human subjects. We cannot generally take brain tissue from living human patients (exceptions include rare instance of removal of brain tissue during a neurosurgical operation for the exturpation of an epileptic focus or a tumour). We are limited to taking urine, blood and possibly cerebrospinal fluid (CSF) samples. (Even the latter is of uncertain ethical status in that it is highly invasive and cannot offer any direct hope of improving that particular patient's condition, although it might ultimately benefit other future patients. Informed consent from a patient capable of understanding these considerations but still willing to undergo discomfort for the sake of assisting medical research is *de rigueur*.) Even CSF samples are not brain tissue and the distribution of transmitters and metabolites from the brain into the CSF is often uncertain and controversial. Thus, in living human patients we generally lack direct access to brain tissue in order to monitor it for changes in neurotransmitter function.

A variant of this approach is to perform direct neurochemical measurements of brain tissue from human patients who have recently died of natural causes. This has the advantage of giving us brain tissue *per se* on which to perform our measurements, but brings with it other difficulties.

Thus, since it is dead, it is by definition in a state different from that which it showed during the period of psychopathology. Considerable time may elapse from the moment of death to the moment of performing the neurochemical measurement. This can be ameliorated by storage of the tissue under refrigeration, but some time will inevitably elapse before relatives or the Coroner permit autopsy and the removal of the brain. The neuroscientist cannot in practice hover by the bedside of the patient, awaiting death, and it would be most unethical if he should. Thus, during the inevitable delay between death and the freezing of the tissue considerable neurochemical changes may occur which will severely distort the picture when assay is finally accomplished. Animal studies on neurotransmitter changes after death can give an idea of what these neurochemical alterations are likely to be, but cannot stop them from occurring. Thus, allowance may perhaps be made for post-mortem changes but they cannot be prevented *in toto*. This is one serious drawback to the use of post-mortem human brain tissue.

Another, more recognised and controlled now than in previous studies, is the effect of previous environment prior to the patient's death. Most so-called normal human brain tissue from which to judge the significance of psychopathological changes is derived from accident victims or others suffering sudden death. Mental patients usually succumb to death only after a long period of hospitalisation, often with other chronic illnesses accompanying the main psychopathology. It is not to be wondered at that long periods of inactivity in bed, on a diet of institutional food, and in a severely regimented and unstimulating environment may cause changes in brain neurotransmitter systems by themselves, unrelated to changes involved in the psychopathology which led to the institutionalisation. The state of the patient's general health is often poor, especially in the aged. These brains are then compared to generally young (although experimenters are now more aware of this confound and try to achieve age-matching), healthy, noninstitutionalised individuals leading active and stimulating lives. Although definite neurochemical changes may be found, and replicated, the problem then becomes that of ascribing to them the psychopathological state as such rather than to the other ways in which the control group unavoidably differs from the experimental.

A final problem has been recently highlighted in the case of a role for dopamine in schizophrenia. When there are drugs available which ameliorate, even if not necessarily cure the condition, it would be unethical to withhold this medication. However, the administration of a drug, usually on a chronic basis, would be expected to cause considerable neurochemical changes in and of itself. Indeed, the biological determinist would argue that it could not be helping the mental condition *without* causing neurochemical alterations. When, on post-mortem assay of human brain tissue, we find changes in a given neurochemical system the question then becomes, are these caused by the psychopathological condition or by the drugs which were administered to try to treat this condition? Much experimental effort is required to attempt to dissociate these two and may not always be successfully accomplished.

If the measurement of neurochemical changes in a given disease state is surrounded by so many difficulties, can we not adopt the other approach mentioned above and use the fact that a drug helps to improve the psychopathology to deduce which systems were originally disturbed? Indeed, if a drug with a single neurochemical action were to produce complete cure of a mental illness this would be very strong evidence. Alas, such things are not so black and white. Thus, drugs often improve the condition but fail to effect a total cure. Was the drug then acting via the system which had malfunctioned initially to cause the psychopathology, or was it modifying an ancillary system to cause the limited improvement observed? Further, no drug has a single neurochemical action. The story of the action of psychoactive drugs is a continual discovery of new, and hitherto unsuspected, neurochemical effects, each of which might justifiably be described as the mechanism of 'cure'. Drugs are not consciously designed by the molecular chemist to affect only one neurochemical system and then tested for an action against mental illness. Indeed, the converse is true. Chemical substances obtained from a wide range of sources such as plants, fungi, and other natural pharmacopoeia, as well as synthetic substances, are screened against mental illness (often in animal models which bear little direct relationship to the human illness) and only then is the successful candidate investigated to ascertain its actions on neurochemical systems. Often it will turn out to affect several such systems and a whole field of research can be required to try to determine which of its many actions is responsible for the improvement in human psychopathology. Again, these efforts need not necessarily be rewarded with success.

Thus, in summary *neither* post-mortem or bodily fluid measurements *nor* the neurochemical mode of action of psychoactive drugs has yet yielded incontrovertible evidence for a role of catecholamines in pathological behaviour (with the exception of Parkinsonism). Taking the two together, however, some suggestive evidence of differing degrees of strength for different mental illnesses has emerged. It is this, admittedly nondefinitive, body of work that I shall now survey.

I: COGNITIVE DISORDERS

Schizophrenia

Introduction

Schizophrenia encompasses a series of diseases characterised by 'a withdrawal into a private fantasy world which is maintained by personal beliefs, idiosyncratic thought patterns and percepts which are not culturally shared' (from Bemporad & Pinsker, 1974; for more detail see Arana (1978) and DSM III). The symptoms of schizophrenia first classified by Kraepelin (1883) include delusions, hallucinations and incoherence, bringing about a deterioration of functioning in work, social relations and self-care. Further subdivision of schizophrenia is often made into the types catatonic, paranoid, disorganised and undifferentiated (DSM III). Paranoid schizophrenia is dominated by persecutory or grandiose delusions, delusional jealousy or similar hallucinations. Catatonic schizophrenia involves marked psychomotor disturbances called 'catatonia'. Disorganised schizophrenia is characterised by incoherence and inappropriate affect while 'undifferentiated' forms a grab-bag category for those cases not fitting into the three earlier diagnoses. Perhaps the most homogeneous category of schizophrenia is that of paranoid schizophrenia and it is precisely here that a more consistent picture is starting to emerge from the neurochemical findings. Innumerable 'causes' of schizophrenia have been suggested on the biochemical level, ranging from cholinergic, GABA-minergic, endorphin systems to prostaglandins, melatonin, gluten and immunological and viral diseases (see Ter Haar, 1979). The following will necessarily be restricted by concentrating on the role of catecholamines in schizophrenia.

Noradrenaline

By far the greatest evidence for a role of noradrenergic systems in the psychopathology of schizophrenia has come from post-mortem assays of noradrenaline markers in human schizophrenic brain tissue. More recently a little, but often contradictory, evidence has started to emerge from the effects of noradrenergic drugs on the schizophrenic condition.

Noradrenergic measures in schizophrenia. Two distinct phases in the story of a noradrenergic role in schizophrenia may be perceived from a historical point of view. The first, based on post-mortem dopamine-β-hydroxylase (DBH) measurements around 1975, suggested a deficiency of brain noradrenaline in schizophrenia. Considerable interest was shown in this idea and numerous attempts to confirm it made in the following few years.

These were generally of a negative conclusion. The second wave of interest in noradrenaline and schizophrenia is still with us and was started around 1978 by measurements of noradrenaline levels *per se* in post-mortem brain tissue. These revealed a picture different from that of the earlier DBH studies. Now, abnormally high levels of noradrenaline were found in limbic areas, especially in paranoid type schizophrenics. Subsequent attempts to confirm this have met with more success than the DBH story, but have served to highlight the importance of the subgrouping of paranoid v. other forms of schizophrenia.

In 1973 Wise & Stein reported reductions in DBH activity in schizophrenic post-mortem brains of between 30 and 50 %. They obtained brain specimens from eight male and ten female schizophrenic patients dying of natural causes at an average age of 71.2 years after hospitalisation for an average of 34.4 years. Six male and six female control brains were obtained via the Medical Examiner's (Coroner's) Office from mentally healthy, normal individuals dying from accidents or heart attack. The latter group had an average age of 57.2 years and of course, had not been hospitalised. Thus, two immediate differences exist between control and experimental groups, beside that of schizophrenia, namely age at death and prior institutionalisation. The authors were thus required to exclude these factors as being causative of their observed changes in brain DBH. They were able to form a subgroup of schizophrenic patients whose average age at death was similar to that of controls. Nonetheless, DBH activity in this subgroup was still only 50 % of control. This would appear to exclude age as a cause. To exclude the effects of hospitalisation, the authors showed that although periods of institutionalisation ranged from 5 to 60 years there was no correlation between duration of hospitalisation and DBH activity. This might exclude a role for institutionalisation in the decreased DBH activity. On the other hand, it might be that even the minimum period of 5 years hospitalisation (in itself a long time) was adequate to change brain DBH. This serves to indicate the possibly insurmountable problems associated with the interpretation of post-mortem human data.

Attempts by Wyatt, Schwartz, Erdelyi & Barchas (1975) to confirm these reductions in brain DBH in schizophrenia were unsuccessful. Thus, nine schizophrenic brains and nine controls failed to differ significantly in their regional DBH activities. Only a nonsignificant decrease in the hippocampus of schizophrenics was seen.

A similar analysis of the DBH activities in samples of cerebrospinal fluid obtained by lumbar puncture from 11 schizophrenic patients reported by Lerner *et al.* (1978) also found no change when comparing patients suffering from other forms of illness such as depression, alcoholism and personality disorders. No normal control group was used here which rather reduces the power of the study. Similar negative results were reported by Cross *et al.* (1978) for brain DBH activities in post-mortem schizophrenic brains.

Thus, in summary, the DBH story has not been generally confirmed with regard to schizophrenia. With negative results in this tortuous clinical field, however, a fair degree of caution is deserved. Given the horrendous difficulties facing the investigators in the collection and subsequent interpretation of post-mortem data, it would be very easy to lose significant differences which were really there. Unlike previous chapters, here there can only be a balance of probabilities' conclusion. In all likelihood, there is no change in brain DBH activity in schizophrenia. But, this conclusion, by the nature of the studies involved, is of its very nature less satisfactory than those enumerated in previous chapters.

The second wave of interest in noradrenaline and schizophrenia came with the report of Farley *et al.* (1978) that the level of noradrenaline *per se* was increased from 50 % to 300 % in regions of the limbic brain in four patients with paranoid schizophrenia compared to age-matched controls (see Table 8.1). This study highlights two important considerations. First, subgroups of schizophrenics may differ. These data were obtained for a

Table 8.1. *Noradrenaline in limbic brain areas of four schizophrenic subjects compared with controls*

	Noradrenaline (micrograms per gram of wet tissue)		
	Controls	Schizophrenics	
Brain region	Mean ± SEM	Mean ± SEM	%
Hypothalamus, total	1.83 ± 0.18	1.86 ± 0.22	102
Hypothalamus, anterior	2.29 ± 0.31	2.07 ± 0.30	90
Hypothalamus, posterior	1.64 ± 0.36	1.88 ± 0.18	115
Hypothalamus, lateral	1.49 ± 0.12	1.63 ± 0.17	109
Nucleus accumbens	1.58 ± 0.16	2.40 ± 0.27*	152
Medial olfactory (preoptic) area	1.49 ± 0.27	1.69 ± 0.19	113
Bed nucleus of stria terminalis	1.23 ± 0.17	2.72 ± 0.26***	221
Ventral septum	0.53 ± 0.11	1.59 ± 0.24**	300
Mammillary body	0.45 ± 0.02	0.69 ± 0.16	153
Paramedian thalamic nuclei	0.48 ± 0.02	0.53 ± 0.06	110

* $P < 0.05$, ** $P < 0.01$, *** $P < 0.001$.
From Farley *et al.*, 1978.

paranoid subgroup. Secondly, different brain regions in the schizophrenic brain may differ in the degree of noradrenaline loss. These data were obtained only on so-called limbic areas (actually, nucleus accumbens, ventral septum, bed nucleus of the stria terminalis and mammillary bodies). Other brain areas showed normal noradrenaline levels.

In support of this result is the finding of Lake *et al.* (1980*a*) that the levels of noradrenaline *per se* in CSF samples from paranoid schizophrenics were higher by 50 % than in controls (age-matched volunteers were admitted to the same hospital wards as the schizophrenics and were on the same diet). Here we start to see the degree of sophistication needed to obtain control values which are really comparable to the experimental groups. All schizophrenics were taken off drug medication for two weeks prior to CSF sampling. Also supporting an elevation of brain noradrenaline in paranoid schizophrenia is the report of Gomes, Shanley, Potgieter & Roux (1980), who found increased noradrenaline content of CSF samples from a group of undifferentiated schizophrenics (i.e. not subgrouped in terms of paranoid features). Finally, recent confirmation of elevated noradrenaline levels in CSF of schizophrenic patients has come from the Italian group of Kemali, Vecchio & Maj (1982). Against this almost surprising unanimity of data must be placed the findings of Bird, Spokes & Iversen (1979*b*), who examined the brains of over 50 schizophrenic patients. No hint of a change was seen in the ventral septal area in which Farley *et al.* (1978) found a 300 % increase in noradrenaline content. Only very small increases were found in the noradrenaline content of the nucleus accumbens and the anterior perforated substance, which failed to reach conventional levels of significance. However, no subgrouping of the 50 patients into those exhibiting paranoid features was carried out, so to some extent this report remains ambiguous.

It has also been questioned whether the increases in CSF noradrenaline seen by Lake *et al.* (1980*a*) might not have been caused by the drug therapy which their paranoid schizophrenics were experiencing (Lipsky, 1980). Although they were taken off all medication for two weeks prior to CSF sampling, many neuroleptic drugs persist in the body for much longer than this. However, in this case (and see the difficulties caused by this problem in the section on dopamine and schizophrenia, p. 398) we may rule out the elevated noradrenaline levels as being iatrogenic (caused by the drug treatment). This is so since neuroleptic drugs are known to *lower* rather than elevate both noradrenaline levels and those of its principal metabolite 3-methoxy-4-hydroxyphenylglycol (MHPG) in brain or CSF samples (see Lake *et al.*, 1980*b*).

Thus, in summary it may be concluded that a remarkable degree of unanimity exists at the moment in published studies that levels of noradrenaline are elevated in some limbic brain parts of a paranoid subgroup of schizophrenic patients (recently reviewed by Hornykiewicz, 1982). Such unanimity may well dissipate like a mirage even before these words appear in print. *Caveat lector.*

Noradrenergic drug effects. A number of drugs which, *inter alia*, affect noradrenergic transmission have been examined for their effect on the psychotic manifestations of schizophrenia. Thus, clonidine, an α-agonist had no beneficial effect in a double-blind trial on the symptoms of schizophrenia but caused an increase in aggressiveness (Jimerson *et al.*, 1980). This generally confirms earlier reports of the lack of effect of clonidine on the schizophrenic symptoms *per se* (Simpson, Kunz-Bartolini & Watts, 1967; Sugerman, 1967).

The β-blocker propranolol has been reported to be beneficial in schizophrenia (Roberts & Amacher, 1978; Lindstrom & Persson, 1980) but not in all studies (see King, Turkson, Liddle & Kinney, 1980). It appears especially effective in a paranoid schizophrenic subgroup (Yorkston *et al.*, 1977) and may be every bit as effective as the more conventional neuroleptic medication with chlorpromazine (Yorkston *et al.*, 1981). The mode of action is unclear since, as well as central β-blocking action, propranolol will block the peripheral sympathetic system, have membrane stabilising effects and may also alter serotonergic transmission (Weinstock & Weiss, 1980).

Recent work suggests that propranolol acts to reduce the liver metabolism of neuroleptic drugs and hence increases brain levels of these agents (Peet, Middlemass & Yates, 1981). This leads to an apparent improvement in the mental condition. Patients not on neuroleptic medication who receive propranolol do *not* show any improvement.

The conventional neuroleptic drugs such as chlorpromazine typically used in the treatment of schizophrenia may have action on noradrenergic systems. They are antagonists of both β- and α_2-receptors (Sulser & Robinson, 1978), antagonise the noradrenaline-induced suppression of cerebellar Purkinje cell firing (Freedman & Hoffer, 1975) and antagonise the noradrenaline-induced accumulation of cAMP (Palmer, Robinson & Sulser, 1971, 1972). Neuroleptic drugs[1] also increase the release of noradrenaline by blockade of inhibition of α_2-receptors (Arbilla, Briley,

1 Recent work has also demonstrated that many neuroleptics block α_1-receptors (binding studies: Robinson, Berney, Mishra & Sulser, 1979; Peroutka & Snyder, 1980; electrophysiology: Marwaha & Aghajanian, 1982). However, not all α_1-blockers are effective antipsychotics (e.g. promazine; Peroutka *et al.*, 1977).

Dubocovich & Langer, 1978; Gross & Schumann, 1980). Of especial interest, then, is the report from Sternberg *et al.* (1981) that the neuroleptic pimozide reduces the abnormally high levels of noradrenaline found in CSF samples from schizophrenic patients. It is possible that one or all of these neurochemical actions of neuroleptic drugs on noradrenergic, rather than dopaminergic, systems may underlie their efficacy in the amelioration of schizophrenia. However, they provide no hard evidence to support a role of noradrenaline in schizophrenia until further neurochemical elucidation of the actual mechanism of importance has been achieved.

In summary, drug studies provide little hard evidence that noradrenaline may have a role in schizophrenia, but they certainly do not rule out this possibility.

Dopamine

It is dopamine which, until very recently, has been the 'traditional' transmitter system in schizophrenia. Initially this was suggested by the neurochemical findings that the neuroleptic drugs, those most commonly effective in the amelioration of schizophrenic conditions, also acted to block or reduce functional activity in the dopaminergic systems of the brain. More recently, actual changes in indices of dopaminergic function have been consistently detected in post-mortem assays of human schizophrenic brain tissue. However, that is not the end of the story. The interpretation of these changes has engendered some of the fiercest controversy in an already highly controversial clinical literature. The present section can only reflect the swings in fashion and interpretation which have occurred up to now; subsequent reinterpretations or new data may change the picture almost overnight.

Dopaminergic measures in schizophrenia. Two distinct changes in dopaminergic measures have generally been claimed. One involves an increase in the levels of dopamine itself in 'limbic' areas of human post-mortem schizophrenic brain tissue, such as the nucleus accumbens. The other involves an increase in the number of receptors for dopamine, usually measured by radioreceptor binding assays.

Thus Bird, Spokes & Iversen (1979*c*) reported a 30 % increase in the dopamine content of the nucleus accumbens in over 50 schizophrenic brains, with a doubling of the dopamine levels in the anterior perforated substance of the same brains. Interestingly, these changes were sharply localised since no increase was seen in the septum or the amygdala of these patients. To exclude the possibility that these changes reflected the chronically ill pre-terminal condition of some of these patients compared to the controls (who usually died from sudden accidents, etc.) these authors divided their patient sample into those schizophrenics succumbing to sudden death versus those dying after a prolonged illness. No differences were seen between these two conditions suggesting that this may be a real change caused by the disease condition.

However, other authors (Crow *et al.*, 1978*b*, 1979) failed to find any increase in dopamine content of the nucleus accumbens in their sample of 16 schizophrenic patients. Interestingly, they did detect an increase in the dopamine content of the area they dissected as caudate nucleus. It has been suggested that one way to reconcile these differences might be that different groups were dissecting out the same brain area into different samples. Thus, Bird *et al.* might include the critical area in their accumbens sample while Crow *et al.* included it in their caudate sample (see Bird *et al.*, 1979*a*). However, an earlier study by Farley *et al.* (1977) reported no increase in dopamine content in *either* the accumbens *or* the caudate samples. However, in this study the low number of patients (3) in the schizophrenic group might mean that the nonsignificant 38 % increase in their caudate sample could, with more subjects, yield something more substantial.

The second change reported by several authors is an increase in the number of dopamine receptors in schizophrenic brains. Thus, Owen *et al.* (1978) report increased binding of the dopamine ligand [^{3}H]spiroperidol to tissue from schizophrenic brains compared to controls. A more than doubling was observed in this measure in the caudate and lesser increases seen in the putamen and nucleus accumbens. There was some overlap between the individual values of the control and schizophrenic groups but the means were highly significantly different.

The difficulties in this study relate to the interpretation of the observed differences. Are they reflections of the underlying disease process or do they show only the neurochemical effect of the neuroleptic drugs used in the treatment of the disease. To answer this Owen *et al.* (1978) extracted a subgroup of two schizophrenics who had never been medicated and yet who still showed a highly significant increase in dopamine binding in the caudate. They further argued that the increase in the number of dopamine receptors seen in animal experiments after chronic neuroleptic treatment is only about 25 % compared to the massive 103 % seen in their human patients. This would appear suggestive but not entirely conclusive.

Similar increases in dopamine receptor numbers in both the accumbens and the caudate were seen by MacKay *et al.* (1980*a*) in a series of 15 patients dying with a case-note diagnosis of schizophrenia. These authors express more uncertainty as to whether the clear-cut changes in dopamine receptors reflect the disease process *per se* or are a consequence of the neuroleptic drug treatment which almost all the patients had been receiving. Finally, although Lee *et al.* (1978) describe two patients who were apparently 'drug-free' prior to death and in whom an increase in dopamine receptor numbers was still found, a different picture has emerged from MacKay *et al.* (1980*b*). They found a 100 % increase in the number of sites of [^{3}H]spiperone binding in both the caudate and, to a lesser extent, in the accumbens of more than 20 schizophrenic patients who had been receiving neuroleptic medication prior to death. However, a further seven patients who had not been receiving drug treatment for up to a month before death failed to show any increase in [^{3}H]spiperone binding in *either* caudate *or* accumbens compared to control brains. This led the authors to suggest that the changes in dopamine receptor number were totally due to drug treatment and did not underlie the basic schizophrenic illness. This interpretation has not been accepted by all authors (Seeman, 1981) despite the fact that Reynolds *et al.* (1980) also report four schizophrenic patients who had received no neuroleptic drugs for three months prior to death and who showed no increase in dopamine receptor numbers as determined by [^{3}H]spiperone binding compared to control brains.

In conclusion, I can only quote the words of S. H. Snyder (1981): 'it appears likely that reported increases in dopamine receptors in brains of schizophrenic patients largely reflect neuroleptic drug effects and are not causally related to the schizophrenic process' (*ibid.*, p. 462).

Dopaminergic drugs and schizophrenia. The original impetus for the dopamine theory of schizophrenia, which at one time seemed the orthodoxy on the subject, was the finding that the neuroleptic drugs introduced for the treatment of schizophrenia in the 1950s were very effective in blocking activity in dopaminergic brain systems. Even the atypical neuroleptics such as thioridazine and clozapine, which do not cause Parkinsonian side-effects in patients, do not block apomorphine emesis, nor elevate the dopamine metabolite homovanillic acid (HVA) in the caudate (Azmita, Durost & Arthurs, 1959; Wang, 1965; O'Keefe, Sharman & Vogt, 1970) turned out to block dopaminergic receptors (Meltzer, Sachar & Frantz, 1975). The absence of Parkinsonian side-effects is often attributed to the anticholinergic activity which these two compounds also possess (Snyder, Greenberg & Yamamura, 1974; but see also Burki, Eichenberger, Sayers & White, 1975). It remains indisputable that neuroleptic agents are the most effective treatment for schizophrenia presently available. The common element present in all such drugs seems to be their dopamine-blocking activity. However, as reviewed above it has proved impossible so far to demonstrate any increase in dopamine receptors *as a consequence of the disease state per se* in human post-mortem tissue. In view of this, the effectiveness of anti-dopaminergic drugs in the treatment of schizophrenia may reflect not the direct reversal of the basic neuropathology but an amelioration achieved only by action of secondary systems.

Alzheimer's disease and (pre)senile dementia

Introduction

Clear changes in functioning of cholinergic systems were reported a number of years ago in patients with Alzheimer's disease (Davies & Maloney 1976). These appeared to correlate with the severity of the disease (Perry *et al.*, 1978), the symptoms of which, first described by Alzheimer (1907), include memory impairment and dementia involving one or more of the following: personality change, impaired judgement, impaired abstract thinking or 'constructional difficulty' (see van Dongen, 1981, for more details on classification). However, patients seemed to show little improvement on drug regimes intended to restore lost cholinergic function (for example Boyd *et al.*, 1977; Christie *et al.*, 1981). Changes in noradrenergic, serotonergic and dopaminergic function have also been reported (see Gottfries, 1980). Since additional studies have often found no change in dopaminergic systems (Davies & Maloney, 1976; Adolfsson, Gottfries, Roos & Winblad, 1979; Davies, 1979; Mann *et al.*, 1980; Yates *et al.*, 1981) the following discussion will concentrate on the role of noradrenaline in Alzheimer's disease.

Noradrenaline

Cell body loss. The first, rather neglected report of changes in noradrenergic systems in Alzheimer's disease came from Forno in 1966. She reported seven cases in which severe loss of cells was seen in the locus coeruleus (LC) (not further quantified, but very striking in Nissl stained material, Fig. 25C in Forno, 1966). More recent studies have quantified the degree of cell loss in the LC of patients dying with Alzheimer's disease. Tomlinson, Irving & Blessed (1981) examined 15 cases of senile dementia of the Alzheimer type and found 8 cases with fewer LC neurones than any of the 25 control patients. The remaining seven Alzheimer patients were in the very lower range of control cell counts. Although cell counts in the LC fall by some 50 % by the seventh decade of life, even in cognitively normal humans (Vijayashankar & Brody, 1979) a far more precipitous decline is seen in age-matched Alzheimer patients (to about 12–10 %). Some correlation with the severity of the cognitive impairment seemed apparent since Bondareff, Mountjoy & Roth (1981) reported that those patients showing greatest dementia and earliest death (some 12 cases) had only 20 % of the usual number of cells in the LC as judged in cresyl violet stained material. A second group of patients with milder dementia and greater survival ages (some eight cases) showed only a decline to 80 % of normal cell numbers.

Losses of cell bodies in the LC appear to be well established in Alzheimer's disease, with Mann, Yates & Hawkes (1982) reporting reductions of 60% and Bondareff *et al.* (1982) losses of 80 %. Perry *et al.* (1981) observed a better than 83 % loss in a group of four patients and Mann *et al.* (1980) figures in excess of 75 % for their most severe group of five patients. Cell losses of this magnitude would be expected to have a most profound effect on cognitive functioning (see section on Selective attention in Chapter 6).

Noradrenaline levels. If more than four-fifths of the cell bodies in the LC of Alzheimer patients are lost, a very marked decrease in the noradrenaline content of areas innervated by the LC system, such as the cortex and hippocampus, would also be expected. Adolfsson *et al.* (1979) assayed the brains of 15–18 patients with senile dementia of the Alzheimer type and found losses of noradrenaline in the frontal gyrus of the cerebral cortex and in the putamen compared to 10–21 control brains. Cortical loss was around 75 % while that in the putamen only 50 %. No losses of dopamine or serotonin were seen in these areas, although some decrease in dopamine in the thalamus and pons was observed. More strikingly, a negative correlation was observed between the level of noradrenaline in the frontal cortex and the severity of the intellectual and emotional dementia (Spearman $\rho = -0.45$, $P < 0.05$). A hint of a similar correlation also appeared for hypothalamic noradrenaline levels. Even more marked noradrenaline loss was shown in the hypothalamus of a small group of patients ($n = 2$) examined by Mann *et al.* (1980). Noradrenaline was reduced by 53 % here with no changes in hypothalamic dopamine. Minor changes (20–26 %) in dopamine, however, were seen in frontal cortex and caudate. Finally, Mann *et al.* (1982) and Yates *et al.* (1981) found losses of 40–50 % in noradrenaline content of hypothalamus, caudate, cerebellum and frontal cortex of patients with senile dementia of the Alzheimer type but not in a group of three patients with dementia of multiple-infarct pathology.

Thus, clear losses of noradrenaline content of areas receiving LC innervation have been demonstrated in Alzheimer patients and may have some specificity to this condition since they did not occur in dementia caused by multiple-infarct pathology.

Dopamine-β-hydroxylase. Cross *et al.* (1981) and Perry *et al.* (1981) examined the activity of the noradrenergic marker enzyme DBH in 8–12 patients with Alzheimer's disease and found lower activities in frontal and temporal cortex as well as the hippocampus. In some cases the losses approached 50 %. That this may be specific to Alzheimer's disease was shown by the absence of any change in similar brain areas in depressive or multiple-infarct dementia patients. Perhaps 50–70 % of patients with 'senile dementia' are of the Alzheimer type, while in only 10–30 % is the dementia due to cerebrovascular insufficiency (van Dongen, 1981).

Noradrenergic drug treatments. Preliminary reports with L-DOPA, which will increase function of both dopamine and noradrenaline systems appeared hopeful. Lewis, Ballinger & Presley (1978) found improvement in ratings of communication and intellectual functioning in a group of 14 patients on doses of 125–875 mg of L-DOPA per day, when compared to those receiving placebo. Further, Johnson, Presley & Ballinger (1978) report that this improvement declined when the patients were taken off the drug and improved again when it was restarted. Some patients maintained this improvement for up to six months on the drug regime. However, Kristensen, Olsen & Theilgaard (1977) found in their study that L-DOPA was *not* effective in pre-senile dementia. It would appear that more specific noradrenergic drug therapies might hold promise.

Conclusion. A number of recent reports come to similar conclusions – that there is a profound malfunction in the LC noradrenergic system in Alzheimer's type senile dementia. This has been shown both in terms of the morphological loss of LC cell bodies and in terms of biochemical loss of noradrenaline content and DBH activity in post-mortem brain tissue of patients dying with this diagnosis. Only erratic evidence has emerged for any change in brain dopaminergic systems in Alzheimer's disease.

However, very clear, reproducible and profound loss of cholinergic function has also been seen. It is thus clear that Alzheimer's disease is not caused by loss of only a single neurochemical system. What aspects of the disease can be ascribed to cholinergic malfunction and which symptoms to noradrenergic loss remains a topic for future research (for some suggestions from the animals experimental literature the reader should see the section on Selective attention in Chapter 6). Even more brain systems than the cholinergic and noradrenergic may turn out to be deficient in Alzheimer's disease. Nonetheless, loss of noradrenergic cells in the LC of 50–80 % would be expected to have a major effect on cognitive functioning and some, if by no means all, of the debilitating characteristics of senile dementia of the Alzheimer type may well be related to the loss of noradrenergic systems.

Korsakoff's psychosis

Introduction

This cognitive disease occurs in nutritionally depleted chronic alcoholic patients and was first described by the Russian neurologist S. S. Korsakoff (1890). Symptoms include severe impairments in learning and memory, with confabulation as an attempt to overcome these deficits.

Both anterograde and retrograde amnesia were originally described by Korsakoff and, since the severity of the two amnesias change in parallel during the development of the disease (Victor, Adams & Collins, 1971), they may both be due to a single underlying pathology. Korsakoff patients also have small but definite impairments in perceptual and cognitive abilities (reviewed in Butters & Cermak, 1976). Motivational deficits may also be seen, with Korsakoff patients being described as apathetic and lacking in spontaneous initiative (Talland, 1965). Although confabulation is a common occurrence in the early stages of the disease as mentioned above, it may fail in the late stages (Talland, 1965; Victor *et al.*, 1971).

Noradrenaline

Post-mortem assays. Although widespread pathological changes in Korsakoff's disease have been known for many years, involving such areas as the mammillary bodies and medial or posterior thalamus (Malamud & Skillicorn, 1956), it is only within the last few years that evidence has emerged to indicate a loss of noradrenergic function as well. McEntee & Mair reported in 1978 that the lumbar CSF of Korsakoff patients showed a much lower concentration of MHPG, a metabolite of noradrenaline, than normal. The extent of the decrease in the nine patients examined was correlated with a psychometric measure of the memory deficits (i.e. Wechsler Adult Intelligence Scale IQ minus Wechsler Memory Scale MG; McEntee & Mair, 1978). Neither the dopamine metabolite HVA nor the serotonin metabolite 5-HIAA (5-hydroxyindole-acetic acid) was lowered in the Korsakoff group compared to controls. However, a later study raised the possibility of slight reductions in CSF HVA in a larger group of Korsakoff patients (R. G. Mair, W. J. McEntee, R. J. Zatorre & P. J. Langlais, personal communication).

Drug effects. If a deficit in central noradrenaline transmission underlies some of the symptoms of Korsakoff's disease, it might be of therapeutic potential to administer aminergic stimulants to these cases. McEntee & Mair (1980*a*) found that the α-agonist drug clonidine produced a significant improvement in memory in a sample of eight Korsakoff patients. Using a double-blind design, clonidine (0.3 mg) caused significant improvement on the memory and visual reproduction tests of the Wechsler Memory Scale. Neither the serotonergic antagonist methysergide nor the mixed noradrenaline/dopamine releasing agent amphetamine were effective on these tests.

Conclusion

Two lines of recent evidence suggest a role for noradrenaline systems in at least some of the diverse symptoms of Korsakoff's psychosis. First, CSF levels of the noradrenaline metabolite MHPG are reduced in Korsakoff patients, with no great change in dopaminergic or serotonergic metabolites. Further, the degree of reduction in MHPG correlates with the severity of the memory impairment. Secondly, administration of clonidine, an α-agonist, partially restores recent memory function in these patients. However, as with other disease (see section on Alzheimer's dementia above) the noradrenaline malfunction is only one of a number of changes in brain neurochemistry and as such will contribute to only a small subset of the overall symptoms of the disease. In view of the evidence linking noradrenaline with selective attention (see Chaper 6), it is likely that the deficits in attention, concentration, ability to change mental set, and visual and verbal abstraction (Victor & Banker, 1978) are those most closely related to the noradrenergic malfunction in Korsakoff's psychosis (McEntee & Mair, 1980*b*; R. G. Mair & W. J. McEntee, personal communication). Nonetheless, a noradrenergic basis and treatment of even a part of the complex cognitive changes seen in Korsakoff's psychosis would be of major importance.

Attentional deficit disorder

Introduction

Initially called minimal brain dysfunction or hyperactive child syndrome this has now been renamed attention deficit disorders (DSM III, 1980). Symptoms of this disease, usually diagnosed in young children, include short attention span, motor hyperactivity, deficits in impulse control and consequent poor performance in learning and educational situations (see Wender, 1978). The disorder is ten times more common in boys than in girls.

Noradrenaline

Shekim, Dekirmenjian & Chapel (1979) found that the concentration of the noradrenaline metabolite MHPG in urine was lower in a sample of 15 hyperactive boys than in their 13 control children. However the experimental sample was very varied, with 8 of the 15 children showing MHPG levels markedly lower than control values while the remaining 7 children diagnosed with attentional deficit disorder showed normal to elevated levels. Further, an earlier study by Wender, Epstein, Kopin & Gordon (1971) failed to find any difference in the MHPG levels in urine from hyperactive compared to control children.

Dopamine

Shaywitz, Cohen & Malcolm (1975*b*) found significantly lower levels of the dopamine metabolite HVA in CSF from hyperactive children after probenecid loading than in controls. No change in the serotonin metabolite 5-HIAA was detected but regretfully the noradrenaline metabolite MHPG was not measured in this study.

An animal model had been reported of minimal brain dysfunction in which rats are treated with 6-OHDA when neonatal and depletions of brain dopamine are affected without loss of noradrenergic systems (Shaywitz, Yager & Klopper, 1976*b*). Increases in locomotor activity are reported during a brief period of early life which return to normal in adulthood. Further, drugs such as amphetamine and methylphenidate are reported to reduce this hyperactivity at nonstereotypic doses (Shaywitz, Klopper, Yager & Gordon, 1976*a*; Shaywitz *et al.*, 1978). Recently evidence has been provided that this animal model may differ in significant features from the human clinical condition of attention deficit disorder. Thieme, Dijkstra & Stoof (1980) found indications of deficits of attention in 6-OHDA-treated rat pups as shown by prolonged locomotion in a novel environment (compared to controls which rapidly habituated) and by more rapid switching between categories of behaviour (grooming and sniffing, in particular, showed shorter duration bouts). However, neither of these 'attentional' deficits were corrected by 0.75 mg kg^{-1} amphetamine, a drug which is effective in human attention deficit disorder. Further, three groups (Eastgate, Wright & Werry, 1978; Pappas *et al.*, 1980; Concannon & Schechter, 1981, 1982) have reported, in contradiction to Shaywitz *et al.* (1976*b*, 1978), that neither amphetamine nor methylphenidate were effective in reducing the locomotor hyperactivity of these 6-OHDA-treated rat pups. Indeed, both drugs further increased such locomotor activity. This is again apparently contrary to the therapeutic effects of these agents in human attentional deficit disorder.

Since it is uncertain how close this animal paradigm comes to being a model of human attentional deficit disorder, it constitutes only slight evidence for a dopaminergic basis of minimal brain dysfunction.

Drug effects

The usual drug prescribed for minimal brain dysfunction (now renamed attentional deficit disorder) is methylphenidate or *d*-amphetamine (Millichamp & Fowler, 1967). Imipramine has also been reported to be of some use (Winsberg, Bialer, Kupietz & Tobias, 1972; Waiser, Hoffman, Polizus & Engelhardt, 1974; Cox, 1982; although some authors call it ineffective, Wender & Klein, 1981). Both amphetamine and methylphenidate release dopamine from presynaptic neurones and amphetamine will additionally release noradrenaline. Imipramine is a noradrenaline uptake inhibitor with some serotonergic effects. It has been reported that the *d*- and *l*-isomers of amphetamine are equally effective in treating minimal brain dysfunction (Arnold, Wender, McCloskey & Snyder, 1972; Arnold *et al.*, 1976) and this might be some evidence in favour of a noradrenergic mediation (see Mason, 1982). However, given the neurochemical impurity of these drugs virtually no conclusion can be drawn from their efficacy in the clinical condition towards the biochemical pathology underlying the disease (Margolin, 1978).

Conclusion

No firm changes in catecholamine function have been demonstrated in children suffering from attentional deficit disorder. Some evidence from the drugs used to treat this disorder would indicate a catecholaminergic involvement but this fails to distinguish noradrenaline from dopamine. Noradrenergic deficits might be expected to be more related to the cognitive and attentional changes of the syndrome while dopaminergic ones might underlie the motor hyperactivity. A dopaminergic animal model of the hyperactivity has indeed been developed using 6-OHDA administration to neonatal rats but may differ on important points from the human clinical condition.

Summary – cognitive disorders

Schizophrenia

(1) Post-mortem assays on brains of schizophrenic patients initially suggested a deficiency in noradrenergic systems on the basis of reduced DBH activities. This does not appear to have stood the test of time and more recently four or more groups have independently found elevated noradrenaline concentrations in limbic brain regions or in CSF of schizophrenics, especially of the paranoid subtype. Neuroleptic drug medication was reported to lower CSF noradrenaline in parallel with an amelioration of the schizophrenic symptoms.

(2) Beta-blockade with propranolol has been reported to be effective in schizophrenia but appears to act by altering liver metabolism of neuroleptic medication and hence elevating brain neuroleptic levels indirectly. This cannot then be invoked as evidence for a noradrenergic overactivity in schizophrenia. Conventional neuroleptics may affect noradrenergic systems but it has not been demonstrated that this is necessary for their clinical effectiveness.

(3) Post-mortem assays on dopaminergic systems in brain tissue from schizophrenics have failed to yield any consistent change with the exception of an elevation in the number of dopamine receptors. This latter change, however, is not generally seen in those few patients who had not been treated with neuroleptics and thus may be iatrogenic – caused by the drug treatment – and *not* related to the neurochemical pathology of the disorder.

(4) The fact that neuroleptic drugs, the most effective treatment for the amelioration of the symptoms of schizophrenia, appear to reduce activity in brain dopaminergic systems, possibly by blocking dopamine receptors, need not in itself mean that the causative pathology of schizophrenia lies in these same dopaminergic systems.

(5) We do not know the cause of schizophrenia but a malfunction of noradrenergic systems may play a part in the alterations in attention and relation to the environment seen in the disease. Other neurochemical systems may well be involved in some of the other symptoms of the condition.

Senile dementia – Alzheimer's disease

(1) Very clear and profound loss of noradrenergic systems has been demonstrated in terms of a 50–80% reduction in the number of cell bodies in the LC in post-mortem brain tissue as well as a loss of noradrenaline in forebrain areas and reduced DBH. Some studies have shown a correlation between the degree of damage to the noradrenergic system and the severity of the dementia.

(2) No widespread examination of the effect of noradrenergic drug treatments in Alzheimer's disease has yet been carried out. Some early results with L-DOPA appeared promising.

(3) No consistent loss of dopaminergic parameters has been seen in Alzheimer's disease although sporadic variations do crop up now and again.

(4) A role for noradrenaline loss in the attentional changes in

cognitive function seen in Alzheimer's disease is strongly indicated. It is certain that other systems (cholinergic) are also deficient and some of the symptoms of the overall disease state would be expected to be related to this, not to the noradrenergic deficiency.

Korsakoff's psychosis

(1) Reductions in CSF MHPG have been reported in Korsakoff patients and seem to be correlated with the severity of the amnesia. Clonidine, an α-agonist drug has been used with some success in improving memory performance of these patients.
(2) Studies on a role for dopamine in Korsakoff's psychosis have not been frequent in the literature; a slight reduction in HVA might be present but this must be regarded as tentative at this stage.
(3) A role for a noradrenergic pathology in the amnesia seen in Korsakoff's psychosis, probably as a consequence of altered attentional filtering, has been suggested. Confirmation is still required.

Attentional deficit disorder

(1) Scant metabolite studies exist for any role of noradrenaline in this disease.
(2) Equally scant metabolite studies exist for a role of dopamine; however, an animal model involving dopamine reduction by 6-OHDA has been proposed. Its relevance and mechanism is yet controversial.
(3) Drug treatment for this disease (amphetamine, methylphenidate) suggests a catecholaminergic basis but fails to differentiate between noradrenaline and dopamine.
(4) A role for noradrenaline in attentional deficit disorder might be suggested by the effects on rat behaviour of 6-OHDA-induced noradrenaline loss (see Chapter 6). Such a suggestion is only tentative at this stage.

II: MOVEMENT DISORDERS

Parkinsonism

Introduction

Parkinson's disease is one of the few instances in which there is firm evidence for the underlying neurochemical pathology. The disease involves progressive loss of motor movements, rigidity of the limbs and resting tremor, first described by James Parkinson (1817). Different categories of 'Parkinsonism' have been outlined (see van Dongen, 1981). These include cases where the movement disorders are the result of encephalitis (postencephalitic Parkinsonism), drug treatments (drug-induced Parkinsonism with, for example, neuroleptic agents), atherosclerosis or infarction of the brain (vascular Parkinsonism) and finally cases, by far the majority, in which no obvious cause is present (idiopathic Parkinsonism). The various motor initiation difficulties are now clearly related to loss of dopamine-containing cells in the nigrostriatal pathway. Other aspects of the disease, especially the dementias and intellectual deterioration seen in long-standing cases, may show some relation to noradrenergic cell loss which also occurs in these conditions.

Dopamine

Post-mortem assays. Perhaps the first studies on the brains of patients dying with Parkinson's disease were reported by Ehringer & Hornykiewicz (1960), who found deficiencies of dopamine in the caudate, putamen, pallidum, substantia nigra and hypothalamus (see Table 8.2). This was confirmed by Birkmayer & Hornykiewicz in 1961. They were soon followed by numerous other authors (Cote *et al.*, 1970; Fahn, Libsch & Cutler, 1971; Greer, Collins & Anton, 1971; Rinne, Sonninen & Hyppa, 1971; Buscaino, Jori & Campanella, 1972). All found marked decreases in dopamine content of striatal nuclei such as the caudate and putamen and that recent treatment with L-DOPA prior to death served to increase dopamine levels in these areas. Similar loss of dopamine content was observed using histochemical fluorescence techniques in the caudate nucleus (Constantinidis, Siegfried, Frigyesi & Tissot, 1974).

Marked reductions in the concentration of the dopamine metabolite HVA were shown in CSF samples from Parkinsonian patients (Bern-

Table 8.2. *Dopamine content in control or Parkinsonian brains ($\mu g\ g^{-1}$ tissue)*

Area	Control	Parkinsonian
Striatum	3.5	0.1
Pallidum	0.1	0.07
Thalamus	0.01	0.01
Substantia nigra	0.46	0.07

From Birkmayer, 1969.

heimer, Birkmayer & Hornykiewicz, 1966). Deficiencies in the dopamine-stimulated adenylate cyclase activity have also been reported in post-mortem Parkinsonian brains (Shibuya, 1979). Urinary excretion of free dopamine is also reduced in Parkinson patients (Barbeau, Murphy & Sourkes, 1961; Hoehn, Crowley & Rutledge, 1977). The number of cell bodies in the substantia nigra of patients dying with Parkinson's disease is markedly reduced, as shown by conventional histology, and, moreover, correlates with the severity of the motor impairments (Alvord *et al.*, 1974). It appears that degeneration in other brain areas (see later for LC) is *not* necessary for the motor symptoms of Parkinsonism to be seen (Adams, van Bogaert & van der Eecken, 1964). Thus, nigrostriatal degeneration is sufficient to cause motor impairments in Parkinsonism. Additonally, in cases of vascular Parkinsonism, degeneration is always seen in the substantia nigra or basal ganglia (De Reuck, de Coster & van der Eeken, 1980).

In summary, a host of studies (see review by Hornykiewicz, 1966) have established that loss of dopaminergic neurones occurs in post-mortem tissue of the nigrostriatal system in Parkinsonian patients. This has been shown by biochemical studies on dopamine concentrations, and also on those of its metabolites, by histofluorescence techniques and by conventional histology. This would seem to be the single case in the study of catecholamines and pathological behaviour in which a neurochemical cause of a disease has been amply demonstrated.

Drug studies. The standard treatment for Parkinsonism, the administration of L-DOPA, was initiated by Birkmayer & Hornykiewicz (1961). However, this acts to increase the brain level of both dopamine *and* noradrenaline and as such fails to distinguish between the two (although noradrenaline levels are raised to a much lesser extent than those of dopamine; Buscaino, Jori & Campanella, 1972). On the other hand, the development of Parkinson motor impairments in patients receiving neuroleptic drugs (see van Rossum, 1967) does constitute strong evidence for a dopaminergic role in motor movement (see Chapter 4).

Noradrenaline

Although loss of noradrenergic systems does not seem to be necessary for the motor symptoms of Parkinsonism, it may contribute to various cognitive changes seen in this disease.

It was reported as long ago as 1953 that brainstem lesions, including the LC, were present in brains of Parkinsonian patients (Greenfield & Bosanquet, 1953). The original report of decreased dopamine in Parkinsonian brains (Ehringer & Hornykiewicz, 1960) also mentioned loss of brain noradrenaline. The largest losses of noradrenaline are seen in nucleus paranigralis and nucleus pigmentosus with more moderate losses in the amygdala, raphe and central gyrus (Farley & Hornykiewicz, 1976). Losses of noradrenaline were found in the caudate, putamen, thalamus, hypothalamus, raphe, substantia nigra, nucleus rubra, cingulate gyrus, amygdala and nucleus accumbens in another series of five Parkinsonian brains (Riederer, Birkmayer, Seemann & Wuketich, 1977). The noradrenaline metabolite MHPG was also slightly reduced in these brains, especially in the basal ganglia, raphe and accumbens. Other authors have also found widespread losses of noradrenaline in Parkinsonian brains (Rinne & Sonninen, 1973; Birkmayer, Danielczyk, Neumayer & Riederer, 1974; Birkmayer, Jellinger & Riederer, 1977). Loss of pigmented cell bodies in the LC has been reported (Bernheimer *et al.*, 1973). Low serum DBH levels have been seen in Parkinsonian patients (Lieberman, Freedman & Goldstein, 1972) but MHPG in CSF samples is apparently unaltered (Chase, Gordon & Ng, 1973). DBH activity in terminal regions of the LC system has been reported to be reduced in Parkinsonian brains (McGeer & McGeer, 1976; Nagatsu *et al.*, 1979).

The effect of the marked losses of LC noradrenaline in Parkinson's disease is uncertain. It does not appear to be important for the motor impairments (see previous section). However, up to 30 % of Parkinsonian patients show clear intellectual impairments (Mindham, 1970). The impairments usually affect orientation, constructions and memory (see review by Boller, 1980), while social behaviour, language and manipulation of old information are less affected. On the Weschler Adult Intelligence Scale tests the most affected item was perceptual organisation (Meier & Martin, 1970; Loranger *et al.*, 1972). The latter authors argue that 'It seems that the Parkinsonian patient's greatest difficulty is in comprehending and analysing novel or unfamiliar stimuli' (*ibid.*, p. 411). As the disease progresses, more severe symptoms such as personality changes, hallucinations and psychosis may arise (Mindham, 1970). These severe forms are seen more often in postencephalitic or vascular Parkinsonian patients. Whether any of these cognitive malfunctions associated with Parkinsonism results from loss of noradrenergic cells in the LC system is unknown. However, the similarities with some aspects of Alzheimer's dementia and the effects of experimental noradrenaline loss in animal models is suggestive.

Conclusion

In Parkinson's disease we have the one example in which both biochemical assays and drug studies have together combined to demonstrate a role for nigrostriatal dopaminergic systems in the motor initiation symptoms. Clear losses of brain noradrenaline also occur, especially in more chronic cases, and may be the cause of intellectual changes seen in the disease. This, however, awaits definitive proof.

Huntington's chorea

A movement disorder often regarded as in some way opposite to Parkinsonism is that of Huntington's chorea. Here, rather than a deficit in the initiation of motor movements, an uncontrollable emission of poorly coordinated movement occurs as first described by George Huntington (1872). Although the condition is often treated with neuroleptic drugs such as haloperidol and pimozide, post-mortem assays on brains of patients dying with Huntington's chorea reveal an essentially intact dopaminergic system in the basal ganglia (McGeer & McGeer, 1976). Considerable degeneration of cells postsynaptic to the dopamine terminals in the caudate and putamen nuclei is seen together with loss of cholinergic and GABAminergic parameters in these nuclei. Thus, in a strict sense, Huntington's chorea does not appear to be caused by a malfunction *per se* of brain dopaminergic systems.

Epilepsy

Introduction

Epilepsy known as the 'sacred disease', has been described since the time of Hippocrates (460–374 B.C.) in classical antiquity (see Adams, 1929). The literature on a catecholaminergic role in epilepsy is unusual in the respect that while considerable evidence has accumulated from animal studies that both brain noradrenergic and dopaminergic systems can alter the sensitivity to seizures induced by various experimental techniques there is virtually no evidence that such a mechanism is actually perturbed in the clinical epilepsies of human patients. Thus, two recent studies which examined the levels of catecholamine metabolites in the CSF of epileptic patients found no change in the noradrenaline metabolite MHPG (Laxer *et al.*, 1979; Peters, 1979), despite the considerable data from animal studies that manipulations of noradrenergic systems can markedly alter seizure susceptibility. One study (Hiramatsu, Fujimoto & Mori, 1982) has reported a rather variable increase in noradrenaline itself in the CSF of epileptics (grand mal). It seems unwise therefore to concentrate too closely, in the following coverage, on the voluminous animal literature. It is perfectly possible from a conceptual viewpoint that, while alterations in brain noradrenergic or dopaminergic systems can affect seizure susceptibility in animals, epilepsy in humans is caused by other neurochemical malfunctions and the catecholaminergic systems are essentially intact. In view of this possibility the review of animal studies on catecholamines and convulsions is deliberately kept brief.

Noradrenaline

Lesion studies. The neurotoxin 6-OHDA has been used extensively to examine a role for noradrenaline in seizure susceptibility. Arnold, Racine & Wise (1973) and Corcoran, Fibiger, McCaughran & Wada (1974) showed that, in rats, intraventricular administration of 6-OHDA facilitated the time course of amygdaloid kindling. This is a paradigm in which electrical stimulation, in this case of the amygdala, is applied on a daily basis. This stimulation initially fails to elicit any overt convulsions. However, with repeated administrations (in the case of the amygdala from 10 to 15 daily stimulations) progressive magnitudes of convulsion develop, culminating in a so-called stage 5 seizure (Racine, 1972), in which convulsive activity of both forelimbs and hindlimbs on both sides of the body and ictal electroencephalographic activity is observed. 6-OHDA treatment depleted both noradrenaline and dopamine, so no differentiation of the catecholamine of importance was possible. Seizures caused by loud sounds (audiogenic seizures) were also potentiated by intraventricular administration of 6-OHDA which depleted both noradrenaline and dopamine (Bourne, Chiu & Picchioni, 1972), as were seizures caused by alcohol withdrawal (Chu, 1978) and electroconvulsive shock (ECS: Browning & Maynert, 1978; London & Buterbaugh, 1978).

More specific lesions using 6-OHDA to destroy only the noradrenergic systems, or only the dopaminergic ones, have been reported recently. Thus, both McIntyre, Saari & Pappas (1979) and Corcoran & Mason (1980) found that depletion of brain noradrenaline with 6-OHDA (either neonatal peripheral administration or adult injection into the fibres of the dorsal noradrenergic bundle) facilitated the time course of amygdaloid kindling. The study of Corcoran & Mason (1980) also included a group of rats with a 60 % depletion of brain dopamine caused by 6-OHDA injection into the fibres of the nigrostriatal dopaminergic bundle and this group showed no potentiation of amygdaloid kindling. The role of amygdaloid noradrenaline in the facilitation of kindling was confirmed by similar 6-OHDA lesions carried out by Mohr & Corcoran (1981) and McIntyre (1980), while McIntyre & Edson (1982) found hippocampal kindling was also facilitated. Bourne *et al.* (1977) used the neonatal 6-OHDA technique to deplete brain noradrenaline and found a marked potentiation of audiogenic seizures, while Mason & Corcoran (1979*a*), using adult dorsal bundle 6-OHDA injections, found a potentiation of convulsions caused by the drug metrazol (also Mason & Corcoran, 1978) or of ECS seizures (Mason & Corcoran, 1979*b*). Depletion of brain dopamine failed to affect ECS or metrazol convulsions. More localised noradrenaline depletions in which 6-OHDA injections were made into the ascending dorsal bundle, the cerebellar noradrenergic innervation of the descending noradrenergic innervation to the spinal cord showed that the potentiation of the metrazol seizure was the result of dorsal bundle noradrenaline loss (Mason & Corcoran, 1979*c*). Thus, the conclusion from the 6-OHDA lesion studies in animals appears to be that considerable evidence exists for a seizure suppressant role of noradrenaline in convulsions induced by many techniques (ECS, metrazol, kindling, audiogenic) while little support has emerged for a role for dopamine in these seizures.

Release/levels measurements in animals. Sato & Nakashima (1975) found that whole brain noradrenaline and dopamine content was significantly reduced in hippocampal-kindled cats killed one week after their last seizure. Since whole brain was analysed, this fails to indicate the brain area of importance. This was rectified by Engel & Sharpless (1977), who found a depletion of both catecholamines in the amygdala, but not in the hippocampus, of kindled rats sacrificed one month after their last seizure. The noradrenaline loss, however, might be caused by electrode implantation, since implanted but not electrically stimulated rats also showed a similar loss of amygdaloid noradrenaline content. Callaghan & Schwark (1976) found a depletion of hypothalamic noradrenaline with no change in dopamine level after amygdaloid kindling. On the other hand, Wilkinson & Halpern (1979) found no change in noradrenaline turnover in the forebrain of rats killed one week after a series of kindled seizures. Dopamine turnover was increased in these rats. Changes in postsynaptic noradrenaline markers have also been reported after kindling. Thus, McNamara (1978*a*) found alterations in β-binding in the forebrain of amygdaloid kindled rats 3 days after their first seizure. These changes took time to develop, not being present 15 h after the first seizure (McNamara, 1978*b*). Schlesinger, Boggan & Freedman (1965) examined genetically seizure-prone DBA/2J mice, and Schaeffer & Kuenzel (1978) examined seizure-prone White Leghorn chicks for a correlation with brain noradrenaline levels. In the chicks an increase in the brainstem and a decrease in the cerebellar noradrenaline were seen when compared to nonseizure-prone controls. Philo, Reiter & McGill (1979) found that pinealectomy caused convulsions in gerbils and simultaneously reduced telencephalic noradrenaline without effect on dopamine. Laguzzi, Acevedo & Izquierdo (1970) found that surgical decortication reduced the concentration of noradrenaline in the hippocampus of rats while increasing the susceptibility to seizures, and Ladisch (1973) found that ouabain administered onto the cortical surface caused convulsions and increased the disappearance of pre-injected [^{3}H]noradrenaline in brainstem and forebrain. A corresponding increase was seen in the labelled noradrenaline metabolites normetanephrine, dihydroxymandelic acid and dihydroxyphenylglycol. Similar results were obtained with ECS-induced convulsions (Ladisch, Steinbauff & Matussek, 1969). However, ouabain convulsions fail to alter the gross levels of noradrenaline in mouse brain (Doggett & Spencer, 1971).

In conclusion, it would appear that noradrenergic systems are activated during convulsions, but so too are many other neurotransmitter pathways. Kindling appears to cause changes in noradrenergic systems which outlast the immediate post-seizure period and may contribute to the mechanism of kindling itself.

Drug studies. Reserpine, which depletes not only noradrenaline and dopamine but also serotonin, has been found to potentiate audiogenic seizures (Lehmann & Busnell, 1963). This effect of reserpine can be reversed by subsequent loading with L-DOPA, suggesting a role for the catecholamines but failing to separate noradrenaline from dopamine (Boggan & Seiden, 1971). L-DOPA on its own may protect against metrazol convulsions (McKenzie & Soroko, 1972) but Dadkar, Dahanukar & Sheth (1979) find that L-DOPA with or without a peripheral decarboxylase inhibitor potentiates metrazol convulsions. The point is not worth labouring since L-DOPA fails to dissociate noradrenaline from dopamine as the catecholamine of importance.

More specific manipulations of noradrenergic or dopaminergic systems have been reported and will be briefly covered here (for noradrenaline) and in the next section (for dopamine). The noradrenaline synthesis inhibitor FLA 63 has been found to potentiate metrazol seizures in mice (Dadkar *et al.*, 1979). Spinal cord seizures, induced by electrical stimulation between C_1 and C_4 of the anaesthetised rat, were potentiated by noradrenaline depletion caused by the benzoquinolizine Ro 4-1284 (Jobe, Ray, Geiger & Bourne, 1977). A converse manipulation of increasing noradrenaline availability by administration of the noradrenaline uptake inhibitors desipramine or nisoxetine significantly shortened the duration of such spinal cord seizures (Jobe & Ray, 1980). Intraventricular administration of isopropylnoradrenaline has been found to protect against convulsions induced by peripheral injection of picrotoxin (Georgiev & Petkova-Radkova, 1976). On the other hand, the putative second messenger of noradrenergic synapses, cAMP, when administered intraventricularly failed to provide protection from, and actually potentiated, picrotoxin seizures. Convulsions caused by bright lights in photosensitive baboons were successfully reduced by intraventricular injection of noradrenaline (Altschuler, Killam & Killam, 1976).

The selective α-agonist clonidine has been reported to decrease susceptibility to audiogenic seizures in mice (Kellogg, 1976), although Dadkar *et al.* (1979) found the drug potentiated metrazol seizures in the same species. Both clonidine and the α_2-agonists oxymetazoline and UK 14, 304 reduced audiogenic seizures in DBA/2 mice upon intraperitoneal administration (Horton, Anlezark & Meldrum, 1980). These effects could be blocked by the selective α_2-antagonists yohimbine and piperoxan. However, yohimbine and piperoxan by themselves failed to potentiate audiogenic convulsions. The α-blocker phenoxybenzamine also failed to affect audiogenic convulsions although it has been reported to protect against metrazol convulsions (Dadkar *et al.*, 1979). Although β-blockers are generally agreed to reduce seizure severity (Murmann, Almirante & Saccani-Guelfi, 1966; Yeoh & Wolf, 1968; Anlezark, Horton & Meldrum, 1979), comparison of the efficacy of the (−) and (+) isomers of propranolol led Anlezark *et al.* (1979) to conclude that this was caused by membrane stabilisation effects rather than β-blockade. However, for metrazol convulsions, Louis, Papanicolaou, Summers & Vajda (1982) came to the opposite conclusion with the two isomers, additionally observing that pindolol, a β-blocker devoid of membrane stabilising effect, also protected against these convulsions.

In summary, general agreement exists that measures which increase noradrenaline availability, such as uptake inhibitors, or which directly activate noradrenaline receptors, such as clonidine or isopropylnoradrenaline or noradrenaline itself, act to reduce seizures of many types (metrazol, audiogenic, electrical spinal cord stimulation). Measures which reduce the availability of noradrenaline, such as synthesis inhibition, potentiate convulsions. Studies with selective α- or β-blockers are less clear, but β-blockade may protect against metrazol, if not audiogenic, convulsions. A similar effect for metrazol may perhaps occur with α-blockade with phenoxybenzamine. However, in multiple drug studies a possibility exists for the metabolism of the convulsive agent to be altered by the receptor blocker in the liver without regard to any putative noradrenergic receptor effects and this has not been adequately controlled in many of the above studies.

Dopamine

Some evidence for a role of dopamine in convulsions has been mentioned above. Sato & Nakashima (1975) found reductions in whole brain dopamine after hippocampal kindling in cats and Engel & Sharpless (1977) found regional loss of dopamine in the amygdala of kindled rats. Increased turnover of dopamine in the forebrain after kindling was reported by Wilkinson & Halpern (1979). Drug studies also support this conclusion with Anlezark & Meldrum (1978) finding that the dopamine agonist apomorphine suppressed audiogenic seizures. This has been extended to the ergot alkaloids, which are also dopamine agonists (Anlezark, Pycock & Meldrum, 1976) and to include protection against photically induced seizures in baboons (Anlezark & Meldrum, 1978). However, Dadkar *et al.* (1979) failed to find any protective effect of apomorphine on metrazol seizures in mice (even at the extremely high dose of 20 mg kg^{-1}). This drug did protect against convulsions induced by electroconvulsive shock in rats (McKenzie & Soroko, 1972). The dopamine agonist *N*,*n*-propylnorapomorphine also protects against both audiogenic and photic seizures (Anlezark, Horton & Meldrum, 1978). Such effects could be blocked by the dopamine antagonist drug haloperidol (Anlezark *et al.*, 1976). However, Kleinrok, Czuczwar, Wojcik & Przegalinski (1978) failed to find any effect of apomorphine on either electroshock or metrazol convulsions. This was supported for metrazol convulsions by the negative findings of Lazarova & Roussinov (1978).

Some clinical studies have reported lower levels of the dopamine metabolite HVA in the CSF of epileptic patients than in that of the control group (Barolin & Hornykiewicz, 1967; Papeschi, Molina-Negro, Sourkes & Erba, 1972; Shaywitz *et al.*, 1975*a*; Leino, MacDonald, Airaksinen & Riekkinen, 1980) and of dopamine itself (Hiramatsu *et al.*, 1982). However, a further six clinical studies failed to find any such difference (Garelis & Sourkes, 1974; Reynolds *et al.*, 1975; Livrea, 1976; Livrea *et al.*, 1976; Laxer *et al.*, 1979; Habel *et al.*, 1981), while some find an increase (Chadwick, Jenner & Reynolds, 1975; Ito, Okuno, Mikawa & Osumi, 1980).

Thus, some slight evidence might implicate dopamine in some, rather limited sorts of convulsions in experimental animals. These would seem to be mainly audiogenic or photically induced seizures. Little evidence exists that dopamine abnormalities play any part in the clinical epilepsies.

Conclusion

In summary, although much evidence exists to implicate brain noradrenaline systems in seizure susceptibility in experimental animals, and some little evidence for a role of dopamine in a few of these paradigms, the failure to detect any change in catecholamine metabolites in human epileptic patients suggests that the catecholamine systems may *not* play any part in the actual pathology of the clinical epilepsies.

Summary – movement disorders

Parkinson's disease

(1) Within the limitations of scientific proof it may be stated that we do indeed know the neurochemical cause of this disease. The movement aspects are the result of loss of dopamine-containing cells from the substantia nigra. This has been shown by post-mortem studies on dopamine levels, dopaminergic cell counts, dopamine metabolite levels and by histofluorescence.

(2) Drug therapy of Parkinson's disease is by administration of L-DOPA or dopaminergic agonist drugs.

(3) Noradrenaline loss also occurs in the more severe or long-standing cases of Parkinson's disease. This has been shown histologically by cell loss in the LC, reduction in noradrenaline content of forebrain areas and lower DBH activities. Such a noradrenergic loss might underlie the cognitive and intellectual dysfunctions seen in chronic cases. A role for noradrenaline in the attentional malfunctions of Parkinson's patients would seem possible, but direct proof is lacking.

Huntington's chorea

(1) Loss of elements postsynaptic to the nigrostriatal dopaminergic system is seen in this disease. The dopaminergic systems themselves appear to be intact.

Epilepsy

(1) No evidence for any role of noradrenaline in the causative pathology or human clinical states has emerged from CSF studies.

(2) The animal literature reveals a wealth of studies indicating that reducing noradrenergic function by 6-OHDA lesion enhances the susceptibility to convulsions of all types, including those caused by kindling, metrazol, ECS and audiogenic means.

(3) Some lowering of brain noradrenaline function may be seen in animals prone to seizures, either genetically or as a result of experimental intervention. This is not completely consistent in all studies.

(4) Increasing noradrenaline availability with uptake inhibitors reduces susceptibility to many types of seizures. Propranolol protects against many sorts of convulsion but this might be because of the membrane stabilising effects of propranolol, not its β-blockade, at least in some cases. Studies with α-receptor selective drugs are less uniform.

(5) Dopaminergic agonists seem to protect against photically induced convulsions and audiogenic seizures but do not have a similar effect on metrazol convulsions.

(6) A few findings have been reported of changes in dopaminergic systems after kindling. These may be larger in magnitude than those seen in noradrenergic systems.

(7) Since no evidence for a role of either catecholamine in human clinical epilepsy has been forthcoming the relevance of the animal data must be questioned.

III. MOOD DISORDERS

Depression

Introduction

Various subdivisions of the affective disorder known as depression may be distinguished (Katz & Hirschfield, 1978). Thus, a separation between unipolar and bipolar may be made, in which the unipolar patients experience only depressed periods while the bipolar group may swing between depressed and manic phases. Another dichotomy often made is between endogenous and nonendogenous depression. Here, endogenous depression is characterised by early morning awakening, anorexia and weight loss and lack of reactivity and severe depression of mood; these symptoms are not brought about by external circumstances, while for nonendogenous depression external causative agents may be identified. So called psychotic depression is seen in schizophrenia-related conditions and schizoaffective disorders (Arana, 1978; Spitzer, Fleiss & Endicott, 1978). This may be extended to the concept of primary and secondary depression. In secondary depression the affective changes are related to a pre-existing psychiatric disease or to some other disease which could cause affective symptoms (Feighner *et al.*, 1972). Other symptoms also found in depressive states include feelings of guilt, worthlessness, hostility, anxiety–tension, loss of interest, somatic and cognitive complaints, motor retardation and bizarre thoughts (Nelson & Charney, 1981).

Noradrenaline

Depression is perhaps the 'classic' instance of widely accepted noradrenergic brain pathology. It was first postulated by various workers in the mid 1960s (Bunney & Davis, 1965; Schildkraut, 1965). Up to perhaps five years ago no serious dispute as to a noradrenergic basis of depression would have been entered. Recently, perhaps as a swing of fashion, various nonnoradrenergic theories of depression have gained vogue. Some of this is due to the advent of clinically effective antidepressant drugs which seem, unlike the early generation of such agents, to have relatively little action on brain noradrenergic systems.

Metabolite levels. The early evidence for a role of brain noradrenergic systems in depression came from studies on the levels of noradrenaline and its metabolites in the urine, and CSF of various classes of depressed patients.

Unipolar patients: This group seems more heterogenous than the bipolar group (see van Dongen, 1981). A subgroup of unipolar patients with low urinary MHPG content can often be distinguished from a second group with urinary MHPG in the normal range (Goodwin, Cowdry & Webster, 1978; Taube *et al*., 1978; Garver & Davis, 1979; Halaris & DeMet, 1979; Schildkraut *et al*., 1979). The second group with normal urinary MHPG levels may often have low CSF 5-HIAA levels, suggesting a serotonergic malfunction. When the symptoms of depression in the low-noradrenaline group have disappeared, it is often found that the levels of urinary MHPG have returned to normal (Shaw *et al*., 1973; Sweeney, Maas & Pichar, 1979).

Paradoxically, activity of the noradrenaline synthesis enzyme DBH is reported to be normal in CSF samples from unipolar depressed patients (Lerner *et al*., 1978). Patients with a history of depression, after dying by suicide, have shown slight changes in noradrenaline levels (amygdala, nucleus ruber) and decreases in MHPG in a few more regions (globus pallidus, hypothalamus, mammillary bodies, substantia nigra, raphe, accumbens: see Riederer & Birkmayer, 1980; van Dongen, 1981). However, studies failing to find any change in brain noradrenaline, its metabolites or enzymes in such post-mortem tissue have also been published (Bourne *et al*., 1968; Pare, Young, Price & Stacey, 1969; Grote *et al*., 1974; Moses & Robins, 1975). Recently, Meyerson *et al*. (1982) reported no change in postsynaptic noradrenaline receptors in brains of suicide victims as revealed by β-binding techniques. Very little evidence for a role of noradrenaline in depression has emerged from post-mortem studies. Most suggestive are the results on CSF and urinary catecholamines and their metabolites.

Bipolar patients: Perhaps better evidence is available here. Levels of both plasma and urinary MHPG correlate with the preceding changes in mood (Bond, Jenner & Sampson, 1972; Jones, Maas, Dekirmenlian & Fawcett, 1973; Schildkraut *et al.*, 1979; Garver & Davis, 1979; Halaris & DeMet, 1979). Urinary MHPG content of the bipolar group is lower than that of controls during the depressed phase and returns to the control values during the manic phase. Again, CSF DBH activities fail to show this relationship (Lerner *et al.*, 1978) with an actual reduction during the manic phase.

Drug effects. Initial evidence for a role of catecholamines in depression came from the clinical effectiveness of the monoamine oxidase inhibitors in depressive illness. More support for a specific role for noradrenaline came with the advent of the tricyclic antidepressants. Of the distinctions made between various subgroups of depressed patients (see above) it is the endogenous group which responds best to tricyclics (Goodwin *et al*, 1973; Katz & Hirschfield, 1978). During successful treatment of such depression, the urinary MHPG content of patients with low initial value (so-called noradrenergic depression) is increased by drugs such as desmethylimipramine or imipramine. The direct β-noradrenaline agonist salbutamol has also been occasionally reported to be an effective antidepressant (Widlocher *et al.*, 1977; Simon, Lecrubier & Jouvent, 1978).

The major effect of all early tricyclic antidepressants was found to be a blockade of the reuptake inactivation of noradrenaline (see Lewi & Colpaert, 1976). However, this effect is immediate, while clinical improvement to antidepressant therapy takes some two weeks or so to occur. Thus, attention has shifted to the chronic effects of tricyclic administration (Vetulani & Sulser, 1975). Here a decrease in the number of central noradrenaline receptors has been reported in experimental animals (Wolfe, Harden, Sporn & Molinoff, 1978) for β-receptors, apparently of the β_2-subtype (Minneman, Dibner, Wolfe & Molinoff, 1979). A decrease in the number of α_2-receptors on LC cells has been shown electrophysiologically (Svensson & Usdin, 1978; McMillen, Warnack, German & Shore, 1980). The secondary tricyclic antidepressants (desmethylimipramine and imipramine, which is metabolised to desipramine (DMI)) are more effective than the tertiary tricyclics (such as amitriptyline, chlorimipramine) in causing receptor alterations (Tang, Seeman & Kwang, 1981) and this parallels the relative clinical efficacy of these agents in low-noradrenaline depressed patient subgroups. More recent antidepressants which initially appeared to interact only weakly with noradrenergic systems (such as the tetracyclic mianserin or zimilidine) have also been reported to decrease the number of β-receptors on chronic administration (Clements-Jewery, 1978), although other workers find reduced sensitivity of the noradrenaline-stimulated adenylate cyclase without reduction in β-binding (Mishra, Janowsky & Sulser, 1980; Sellinger-Barnette, Mendels & Frazer, 1980). Evidence that the decrease in receptor binding (of either β- or α_2-type) may be related to the therapeutic mechanism of action of the antidepressants comes from the similarity in the time course for DMI administration in reducing β-binding with the time course of its clinical response (Huang, Maas & Hu, 1980) and from DMI's ability to reduce behavioural effects of the α_2-agonist clonidine with a time course similar to its clinical response (Checkley, Slade, Shur & Dawling, 1981).

At the moment, the field is in a state of flux, with it no longer being clear that depression is caused by a deficiency of brain noradrenaline or that antidepressants work by increasing the availability of noradrenaline at the synapse as a consequence of reuptake inhibition (see outstanding review by Green & Nutt, 1983). Reductions in numbers of postsynaptic receptors, either as a consequence of increased synaptic noradrenaline or as a direct action of the drug itself (mianserin, zimilidine), may instead be the mechanism of clinical actions. This implies the reverse cause of depression, namely too much noradrenaline activity. During this period of uncertainty it has even been suggested that noradrenaline might not be the amine of critical importance in depression and I will now consider suggestions of a role for dopamine.

Dopamine

Metabolite studies. Although until recent years the emphasis in metabolite determinations in the urine and CSF of depressed patients has been on those of noradrenergic origin, some slight indication of changes in those of dopaminergic origin has also accumulated (along with considerable evidence for a subgroup of serotonergic nature). While the baseline of levels of the dopamine metabolite HVA may be unchanged in CSF, it has been reported to be below normal using the probenecid accumulation technique (van Praag & Korf, 1971, 1975; Sjostrom & Roos, 1972; Goodwin, Post, Dunner & Gordon, 1973; Banki, 1977). However, some caution is needed with these results since they were obtained on patients with marked motor retardation as a symptom of their depression and the reduced dopamine metabolism might be related to the lesser activity and movement seen in these patients (see Chapter 4). That the post-probenecid HVA levels rise to normal as the depression abates (van Praag, Korf & Lakke, 1975) is no evidence of its causative nature, since so too will the motor retardation abate. This indeed is the conclusion of most investigators in this area, namely that the low dopamine metabolism reflects the reduced motor activity seen as a result of depression rather than the depressive pathology itself (van Praag, 1980).

Drug studies. More supportive evidence for a role of dopamine in depression comes from some of the newer antidepressant drugs and their neurochemical effects. L-DOPA, which is converted into both noradrenaline and dopamine (although it may form rather more dopamine than noradrenaline *in vivo*) has been reported to improve the motor retardation symptoms of depressed patients *but no improvement in mood per se* was seen (Goodwin, Murphy & Brodie, 1970; van Praag & Korf, 1975). This would seem to make it more likely that the dopaminergic effect in depression is a consequence of reduced motor activity rather than its role in the affective pathology. On the other hand, the direct dopamine receptor agonist piribedil has been found to be clinically effective in improving mood in both a subgroup of depressed patients who had low HVA accumulation after probenecid (Post *et al.*, 1978) and in a group of unselected patients (Reus, Lake & Post, 1980). The dopaminergic agonist bromocryptine and the dopamine/noradrenaline uptake inhibitor nomifensine are also clinically effective antidepressants (van Scheyden, van Praag & Korf, 1977), as is the pure dopamine uptake inhibitor buproprion.

Although the classic antidepressants, the tricyclic agents, have little or no effect on dopaminergic systems in the acute phase, evidence has recently emerged that they may change dopaminergic parameters upon chronic dosage. Thus, repeated administration of amitriptyline, imipramine, chlorimipramine, iprindole and mianserin will cause a putative subsensitivity of dopamine autoreceptors as shown by reduction in the inhibitory effect of small doses of apomorphine on locomotor activity (see Chapter 4) in rats, on striatal 3,4-dihydroxyphenylglycol levels (Serra, Argiolas, Fadda & Gessa, 1980, but see Holcomb, Bannon & Roth, 1982) and on nigrostriatal dopaminergic neuronal firing rates (Chiodo & Antelman, 1980) although this latter has been questioned (MacNeil & Gower, 1982; Welch, Kim & Liebman, 1982; see reply by Chiodo & Antelman, 1982). There is also one report of postsynaptic changes induced by chronic desipramine or imipramine treatment in the rat on [^{3}H]spiperone binding in the striatum (reduction, Koide & Matsushita, 1981), although this is not universally agreed.

Thus, in summary there is suggestive evidence that the therapeutic response to both the classic tricyclic antidepressant drugs and to the newer agents may involve dopaminergic systems. It is unclear precisely what change in these systems is necessary for clinical efficacy (i.e. increase in activity, decrease in postsynaptic receptors). The evidence from metabolite studies suggests a subgroup with low probenecid accumulation of HVA but most authors view this as a consequence, not a cause, of the depressive pathology.

Anxiety

Introduction

Klein, Zitrin & Woerner (1978) use the term 'anxiety' as a state of 'uneasy, apprehensive, or worried about what may happen'. Clearly, a distinction is normally made between situational anxiety for which a good reason is present (some stress, disease or justifiable anxious expectations) and so-called 'pathological' anxiety for which no reason can be discerned but which is nonetheless present for periods of time ('phobic' or 'anxiety disorders' – see DSM III and Feighner *et al.*, 1972).

Noradrenaline

Animal models. A role for the LC noradrenergic system in fear and anxiety has been advanced by Redmond and colleagues (Redmond *et al.*, 1976; Redmond, 1977; Redmond & Huang, 1979). This has been described in more detail in Chapter 6. Essentially, the findings are that electrolytic lesion to the LC in the monkey leads to a placid animal which fails to show the usual fear responses to threat from the human experimenter or to a toy snake. Electrical stimulation in the vicinity of LC caused the emission of behaviours such as scratching and grimacing often associated with fear situations in the wild. However, as reviewed earlier (pp. 287–90) more selective 6-OHDA lesions in the rat of either the LC (Crow *et al.*, 1978*a*; File, Deakin, Longden & Crow, 1979) or of the ascending noradrenergic bundle (see Mason, 1982, for review) failed to reduce fear behaviours as assessed by social interaction, open field, passive avoidance, one- and two-way active avoidance, Sidman avoidance, conditioned emotional response and taste neophobia. Furthermore, electrical stimulation of the LC in the rat has been reported to support self-stimulation – an indication of positive reward rather than fear (Crow, Spear & Arbuthnott, 1972). Van Dongen (1980*a*,*b*) has mapped the region of the cat brain in which injections of the cholinergic agonist carbachol would elicit fear responses and shown that it is not the LC but the adjacent rostral pontine reticular formation – a nonnoradrenergic region. This would suggest that the fear-related effects reported by Redmond and colleagues may reflect activation of adjacent *non*noradrenergic neural circuits and so provide little direct evidence for a role of noradrenaline in human anxiety states. In summary (and see Chaper 6, for more detail) the current state of the animal literature does not support a role for noradrenaline in fear.

Clinical and drug effects. It is reported that electrical stimulation in the vicinity of the LC in patients undergoing brain surgery induces feelings of fear and death (Nashold, Wilson & Slaughter, 1974). However, the localisation appears to have been in the central grey rather than the LC proper (F. G. Graeff, personal communication). This, then, could not be claimed as evidence for a role of noradrenergic systems in anxiety.

Beta-blocking drugs such as propranolol have been used to treat anxiety states (Floru, 1977; Greenblatt & Shader, 1978). However, the majority of studies indicate that it is the peripheral actions of β-blockade on the somatic manifestations of fear, rather than any direct central effect, which are of importance (Gottschalk, Stone & Gleser, 1974; Bernardt, Silverstone & Singleton, 1980). Thus, the anti-anxiety effects of propranolol are weak evidence at best for a role of noradrenaline in anxiety.

Tricyclic antidepressants have been used to treat pathological anxiety (Klein *et al.*, 1978) but these would be expected to increase the availability of brain noradrenaline, not reduce it as would be required by the Redmond hypothesis.

The major class of anti-anxiety agents, the benzodiazepines, do indeed decrease the firing of LC cells and hence reduce the availability of forebrain noradrenaline (Grant, Huang & Redmond, 1980). However, van Dongen (1981) makes the point that the effect of benzodiazepines on noradrenaline turnover shows tolerance with repeated administration while the anti-anxiety effect of these drugs shows sensitisation rather than tolerance (Haefely, 1978). This would imply that the anti-anxiety action of benzodiazepines does not depend on any effect that they might have on noradrenergic systems. Finally, Sanghere *et al.* (1982) found that the clinically effective anti-anxiety agent buspirone failed to decrease LC activity in rats.

Conclusion. The noradrenergic theory of anxiety, initially suggested from animal experiments, seems to have received very little support from the clinical literature and should now be abandoned.

Mania

Noradrenaline

Some of the evidence implicating noradrenaline in mania has been covered in passing in the section of Depression (see above). I will merely note here that, during the manic phase of patients suffering from bipolar depression, both plasma and urinary levels of the noradrenaline metabolite MHPG are increased as compared to those during the depressive phase (Bond *et al.*, 1972; Jones *et al.*, 1973; Garver & Davis, 1979). However, this is partially circular since even during the manic phase the MHPG levels do not exceed those of matched control subjects. However, levels of the noradrenergic enzyme DBH in CSF are *lower* in the manic phase (Lerner *et al.*, 1978). Although increases in locomotor activity and agitation might be expected to increase catecholamine metabolites, this may be excluded by the demonstration that the increase in MHPG content *precedes* the subsequent change in mood and activity (Taube *et al.*, 1978).

Manic symptoms may be diminished by treatment with the β-blocker propranolol (Emrich *et al.*, 1979) or by DBH inhibitors such as fusaric acid (Sacks & Goodwin, 1974). The most common treatment, with lithium, has among many other effects the action of antagonising noradrenaline-stimulated adenylate cyclase (see van Dongen, 1981). This is all too preliminary to allow any firm conclusions.

Summary – mood disorders

Depression

(1) Reduction in urine and CSF MHPG levels are found in some depressed patients and changes in these levels seem to precede mood changes indicating a causative role. A second subgroup of patients with normal MHPG levels may show lowered serotonergic metabolites. Post-mortem assays on brain tissue from suicides have failed to demonstrate any consistent change in noradrenergic systems.

(2) Conventional antidepressant therapy with tricyclics was believed for a long time to act by inhibiting the reuptake of noradrenaline and hence elevating the availability of noradrenaline in the synapse. Newer antidepressant agents do not have this effect. Attention has shifted to the effects of these drugs on postsynaptic β-receptor number and on α-receptor density. This field is still in a state of flux.

(3) Depressed patients with motor retardation show reduced HVA levels in the CSF suggesting a dopaminergic malfunction. Most authors regard this as a consequence of the lesser motor activity seen as a symptom of depression and *not* as a cause of the mood alteration *per se*.

(4) Both the classic tricyclic drugs and the newer antidepressants may act on dopaminergic systems. The direction or mechanism of their

effect is not clear as yet and this area remains suggestive rather than definitive.

(5) No global conclusion on the neurochemical cause of depression is warranted by the state of the present evidence. It is possible that there may be different types of depression related to low noradrenaline, low dopamine and low serotonin.

Anxiety

(1) An animal model of anxiety initially suggested a possible role of noradrenaline in this human clinical state. Further studies with 6-OHDA have indicated that the anxiety effects were *not* related to noradrenergic systems.

(2) The anti-anxiety effects of propranolol seem to be the result of its peripheral somatic rather than its central effects and cannot be invoked as evidence for a noradrenergic basis of anxiety. This effect of the classic anti-anxiety agents, the benzodiazipines, on noradrenaline turnover shows tolerance while their effects on anxiety shows sensitisation. This would imply that the benzodiazepines do *not* have their anti-anxiety effect via noradrenergic systems.

(3) The noradrenergic theory of anxiety should be abandoned.

Mania

(1) The evidence for a causative role of noradrenaline in mania is disappointingly slim. No great emphasis should be placed upon it.

Conclusions

Disease or symptoms

As can be seen from the above summary there is no one disease which can be said to result in its entirety from a noradrenergic or a dopaminergic malfunction. Even Parkinson's disease cannot be said to be entirely caused by the well-proven dopamine deficiency. The motor impairment may be caused by dopamine loss but the intellectual perturbations seen in long-standing cases may owe their origin to noradrenergic pathology. Even the whole motor impairment is not the result of dopamine depletion, since the tremor often fails to respond to L-DOPA therapy whereas the difficulties in motor initiation are alleviated by such treatment. It would seem unprofitable to expect all of one disease state to be caused by malfunction of a single neurochemical system. A more useful heuristic approach is to examine the symptoms (van Dongen, 1980*b*). Here, a symptom common to a number of disease conditions, but by no means constituting the entirety of any of these diseases, may be seen to result from malfunction in a given neurochemical system. It may be seen that the above survey of catecholamines and pathological behaviour raises the possibility that a noradrenergic malfunction might cause a change in attentional processes. This is suggested from the animal 6-OHDA lesion studies (see Chapter 6). Such changes in attention and relation-to-the-environment are seen in the diverse disease categories of schizophrenia, Korsakoff's psychosis, Alzheimer's disease, attentional deficit disorder and in the intellectual changes of Parkinson's disease. It is not suggested that any of these diseases is a pure noradrenergic disease. Most will involve the simultaneous failure of one or more nonnoradrenergic neurotransmitter systems. However, the noradrenergic pathology may give rise to the attentional symptoms seen as a part of *all* of the above conditions. The presence of other deficiencies will give rise to those features of the disease states which markedly separate them one from another, and allow differential clinical diagnosis. Such may prove to be the most productive of future research strategies.

References

Adams, F. (1929). *The genuine works of Hippocrates*. William Wood & Co.: New York.

Adams, R. D., van Bogaert, L. & van der Eecken, H. (1964). Striato-nigral degeneration. *Journal of Neuropathology and Experimental Neurology*, **3,** 584–608.

Adolfsson, R. H., Gottfries, C. G., Roos, B. E. & Winblad, B. (1979). Changes in the brain catecholamines in patients with dementia of Alzheimer type. *British Journal of Psychiatry*, **135,** 216–23.

Altschuler, H. L., Killam, E. K. & Killam, K. F. (1976). Biogenic amines and the photomyoclonic syndrome in the baboon, *Papio papio*. *Journal of Pharmacology and Experimental Therapeutics*, **196,** 156–66.

Alvord, E. C., Forno, L. S., Kusse, J. A., Kaufman, R. J., Rhodes, J. S. & Goetowski, C. R. (1974). The pathology of Parkinsonism: a comparison of degenerations in cerebral cortex and brain stem. *Advances in Neurology*, **5,** 179–93.

Alzheimer, A. (1907). A new disease of the cortex. *Allgemeine Zeitschrift für Psychiatrie und ihre Grenzgebiete*, **64,** 146–8.

Anlezark, G. M., Horton, R. W. & Meldrum, B. S. (1978). Dopamine agonists and reflex epilepsy. *Advances in Biochemical Psychopharmacology*, **19,** 383–8.

Anlezark, G. M., Horton, R. W. & Meldrum, B. S. (1979). The anticonvulsive action of the (−) and (+) enantiomers of propranolol. *Journal of Pharmacy and Pharmacology*, **31,** 482–3.

Anlezark, G. M. & Meldrum, B. S. (1978). Blockade of photically induced epilepsy by 'dopamine agonist' ergot alkaloids. *Psychopharmacology*, **57,** 57–62.

Anlezark, G. M., Pycock, C. & Meldrum, B. S. (1976). Ergot alkaloids as dopamine agonists: comparison in two rodent models. *European Journal of Pharmacology*, **37,** 295–302.

Arana, J. D. (1978). Schizophrenic psychoses. In *Clinical psychopathology*, ed. G. U. Balis, L. Wurmser & E. McDaniel, pp. 123–35. Butterworth: Boston.

Arbilla, S., Briley, M. S., Dubocovich, M. L. & Langer, S. Z. (1978). Neuroleptic binding and their effects on the spontaneous and potassium-evoked release of ^{3}H-dopamine from the striatum and of ^{3}H-noradrenaline from the cerebral cortex. *Life Sciences*, **23,** 1775–80.

Arnold, L. E., Huestis, R. D., Smeltzer, D. J., Scheib, J., Wemmer, D. & Colner, G. (1976). Levoamphetamine vs dextroamphetamine in minimal brain dysfunction. *Archives of General Psychiatry*, **33,** 292–301.

Arnold, L. E., Wender, P. H., McCloskey, K. & Snyder, S. H. (1972). *Levo*amphetamine and *dextro*amphetamine: comparative efficacy in the hyperkinetic syndrome. *Archives of General Psychiatry*, **27,** 826–32.

Arnold, P. S., Racine, R. J. & Wise, R. A. (1973). Effects of atropine, reserpine, 6-hydroxydopamine and handling on seizure development in the rat. *Experimental Neurology*, **40,** 457–70.

Azmita, H., Durost, H. & Arthurs, D. (1959). The effect of thioridazine (Mellaril) on mental symptoms. Comparison with chlorpromazine and promazine. *Canadian Medical Association Journal*, **81,** 549–53.

Banki, C. M. (1977). Correlation between cerebrospinal fluid amine metabolites and psychomotor activity in affective disorders. *Journal of Neurochemistry*, **29,** 255–7.

Barbeau, A., Murphy, G. F. & Sourkes, T. L. (1961). Excretion of dopamine in diseases of the basal ganglia. *Science*, **133,** 1706–7.

Barolin, G. S. & Hornykiewicz, O. (1967). Zür diagnostischen Wertigkeit der homovanillin saure im liquor cerebrospinalis. *Wiener klinische Wochenschrift*, **79,** 815–18.

Bemporad, J. R. & Pinsker, H. (1974). Schizophrenia; the manifest symptomatology. In *American Handbook of Psychiatry*, vol. 3, ed. S. Arietti & E. B. Brodym, pp. 524–50. Basic Books: New York.

Bernardt, M. W., Silverstone, T. & Singleton, W. (1980). Behavioural and subjective effects of beta-adrenergic blockade in phobic subjects. *British Journal of Psychiatry*, **137,** 452–7.

Bernheimer, H., Birkmayer, W. & Hornykiewicz, O. (1966). Homovanillinsaure im liquor cerebrospinalis. Untersuchungen beim Parkinson syndrom und anderen erkrankungen des Zentralnervousystems. *Wiener klinische Wochenschrift*, **78,** 417–19.

Bernheimer, H., Birkmayer, W., Hornykiewicz, O., Jellinger, K. & Seitelberger, F. (1973). Brain dopamine and the syndromes of Parkinson and Huntington. *Journal of Neurological Sciences*, **20,** 415–55.

Bird, E. D., Crow, T. J., Iversen, L. L., Longden, A., Mackay, A. V. P., Riley, G. J. & Spokes, E. G. (1979*a*). Dopamine and homovanillic acid concentrations in the post-mortem brain in schizophrenia. *Journal of Physiology*, **293,** 360P.

Bird, E. D., Spokes, E. G. & Iversen, L. L. (1979*b*). Brain norepinephrine and dopamine in schizophrenia. *Science*, **204,** 93–4.

Bird, E. D., Spokes, E. G. S. & Iversen, L. L. (1979*c*). Increased dopamine concentration in limbic areas of brain from patients dying with schizophrenia. *Brain*, **102,** 347–60.

Birkmayer, W. (1969). The importance of monoamine metabolism for the pathology of the extrapyramidal system. *Journal of Neurovisceral Relations, suppl.*, **9,** 297–308.

Birkmayer, W., Danielczyk, W., Neumayer, E. & Riederer, P. (1974). Nucleus ruber and L-DOPA psychosis: biochemical post-mortem findings. *Journal of Neural Transmission*, **35,** 93–116.

Birkmayer, W. & Hornykiewicz, O. (1961). Der L-3,4-dihydroxyphenylalanine (Dopa) effekt bei der Parkinson – akinese. *Wiener klinische Wochenschrift*, **73**, 787–8.

Birkmayer, W., Jellinger, K. & Riederer, P. (1977). Striatal and extrastriatal dopaminergic function. In *Psychobiology of the striatum*, ed. A. R. Cools, A. H. M. Lohman & J. H. L. van der Bercken, pp. 141–53. North-Holland Publishing Co.: Amsterdam.

Boggan, W. O. & Seiden, L. S. (1971). DOPA reversal of reserpine enhancement of audiogenic seizure susceptibility in mice. *Physiology and Behavior*, **6**, 215–17.

Boller, F. (1980). Mental status of patients with Parkinson's disease. *Journal of Clinical Neuropsychology*, **2**, 157–72.

Bond, P. A., Jenner, F. A. & Sampson, G. A. (1972). Daily variations in the urine content of 3-methoxy-4-hydroxyphenylglycol in two manic-depressive patients. *Psychological Medicine*, **2**, 81–5.

Bondareff, W., Mountjoy, C. Q. & Roth, M. (1981). Selective loss of neurones of origin of adrenergic projection to cerebral cortex (nucleus locus coeruleus) in senile dementia. *Lancet*, **i**, 783–4.

Bondareff, W., Mountjoy, C. Q. & Roth, M. (1982). Loss of neurons of origin of the adrenergic projection to cerebral cortex (nucleus locus coeruleus) in senile dementia. *Neurology*, **32**, 164–8.

Bourne, H. R., Bunney, W. E., Colburn, J. M., Davis, J. M., Davis, J. N., Shaw, D. M. & Coppen, A. J. (1968). Noradrenaline, 5-hydroxytryptamine and 5-hydroxyindole acetic acid in hindbrains suicidal patients. *Lancet*, **ii**, 805–8.

Bourne, W. M., Chin, L. & Picchioni, A. L. (1972). Enhancement of audiogenic seizure by 6-hydroxydopamine. *Journal of Pharmacy and Pharmacology*, **24**, 913–14.

Bourne, W. M., Chin, L. & Picchioni, A. L. (1977). Effects of neonatal 6-hydroxydopamine treatment on audiogenic seizures. *Life Sciences*, **21**, 701–6.

Boyd, W. O., Graham-White, J., Blackwood, G., Glen, I. & McQueen, J. (1977). Clinical effects of choline in Alzheimer senile dementia. *Lancet*, **ii**, 711.

Browning, R. A. & Maynert, E. W. (1978). Effect of intracisternal 6-hydroxydopamine on seizure susceptibility in rats. *European Journal of Pharmacology*, **50**, 97–101.

Bunney, W. E. & Davis, J. W. (1965). Norepinephrine in depressive reactions. A review. *Archives in General Psychiatry*, **13**, 483–94.

Burki, H. R., Eichenberger, E., Sayers, A. C. & White, T. G. (1975). Clozapine and the dopamine hypothesis of schizophrenia, a critical appraisal. *Pharmakopsychiatrie–Neuro-Psychopharmakologie*, **8**, 115–21.

Buscaino, G. A., Jori, A. & Campanella, G. (1972). Brain dopamine and noradrenaline in 3 Parkinsonians deceased during L-DOPA treatment. *Acta neurologica (Napoli)*, **27**, 537–44.

Butters, N. & Cermak, L. S. (1976). Neuropsychologic studies of alcoholic Korsakoff patients. In *Empirical studies of alcoholism*, ed. A. Goldstein & M. Neuringer, pp. 113–30. Ballinger: Cambridge, Mass.

Callaghan, D. A. & Schwark, W. S. (1976). Neurochemical changes and drug effects in a model of epilepsy in the rat. *Neuroscience Abstracts*, **2**, 257.

Chadwick, D., Jenner, P. & Reynolds, E. H. (1975). Amines, anticonvulsants and epilepsy. *Lancet*, **i**, 473–6.

Chase, T. N., Gordon, E. K. & Ng, L. K. Y. (1973). Norepinephrine metabolism in the central nervous system of man: studies using 3-methoxy-4-hydroxyphenylglycol levels in cerebrospinal fluid. *Journal of Neurochemistry*, **21**, 581–7.

Checkley, S. A., Slade, A. P., Shur, E. & Dawling, S. (1981). A pilot study on the mechanism of action of desipramine. *British Journal of Psychiatry*, **138**, 248–51.

Chiodo, L. A. & Antelman, S. M. (1980). Repeated tricyclics induce a progressive dopamine autoreceptor subsensitivity independent of daily drug treatment. *Nature*, **287**, 451–4.

Chiodo, L. A. & Antelman, S. M. (1982). Do antidepressants induce dopamine autoreceptor subsensitivity? *Nature*, **398**, 302–3.

Christie, J. E., Shering, A., Ferguson, J. & Glen, A. I. M. (1981). Physostigmine and arecoline: effects of intravenous infusions in Alzheimer presenile dementia. *British Journal of Psychiatry*, **138**, 46–50.

Chu, N. S. (1978). Enhancement of alcohol withdrawal seizures by 6-hydroxydopamine. *Epilepsia*, **19**, 603–9.

Clements-Jewery, S. (1978). The development of cortical beta-adrenoreceptor sub-sensitivity in the rat by chronic treatment with trazodone, doxepin and mianserin. *Neuropharmacology*, **17**, 779–81.

Concannon, J. T. & Schechter, M. D. (1981). Hyperactivity in developing rats; sex differences in 6-hydroxydopamine and amphetamine effects. *Pharmacology, Biochemistry & Behavior*, **14**, 5–10.

Concannon, J. T. & Schechter, M. D. (1982). Failure of amphetamine isomers to decrease hyperactivity in developing rats. *Pharmacology, Biochemistry & Behavior*, **17**, 5–9.

Constantinidis, J., Siegfried, J., Frigyesi, T. L. & Tissot, R. (1974). Parkinson's disease and striatal dopamine: *in vivo* morphological evidence for the presence of dopamine in the human brain. *Journal of Neural Transmission*, **35**, 13–22.

Corcoran, M. E., Fibiger, H. C., McCaughran, J. A. & Wada, J. A. (1974). Potentiation of amygdaloid kindling and metrazol-induced seizures by 6-hydroxydopamine in rats. *Experimental Neurology*, **45**, 118–33.

Corcoran, M. E. & Mason, S. T. (1980). Role of forebrain catecholamines in amygdaloid kindling. *Brain Research*, **190**, 473–84.

Cote, L., Yahr, M. D., Duvoisin, R. C., Wolf, A. & Marksberry, W. (1970). Analysis of deaths in Parkinson patients on long-term levodopa therapy. *Transactions of the American Neurological Association*, **95**, 73.

Cox, W. H. (1982). An indication for use of imipramine in attention deficit disorders. *American Journal of Psychiatry*, **139**, 1059–60.

Cross, A. J., Crow, T. J., Killpack, W. S., Longden, A., Owen, F. & Riley, G. R. (1978). The activities of brain dopamine-beta-hydroxylase and catechol-*o*-methyltransferase in schizophrenics and control. *Psychopharmacology*, **59**, 117–22.

Cross, A. J., Crow, T. J., Perry, E. K., Perry, R. H., Blessed, G. & Tomlinson, B. E. (1981). Reduced dopamine-beta-hydroxylase activity in Alzheimer's disease. *British Medical Journal*, **282**, 93–4.

Crow, T. J., Baker, H. F., Cross, A., Joseph, M. H., Lofthouse, R., Longden, A., Owen, F., Riley, G. J., Glover, V. & Killpack, W. S. (1979). Monoamine mechanisms in chronic schizophrenia: post-mortem neurochemical findings. *British Journal of Psychiatry*, **134**, 249–56.

Crow, T. J., Deakin, J. F. W., File, S. E., Longden, A. & Wendlandt, S. (1978*a*). The locus coeruleus noradrenergic system – evidence against a role of attention, habituation, anxiety and motor activity. *Brain Research*, **155**, 249–61.

Crow, T. J., Owen, F., Cross, A. J., Lofthouse, R. & Longden, A. (1978*b*). Brain biochemistry in schizophrenia. *Lancet*, **i**, 36–7.

Crow, T. J., Spear, P. J. & Arbuthnott, G. W. (1972). Intracranial self-stimulation with electrodes in the region of the locus coeruleus. *Brain Research*, **36**, 275–87.

Dadkar, V. N., Dahanukar, S. A. & Sheth, U. K. (1979). Role of dopaminergic and noradrenergic mechanisms in Metrazole convulsions in mice. *Indian Journal of Medical Research*, **70**, 492–4.

Davies, P. (1979). Neurotransmitter-related enzymes in senile dementia of Alzheimer type. *Brain Research*, **171**, 319–21.

Davies, P. & Maloney, A. J. F. (1976). Selective loss of central cholinergic neurones in Alzheimer's disease. *Lancet*, **ii**, 1403.

De Reuck, J., de Coster, W. & van der Eeken, H. (1980). Parkinsonism in patients with cerebral infarcts. *Clinical Neurology and Neurosurgery*, **82**, 117–85.

Doggett, N. S. & Spencer, P. S. J. (1971). Pharmacological properties of centrally administered ouabain and their modification by other drugs. *British Journal of Pharmacology*, **42**, 242–53.

DSM III (1980). *Diagnostic and statistical manual of mental disorders*, 3rd edn. American Psychiatric Association: Washington, D.C.

Eastgate, S. M., Wright, J. J. & Werry, J. S. (1978). Behavioral effects of methylphenidate in 6-hydroxydopamine-treated neonatal rats. *Psychopharmacology*, **58**, 157–8.

Ehringer, H. & Hornykiewicz, O. (1960). Verteilung von noradrenaline und dopamine (3-hydroxytyramine) in gehirn des menschen und ihr verhalten bei erkrankungen des extrapyramidalen systems. *Wiener klinische Wochenschrift*, **38**, 1236–9.

Emrich, H. M., von Zerssen, D., Mohler, H. J., Kissling, W., Cording, C., Schietsch, H. J. & Riedel, E. (1979). Action of propranolol in mania; comparison of the effects of the *d*- and *l*-stereoisomer. *Pharmakopsychiatrie-Neuro-psychopharmakologie*, **12**, 295–304.

Engel, J. & Sharpless, N. S. (1977). Long-lasting depletion of dopamine in rat amygdala induced by kindling stimulation. *Brain Research*, **136**, 381–6.

Fahn, S., Libsch, L. R. & Cutler, R. W. (1971). Monoamines in the human neostriatum: topographic distribution in normals and in Parkinson's disease and their role in akinesia, rigidity, chorea and tremor. *Journal of Neurological Sciences*, **14**, 427.

Farley, I. J. & Hornykiewicz, O. (1976). Noradrenaline in subcortical brain regions of patients with Parkinson's disease and control subjects. In *Advances in Parkinsonism*, ed. W. Birkmayer & O. Hornykiewicz, pp. 178–85. Roche: Basle.

Farley, I. J., Price, K. S. & Hornykiewicz, O. (1977). Dopamine in the limbic regions of the human brain: normal and abnormal. *Advances in Biochemical Psychopharmacology*, **16**, 57–64.

Farley, I. J., Price, K. S., McCullough, E., Deck, J. H. N., Hordynski, W. & Hornykiewicz, O. (1978). Norepinephrine in chronic paranoid schizophrenia: above-normal levels in limbic forebrain. *Science*, **200**, 456–8.

Feighner, J. P., Robins, E., Gruze, S. B., Woodruff, R. A., Winokur, G. & Munoz, R. (1972). Diagnostic criteria for use in psychiatric research. *Archives of General Psychiatry*, **26**, 57–63.

File, S. E., Deakin, J. F. W., Longden, A. & Crow, T. J. (1979). An investigation of the role of the locus coeruleus in anxiety and agonistic behaviour. *Brain Research*, **169**, 411–20.

Floru, L. (1977). Use of beta-blocking agents in psychiatry and neurology. *Fortschritte der Neurologie-Psychiatrie und ihrer Grenzgebiete*, **45**, 112–27.

Forno, L. S. (1966). Pathology of Parkinsonism. *Journal of Neurosurgery*, **24**, 266–71.

Freedman, R. & Hoffer, B. J. (1975). Phenothiazine antagonism of the noradrenergic inhibition of cerebellar Purkinje cells. *Journal of Neurobiology*, **6**, 277–88.

Garelis, E. & Sourkes, T. L. (1974). Use of cerebrospinal fluid drawn at pneumoencephalography in the study of monoamine metabolism in man. *Journal of Neurology, Neurosurgery and Psychiatry*, **37**, 704–10.

Garver, D. L. & Davis, J. M. (1979). Biogenic amine hypotheses of affective disorders. *Life Sciences*, **24**, 383–94.

Georgiev, V. P. & Petkova-Radkova, B. P. (1976). Effect of isopropyl-noradrenaline, adenosine, cyclic 3′,5′-adenosine monophosphate and dibutyryl cyclic adenosine monophosphate on the picrotoxin convulsive-seizure threshold. Interaction with gamma-amino-butyric acid. *Acta physiologica et pharmacologica bulgarica*, **2**, 52–8.

Gomes, U. C. R., Shanley, B. C., Potgieter, L. & Roux, J. T. (1980). Noradrenergic overactivity in chronic schizophrenia: evidence based on cerebrospinal fluid noradrenaline and cyclic nucleotide concentrations. *British Journal of Psychiatry*, **137**, 346–51.

Goodwin, F. K., Cowdry, R. W. & Webster, M. H. (1978). Predictors of drug response in the affective disorders: towards an integrated approach. In *Psychopharmacology – a generation in progress*, ed. M. A. Lipton, A. DiMascio & K. F. Killam, pp. 1277–88. Raven Press: New York.

Goodwin, F. K., Murphy, D. C. & Brodie, H. K. H. (1970). L-Dopa, catecholamines and behavior; a clinical and biochemical study in depressed patients. *Biological Psychiatry*, **2**, 341–66.

Goodwin, F. K., Post, R. M., Dunner, D. L. & Gordon, E. K. (1973). Cerebrospinal fluid amine metabolism in affective illness; the probenecid technique. *American Journal of Psychiatry*, **130**, 73–84.

Gottfries, C. G. (1980). Biochemistry of dementia and normal ageing. *Trends in Neuroscience*, **3**, 55–7.

Gottschalk, L. A., Stone, W. N. & Gleser, G. C. (1974). Peripheral versus central mechanisms accounting for antianxiety effects of propranolol. *Psychosomatic Medicine*, **36**, 47–56.

Grant, S. J., Huang, Y. H. & Redmond, D. E. (1980). Benzodiazepenes attenuate single unit activity in the locus coeruleus. *Life Science*, **27**, 2231–6.

Green, A. R. & Nutt, D. J. (1983). Antidepressants. In *Psychopharmacology 1:* Part 1 *Preclinical psychopharmacology*, eds. D. G. Grahame-Smith & P. J. Cowan, pp. 1–37. Excerpta Medica: Amsterdam.

Greenblatt, D. J. & Shader, R. I. (1978). Pharmacotherapy of anxiety with benzodiazepenes and beta adrenergic blockers. In *Psychopharmacology – a generation of progress*, ed. M. A. Lipton, A. DiMascio & K. F. Killam, pp. 1381–90. Raven Press: New York.

Greenfield, J. G. & Bosanquet, F. D. (1953). The brain stem lesions in Parkinsonism. *Journal of Neurology, Neurosurgery and Psychiatry*, **16**, 213.

Greer, M., Collins, G. H. & Anton, A. H. (1971). Cerebral catecholamines after levodopa therapy. *Archives of Neurology*, **25**, 461.

Gross, G. & Schumann, H. J. (1980). Enhancement of noradrenaline release from rat cerebral cortex by neuroleptic drugs. *Naunyn-Schmiedeberg's Archives of Pharmacology*, **315**, 103–10.

Grote, S. S., Moses, S. G., Robins, E., Hudgens, R. W. & Croninger, A. B. (1974). A study of selected catecholamine metabolizing enzymes: a comparison of depressive suicides and alcoholic suicides with controls. *Journal of Neurochemistry*, **23**, 791–802.

Habel, A., Yates, C. M., McQueen, J. K., Blackwood, D. & Elton, R. A. (1981).

Homovanillic acid and 5-hydroxyindoleacetic acid in lumbar cerebrospinal fluid in children with afebrile and febrile convulsions. *Neurology*, **31**, 488–91.

Haefely, W. E. (1978). Behavioral and neuropharmacological aspects of drugs used in anxiety and related states. In *Psychopharmacology – a generation of progress*, ed. M. A. Lipton, A. DiMascio & K. F. Killam, pp. 1359–74. Raven Press: New York.

Halaris, A. E. & DeMet, E. M. (1979). Effects of the potential antidepressant AHR-1118 on amine uptake and on plasma 3-methoxy-4-hydroxyphenylglycol in manic-depressive illness. *Psychopharmacology Bulletin*, **15**, 95–7.

Hiramatsu, M., Fujimoto, N. & Mori, A. (1982). Catecholamine level in cerebrospinal fluid of epileptics. *Neurochemical Research*, **7**, 1299–305.

Hoehn, M. M., Crowley, T. J. & Rutledge, C. O. (1977). The Parkinsonian syndrome and its dopamine correlates. *Advances in Experimental Medicine and Biology*, **90**, 243–54.

Holcomb, H. H., Bannon, M. J. & Roth, R. H. (1982). Striatal dopamine autoreceptors uninfluenced by chronic administration of anti-depressants. *European Journal of Pharmacology*, **82**, 173–8.

Hornykiewicz, O. (1966). Dopamine (3-hydroxytryptamine) and brain function. *Pharmacological Reviews*, **18**, 925–64.

Hornykiewicz, O. (1982). Brain catecholamines and schizophrenia –a good case for noradrenaline. *Nature*, **299**, 484–6.

Horton, R., Anlezark, G. & Meldrum, B. (1980). Noradrenergic influences on sound-induced seizures. *Journal of Pharmacology and Experimental Therapeutics*, **214**, 437–42.

Huang, H. Y., Maas, J. W. & Hu, G. H. (1980). The time-course of noradrenergic pre- and post-synaptic activity during chronic desipramine treatment. *European Journal of Pharmacology*, **68**, 41–8.

Huntington, G. (1872). On chorea. *Medical and Surgical Reports*, **26**, 317–21.

Ito, M., Okuno, T., Mikawa, H. & Osumi, Y. (1980). Elevated homovanillic acid in cerebrospinal fluid of children with infantile spasms. *Epilepsia*, **21**, 387–92.

Jimerson, D. C., Post, R. M., Stoddard, F. J., Gillin, J. C. & Bunney, W. E. (1980). Preliminary trial of the noradrenergic agonist clonidine in psychiatric patients. *Biological Psychiatry*, **15**, 45–57.

Jobe, P. C. & Ray, T B. (1980). Effects of norepinephrine and 5-hydroxy-tryptamine reuptake inhibitors on electrically-induced spinal cord seizures in rats. *Research Communications of Chemical Pathology and Pharmacology*, **30**, 185–8.

Jobe, P. C., Ray, T. B., Geiger, P. F. & Bourne, W. M. (1977). Effects of Ro 4-1284 on electrically-induced spinal cord seizures and on spinal cord norepinephrine and 5-hydroxytryptamine levels. *Research Communications in Chemical Pathology and Pharmacology*, **18**, 601–12.

Johnson, K., Presley, A. S. & Ballinger, B. R. (1978). Levodopa in senile dementia. *British Medical Journal*, **i**, 1625.

Jones, F., Maas, J. W., Dekirmenlian, H. & Fawcett, J. A. (1973). Urinary catecholamine metabolites during behavioral changes in a patient with manic-depressive cycles. *Science*, **179**, 300–2.

Katz, M. M. & Hirschfield, R. M. A. (1978). Phenomenology and classification of depression. In *Psychopharmacology – a generation of progress*, ed. M. A. Lipton, A. DiMascio & K. F. Killam, pp. 1185–95. Raven Press: New York.

Kellogg, C. (1976). Audiogenic seizures: relation to age and mechanisms of monoamine neurotransmission. *Brain Research*, **106**, 87–103.

Kemali, D., Vecchio, M. D. & Maj, M. (1982). Increased noradrenaline levels in CSF and plasma of schizophrenic patients. *Biological Psychiatry*, **17,** 711–19.

King, D. J., Turkson, S. N. A., Liddle, J. & Kinney, C. D. (1980). Some clinical and metabolic aspects of propranolol in chronic schizophrenia. *British Journal of Psychiatry*, **137,** 458–68.

Klein, D. F., Zitrin, C. M. & Woerner, M. (1978). Antidepressants, anxiety, panic and phobias. In *Psychopharmacology – a generation of progress*, ed. M. A. Lipton, A. DiMascio & K. F. Killam, pp. 1401–10. Raven Press: New York.

Kleinrok, Z., Czuczwar, S., Wojcik, A. & Przegalinski, E. (1978). Brain dopamine and seizure susceptibility in mice. *Polish Journal of Pharmacology and Pharmacy*, **30,** 513–19.

Koide, T. & Matsushita, H. (1981). An enhanced sensitivity of muscarinic cholinergic receptor associated with dopaminergic receptor subsensitivity after chronic antidepressant treatment. *Life Sciences*, **28,** 1139–45.

Korsakoff, S. S. (1890). A psychic disturbance associated with multiple neuritis. *Allgemeine Zeitschrift für Psychiatrie und ihre Grenzgebiete*, **46,** 475–85.

Kraepelin, E. (1883). *Compendium of psychiatry*, 1st edn. A. Abel: Leipzig. English translation from 8th edn by R. M. Barclay (1919), *Dementia praecox and paraphrenia*. E. & S. Livingstone: Edinburgh.

Kristensen, V., Olsen, M. & Theilgaard, A. (1977). Levodopa treatment of pre-senile dementia. *Acta psychiatrica scandinavica*, **55,** 41–51.

Ladisch, W. (1973). Ouabain induced convulsions and ^{3}H-norepinephrine metabolism in the rat brain. *Journal of Neural Transmission*, **34,** 235–8.

Ladisch, W., Steinbauff, N. & Matussek, N. (1969). Chronic administration of electroconvulsive shock and norepinephrine metabolism in the rat brain. II. ^{3}H-NE metabolism after intracisternal injection with and without the influence of drugs in different brain regions and ^{3}H-NE uptake *in vitro*. *Psychopharmacologia*, **15,** 196–304.

Laguzzi, R. F., Acevedo, C. & Izquierdo, J. A. (1970). Seizure activity and hippocampal norepinephrine content. *Arzneimittel-Forschung*, **20,** 1904–5.

Lake, C. R., Sternberg, D. E., van Kammen, D. P. Ballenger, J. C., Ziegler, M. G., Post, R. M., Kopin, I. J. & Bunney, W. E. (1980*a*). Schizophrenia: elevated cerebrospinal fluid norepinephrine. *Science*, **207**, 331–3.

Lake, C. R., Sternberg, D. E., van Kammen, D. P., Ballenger, J. C., Ziegler, M. G., Post, R. M., Kopin, I. J. & Bunney, W. E. (1980*b*). Elevated cerebrospinal fluid norepinephrine in schizophrenics: confounding effects of treatment drugs. *Science*, **210,** 97.

Laxer, K. D., Sourker, T. L., Fang, T. Y., Young, S. N., Gouther, S. G. & Missala, K. (1979). Monoamine metabolites in the CSF of epileptic patients. *Neurology*, **29,** 1157–60.

Lazarova, M. B. & Roussinov, K. S. (1978). On certain effects of dopaminergic agents in pentylenetetrazol convulsions. *Acta physiologica et pharmacologica bulgarica*, **4,** 50–5.

Lee, T., Seeman, P., Towtelotte, W. W., Farley, I. J. & Hornykiewicz, O. (1978). Binding of ^{3}H-apomorphine in schizophrenic brains. *Nature*, **274,** 897–900.

Lehmann, A. & Busnell, R. G. (1963). A study of audiogenic seizure. In *Acoustic behavior in animals*, ed. R. G. Busnell, pp. 244–74. Elsevier: New York.

Leino, E., MacDonald, E., Airaksinen, M. & Riekkinen, P. J. (1980). Homovanillic acid and 5-hydroxyindoleacetic acid levels in cerebrospinal fluid of patients with progressive myoclonus epilepsy. *Acta neurologica scandinavica*, **62,** 41–54.

Lerner, P., Goodwin, F. K., van Kammen, D. P., Post, R. M., Major, L. F., Ballenger, J. C. & Lovenberg, W. (1978). Dopamine-beta-hydroxylase in the cerebrospinal fluid of psychiatric patients. *Biological Psychiatry*, **13,** 685–94.

Lewi, P. J. & Colpaert, F. C. (1976). On the classification of anti-depressant drugs. *Psychopharmacology*, **49**, 219–24.

Lewis, C., Ballinger, B. R. & Presley, A. S. (1978). Trial of levodopa in senile dementia. *British Medical Journal*, **i**, 550.

Lieberman, A. N., Freedman, L. S. & Goldstein, M. (1972). Serum dopamine-beta-hydroxylase activity in patients with Huntington's chorea and Parkinson's disease. *Lancet*, **i**, 153–4.

Lindstrom, L. H. & Persson, E. (1980). Propranolol in chronic schizophrenia: a controlled study in neuroleptic-treated patients. *British Journal of Psychiatry*, **137**, 126–30.

Lipsky, J. J. (1980). Elevated cerebrospinal fluid norepinephrine in schizophrenics: confounding effects of treatment drugs. *Science*, **210**, 97.

Livrea, P. (1976). Dosaggi di acido omovanillicoet acids 5-idrossindoleacetico liquorali in disturbi del sonno, epilepsia, corea degenerativa et vasculopatie cerebrali acute. *Acta neurologica (Napoli)*, **31**, 580–600.

Livrea, P., DiReda, L. & Papagno, G. (1976). Livelli liquorali de HVA e de 5-HIAA in soggetti epilettica: Effecto della privazione totale de sonno. *Acta neurologica (Napoli)*, **31**, 632–6.

London, E. D. & Buterbaugh, G. G. (1978). Modification of electroshock convulsive responses and thresholds in neonatal rats after brain monoamine reduction. *Journal of Pharmacology and Experimental Therapeutics*, **206**, 81–90.

Loranger, A. W., Goodell, H., McDowell, F. H., Lee, J. E. & Sweet, R. D. (1972). Intellectual impairment in Parkinson's syndrome. *Brain*, **95**, 405–12.

Louis, W. J., Papanicolaou, J., Summers, R. J. & Vajda, F. J. E. (1982). Role of central beta-adrenoceptors in the control of pentylenetetrazol-induced convulsions in rats. *British Journal of Pharmacology*, **75**, 441–6.

McEntee, W. J. & Mair, R. G. (1978). Memory impairment in Korsakoff's psychosis: a correlation with brain noradrenergic activity. *Science*, **202**, 905–7.

McEntee, W. J. & Mair, R. G. (1980*a*). Memory enhancement in Korsakoff's psychosis by clonidine: further evidence of a noradrenergic deficit. *Annals of Neurology*, **7**, 466–70.

McEntee, W. J. & Mair, R. G. (1980*b*). Korsakoff's amnesia: a noradrenergic hypothesis. *Psychopharmacology Bulletin*, **16**, 22–4.

McGeer, P. L. & McGeer, E. G. (1976). Enzymes associated with the metabolism of catecholamines, acetylcholine and GABA in human controls and patients with Parkinson's disease and Huntington's chorea. *Journal of Neurochemistry*, **26**, 65–76.

MacKay, A. V. P., Bird, E. D., Iversen, L. L., Spokes, E. G., Creese, I. & Snyder, S. H. (1980*a*). Dopaminergic abnormalities in postmortem schizophrenic brain. *Advances in Biochemical Psychopharmacology*, **24**, 325–33.

MacKay, A. V. P., Bird, E. D., Spokes, E. G., Rossor, M., Iversen, L. L., Creese, I. & Snyder, S. H. (1980*b*). Dopamine receptors and schizophrenia: drug effect or illness? *Lancet*, **ii**, 915–16.

McIntyre, D. C. (1980). Amygdala kindling in rats: facilitation after local amygdala norepinephrine depletion with 6-hydroxy-dopamine. *Experimental Neurology*, **69**, 395–407.

McIntyre, D. N. & Edson, N. (1982). Effect of norepinephrine depletion on dorsal hippocampus kindling in rats. *Experimental Neurology*, **77**, 700–4.

McIntyre, D. C., Saari, M. & Pappas, B. A. (1979). Potentiation of amygdala kindling in adult or infant rats by injections of 6-hydroxydopamine. *Experimental Neurology*, **63**, 527–44.

McKenzie, G. M. & Soroko, F. E. (1972). The effect of apomorphine, *d*-amphetamine and L-DOPA on MES convulsions – a comparative study in the rat and mouse. *Journal of Pharmacy and Pharmacology*, **24**, 696–7.

McMillen, B. A., Warnack, W., German, D. C. & Shore, P. A. (1980). Effects of chronic desipramine treatment on rat brain noradrenergic responses to alpha-adrenergic drugs. *European Journal of Pharmacology*, **61**, 239–46.

McNamara, J. O. (1978*a*). Selective alterations of regional beta-adrenergic receptor binding in the kindling model of epilepsy. *Experimental Neurology*, **61**, 582–91.

McNamara, J. O. (1978*b*). Muscarinic cholinergic receptors participate in the kindling model of epilepsy. *Brain Research*, **154**, 415–20.

MacNeil, D. A. & Gower, M. (1982). Do antidepressants induce dopamine autoreceptor subsensitivity? *Nature*, **298**, 301.

Malamud, N. & Skillicorn, S. A. (1956). Relationship between the Wernicke and Korsakoff syndrome: a clinicopathologic study of seventy cases. *American Medical Association Archives of Neurology and Psychiatry*, **76**, 585–96.

Mann, D. M. A., Lincoln, J., Yates, P. O., Stamp, J. E. & Toper, S. (1980). Changes in the monoamine containing neurones of the human C.N.S. in senile dementia. *British Journal of Psychiatry*, **136**, 533–41.

Mann, D. M., Yates, P. O. & Hawkes, J. (1982). The noradrenergic system in Alzheimer and multiple infarct dementias. *Journal of Neurology, Neurosurgery and Psychiatry*, **45**, 113–19.

Margolin, D. I. (1978). The hyperkinetic child syndrome and brain monoamines: pharmacology and therapeutic implications. *Journal of Clinical Psychiatry*, **99**, 120–30.

Marwaha, J. & Aghajanian, G. K. (1982). Typical and atypical neuroleptics are potent antagonists at alpha$_1$-adrenoceptors of the dorsal lateral geniculate nucleus. *Naunyn-Schmiedeberg's Archives of Pharmacology*, **321**, 32–7.

Mason, S. T. (1982). Noradrenaline in the brain: progress in theories of behavioural function. *Progress in Neurobiology*, **16**, 263–303.

Mason, S. T. & Corcoran, M. E. (1978). Forebrain noradrenaline and Metrazol-induced seizures. *Life Sciences*, **23**, 167–72.

Mason, S. T. & Corcoran, M. E. (1979*a*). Catecholamines and convulsions. *Brain Research*, **170**, 497–507.

Mason, S. T. & Corcoran, M. E. (1979*b*). Depletion of brain noradrenaline but not dopamine by intracerebral 6-OHDA potentiates ECS convulsions. *Journal of Pharmacy and Pharmacology*, **31**, 209–11.

Mason, S. T. & Corcoran, M. E. (1979*c*). Seizure susceptibility after depletion of spinal or cerebellar noradrenaline with 6-OHDA. *Brain Research*, **166**, 418–21.

Meier, M. J. & Martin, W. F. (1970). Intellectual changes associated with levodopa therapy. *Journal of American Medical Association*, **213**, 465–6.

Meltzer, H. Y., Sachar, E. J. & Frantz, A. G. (1975). Dopamine antagonism by thioridazine in schizophrenia. *Biological Psychiatry*, **10**, 53–7.

Meyerson, L. R., Wennogle, L. P., Abel, M. S., Coupet, J., Lippa, A. S., Rauh, C. E. & Beer, B. (1982). Human brain receptor alterations in suicide victims. *Pharmacology, Biochemistry & Behavior*, **17**, 169–63.

Millichamp, J. G. & Fowler, G. W. (1967). Treatment of 'minimal brain dysfunction' syndromes. *Pediatric Clinics of North America*, **14**, 767–77.

Mindham, R. H. S. (1970). Psychiatric symptoms in Parkinsonism. *Journal of Neurology, Neurosurgery and Psychiatry*, **33**, 188–91.

Minneman, K. P., Dibner, M. D., Wolfe, B. B. & Molinoff, P. B. (1979). Beta$_1$ and beta$_2$ adrenergic receptors in rat cerebral cortex are independently regulated. *Science*, **204,** 866–8.

Mishra, R., Janowsky, A. & Sulser, F. (1980). Action of mianserin and zimelidine on the norepinephrine receptor coupled adenylate cyclase system in brain; subsensitivity without reduction in beta-adrenergic receptor binding. *Neuropharmacology*, **19,** 983–8.

Mohr, E. & Corcoran, M. E. (1981). Depletion of noradrenaline and amygdaloid kindling. *Experimental Neurology*, **72,** 507–11.

Moses, S. G. & Robins, E. (1975). Regional distribution of nor-epinephrine and dopamine in brains of depressive suicides and alcoholic suicides. *Psychopharmacology Communications*, **1,** 327–37.

Murmann, W., Almirante, L. & Saccani-Guelfi, M. (1966). Central nervous system effects on four beta-adrenergic receptor blocking agents. *Journal of Pharmacy and Pharmacology*, **18,** 317–18.

Nagatsu, T., Kato, T., Nagatsu, I., Kondo, Y., Inagaki, S., Iizuka, R. & Narabayashi, H. (1979). Catecholamine-related enzymes in the brain of Parkinsonian patients. In *Catecholamines: basic and clinical frontiers*, ed. E. Usdin, I. J. Kopin & J. Barchas, pp. 1587–9. Pergamon Press: New York.

Nashold, B. S., Wilson, W. P. & Slaughter, G. (1974). The midbrain and pain. In *Advances in neurology*, vol. 4, pp. 191–6. Raven Press: New York.

Nelson, J. C. & Charney, D. S. (1981). The symptoms of major depressive illness. *American Journal of Psychiatry*, **138,** 1–13.

O'Keefe, R., Sharman, D. F. & Vogt, M. (1970). Effects of drugs used in psychoses on cerebral dopamine metabolism. *British Journal of Pharmacology*, **38,** 287–91.

Owen, F., Crow, T. J., Poulter, M., Cross, A. J., Longden, A. & Riley, G. J. (1978). Increased dopamine-receptor sensitivity in schizophrenia. *Lancet*, **ii,** 221–6.

Palmer, G. C., Robinson, G. A. & Sulser, F. (1971). Modification by psychotropic drugs of the cyclic adenosine monophosphate response to norepinephrine in rat brain. *Biochemical Pharmacology*, **20,** 236–9.

Palmer, G. C., Robinson, G. A. & Sulser, F. (1972). Modification by psychotropic drugs of the cyclic AMP response to norepinephrine in the rat brain *in vitro*. *Psychopharmacology*, **23,** 201–11.

Papeschi, R., Molina-Negro, P., Sourkes, T. L. & Erba, G. (1972). The concentration of homovanillic and 5-hydroxyindoleacetic acids in ventricular and lumbar CSF. *Neurology*, **22,** 1151–9.

Pappas, B. A., Gallivan, J. V., Dugan, T., Saari, M. & Ings, R. (1980). Intraventricular 6-hydroxydopamine in the newborn rat and locomotor responses to drugs in infancy. No support for the dopamine depletion model of minimal brain dysfunction. *Psychopharmacology*, **70,** 41–6.

Pare, C. M. B., Young, D. P. H., Price, K. & Stacey, R. S. (1969). 5-Hydroxytryptamine, noradrenaline and dopamine in brainstem, hypothalamus and caudate nucleus of controls and patients committing suicide by coal-gas poisoning. *Lancet*, **ii,** 133–5.

Parkinson, J. (1817). *An essay on the shaking palsy*. Sherwood, Neely & Jones: London.

Peet, M., Middlemass, D. A. & Yates, R. A. (1981). Propranolol and schizophrenia. 2. Clinical and biochemical effects of combining propranolol and chlorpromazine. *British Journal of Psychiatry*, **139,** 112–17.

Peroutka, S. J. & Snyder, S. H. (1980). Relationship of neuroleptic drug effects at brain dopamine, serotonin, alpha-adrenergic and histamine receptors to clinical potency. *American Journal of Psychiatry*, **137,** 1517–22.

Peroutka, S. J., U'Pritchard, D. C., Greenberg, S. A. & Snyder, S. H. (1977). Neuroleptic drug interactions with norepinephrine alpha receptor binding sites in rat brain. *Neuropharmacology*, **16,** 549–56.

Perry, E. K., Tomlinson, B. E., Blessed, G., Bergmann, K., Gibson, P. H. & Perry, R. H. (1978). Correlation of cholinergic abnormalities with senile plaques and mental test scores in senile dementia. *British Medical Journal*, **ii,** 1457–9.

Perry, E. K., Tomlinson, B. E., Blessed, G., Perry, R. H., Cross, A. J. & Crow, T. J. (1981). Noradrenergic and cholinergic systems in senile dementia of Alzheimer type. *Lancet*, **ii,** 149.

Peters, J. G. (1979). Noradrenaline and serotonin spinal fluid metabolites in temporal lobe epileptic patients with schizophrenic symptomatology. *European Neurology*, **18,** 15–18.

Philo, R., Reiter, R. J. & McGill, J. R. (1979). Changes in brain dopamine, norepinephrine and serotonin associated with convulsion induced by pinealectomy in the gerbil. *Journal of Neural Transmission*, **46,** 239–52.

Post, R. M., Gerner, R. H. & Carmen, J. S. (1978). Effect of a dopamine agonist piribedil in depressed patients. *Archives of General Psychiatry*, **35,** 609–15.

Racine, R. J. (1972). Modification of seizure activity by electrical stimulation. II. Motor seizure. *Electroencephalography and Clinical Neurophysiology*, **32,** 281–94.

Redmond, D. E. (1977). Alteration in the function of the nucleus locus coeruleus: a possible model for studies of anxiety. In *Animal models in psychiatry and neurology*, ed. I. Hanin & E. Usdin, pp. 293–306. Pergamon Press: New York.

Redmond, D. E. & Huang, Y. H. (1979). Current concepts. 2. New evidence for a locus coeruleus–norepinephrine connection with anxiety. *Life Sciences*, **25,** 2149–62.

Redmond, D. E., Huang, Y. H., Snyder, D. R. & Maas, J. W. (1976). Behavioral effects of stimulation of the nucleus coeruleus in the stump-tailed monkey (*macaca arctoides*). *Brain Research*, **116,** 502–10.

Reus, V. I., Lake, R. & Post, R. M. (1980). Effect of piribedil (ET-495) on plasma norepinephrine: relationship to antidepressant response. *Communications in Psychopharmacology*, **4,** 207–13.

Reynolds, E. H., Chadwick, D. & Jenner, P. (1975). Folate and monoamine metabolism in epilepsy. *Journal of the Neurological Sciences*, **26,** 605–15.

Reynolds, G. P., Reynolds, L. M., Riederer, P., Jellinger, K. & Gabriel, E. (1980). Dopamine receptors and schizophrenia: drug effect or illness? *Lancet*, **ii,** 1251.

Riederer, P. & Birkmayer, W. (1980). A new concept: brain area specific imbalance of neurotransmitters in depression syndrome – human brain studies. In *Enzymes and neurotransmitters in mental disease*, ed. E. Usdin, T. L. Sourkes & M. B. H. Youdim, pp. 261–80. John Wiley & Sons: Chichester.

Riederer, P., Birkmayer, W., Seeman, D. & Wuketich, S. (1977). Brain noradrenaline and 3-methoxy-4-hydroxyphenylglycol in Parkinson's syndrome. *Journal of Neural Transmission*, **41,** 241–51.

Rinne, U. K. & Sonninen, V. (1973). Brain catecholamines and their metabolites in Parkinsonian patients. *Archives of Neurology*, **28,** 107–10.

Rinne, U. K., Sonninen, V. & Hyppa, M. (1971). Effect of L-dopa on brain monoamines and their metabolites in Parkinson's disease. *Life Sciences*, **10,** 549–53.

Roberts, P. & Amacher, E. (1978). *Propranol and schizophrenia*. Liss: New York.

Robinson, S. E., Berney, S., Mishra, R. & Sulser, F. (1979). The relative role of dopamine and norepinephrine receptor blockade in the action of antipsychotic drugs: metoclopramide, thiethylperazine and molindone as pharmacological tools. *Psychopharmacology*, **64,** 141–7.

Sacks, R. L. & Goodwin, F. K. (1974). Inhibition of dopamine-beta-hydroxylase in manic patients. *Archives of General Psychiatry*, **31,** 649–54.

Sanghere, M. K., McMillen, B. A. & German, D. C. (1982) Buspirone, a non-benzodiazepene anxiolytic, increases locus coeruleus noradrenergic neuronal activity. *European Journal of Pharmacology*, **86,** 107–17.

Sato, M. & Nakashima, T. (1975). Kindling: secondary epileptogenesis, sleep and catecholamines. *Canadian Journal of Neurological Sciences*, **2,** 439–46.

Schaeffer, M. M. & Kuenzel, W. J. (1978). Brain noradrenaline concentration in seizure-prone chicks, *Gallus domesticus*. *Poultry Science*, **57,** 1052–5.

Schildkraut, J. J. (1965). The catecholamine hypothesis of affective disorders; a review of the supporting evidence. *American Journal of Psychiatry*, **122,** 509–22.

Schildkraut, J. J., Orsulak, P. J., Schatzber, A. F., Gudeman, J. E., Cole, J. O., LaBrie, R. A. & Rhode, W. A. (1979). Biochemical discrimination of subtypes of depressive disorders. In *Catecholamines: basic and clinical frontiers*, ed. E. Usdin, I. J. Kopin & J. Barchas, pp. 1860–2. Pergamon Press: New York.

Schlesinger, K., Boggan, W. & Freedman, D. (1965). Genetics of audiogenic seizures. 1. Relation to brain serotonin and norepinephrine in mice. *Life Sciences*, **4,** 2345–51.

Seeman, P. (1981). Dopamine receptors in post-mortem schizophrenic brains. *Lancet*, **i,** 1103.

Sellinger-Barnette, M. M., Mendels, J. & Frazer, A. (1980). The effect of psychoactive drugs on beta-adrenergic binding sites in rat brain. *Neuropharmacology*, **19,** 447–54.

Serra, G., Argiolas, A., Fadda, F. & Gessa, G. L. (1980). Hyposensitivity of dopamine autoreceptors by chronic administration of tricyclic antidepressants. *Pharmacology Research Communications*, **12,** 619–24.

Shaw, D. M., O'Keefe, R. O., MacSweeney, S. A., Brookbank, B. W. L., Noguera, R. & Coppen, A. (1973). 3-Methoxy-4-hydroxyphenyl-glycol in depression. *Psychological Medicine*, **3,** 333–6.

Shaywitz, B. A., Cohen, D. J. & Bowers, M. B. (1975*a*). Reduced cerebrospinal fluid 5-hydroxyindoleacetic acid and homovanillic acid in children with epilepsy. *Neurology*, **25,** 72–9.

Shaywitz, B. A., Cohen, D. J. & Malcolm, B. B. (1975*b*). CSF amine metabolites in children with minimal brain dysfunction – evidence for alteration of brain dopamine. *Pediatric Research*, **9,** 385.

Shaywitz, B. A., Klopper, J. H. & Gordon, J. W. (1978). Methylphenidate in 6-hydroxydopamine-treated developing rat pups. *Archives of Neurology*, **35,** 463–9.

Shaywitz, B. A., Klopper, J. H., Yager, R. D. & Gordon, J. W. (1976*a*). A paradoxical response to amphetamine in developing rats treated with 6-hydroxydopamine. *Neurology*, **26,** 363.

Shaywitz, B. A., Yager, R. E. & Klopper, J. H. (1976*b*). Selective brain dopamine depletion in developing rats: an experimental model of minimal brain dysfunction. *Science*, **191,** 305–8.

Shekim, W. O., Dekirmenjian, H. & Chapel, J. L. (1979). Urinary MHPG excretion in minimal brain dysfunction and its modification by *d*-amphetamine. *American Journal of Psychiatry*, **136,** 667–71.

Shibuya, M. (1979). Dopamine-sensitive adenylate cyclase activity in the striatum in Parkinson's disease. *Journal of Neural Transmission*, **44**, 287–95.

Simon, P., Lecrubier, R. & Jouvent, R. (1978). Experimental and clinical evidence of the antidepressant effect of a beta-adrenergic stimulant. *Psychological Medicine*, **8**, 335–8.

Simpson, G. M., Kunz-Bartolini, E. & Watts, T. P. S. (1967). A preliminary evaluation of the sedative effects of Catapres a new anti-hypertensive agent in chronic schizophrenic patients. *Journal of Clinical Pharmacology*, **7**, 221–9.

Sjostrom, R. & Roos, B. E. (1972). 5-Hydroxyindoleacetic acid and homovanillic acid in cerebrospinal fluid in manic-depressive psychosis. *European Journal of Clinical Pharmacology*, **4**, 170–8.

Snyder, S. H. (1981). Dopamine receptors, neuroleptics and schizophrenia. *American Journal of Psychiatry*, **138**, 460–4.

Snyder, S. H., Greenberg, D. & Yamamura, H. I. (1974). Anti-schizophrenic drugs and brain cholinergic receptors. *Archives of General Psychiatry*, **31**, 58–64.

Spitzer, R. L., Fleiss, J. L. & Endicott, J. (1978). Problems of classification; reliability and validity. In *Psychopharmacology – a generation of progress*, ed. M. A. Lipton, A. DiMascio & K. F. Killam, pp. 857–69. Raven Press: New York.

Sternberg, D. E., van Kammen, D. P., Lake, C. R., Ballenger, J. C., Marder, S. R. & Bunney, W. E. (1981). The effect of pimozide on CSF norepinephrine in schizophrenia. *American Journal of Psychiatry*, **138**, 1045–50.

Sugerman, A. A. (1967). A pilot study of ST-155 (Catapres) in chronic schizophrenia. *Journal of Clinical Pharmacology*, **7**, 226–32.

Sulser, F. & Robinson, S. E. (1978). Clinical implications of pharmacological differences among antipsychotic drugs (with particular emphasis on biochemical central synaptic adrenergic mechanisms). In *Psychopharmacology – a generation of progress*, ed. M. A. Lipton, A. DiMascio & K. F. Killam, pp. 943–54. Raven Press: New York.

Svensson, T. H. & Usdin, T. (1978). Feedback inhibition of brain noradrenaline neurons by tricyclic antidepressants; alpha-receptor mediation. *Science*, **202**, 1089–91.

Sweeney, D. R., Maas, J. W. & Pichar, D. (1979). Urinary 3-methoxy-4-hydroxyphenylglycol and state variables in affective disorder. In *Catecholamines: basic and clinical frontiers*, ed. E. Usdin, I. J. Kopin & J. Barchas, pp. 1917–19. Pergamon Press: New York.

Talland, G. A. (1965). *Deranged memory*. Academic Press: New York.

Tang, S. W., Seeman, P. & Kwang, S. (1981). Differential effects of chronic desipramine and amitriptyline treatment on rat brain adrenergic and serotonergic receptors. *Psychiatric Research*, **4**, 129–38.

Taube, S. L., Kirkstein, L. S., Sweeney, D. R., Heninger, G. R. & Maas, J. W. (1978). Urinary 3-methoxy-4-hydroxyphenylglycol and psychiatric diagnosis. *American Journal of Psychiatry*, **135**, 78–81.

Ter Haar, M. B. (1979). Schizophrenia. *Trends in Neuroscience*, **2**, xiii.

Thieme, R. E., Dijkstra, H. & Stoof, J. C. (1980). An evaluation of the young dopamine-lesioned rat as an animal model for minimal brain dysfunction (MBD). *Psychopharmacology*, **67**, 165–9.

Tomlinson, B. E., Irving, D. & Blessed, G. (1981). Cell loss in the locus coeruleus in senile dementia of Alzheimer type. *Journal of the Neurological Sciences*, **49**, 419–28.

van Dongen, P. A. M. (1980*a*). Locus coeruleus region: effects on behaviour of cholinergic, noradrenergic and opiate drugs injected intra-cerebrally into freely moving cats. *Experimental Neurology*, **67**, 52–78.

van Dongen, P. A. M. (1980*b*). *The noradrenergic locus coeruleus. Behavioural effects of intra-cerebral injections and a survey of its structure, functions and pathology.* Krips Repro: Meppel, The Netherlands.
van Dongen, P. A. M. (1981). The human locus coeruleus in neurology and psychiatry. *Progress in Neurobiology*, **17**, 97–137.
van Praag, H. M. (1980). Central monoamine metabolism in depressions. II. Catecholamines and related compounds. *Comparative Psychiatry*, **31**, 44–54.
van Praag, H. M. & Korf, J. (1971). Retarded depression and the dopamine metabolism. *Psychopharmacologia*, **19**, 199–203.
van Praag, H. M. & Korf, J. (1975). Central monoamine deficiency in depressions; causative or secondary phenomenon? *Pharmakopsychiatrie Neuro-Psychopharmakologie*, **8**, 322–6.
van Praag, H. M., Korf, J. & Lakke, J. P. W. F. (1975). Dopamine metabolism in depressions, psychoses and Parkinson's disease; the problem of the specificity of biological variables in behaviour disorders. *Psychological Medicine*, **5**, 138–46.
van Rossum, J. M. (1967). The significance of dopamine receptor blockade for the action of neuroleptic drugs. In *Neuropsychopharmacology*, ed. H. Brill, pp. 321–9. Excerpta Medica: Amsterdam.
van Scheyden, J. D., van Praag, H. M. & Korf, J. (1977). A controlled study comparing nomifensine and clomipramine in unipolar depression, using the probenecid technique. *British Journal of Clinical Pharmacology*, **4**, 179S–84S.
Vetulani, J. & Sulser, F. (1975). Action of various antidepressant treatments reduces reactivity of the noradrenergic cyclic AMP-generating system in limbic forebrain. *Nature*, **257**, 495–6.
Victor, M., Adams, R. D. & Collins, G. H. (1971) *The Wernicke–Korsakoff syndrome*. F. A. Davis: Philadelphia.
Victor, M. & Banker, B. Q. (1978). Alcohol and dementia. *Aging*, **7**, 149–72.
Vijayashankar, N. & Brody, H. (1979). A quantitive study of the pigmented neurons in the nuclei locus coeruleus and subcoeruleus in man as related to ageing. *Journal of Neuropathology and Experimental Neurology*, **38**, 490–8.
Waiser, J., Hoffman, S. P., Polizus, P. & Engelhardt, D. M. (1974). Outpatient treatment of hyperactive schoolchildren with imipramine. *American Journal of Psychiatry*, **131**, 587–91.
Wang, S. C. (1965). Emetic and antiemetic drugs. In *Physiological pharmacology*, ed. W. A. Root & F. G. Hoffman, pp. 256–85. Academic Press: New York.
Weinstock, M. & Weiss, C. (1980). Antagonism by propranolol of isolation-induced aggression in mice: correlation with 5-hydroxytryptamine blockade. *Neuropharmacology*, **19**, 653–6.
Welch, J., Kim, H. S. & Liebman, J. (1982). Do antidepressants induce dopamine autoreceptor subsensitivity? *Nature*, **298**, 301–2.
Wender, P. H. (1978). Minimal brain dysfunction: an overview. In *Psychopharmacology – a generation of progress*, ed. M. A. Lipton, A. DiMascio & K. F. Killam, pp. 41–50. Raven Press: New York.
Wender, P. H., Epstein, R. S., Kopin, I. J. & Gordon, E. K. (1971). Urinary metabolites in children with minimal brain dysfunction. *American Journal of Psychiatry*, **127**, 147–51.
Wender, P. H. & Klein, D. F. (1981). *Mind, mood and medicine*, p. 215. Farrar, Straus & Giroux: New York.

Widlocher, D., Lecrubier, Y., Jouvent, R., Puech, A. J. & Simon, P. (1977). Antidepressant effect of salbutamol. *Lancet*, **ii,** 767–8.

Wilkinson, D. M. & Halpern, L. M. (1979). Turnover kinetics of dopamine and norepinephrine in the forebrain after kindling in rats. *Neuropharmacology*, **18,** 219–22.

Winsberg, B. G., Bialer, I., Kupietz, S. & Tobias, J. (1972). Effects of imipramine on behavior of neuropsychiatrically impaired children. *American Journal of Psychiatry*, **128,** 109–15.

Wise, C. D. & Stein, L. (1973). Dopamine-beta-hydroxylase deficits in the brains of schizophrenic patients. *Science*, **181,** 344–7.

Wolfe, B. B. Harden, T. K., Sporn, J. R. & Molinoff, P. B. (1978). Presynaptic modulations of beta adrenergic receptors in rat cerebral cortex after treatment with antidepressants. *Journal of Pharmacology and Experimental Therapeutics*, **207,** 446–57.

Wyatt, R. J., Schwartz, M. A., Erdelyi, E. & Barchas, J. D. (1975). Dopamine beta-hydroxylase activity in brains of chronic schizophrenic patients. *Science*, **187,** 368–70.

Yates, C. M., Ritchie, I. M., Simpson, J., Maloney, A. F. J. & Gordon, A. (1981). Noradrenaline in Alzheimer-type dementia and Down's syndrome. *Lancet*, **ii,** 39–40.

Yeoh, P. N. & Wolf, H. H. (1968). Effects of some adrenergic agents on low frequency electroshock seizures. *Journal of Pharmacological Sciences*, **57,** 340–2.

Yorkston, N. J., Zaki, S., Pitcher, D. R., Gruzelier, J., Hollander, J. H. & Serjeant, H. G. S. (1977). Propranolol as an adjunct to the treatment of schizophrenia. *Lancet*, **ii,** 575–8.

Yorkston, N. J., Zaki, S. A., Weller, M. P., Gruzelier, J. H. & Hirsch, S. R. (1981). DL-propranolol and chlorpromazine following admission for schizophrenia. A controlled comparison. *Acta psychiatrica scandinavica*, **63,** 13–27.

Epilogue

'for there are many things that hinder sure knowledge, the obscurity of the subject and the shortness of human life.'

Protagoras, fragment 4

It will be seen from the foregoing that 20 years of active research on the brain catecholaminergic systems have yielded many data, some discoveries and a lot of unanswered questions. The readers who have worked their way through the preceding text will be in a better position to decide what is known, what is quite likely and what is merely guessed at, concerning the functions of brain catecholamines in behaviour. Many questions remain unanswered, needing only a little dedicated research effort to yield considerable fruit. When the fashion in neuroscience swings from the currently in-vogue peptides back to the catecholamines once more it may be hoped that that effort will be forthcoming. In the meantime, it has been the function of this book to draw together the knowledge gained in the last 20 years into a systematised overview so that it will be available for the guidance of the next generation of 'catecholaminologists'. In this way, profitable lines of research are highlighted and those experimental dead-ends already tried marked with no-entry signs. The book will have served its purpose, if, in the words of Sir Samuel Dill in another context[1] 'These now forgotten pedants . . . softened the impact of barbarism, and kept open for coming ages the access to the distant sources of our intellectual life'.

[1] Dill, S. (1899). *Roman society in the last century of the Western Empire*. Meriden Press: New York.

Appendix: Behavioural terms and paradigms

Conditioned suppression. This refers to the cessation or reduction in ongoing behaviour caused by presentation of a stimulus previously paired with an aversive event, such as electric footshock. Typically, the ongoing behaviour is lever-pressing for food reward and presentation of a light or tone previously paired with shock (either in the same apparatus as the lever-pressing, called 'on-the-baseline' or in a separate apparatus, called 'off-the-baseline') causes a reduction in the frequency of lever-pressing. Often used as a measure of fear.

Continuous reinforcement (CRF). Every time the animal performs the correct response, reward is presented. This contrasts with partial reinforcement (PRF) in which not every response yields reward. Continuously reinforced tasks may include CRF lever-pressing in a Skinner box in which every lever-press yields a food pellet or CRF alley-way running in which every time the animal runs along the runway to the goal box, it finds one or more food pellets there. In the first case the number of lever-presses per session and in the latter the speed of running to the goal box are the measures recorded to indicate learning.

Differential reinforcement of low rates of responding (DRL). This task requires the animal to withhold a response for a specified period of time. It may be regarded either as an index of response inhibition or of time perception on the part of the animal. It is most often studied in a lever-press situation in a Skinner box. Once the first lever-press has been made by the animal, it must wait a specified period of time (between 5 s and several minutes) before a second response is rewarded. Responses made before this DRL interval are up are not rewarded and will reset the timer, so requiring the animal to wait even longer for reward. Animals typically indulge in chains of displacement behaviour during the DRL interval to keep themselves away from the lever, and so disruption of DRL behaviour may come about as a result of any one of many ways of disrupting behavioural sequencing and structuring.

Discriminated escape from shock. The animal is typically placed in a Y-maze and receives continuous or pulsed electric footshock until it escapes to the safe goal box. This safe goal box is signalled by a discriminative stimulus, such as position (always being the left goal box) or brightness (always being the brighter lit goal box despite being equally often left as right). Running into the unsafe goal box results only in continued footshock until the animal retraces its path and enters the safe goal box. The number of times it enters the unsafe goal box prior to reaching the safe one is recorded as a measure of the degree of learning on this task.

Extinction. The omission of a previously presented reward. Usually such sessions would follow immediately after acquisition sessions in which a response was trained and would appear identical to the animal except that no reward would ever be presented during extinction. Typically, animals cease to emit the previously reinforced behaviour after variable periods of time on extinction. Extinction of CRF schedules is quicker than that of partially reinforced schedules and extinction of ICSS may be extremely rapid, whatever the schedule.

Fixed interval (FI). Typically a lever-pressing schedule in a Skinner box. Reward becomes available in response to a lever-press after the passage of a fixed amount of time. Unlike a DRL schedule, responses prior to completion of the fixed interval do not reset the timer and a 'scalloping' pattern emerges in well-trained animals in which more and more lever-presses are emitted as the period of the fixed interval draws near completion.

Fixed ratio (FR). Typically a lever-pressing schedule in a Skinner box. Not every lever-press is rewarded. Thus, on a FR10, every tenth lever-press yields one food pellet. On a FR schedule a marked suppression of responding (called the postreinforcement pause) occurs immediately after delivery of each food reward. During this time no lever-presses are made and then, once the animal resumes pressing, a steady stream is emitted until the FR is completed. CRF is, by definition, FR1.

Freezing avoidance. A form of passive avoidance in which the rat has to become immobile upon presentation of the warning stimulus in order to avoid subsequent electric footshock. This requires the suppression or withholding of the animal's innate tendency to flee from an aversive event.

One-way active avoidance. The animal is placed in one half of an apparatus divided into two chambers. A warning stimulus, often a buzzer or tone, is presented for 5 to 20 s. Upon completion of this period, electric footshock comes on in the half of the apparatus where the animal has been deposited. Escape into the safe, unshocked, other half of the apparatus occurs (this is an escape response) terminating shock and warning stimulus. The animal is allowed to remain in the safe half for an intertrial interval and is then removed by hand and placed in the first half of the apparatus. The procedure is repeated. Eventually the animal will run to the safe half as soon as the warning stimulus comes on (this is an avoidance response). The warning stimulus is immediately terminated and no footshock is presented. The animal is replaced into the first half of the apparatus after an intertrial interval and the procedure continues. Well-trained animals which run to the safe side as soon as the warning stimulus comes on will maintain this behaviour for many trials with no footshock now being presented. The number of shocks to reach such a level of performance is typically taken as the measure of learning ability.

Partial reinforcement (PRF). In this situation not every correct response made by the animal yields reward. Thus, on a lever-pressing task in a Skinner box, more than one lever-press must be emitted in order to yield food delivery. FR, VR or VI schedules (see above and below) are typically regarded as partially reinforced. In a discrete trial paradigm in a runway, not every run up to the goal box yields food pellets in the goal box. The order of rewarded and nonrewarded trials is typically randomised, often with the restriction that not more than three rewarded or three unrewarded trials may occur consecutively. Animals trained on PRF typically take longer to extinguish than animals trained on CRF.

Passive avoidance – Step-down. The animal is placed on a platform in the middle of an arena, the floor of which can be electrified. Upon step-down (which occurs naturally within 15 to 30 s) electric footshock is experienced. The animal may now be removed and re-tested 24 h or more later. Upon re-test it will remain on the platform much longer than it did initially (often up to 3 min). This increased step-down latency on 24 h re-test is a measure of both the initial learning and its subsequent retention over 24 h.

Separation of initial learning and subsequent retention may be effected by requiring the animal to climb back up to the platform to escape electric footshock upon having stepped down initially. The trial continues until the animal has remained on the platform for three consecutive minutes. The number of step-downs and the time in the apparatus constitute a measure of initial acquisition learning. The step-down latency when re-tested 24 h later is then a pure measure of retention, since all rats were trained to the same criterion during acquisition.

Passive avoidance – Step-through. This is similar to the above except that the response is to step from one compartment to another of a two-compartment apparatus. The two compartments are usually distinct, one being dark and the other lit, for example. Re-test at 24 h examines the latency to re-enter the second compartment when placed again in the first after the retention interval.

Pole-jump avoidance. This is similar to one-way active avoidance except that the required response to escape or avoid electric footshock is to jump on to a vertical pole hanging down from the roof of the apparatus. This is consequently more demanding on the motor abilities of the animal.

Sidman avoidance. Electric footshocks occur at regular intervals via the metal floor of the apparatus. A response, typically lever-pressing in a Skinner box, will postpone the next shock for a fixed period. A second response will postpone shock for a further period. Failure to respond for a certain period will allow shock to re-occur. Thus, a well-trained animal will constantly press the lever at a rate just high enough to ensure that shock is never presented. Parameters of this schedule, which is quite difficult to learn, are the shock–shock interval (S–S) at which shocks occur normally in the absence of a response and the response–shock interval (R–S) being the period for which the response postpones the shock. No warning signal, other than the passage of time, occurs to warn the animal of the imminent presentation of shock.

Signalled shock-avoidance lever-press. This is the Skinner box analogue of one-way active avoidance task. Upon presentation of a warning stimulus, the animal has a certain period of time in which to press the lever before a shock occurs. A lever response counts as an avoidance response and shock is not presented.

Spatial discrimination (T-maze). The animal is placed in the stem of a

T-maze and allowed to run to the choice point. The correct response which leads to food or escapes electric shock is to turn into one of the arms, say the left one. The cue is thus the position in space of the arm which leads to food etc. This task is very simple and is rapidly learned by the animal. The number of errors (entries into the other arm) and the running speed are two measures which may be used to chart learning.

Spontaneous alternation. Animals allowed to choose one arm of a T-maze on the first trial will most often select to visit the other arm on a second trial. This is called spontaneous alternation and may reflect processes involved in learning about, or habituating to, novel stimuli.

Step-up avoidance. Similar to one-way active avoidance except that the response required to avoid or terminate electric footshock is to step up onto a shelf. At the start of the next trial this shelf can be made to retract into the wall, thus depositing the rat back down onto the electrifiable grid.

Successive visual discrimination. In a Skinner box successive periods of time are arbitarily designated S+ and S−. During the S+, one stimulus is present which indicates that lever-pressing will yield food on a CRF or PRF schedule. At the completion of this period of time, the S− period occurs in which a second stimulus, or the absence of the first, indicates that lever-pressing does not yield food. The S+ period may then alternate with the S− or they may occur on a random basis. Typically, animals soon come to confine their responding to the S+ periods only and show very few responses during the S−.

Taste aversion. Animals are allowed to consume a novel tasting solution (often saccharin) by mouth in a free-drinking situation for one session. They are immediately injected with an agent such as lithium chloride which will cause illness. Upon presentation of the solution for a second time some days later, animals are found to consume very little of it as a result of the association between the taste of the solution and the illness consequent to the lithium chloride injection.

Three-compartment linear maze. This is essentially three T-mazes placed end to end so that the animal is faced with three choice points. These may be cued on the basis of position (always take the left arm) or on brightness (always take the brighter lit arm) or an alternation strategy may be required (i.e. left, right, left).

Trace avoidance. This is similar to one-way active or two-way active avoidance but instead of the warning signal remaining on until the shock comes on, or until the animal performs the avoidance response, the signal comes on only briefly and shock occurs some time after it has been terminated. It is a more difficult task due to the temporal separation of the offset of the warning stimulus and the onset of shock.

Two-lever reset paradigm. A rate-free intracranial self-stimulation procedure in which responding on one lever yields electrical stimulation of the brain which decreases in intensity every so many presentations. A response on the second lever is required in order to reset the intensity of the electrical brain stimulation to its original value. Typically, rats emit so many responses on lever 1 and, when the stimulation becomes less rewarding as it weakens, then switch to lever 2. Changes in the rewarding nature of the electrical brain stimulation as a result of drug or lesion manipulation may thus be 'titrated' using this paradigm.

Two-way active avoidance. This is similar to one-way active avoidance except that, after the animal has moved into the second half of the apparatus, a new trial is initiated in this half (after an intertrial interval) *without* returning the animal by hand to the first half of the apparatus. It is a more difficult task than one-way avoidance, since shock is experienced in both halves of the apparatus and there is consequently no safe side of the apparatus. Animals are thus required during the task to re-enter a compartment in which they have sometimes experienced shock.

Variable interval (VI). This is similar to a fixed-interval schedule except that the period after which a lever-press will yield food is varied from one time to another. There is thus no patterning of responses near the time of completion of the interval but a low and steady emission of responses on a regular ongoing basis.

Variable ratio (VR). This is similar to a fixed ratio but the number of lever-presses required before food is presented is varied. No fixed number of responses is thus adequate to complete the ratio and typically a high and steady rate of responding emerges.

Variable time schedule (VT). Here food is presented independently of lever-pressing after a variable period. Lever-pressing neither shortens nor prolongs the period after which food will occur. Food delivery occurs even without a response to trigger it.

Index